H • O • L • T

EARTH SCIENCE

WILLIAM L. RAMSEY
LUCRETIA A. GABRIEL
JAMES F. McGUIRK
CLIFFORD R. PHILLIPS
FRANK M. WATENPAUGH

HOLT, RINEHART AND WINSTON, PUBLISHERS
New York • Toronto • Mexico City • London • Sydney • Tokyo

THE AUTHORS

William L. Ramsey
Former Head of the Science Department
Helix High School
La Mesa, California

Lucretia A. Gabriel
Science Consultant
Guilderland Central Schools
Guilderland, New York

James F. McGuirk
Head of the Science Department
South High Community School
Worcester, Massachusetts

Clifford R. Phillips
Former Head of the Science Department
Monte Vista High School
Spring Valley, California

Frank M. Watenpaugh
Former Science Department Chairman
Helix High School
La Mesa, California

About the Cover:

The photograph on the cover shows mesas and buttes in Monument Valley. The mesas and buttes are traces of what was once a broad plateau. They rise more than 300 meters above the valley floor. Monument Valley is located on the Utah–Arizona border on the Navajo Indian Reservation.

Photo Credits appear on page 563.
Cover photograph by Shelly Grossman/Woodfin Camp & Assoc.

Art Credits:

Andres Acosta, Robin Brickman, Ebet Dudley, Leslie Dunlap, General Cartography Inc., Jean Helmer, John Jones, Kathie Kelleher, Network Graphics, James Watling, Craig Zuckerman

ISBN 0-03-001904-4
789 039 9876

ACKNOWLEDGMENTS

Teacher Consultants

Pat Courtade
Gulf Comprehensive High School
New Port Richey, Florida

Sandy Meyer
Caledonia Junior/Senior High School
Caledonia, Minnesota

Julia Harlan
Wangenheim Junior High School
San Diego, California

William A. Moore Jr.
Director of Science Exploration
and Experimentation Center
Junior High School 240
Brooklyn, New York

Larry Nilson
Wayland Junior High School
Wayland, Massachusetts

J. Peter O'Neil
Waunakee Junior High School
Waunakee, Wisconsin

Bob Whitney
Dover Plains Junior/Senior High School
Dover Plains, New York

Content Critics

Dr. John E. Callahan
Professor of Geology
Appalachian State University
Boone, North Carolina

Dr. William Imperatore
Professor of Meteorology
Appalachian State University
Boone, North Carolina

Dr. Frank McKinney
Professor of Oceanography
Appalachian State University
Boone, North Carolina

Dr. Thomas Rokoske
Professor of Astronomy
Appalachian State University
Boone, North Carolina

Safety Consultant

Franklin D. Kizer
Executive Secretary
Council of State Science Supervisors
Lancaster, Virginia

Computer features written by

Anthony V. Sorrentino, D.Ed.
District Coordinator of Computer
Instruction
Monroe-Woodbury School District
Central Valley, New York

Computer Consultant

Nicholas Paschenko, M.Ed.
Computer Coordinator
Englewood Cliffs School System
Englewood Cliffs, New Jersey

Readability Consultant

Jane Kita Cooke
Assistant Professor of Education
College of New Rochelle
New Rochelle, New York

TO THE STUDENT

This book is about your home. Every city, state, or country where you might live is located somewhere on the planet Earth. The earth is not only the planet on which you live. It also provides you with air, water, food, and most of the other materials you need to live. As you study this book, you will learn how scientists have come to understand much about the way the earth works. This scientific knowledge will then help you to understand and appreciate your home planet.

How to Use This Book

HOLT EARTH SCIENCE consists of 19 chapters organized into five units. In the first chapter, What Is Science?, you will learn about science and some of the tools scientists use to solve problems. The two units following Chapter 1 are concerned with geology, the study of the solid part of the earth. The third unit deals with meteorology, the study of the layer of gases that surrounds the earth. In unit four, oceanography, the study of the oceans and the earth's surface below the oceans, is explored. In the fifth and final unit, astronomy, the study of the universe beyond the earth, is presented. Each unit is divided into chapters. Chapter Goals at the beginning of each chapter tell you what you will learn in that chapter. At the end of the chapter, a Chapter Review made up of Vocabulary, Questions, Applying Science, and a Bibliography will help you to remember what you learned in that chapter and to explore some applications and extensions of that knowledge. Each chapter is further divided into sections that begin with a list of section objectives. You will find many interesting and helpful diagrams and photographs in each section. Science words that may be new to you are printed in boldface type. Their definitions are printed alongside in the page margin. Each section ends with a brief Summary of the important ideas developed in the section. Questions at the end of the section will help you to test if you have accomplished the objectives for that section. Most sections are followed by an Investigation or a Skill-Building Activity. These Investigations and Activities will help you to develop good laboratory techniques and to practice basic science skills. Three features of special interest called Careers, Technology, and ¡Compute! are found throughout the book. These features will tell you about careers in earth science, about the ways technology affects our lives, and about the ways computers are used in science.

Safety

In your study of earth science, you will learn many fascinating things about the natural world. However, the study of science can also involve many potential dangers. Science classrooms and laboratories contain equipment and chemicals that can be dangerous if not handled properly. You should always follow the directions and cautions in the book when doing an Investigation. In addition, your teacher will explain to you the precautions and rules that must be followed to insure the safety of you and your classmates. Your responsibility is to follow these rules carefully and to demonstrate a positive attitude toward laboratory safety.

Work Habits

- Never work alone in a science laboratory.
- Never eat, drink, or store food in a science laboratory.
- Wash hands before and after work and after spill cleanups.
- Restrain loose clothing, long hair, and dangling jewelry.
- Never leave heat sources unattended (e.g. gas burners, hot plates, etc.).
- Do not store chemicals and/or apparatus on lab bench, and keep lab shelves organized.
- Never place glassware or chemicals (in bottles, beakers/flasks, wash bottles, etc.) near the edges of a lab bench.
- Read all lab procedures before doing laboratory work.
- Analyze accidents to prevent repeat performances.
- Protection should be provided for the lab worker and lab partner.
- Do not mix chemicals in the sink drain.
- Have actions pre-planned in case of an emergency (e.g., what devices should be turned off, which escape route to use, and how to use a fire extinguisher).

Safety Wear

- Approved eye or face protection should be worn continuously.
- Gloves should be worn which will resist penetration by the chemical being handled and which have been checked for pinholes, tears, or rips.
- Wear a laboratory coat or apron to protect skin and clothing from chemicals.
- Footwear should cover feet completely; no open-toe shoes.

CONTENTS

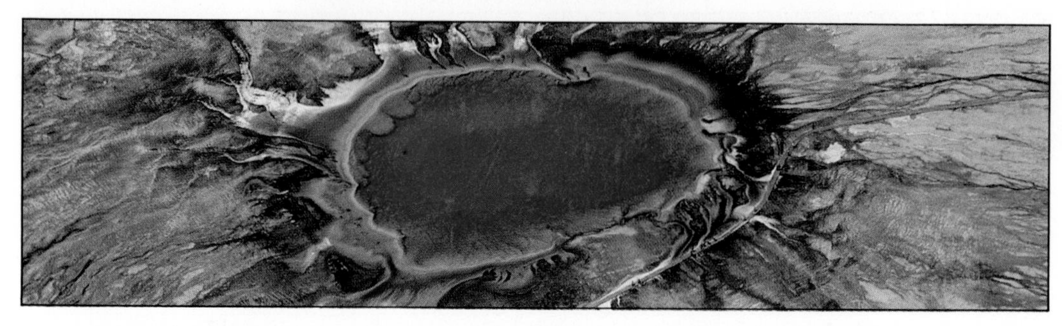

CHAPTER

1

This volcanologist is wearing protective clothing and a mask to shield him from the intense heat of the lava. He is using a steel pipe in order to take a sample of the molten lava.

WHAT IS SCIENCE?

CHAPTER GOALS

1. Describe the branches of earth science and their importance in our lives.
2. Explain how scientists find answers to questions.
3. Explain why making careful observations and measurements is important in science.

1–1. What Is Earth Science?

At the end of this section you will be able to:

- ☐ Explain how scientific study of the earth began.
- ☐ Describe four main branches of earth science.
- ☐ Explain why earth science is important.

Job: Collect some lava. Study the lava to find the source of the melted rock. Such a job is one type of work an earth scientist does. The picture on the left page shows an earth scientist taking a sample of lava. Lava is the melted rock that comes from volcanoes. By studying these samples, earth scientists can learn much about volcanoes.

THE SCIENCE OF THE EARTH

A volcano can be one of nature's most frightening shows. In a few minutes a volcanic eruption can throw out millions of kilograms of rocks and deadly clouds of hot gases. Whole mountains or islands can be blasted away. There are records of cities that have been destroyed by volcanoes. About 1450 B.C. a volcanic eruption took place on an island near Greece. In one day, most of the island and its people disappeared. This event may have been the basis of the story of the lost land of Atlantis.

In Mexico in 1943, a farmer discovered a crack about 25 meters long in his cornfield. The crack had opened during the night. Red-hot stones and ash blew out of the crack. By the next morning, a volcano ten meters high had appeared. Within a year, the volcano grew to a height of over 300 meters. Ash from the volcano covered much of the countryside. The farmer lost his cornfield. In the years that followed, thousands of people lost their homes.

Fig. 1-1 The final moments of some victims in Pompeii. Their bodies left openings in the hardened volcanic material. The openings were then filled with plaster to make these molds.

Perhaps the best-known volcanic disaster took place in A.D. 79. It was the loss of the Roman city of Pompeii (Pom-**pay**). The city, with at least 2,000 people, was buried by volcanic material. See Fig. 1-1. Volcanic material has kept the city just as it was before being buried.

Stories have been made up to explain volcanoes. For example, the Romans believed that volcanoes were caused by a god. They thought the Roman god Vulcan had huge ovens under the ground. Volcanoes (named after Vulcan) were the chimneys of those huge hidden ovens. Scientific study of volcanoes began when people started to make careful **observations** (ob-sur-**vay**-shuns) of the earth. *Observations* are any information that we gather by using our senses. For hundreds of years, people had lived in many places where old volcanoes quietly rested. They did not know that the mountains and fields were once the scene of fire and explosion. Toward the end of the 18th century scientists studying the earth began to find that volcanoes were not unusual geographic features. They found rocks believed to have formed from lava. Some mountains were found to be old volcanoes. Scientists began to see that much of the earth's surface was shaped by volcanoes. See Fig. 1-2. For the first time, scientists became aware that the earth's surface could change. If volcanoes could build mountains, then the shape of the earth's surface must not always have been the same. By observing the changes in the earth's surface, scientists believed they could find out how the earth works. This was the beginning of the scientific study of the earth.

Observations Any information that we gather by using our senses.

Fig. 1-2 How do you think this mountain was formed?

THE EARTH SCIENCES

Earth is the planet we live on, and earth science is the study of our planet. Earth science can be divided into four branches. One branch is called **geology.** *Geology* is the study of the solid part of the earth. Scientists called geologists study the matter and the structure of the planet. They do this to find out how the earth formed and how it changes. Some geologists study volcanoes because volcanoes change the surface of the earth.

Geology The branch of earth science that studies the solid part of the earth.

The earth is surrounded by a layer of gases that make up its atmosphere. **Meteorology** is the branch of earth science that studies the atmosphere. The scientists who work in the field of *meteorology* are called meteorologists. They also observe weather and study climates. See Fig. 1-3.

Meteorology The branch of earth science that studies the atmosphere.

Oceans cover most of the earth's surface. **Oceanography** is the branch of science that studies oceans. *Oceanography* also includes the study of the earth's surface below the oceans. The scientists who make these studies are called oceanographers, or *marine scientists*. See Fig. 1-4.

Oceanography The branch of earth science that studies the oceans and the ocean floor.

Fig. 1-3 Meteorologists study the weather.

Fig. 1-4 Oceanographers study the oceans.

Astronomy The branch of earth science that studies the universe.

Our planet is a part of the universe. For this reason, **astronomy** is a part of earth science. *Astronomy* is the study of the universe beyond the earth. Astronomers study other planets, the sun, and the stars. See Fig. 1-5. In this book, you will learn about geology, meteorology, oceanography, and astronomy. Each will be studied in a different unit.

WHY EARTH SCIENCE IS IMPORTANT

Our planet is always being changed by powerful forces. Almost every day the news tells you about volcanoes, earthquakes, storms, floods, and drought. At some time, these forces

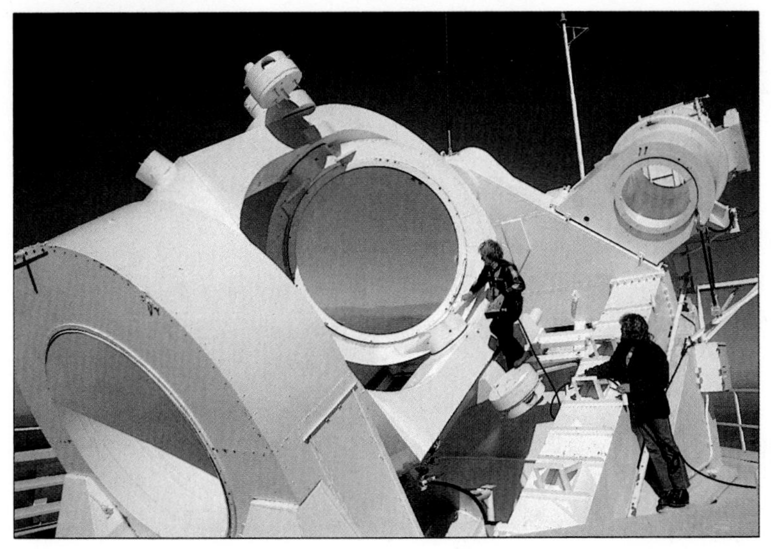

Fig. 1-5 Astronomers study the universe.

will touch the life of every person. What causes earthquakes and weather? Why do volcanoes suddenly erupt? Finding the answers to questions like these can save lives and property.

Understanding the earth also can improve the way we live. Exploring the land and oceans will help find oil, coal, and metals. Almost all of the things we use every day come from the earth. Human-made wastes cause pollution of air and water. We must understand the atmosphere and oceans to control pollution and its effects. From space, we can see the earth as never before. Already pictures of the earth from space have helped to locate valuable deposits of metals and sources of pollution.

SUMMARY

Scientific study of the earth began with observations. The four main branches of earth science are geology, meteorology, oceanography, and astronomy. Scientific study of the earth can improve our lives on this planet.

QUESTIONS

Use complete sentences to write your answers.

1. Explain how observations of volcanoes led to scientific study of the earth.
2. Describe the four main branches of earth science.
3. Give one example of why the study of earth science could be important to you.

INVESTIGATION

OBSERVATIONS

PURPOSE: To make observations of a very common event. Then to ask a question that requires more facts to be gathered through experiment in order to arrive at a conclusion.

MATERIALS:

7 index cards   2 sheets of paper
stapler

PROCEDURE:

A. Staple two index cards together using one staple near the center of the cards. Staple four other index cards together in the same way.

B. Compare the weight of one card to the two stapled together.

   1. Which is heavier?

C. Hold one card in one hand and the two stapled cards in the other. With your hands at waist height, drop them both at the same instant. Note which hits the floor first. Do this several times.

   2. Which falls through the air faster?

D. Pick up the two-card pile in one hand and the four-card pile in the other.

   3. Which is heavier?
   4. Which pile do you think will fall through the air faster?

E. Holding the cards waist high, drop them at the same instant. Note which pile strikes the floor first. Do this several more times.

   5. Which pile fell through the air faster?
   6. Was your answer for question 4 above correct?

F. At this point you could conclude that weight determines how fast an object falls through air. A scientist would ask, "Is the speed at which an object falls through the air determined only by its weight?" Another experiment is needed to answer this question. Crumple one sheet of paper into a ball.

   7. Which is heavier, the ball or the flat sheet?
   8. On the evidence you have at this time, which should fall through air faster?

G. Hold the two pieces of paper waist high and drop them at the same instant. Do this several times.

   9. Do objects having the same weight always fall at the same speed?
   10. What explanation can you give for this result?

H. Fold the flat sheet of paper into quarters. Drop the folded paper from waist height and observe its speed.

   11. What do you observe about the folded paper's speed compared to the flat sheet?

CONCLUSIONS:

1. What two factors affect the speed at which objects fall through the air?
2. What do you think would happen to the speed of falling objects if air were not present?
3. What must you do to be certain of your answer to question 2?

1-2. Scientific Thinking

At the end of this section you will be able to:

- ☐ Explain how asking questions can lead to scientific thinking.
- ☐ List and explain the steps in scientific problem solving.
- ☐ Distinguish between a *hypothesis* and *scientific theory*.

It was a place for camping, hunting, and fishing. There were green forests, meadows, and lakes. This was the area around Mount St. Helens in the state of Washington. At 8:32 A.M. on May 18, 1980, the mountain blew its top. About 250 billion kilograms of rock and dust were blasted out by the eruption. The forests were turned into a landscape that looked like the surface of the moon. See Fig. 1-6.

Fig. 1-6 (left) Mount St. Helens before it erupted. (right) Mount St. Helens after the eruption.

ASKING QUESTIONS

Since that May morning in 1980, scientists have been studying Mount St. Helens. They are searching for answers to questions about the eruption. One important question is: Does a volcano give warning signs that it is going to erupt? Scientific thinking begins with such a question. Finding an answer could help to save many lives in the future.

Fig. 1-7 Oil geologists use the scientific method to find oil.

SCIENTIFIC PROBLEM SOLVING

Research is a way scientists look for answers to questions. Scientists who want to find a way to tell when a volcano will erupt must gather as much information as possible about volcanoes. Scientific research is often done by following a number of steps. The steps are called the **scientific method.** The *scientific method* is a guide to scientific problem solving. It is not a set of instructions that will always result in an answer. See Fig. 1-7. Applying the scientific method, research on the eruption of Mount St. Helens might take the following steps.

Scientific method A guide scientists use in solving problems.

1. State the problem. Sometimes hot liquid rock called lava forces its way to the surface of a volcano. When this happens, it is called a volcanic eruption. An erupting volcano also gives off gases. Sometimes the lava flows out quietly. Other times there are explosions. The lava flow and explosions can damage property and injure or kill people in the area.

The first step of the scientific method is to identify and state the problem so that it can be researched. This is not always easy. The problem with Mount St. Helens is to find a way of telling when an eruption will happen. In science, telling what is likely to happen is called predicting. The problem is then stated in the form of a question. The question could be: "Can the eruptions of Mount St. Helens be predicted?" The question must clearly state the problem. Then the research can begin.

2. Gather information. Research usually begins by gathering all information that is already known. For example, scientists had records that showed earthquakes happening before the eruption of Mount St. Helens. These records came from instruments placed on the volcano. See Fig. 1-8. The records showed that earthquakes began around Mount St. Helens several months before it erupted. Most of the earthquakes were small. Sometimes as many as fifteen small earthquakes were recorded in one hour.

Fig. 1-8 A seismograph on Mount St. Helens. It records earthquakes.

3. Form a **hypothesis.** A *hypothesis* (hie-**poth**-uh-sus) is a possible answer to a question. The question to be answered here is, "Can the eruptions of Mount St. Helens be predicted?" A hypothesis often comes from first seeing some kind of pattern in observations. For example, the earthquake records seem to show that groups of small earthquakes come before eruptions of Mount St. Helens. Thus a hypothesis could be stated as: Mount St. Helens is likely to erupt shortly after many small earthquakes are observed

Hypothesis A statement that explains a pattern in observations.

4. Test the hypothesis. Scientists think of a hypothesis as only one of many possible answers to a question. They do not consider a hypothesis to be correct until it has been tested. A hypothesis is tested by doing one or more **experiments.** An *experiment* is a setup that enables observations to be made. Ideally, *experiments* are set up so that they cannot be disturbed. In this way, scientists can be sure that the results of the experiment are correct. These experiments are called *controlled* experiments. In these experiments, scientists are able to *control* anything that may affect the experiment. Therefore, many controlled experiments are done in laboratories. For example, suppose light affected an experiment. In the laboratory scientists could block out the light. However, many experiments in earth science cannot be set up under controlled conditions. For example, earthquakes and volcanoes cannot be made to happen. So earthquakes and volcanoes must be tested by a different kind of experiment. In this kind of experiment, scientists make as many observations as they can. From these observations, they then try to find a way of telling when an

Experiment A setup to test a particular hypothesis.

eruption will take place. For these scientists, the volcano has become their laboratory.

5. Accept or change the hypothesis. Experiments will show if the hypothesis is correct. Since 1980, scientists have used earthquakes as early warnings of volcanic eruptions. They have been able to predict many Mount St. Helens eruptions. Some were predicted up to three weeks before they happened. Often other experiments will show how the hypothesis can be improved. For example, there were other kinds of observations made before the Mount St. Helens eruption. Scientists observed that parts of the mountain moved out slightly. These movements were measured by instruments placed on the mountain. See Fig. 1-9. Scientists also observed that certain gases were given off just before the eruption. These other ob-

Fig. 1-9 A laser range finder is used to measure the movements of parts of Mount St. Helens. Measurements can be made to an accuracy of about one part per million.

servations, along with earthquake activity, can be used to make better predictions. When scientists find that their predictions are correct, they then can accept the hypothesis as correct.

SCIENTIFIC THEORIES

The hypothesis that used earthquakes as early warnings of volcanic eruptions was tested many times and found to be correct. Other observations were also used to help predict volcanic eruptions. Very often several or many hypotheses will lead to the development of a **scientific theory.** A *scientific theory* is a general explanation made up of many hypotheses, not just one. A scientific theory that only predicts eruptions on Mount St. Helens is not complete. It must answer a wider range of questions. For example, can the theory predict eruptions on other volcanoes? Why do earthquakes and volcanoes happen in the same places? A good scientific theory is able to answer these questions. However, even good theories may change as new observations are made.

Scientific theory A general explanation that includes many hypotheses that have been tested and found to be correct.

Some scientific theories have been tested many times and found to be correct. A theory that has been found to be correct every time it is tested is called a *scientific law.* An example of a scientific law is the law of gravity. The force of gravity on the earth obeys this scientific law. Even bodies in space, such as the sun and planets, obey this law.

SUMMARY

Scientific thinking begins with asking questions. The scientific method is a guide that scientists use to find answers. Scientific research may lead to a possible answer or hypothesis. After a hypothesis has been tested and found to be correct, it may become part of a scientific theory. When a theory has been found to be correct, it becomes a scientific law.

QUESTIONS

Use complete sentences to write your answers.

1. Give an example of how asking a question can lead to scientific thinking.
2. List the steps that can be used as a guide to scientific problem solving.
3. Explain the difference between a scientific theory and a hypothesis.

INVESTIGATION

AN EXPERIMENT

PURPOSE: Make a hypothesis and carry out an experiment to check the hypothesis.

MATERIALS:

2 thermometers (Celsius)
2 pieces of dry cloth
1 medicine dropper

PROCEDURE:

A. Copy the table shown below.

Condition	Temperature (°C)	
	First Reading	Second Reading
Bulbs Bare		
Bulbs Covered		

B. Place both thermometers side by side on the table. Take care that they will not roll off the table. Place the bulbs of the thermometers 2 cm apart. See Fig. 1–10.

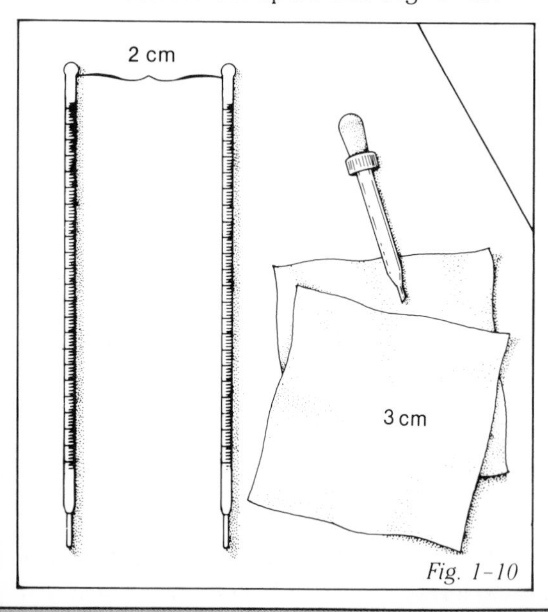

Fig. 1–10

C. Allow the thermometers to remain undisturbed for two minutes. Then read both thermometers and record their temperatures in your table. The reading on the two thermometers may differ by a degree or two, even though they are at the same temperature.

D. Wrap a piece of dry cloth around the bulb of each thermometer. Put them back in the same place as before. Again, after two minutes, read each thermometer and record in your data table.
   1. Was there a change in the reading for either thermometer?
   2. What effect would you predict there would be if you wet the cloth on one of the thermometers with water?

E. The answer you gave for question 2 is called a "hypothesis." A hypothesis needs to be tested. What you do in testing the hypothesis is called an "experiment."
   3. Design an experiment that will test your hypothesis.

F. Carry out your experiment. Record your results in the data table. Record the temperature of both thermometers. Describe the condition of each in the data table.

CONCLUSIONS:

1. Using complete sentences, describe your hypothesis.
2. Using the results of the experiment, tell whether or not your hypothesis was correct.
3. Does the scientific method always give a correct hypothesis? Explain.

1-3. Measuring

At the end of this section you will be able to:

- ☐ Explain why measurements are important in making scientific observations.
- ☐ Name and use the SI units of measurement.
- ☐ Explain why measurements are likely to contain error.

Look at Fig. 1-11. What would happen to the ball held in each dragon's mouth during an earthquake? Even a small earthquake could cause a ball to fall into the mouth of the toad below. This instrument was used by the Chinese around A.D. 130 to detect earthquakes. The strength of the earthquake could also be measured. It was measured by the number of balls dropped.

MEASUREMENTS IN SCIENCE

An earthquake takes place very quickly. It leaves no record except in the damage that is done. However, the ancient Chinese could make a record of earthquakes using the instrument shown in Fig. 1-11.

Fig. 1-11 The ancient Chinese used this instrument to detect and measure earthquakes.

Measurement An observation made by counting something.

Like the ancient Chinese, modern scientists also try to make observations that can be kept as a record. This is a reason for scientific observations often being made in the form of **measurements.** A *measurement* is an observation that can be made by counting something. Measuring may be as simple as counting the number of balls falling from the mouths of the dragons. However, modern earthquake measurements are not so simple. Instruments now used to record earthquakes measure their strength by traces on paper held by turning drums. See Fig. 1–12. Scientists make as many observations as possible in the form of measurements. A measurement is written as a number value. Then it can be understood because numbers have the same meaning for everyone.

All measurements are made by using a number and a unit. The number tells how many. The unit tells what is being counted. For example, you might tell someone your height in feet and inches. Feet and inches are units commonly used in the United States for measuring length. However, people in other countries would have a problem understanding feet and inches. Units of measurement used in almost all parts of the world are based on the metric system. The modern form of the metric system is called the International System of Units, which is usually written as SI. SI stands for the French name *le Système International.* Common metric units are the same as SI units.

Fig. 1–12 Earthquakes from all over the United States are recorded by these instruments at the National Earthquake Information Service in Golden, Colorado.

Scientists use SI units to make measurements. One reason is that SI units are easy to use. This is because they are based on the number ten. All SI units can be changed into larger or smaller units by using the number ten. For example, the kilometer is equal to 1,000 (10 × 10 × 10) meters. One meter is equal to 100 (10 × 10) centimeters. Changing units such as feet into larger or smaller units is not so simple. To change feet into inches, you must first know that 1 foot equals 12 inches. To change feet into yards, you must know that there are 3 feet in 1 yard. With SI units, you only must remember that the size of all units can be changed by using the number ten.

MAKING MEASUREMENTS

Measurements are usually made with the use of tools or instruments. Instruments can improve our observations. For example, telescopes are instruments that help us to see deep into

Fig. 1-13 Radiotelescopes have detected some of the most distant objects in the universe.

space. See Fig. 1-13. Microscopes help us to see objects that are too small to be seen with just our eyes. Measurements also make observations more useful. To say something is large is not enough. Measurements are needed to show how large.

1. Length or distance. The basic SI unit used to measure length or distance is the **meter** (m). A *meter* is slightly longer than a yard. Lengths or distances that are much longer than one meter are measured in larger units. The unit most often used to measure distances much longer than one meter is the *kilometer* (km). A kilometer is equal to 1,000 meters. "Kilo-" means 1,000.

Meter The basic SI unit of length. A meter is equal to 39.4 inches.

Units smaller than a meter are used when measuring lengths shorter than a meter. A unit equal to 1/100 (0.01) of a meter is called a *centimeter* (cm). A unit smaller than a centimeter is called a *millimeter* (mm). A millimeter is equal to 1/1,000 (0.001) meter.

A basic unit such as a meter can be changed into a larger or smaller unit by using a prefix. A prefix is always put in front of another word. For example, "kilo-" and "milli-" are prefixes added to the word "meter." Each of these prefixes tells how much larger or smaller the basic unit has been made. Table 1-1 shows the prefixes that can be used with all basic SI units. Lengths can be measured in SI units using a metric ruler. Metric rulers are often marked in centimeters, which are numbered.

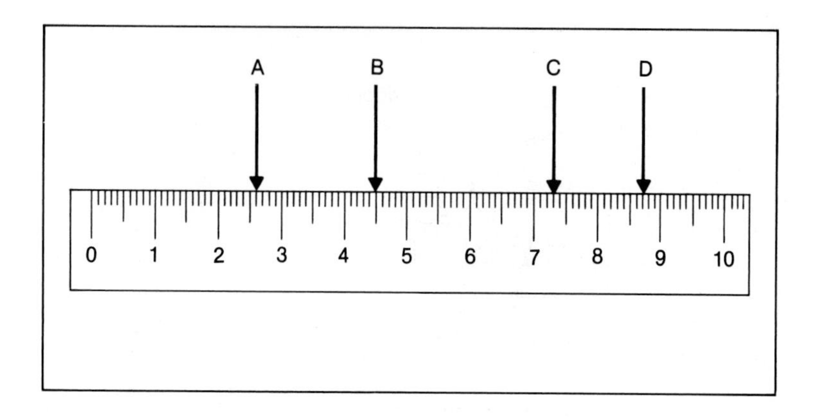

Fig. 1–14 The numbered divisions on this metric ruler are centimeters. What are the readings at A, B, C, and D in centimeters?

Each centimeter is divided into millimeters, which are not numbered. See Fig. 1–14. A common metric ruler is one meter in length and is called a meter stick.

Prefix	Meaning	Example
Milli-	1/1,000 (.001)	Millimeter
Centi-	1/100 (.01)	Centimeter
Deci-	1/10 (.1)	Decimeter
Deca-	10 times	Decameter
Hecto-	100 times	Hectometer
Kilo-	1,000 times	Kilometer

Table 1–1

2. Mass. Mass is the amount of matter an object has. Measuring mass means finding the amount of matter in an object. Measuring how much an object weighs is one way to find its mass. The weight of an object usually depends on its mass. Thus mass and weight are often used as if they were the same. However, they are not the same. The weight of an object can change. This is because weight also depends on the object's distance from the earth. Thus someone in space, far from the earth, is nearly weightless. But the mass or amount of matter in the person's body cannot change. The weight of an object remains the same only when measured on the earth.

Kilogram An SI unit of mass equal to 1,000 grams.

The SI unit for mass is the **kilogram** (kg). One *kilogram* of mass weighs about 2.2 pounds. A smaller SI unit of mass is a gram (g). One gram is 1/1,000 (0.001) kg. A nickel coin weighs about 5 g. There are 28.35 g in 1 ounce.

Mass can be measured by using a balance. See Fig. 1–15. The balance works like a seesaw. The mass of the object on the left pan is balanced by moving the sliding weights, or riders, on the

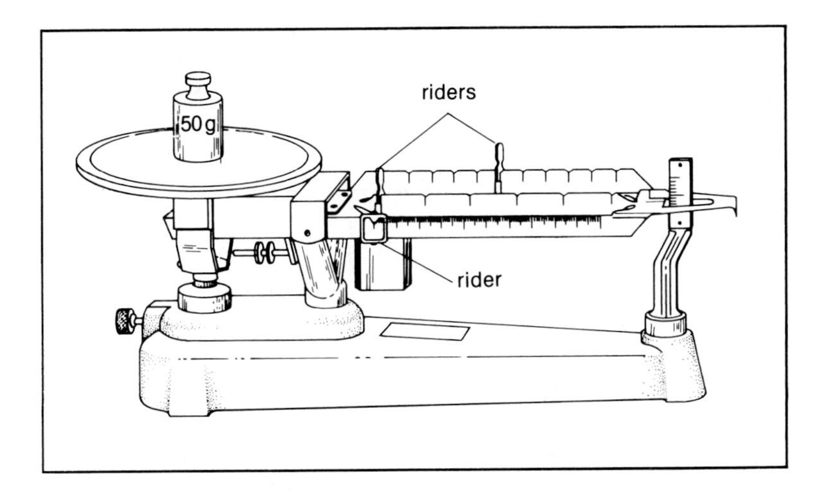

Fig. 1-15 A balance is used to measure mass.

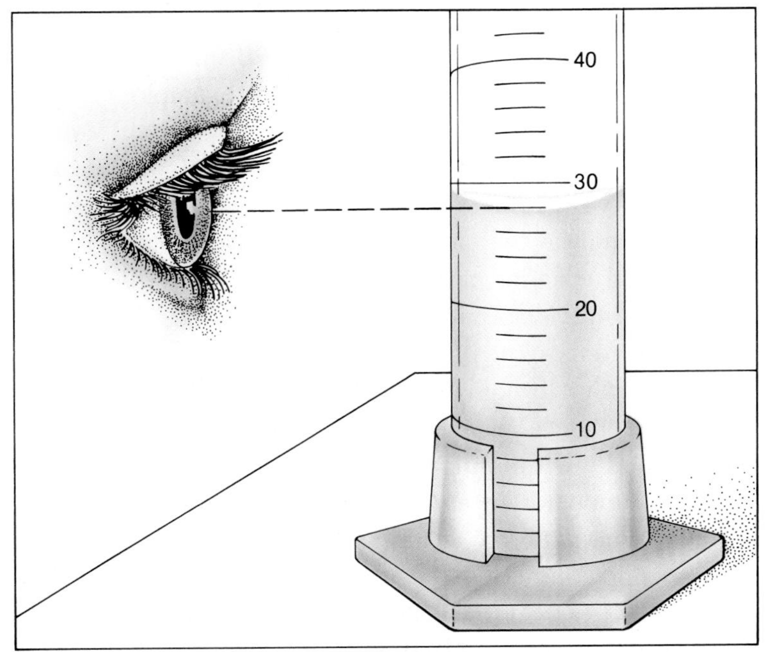

Fig. 1-16 To read a graduated cylinder, hold up the cylinder so that the surface of the liquid is level with your eye. Read the mark closest to the bottom of the curve. Numbered marks are ml or cm. What is the reading shown?

right. Each rider gives a reading. The readings are then added to get the mass of the object.

3. Volume. All materials take up space. Rocks, water, and even air take up space. Liquids usually are measured by their volume. The volume of a liquid is the amount of space the liquid takes up. The volume of a liquid can be measured with a graduated cylinder. See Fig. 1-16. The liquid is poured into the cylinder. Holding the cylinder with the liquid surface at eye level, read the mark closest to the surface. Since the liquid surface is curved, use the mark closest to the bottom of that curve.

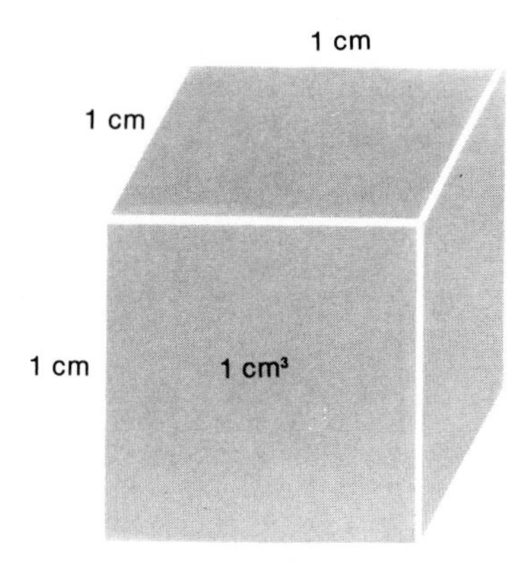

Fig. 1-17 A cube that measures 1 cm on each side has a volume of 1 cm³. A sugar cube has a volume of about 1 cm³.

The unit often used when measuring the volume of liquids is the **liter** (L). A *liter* is a little more than a quart. Small volumes are measured in milliliters (mL). One mL is 1/1,000 (0.001) liter. One mL is equal in volume to a cube that is 1 cm on each side. Such a cube has a volume of one cubic cm. This is written as 1 cm^3. See Fig. 1-17. Thus, 1 mL and 1 cm^3 are equal and are used in the same way. The table below shows the relationship between the units commonly used for measuring liquids.

Unit	Symbol	Value
liter	L	1,000 mL
milliliter	mL	0.001 L

The volumes of solids can also be measured. However, the volumes of solids are measured in cm^3. When a solid has a *regular* shape, such as a cube, its volume can be measured by multiplying the lengths of its three sides. For example, the cube shown in Fig. 1-17 has a volume of 1 cm $\times$ 1 cm $\times$ 1 cm or 1 cm^3. If a solid has an *irregular* shape, such a common rock, its volume can be measured by finding out how much water it can displace. One way this can be done with a small object is to fill a graduated cylinder partially with water. Read the volume. Then lower the solid into the water. The solid will cause the water level to rise. The difference between the new water level and the original level is equal to the volume of the solid.

Liter A unit of volume. A liter is equal to 1.05 quarts.

4. Temperature. The most widely used temperature scale in the world is the **Celsius** (C) scale. Most scientific measurements of temperature are based on the *Celsius* scale. The freezing temperature of water on the Celsius scale is 0°C. The boiling temperature of water is 100°C. You are probably used to temperatures being given in the Fahrenheit scale. The Celsius and Fahrenheit scales are shown in Fig. 1-18.

Celsius The temperature scale used most in science. It is based on the freezing point of water, 0°C, and the boiling point of water, 100°C.

5. Density. Sometimes measurements are made by using two kinds of units. For example, the measurement of **density** uses units of mass and volume. *Density* tells how much mass there is in a certain volume of material. The units for density are grams per cubic centimeter (g/cm^3). For example, the density of lead is 11.3 g/cm^3. This means that 1 cm^3 of lead has a mass of 11.3 g. Aluminum has a density of 2.7 g/cm^3. This means that 1 cm^3 of aluminum has a mass of 2.7 g. Thus a piece of lead has a much larger mass than a piece of aluminum with the same volume. Therefore, lead is said to be much more dense than aluminum.

Density The amount of mass in a given volume of a substance; density = mass/volume.

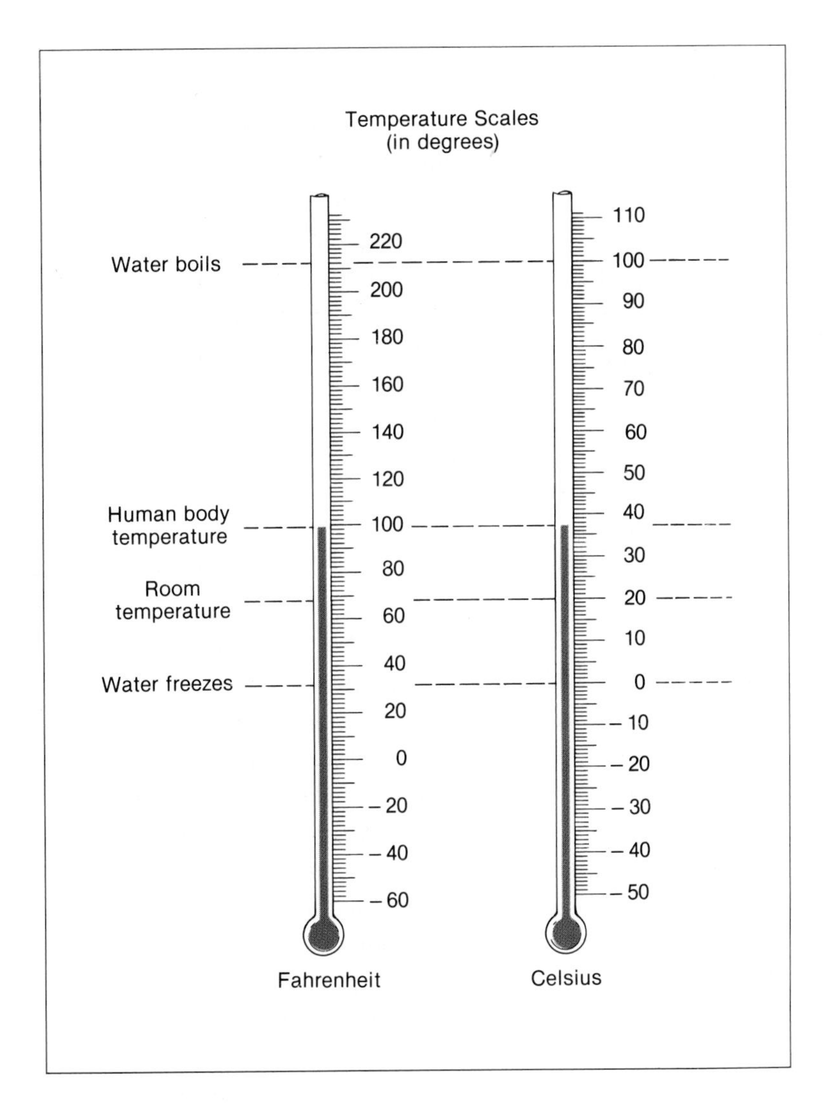

Fig. 1-18 A comparison of the Celsius and Fahrenheit temperature scales. You can use this graph to change from one scale to the other. Find the temperature on one scale. Then use a ruler to move across to the other scale to find the equivalent temperature.

ESTIMATING

Each time a measuring instrument is used, you will need to estimate the final answer. For example, suppose that you want to measure the length of this book. The edge of the book will not be exactly even with the smallest mark on the ruler. You probably will choose the mark that is closest to the edge of the book. The smallest mark on most metric rulers measures millimeters. Therefore, your measurement could easily be off by almost as much as 1 mm. See Fig. 1-19. There is no way you can remove this error. No measurement using instruments can be exact. This is because all measurements made by instruments need some estimating.

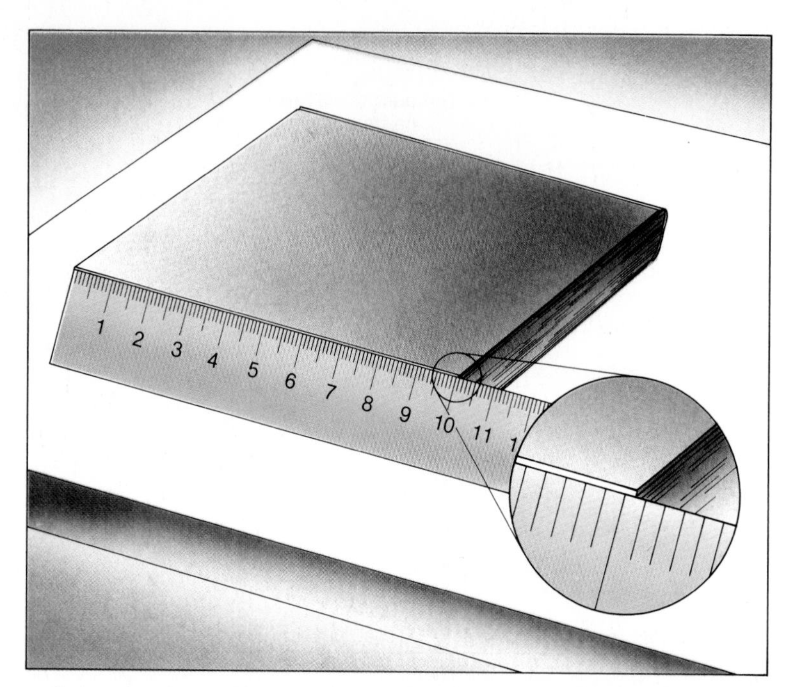

Fig. 1-19 When measuring length using a metric ruler, you will have to estimate to the nearest mm.

Scientists know that their measurements contain some error. They try to keep the errors as small as possible. One method of keeping the error small is to make the same measurement many times. Some errors make measurements too large. Other errors make measurements too small. These errors will tend to balance each other. The measurements are then averaged into a final measurement. Averaging is one way scientists try to make measurement error as small as possible.

SUMMARY

Measurements are important because they make observations more useful and have the same meaning for everyone. The basic SI units are the meter, liter, and kilogram. All measurements made by instruments need some estimating.

QUESTIONS

Use complete sentences to write your answers.

1. Why are measurements used in science?
2. Describe each of the SI units of measurement for (a) length, (b) volume, and (c) mass. Compare each to a common measure.
3. Describe two metric units of length smaller than a meter.
4. Explain what causes measurements to contain error.

INVESTIGATION

MEASURING WITH THE SI SYSTEM

PURPOSE: To use the SI system in making measurements of length, mass, and volume.

MATERIALS:

ruler
10 index cards
balance
graduated cylinder
rubber stopper

PROCEDURE:

A. Copy the following table. Record all your measurements in your copy of the table.

Measurement	Index Card	Rubber Stopper
Length (cm)		———
Width (cm)		———
Thickness (cm)		———
Mass (g)		
Volume (cm^3)		

B. Use the metric side of the ruler to measure the length and width of one index card. Record your results.

C. Measure the thickness of one index card to the nearest 0.1 cm.
   1. What is your measurement?

D. Use your ruler to measure the thickness of ten index cards that are held tightly together. Find the thickness of one card by dividing by ten. Record this in your table to the nearest hundredth of a cm.

E. Find the volume of one index card (multiply its length × width × thickness). Use the thickness obtained in step D. Record this.

F. Determine the mass of one index card. Record in the table.

G. Determine the mass of the stopper and record in your table.

H. Find the volume of the rubber stopper by doing the following:
   2. Fill the graduated cylinder about half-full with water. Note and record the volume of water. Remember to read to the bottom of the curved surface of the water.
   3. Carefully lower the stopper into the cylinder until it is completely underwater. Note and record the new volume of water.
   4. The difference in the two volumes is the volume of the solid. The measurements on the graduated cylinder are in milliliters (mL). Each 1.0 mL is equal to 1.0 cm^3. Record the volume of the irregular solid in your table to the nearest cm^3.

CONCLUSIONS:

1. Name and give the symbols for the units of length, mass, and volume of the SI system used in this activity.
2. Explain why ten index cards were used for the measurements made in steps C and E.
3. Why do you think the volume of the stopper was measured using a graduated cylinder rather than a ruler?
4. If you were given a cube and a small ball, explain how you would measure the volume of each.

SKILL-BUILDING ACTIVITY

USING A BALANCE

PURPOSE: A balance is an instrument used in determining the mass of an object. You need to know how to use the balance properly to measure mass accurately.

MATERIALS:

balance   objects to weigh

PROCEDURE:

A. A balance must always be checked with no mass on it. It should be placed on a level surface and adjusted until the indicator shows that it is level. When the pointer moves equally on both sides of the center of the scale, it is in balance. See Fig. 1–20. An equal-arm balance, like the one in Fig. 1–20, has right and left sides that are the same size and shape. If your balance is an equal-arm balance, find the mass of each of two different objects by doing the following:
   a. Place the object of unknown mass on the left pan of the balance.
   b. Place known masses on the right pan until the pointer again moves equally on both sides of the center of the scale.
   c. If your balance has a numbered scale and a rider on that scale, you can use the rider to help balance the unknown mass.
   d. The unknown mass is equal to the total of the known masses plus the rider reading.

B. Another kind of balance is the triple-beam balance shown in Fig. 1–15. It has one pan and three movable riders. If your balance is a triple-beam balance, weigh two objects by doing the following:
   e. Place the object of unknown mass on the pan.
   f. Move the riders until the pointer moves equally on both sides of the center of the scale.
   g. The unknown mass is equal to the total of the readings of the riders. Always report mass to the smallest unit on the scales. For example: back rider = 10 g, middle rider = 100 g, front rider = 8.6 g; the mass is 118.6 g.

CONCLUSIONS:

1. Describe how you would use an equal-arm balance to find the mass of an object.
2. Describe how you would use a triple-beam balance to find the mass of an object.

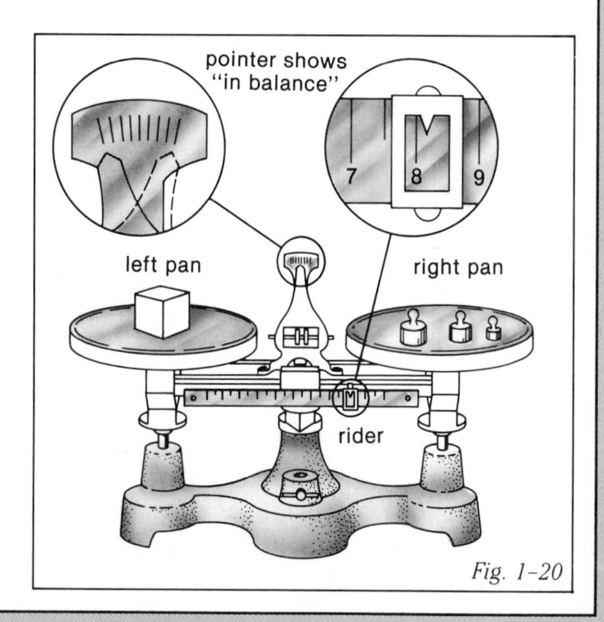

Fig. 1–20

TECHNOLOGY

INTRODUCTION

Welcome to the future, a world created by scientific research and development, a world of questions and possibilities. Science is much more than knowledge of facts that have already been discovered. Science is an activity for answering questions and seeking new knowledge. It is also an attempt to explain what new facts mean and how these facts can be used. Science is also the foundation upon which new technology is built. In the Special Features sections, we have tried to inform you of current research and technological developments that have actually extended the subject matter of science or changed the nature of what you must learn in order to live and work in modern society. The Special Features sections go beyond textbook material to encourage you to explore, on your own, events in the scientific world—events that are constantly changing and that often raise more questions than they answer. For example, you will read about weather satellites helping to save crops and lives, about new technology creating another gold rush in California, about computerized images from heat-sensing satellites charting warm ocean currents, and of the possibility of harnessing the wind energy to supply a share of our electrical energy.

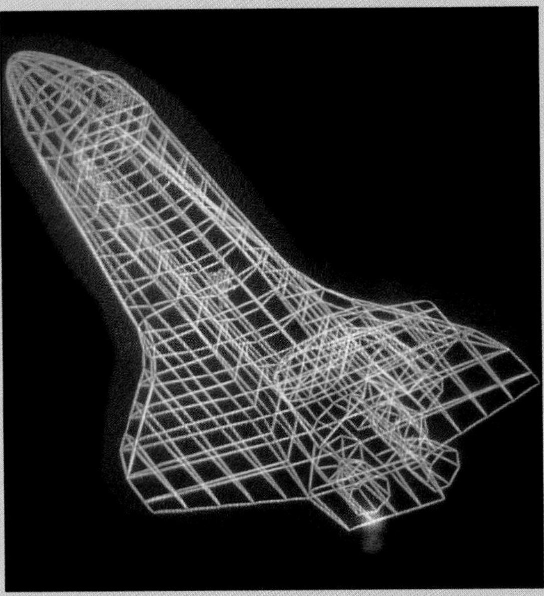

The earth and planetary sciences are undergoing amazing changes because of the information available through new technologies. The new technologies, the result of the application of scientific thought and research, have also created a society of different jobs and careers. To find work in today's job market, you have to understand the changes brought about by the addition of new technologies and the elimination of old ones. For example, typographers used to set books into print using physical type —that is, actual letters engraved on metal. They arranged the letters line by line for each page of print. The pages of this book were not printed that way. Instead, the words were entered into a word processor, stored on a floppy disk, and became type when a compositor programmed a computer to translate the words on the floppy disk onto printed pages. Even the owl in the computer section was not drawn by a human artist; it was made by a computer. Changes such as these mean learning skills different from those that people had to learn even as recently as ten years ago. Computers are becoming a fact of daily life, and learning to use them is essential. The section titled Compute! will help you to learn science through computers and to learn more about computers while applying information in your text to the programs. That section includes points of interest for both the student who wants an introduction to computers and one who is looking for new software to try.

In each Special Features section, you will find words that may not be completely familiar to you, such as *rotor, oscillation,* and *sensor.* As you learn to use these words comfortably, your understanding will increase. Be patient with yourself; this is only the beginning of a fascinating journey into a future of many possibilities.

CHAPTER REVIEW

VOCABULARY

On a separate piece of paper, match the number of each sentence with the term that best completes it. Use each term only once.

observation	scientific theory	liter (L)	kilogram (kg)
experiment	scientific method	meteorology	Celsius
hypothesis	measurement	oceanography	astronomy
geology	meter (m)	density	

1. The branch of science dealing with the universe beyond the earth is ______.
2. A scientist who studies volcanoes works in the branch of earth science called ______.
3. A(n) ______ is made by counting something.
4. Hypotheses are tested by planning a(n) ______.
5. A(n) ______ is something you know from seeing, hearing, or feeling.
6. The basic unit of length in the metric system is called the ______.
7. A(n) ______ is a possible answer to a question.
8. The ______ is a metric unit of volume.
9. When a hypothesis is found to be correct, it becomes part of a(n) ______.
10. The ______ is the SI unit of mass.
11. The temperature scale widely used in science is the ______ scale.
12. ______ is the study of the oceans and the ocean floor.
13. A guide used by scientists in solving problems is called the ______.
14. One who studies weather works in the field of ______.
15. ______ is the amount of mass in a given volume of a substance.

QUESTIONS

Give brief but complete answers to each of the following questions. Unless otherwise indicated, use complete sentences to write your answers.

1. What is earth science?
2. How did the scientific study of the earth begin?
3. Why are volcanoes a part of geology?
4. Why is astronomy a branch of earth science?
5. How can understanding earth science save lives?

6. Describe how scientific thinking starts.
7. In your own words, describe the five steps that may be used as a guide to scientific problem solving.
8. When does a scientific hypothesis become a part of a scientific theory?
9. When does a scientific theory become a scientific law? Give an example of a scientific law.
10. How are experiments usually done in earth science?
11. Why are controlled experiments considered ideal experiments?
12. Why do scientists use SI units for measuring?
13. What is the purpose of making the same measurement many times?
14. Convert: (a) 80 cm to meters; (b) 30 cm to millimeters; (c) 12 mm to centimeters.
15. Convert: (a) 512 g to kilograms; (b) 6.12 L to milliliters; (c) 450 ml to liters.

APPLYING SCIENCE

1. Write a job description for (a) a geologist, (b) an astronomer, (c) an oceanographer, and (d) a meteorologist.
2. Write at least six observations of your classroom. Then compare your list with the lists of your classmates. (a) Were your observations written in the same or in a different way? (b) Did any of the observations use numbers? Were these observations easier to understand? Why?
3. Collect labels, containers, etc., that show SI units. Show your collection to the class.

BIBLIOGRAPHY

American Metric Council. *The Metric System Day by Day.* Washington, D.C.: American National Metric Council, 1977.

Campbell, Norman Robert. *What is Science?* New York: Dover.

Davies, Paul. *The Edge of Infinity.* New York: Simon & Schuster, 1982.

Gore, Rick. "Eyes of Science." *National Geographic,* March 1978.

National Bureau of Standards. *The International System of Units (SI).* Special Publication 330. Washington, D.C.: U.S. Dept. of Commerce, 1981.

Zubrowski, Bernie. *Messing Around with Drinking Straw Construction.* Boston: Little, Brown, 1981.

UNIT 1

THE ACTIVE EARTH

CHAPTER

2

This Landsat photo from an orbiting spacecraft shows the beauty and variety of the earth's surface. Urban areas (red), forests (green), marshes (blue) are 3 of 38 identifiable areas.

A UNIQUE PLANET

CHAPTER GOALS

1. Develop a hypothesis that explains present observations of the solar system and how it began.
2. Explain how information about the inside of the earth is obtained.
3. Develop models to represent the earth's size, shape, and surface.
4. Describe the location of a place on the earth.
5. Describe the earth's motions using observations of the sun and stars.

2-1. Origin of the Earth

At the end of this section you will be able to:

- Use present observations of the solar system to explain how it began.
- Explain the collision and dust cloud hypotheses.
- Describe why the dust cloud hypothesis is used to explain how the solar system formed.

The photo at the left is a view of the surface of the earth. It was taken by *Landsat*, a spacecraft above the earth. Looking down on the earth from space has given scientists a new view of the planet we live on.

THE EARTH AND PLANETS

Any hypothesis that explains how the earth formed must also explain certain observations. First, the earth is one of nine planets and other bodies that move around the sun. The earth, planets, sun, and smaller bodies are called the **solar system.** Each planet in the *solar system* follows a path or *orbit* around the sun. A hypothesis that explains how the earth formed should explain why the earth is part of the solar system. Second, the hypothesis should explain why the orbits of the planets fall on nearly the same plane. The planets move as if they were on a flat surface such as the top of a desk. Third, the hypothesis also should explain why the planets are different. Four of the planets closest to the sun are like the earth. They are made

Solar system The sun with its nine planets and smaller bodies that move around it.

Fig. 2-1 Formation of the solar system: (a) a slowly rotating gas-and-dust cloud begins to shrink; (b) rotation accelerates, forming a disk with matter collecting at the center; (c) shrinking continues as the sun forms, with rings of material circling the new star; (d) the planets grow from the rings, following paths around the sun.

mostly of heavy solids. However, four of the planets far from the sun are made mostly of light gases.

THE COLLISION HYPOTHESIS

In science a hypothesis is often changed. Sometimes a better hypothesis is developed. The new hypothesis might be able to explain observations better. One hypothesis tried to explain the origin of the earth and other planets by saying that the planets formed when another star passed very close to the sun. The passing star then tore large pieces of matter away from the sun. These pieces of the sun cooled and formed bodies that became the planets. This hypothesis has at least two problems. Material drawn out of the sun would be so hot that it would expand instead of contract. Thus solid planets could not form. The hypothesis also says that the sun and another star would have to come very close. Observations show that stars are spread very far apart in space. The chance of such a near collision is very small. These problems, along with others, led scientists to search for a better hypothesis.

THE DUST CLOUD HYPOTHESIS

The hypothesis accepted by most scientists today is the dust cloud hypothesis. See Fig. 2-1. It says that the solar system came from a huge cloud of dust and gas in space. Such clouds can be seen with telescopes. The cloud of dust and gas was much larger than the solar system is today. This hypothesis says that the solar system started to form when this cloud of gas and dust began to shrink and spin slowly. The explosion of a nearby star may have caused the cloud to shrink. As it became smaller, it turned faster. The same thing happens to spinning skaters. When they pull their arms in, they spin faster. See Fig. 2-2.

As the cloud turned faster, it also became flatter. The shape began to look like a giant wheel or disk. This became the plane of the solar system. At the same time, most of the matter was gathering into a body around the center of the disk. As this body grew, it became hotter. Thus a hot new star, the sun, was formed. As the disk of gas and dust was still turning around the sun, smaller bodies were also beginning to grow. These smaller bodies were too small to have enough heat to become stars. They became the planets and smaller bodies of the solar system. Thus the sun came to have a system of large and small

Fig. 2-2 When a skater draws her arms closer to her body, she spins faster.

bodies around it. They moved in orbits on the plane of the disk from which they grew. Fig. 2-1 shows how the solar system formed from a cloud of gas and dust.

After the planets formed, they began to change. In the hot inner part of the young solar system, the heat from the sun caused the planets to lose most of the lighter gases. The heavier solids remained. The planets far from the sun were in the cold outer parts of the solar system. They were able to hold the lighter gases. This is one reason why the planets closest to the sun are different from the planets far from the sun.

At the present time, the dust cloud hypothesis is able to explain many observations. Future observations may show this hypothesis to be correct. Or the hypothesis may be changed as new observations are made.

SUMMARY

The collision and dust cloud hypotheses try to explain many observations of the solar system. The dust cloud hypothesis is accepted by most scientists today. It says that the solar system formed from a spinning cloud of dust and gas.

QUESTIONS

Use complete sentences to write your answers.

1. What observations must a hypothesis explain for how the earth formed?
2. Using the collision hypothesis, explain how the earth and other planets formed.
3. Why is the collision hypothesis no longer accepted by most scientists?
4. Using the dust cloud hypothesis explain how the solar system formed.

INVESTIGATION

A MODEL OF THE DUST CLOUD HYPOTHESIS

PURPOSE: To make and study a model of the dust cloud hypothesis.

MATERIALS:

celery seeds — round pan or dish
small spoon — container for water

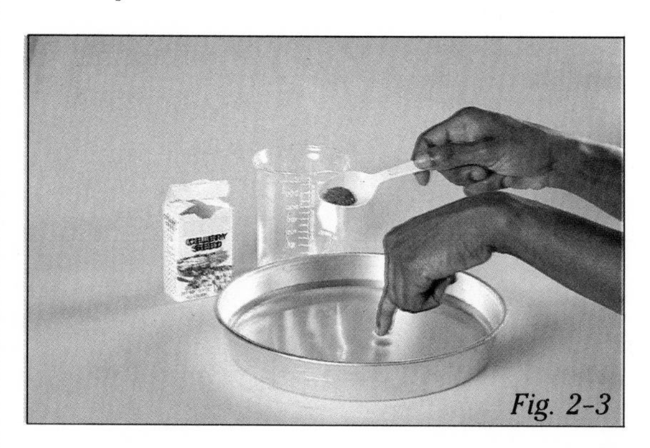

Fig. 2-3

PROCEDURE:

A. Place the pan on a flat surface. Fill it about half-full of water. Obtain a half-teaspoon of celery seeds. Sprinkle the seeds over the surface of the water as evenly as possible. See Fig. 2–3. Have your partner observe what happens as this is done. Observe only the seeds that float on the water.
   1. What was the very first thing that happened to the floating seeds?
   2. What happened to the floating seeds after a few minutes?
   3. Which part of the dust cloud hypothesis would this show?

B. Observe what happens to the floating seeds for a few more minutes.
   4. What do you observe happening to the seeds?
   5. Which part of the dust cloud hypothesis would this show?

C. Observe the seeds until there is no longer any movement. The forces that are causing this reaction allow you to see, in a very short period of time, what would take billions of years for gravity to achieve on the scale of our solar system.
   6. Describe the pattern of seeds.
   7. What do you think the pattern will be like in a few hours?

D. Pour out the water and seeds into the container given by the teacher.

E. Once again, place the pan on a flat surface. Then fill the pan about half-full of water. Obtain another half-teaspoon of celery seeds. Use your finger to stir the water in the pan to make it move in a circle. Remove your finger and then, immediately but slowly, sprinkle the celery seeds over the moving water surface.
   8. In the very beginning, what happens to the floating seeds?

F. Watch the seeds until all the motion has stopped.
   9. Describe the final pattern of seeds after motion has stopped.

CONCLUSIONS:

1. In what way does this model act like the dust cloud hypothesis?
2. In what ways does this model differ from the dust cloud hypothesis?
3. How is the second model a better one for the dust cloud hypothesis?

2-2. Birth of a Planet

When you finish this section you will be able to:

- Explain how scientists study the inside of the earth.
- Describe the inside of the earth using the terms "core," "mantle," and "crust."
- Explain how the earth might have formed continents, oceans, and an atmosphere.

The moon is the earth's closest neighbor in space. Using the moon as a laboratory has helped scientists learn about the earth's early history. For example, a rock taken from the moon looks like many rocks you can pick up on the earth. See Fig. 2-4. But moon rocks tell a different story. Almost all rocks on the earth's surface have been changed many times during the earth's history. However, moon rocks have hardly changed since the time the earth began. Unlike earth rocks, moon rocks give scientists information about the early solar system. Since the earth formed as a part of the solar system, the moon's story is also about the earth's beginning. From what we know about the moon, we can make a model of the early earth.

Fig. 2-4 An astronaut uses a special rake to collect rock chips from the surface of the moon.

EXPLORING THE INSIDE OF THE EARTH

Somehow, in its early history, the inside of the earth must have formed layers. Earthquake waves show the inside of the earth to be made up of layers. You probably have felt vibrations caused by a passing train or truck. The vibrations pass through the ground as a kind of wave. Earthquakes also produce waves. Large earthquakes can produce some waves that pass through the earth. When those waves return to the earth's surface, they carry information about the materials inside the earth. This is similar to the way X-rays can give information about the inside of your body. It has helped scientists to "see" the inside of the earth.

THE EARTH'S LAYERS

Earthquake waves show that the inside of the earth is made up of three main layers. Covering the outside of the earth is a thin, solid layer called the **crust.** The thickness of the *crust* differs from place to place. The parts of the crust that make up the sea floor are from 4 to 7 km thick. Under the continents, the crust is about 35 km in thickness. Beneath some mountains, the crust is as much as 70 km thick. Since we can study the crust directly, we know more about it than we do about the other layers. However, the crust makes up only 1 percent of the earth's volume.

Crust The thin, solid outer layer of the earth.

In the early 1900's a Yugoslavian scientist named A. Mohorovicic observed that earthquake waves changed their speed. They changed their speed a short distance beneath the earth's surface. He believed that this was because the waves had entered a kind of rock different from the crust. Below the crust, another layer of rock had been found. This layer was called the **mantle.** The boundary between the crust and the *mantle* is now called the *Moho,* after Mohorovicic. The mantle is a very thick layer of about 2,870 km. It makes up about 80 percent of the earth's volume and almost two-thirds of the earth's mass.

Mantle The layer of earth below the crust having a thickness of about 2,870 km.

Below the mantle, deep in the earth's center, is a very heavy **core.** The *core* has a radius of about 3,500 km. The outer core is liquid. The outer core's thickness is about 2,000 km. The inner core is solid. The inner core's radius is about 1,500 km. The core makes up 19 percent of the earth's volume and nearly one-third of its mass. Only rocks made of iron and nickel are heavy enough to compare to the mass of the core. Most of the

Core The center layer of the earth, made up of an inner and an outer part, having a radius of about 3,500 km.

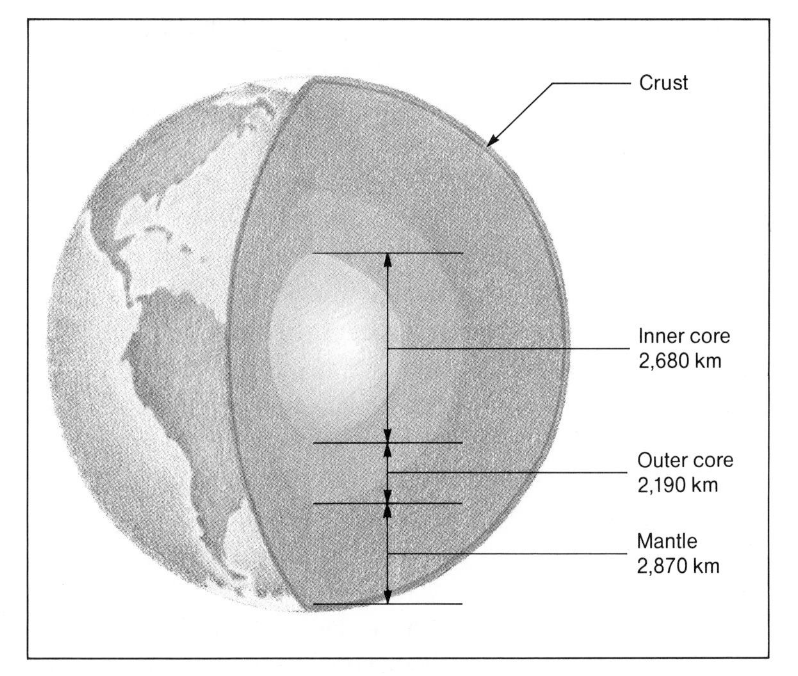

Fig. 2-5 The interior of the earth is made up of layers, as shown in this diagram. The thickness of the crust is thinner than can be shown correctly in a diagram this size.

core is thought to be made up of these two metals. Fig. 2-5 shows the layers of the earth.

People who drill into the earth for minerals and oil find that the rocks deep in the crust are hotter than those on the surface. The deeper they drill into the earth, the higher the temperature. Measurements in the upper part of the earth's crust show that the temperature goes up about 30°C for each kilometer in depth. Scientists believe that the temperature in the mantle increases to about 3,000°C. Thus scientists find the earth to be made up of three main layers and a very hot interior. How can this be explained?

THE EARTH'S BEGINNING

Moon rocks show that the moon and planets of our solar system formed about 4.5 billion years ago. The moon and earth share much of their early history. Most moon rocks are similar to rocks found around volcanoes on the earth. This means that the moon was once very hot. During that time, the inside of the moon formed layers. This may be similar to the way the earth formed its layers. Planets, such as the earth, that are made up of layers are said to be **differentiated.** Scientists are not completely sure about how the earth became *differentiated* because the process took place so long ago. But layers and heat in

Differentiation The process by which planets became separated into layers.

the earth show that the earth once was very hot. Because of the heat, the entire earth was made up of melted materials. Then heavy substances such as iron sank and formed the core. Lighter materials floated and became the mantle and crust. In time, the young planet cooled. The crust and much of the mantle became solid. However, some of the heat still remains inside the earth.

Among the materials that formed was water. This water was locked up in certain minerals. As the earth began to cool, hot, melted rock poured out from the inside of the earth. Continents probably formed from the lightest rocks that cooled. The light rocks formed solid islands that floated on the still-soft crust. Water vapor escaped and turned into liquid water. In time, enough water gathered on the surface to form the world's first oceans. At the same time that water vapor was escaping from the earth's interior, other gases were also given off. The lightest of these gases escaped into space. Others were held by the earth's gravity. Thus a thin covering of gases remained over the surface to form the earth's first atmosphere.

Scientists believe that about four billion years ago the earth ended its development. All of the forces still working today to change the planet were set into motion. Today's earth was shaped by these forces.

SUMMARY

Most of what scientists know about the inside of the earth is based on earthquake waves. The earth is divided into three main layers—the core, mantle, and crust. The earth's continents, oceans, and atmosphere were formed because of heating and melting.

QUESTIONS

Use complete sentences to write your answers.

1. How do earthquakes give information about the inside of the earth?
2. Describe each layer of the earth.
3. Describe how the earth might have formed layers.
4. Explain how the continents, oceans, and atmosphere formed on the earth.

INVESTIGATION

THE ORIGIN OF A LAYERED EARTH

PURPOSE: To use observations from two experiments to show how layers were formed in the earth.

MATERIALS:

250-mL beaker
soil sample
heavy small rock
light small rock
water

PROCEDURE:

A. Fill the beaker about half-full of water. Then obtain 60 mL of the soil sample. Add the soil sample to the water in the beaker and stir. Allow the mixture to sit undisturbed for about five minutes.
   1. What evidence of layering do you find in the beaker?

B. Look closely at the material that has settled to the bottom.
   2. Compare the size of the particles at the bottom with those found at the top.

C. Look closely at the top of the water in the beaker.
   3. Describe any particles floating on the surface.
   4. How does the size of these particles compare with those that sank to the bottom?

D. Pick up the two rock samples one at a time.
   5. Which rock, the heavy one or the light one, would you predict to fall faster in the water?

E. Test your prediction by experiment. Use the same container of water and soil from before. Drop both rocks at the same time from just above the surface of the water. Your partner will observe which hits bottom first and what happens when they strike bottom.
   6. Was your prediction correct?

F. The table below shows the average density of the layers of the earth.

Layer	Density (g/cm^3)
Crust	2.8
Mantle	4.5
Core	10.7

   1. What is the density of the crust?
   2. The density of the earth is 5.5 g/cm^3. Compare the density of the crust with the density of the earth.
   3. How do you explain the difference?
   4. Explain how scientists were able to determine the density of the mantle and the core from knowing the density of the crust and the earth.

CONCLUSIONS:

1. If the earth was melted at one time, would this have helped layers to form? Explain your answer.
2. If the earth was melted, would light material or heavy material end up in the center as a core?
3. If there are layers inside the earth, would the materials of which they are made be similar?
4. Why are sources of heavy metals most often found in those areas where volcanoes once were active?

2–3. Models of the Earth

When you finish this section you will be able to:

- ☐ Show how the size and shape of the earth can be found.
- ☐ Describe how maps are made from globes.
- ☐ Read information from maps.

In 1492, when Columbus sailed to find the Far East, some of the crew nearly forced him to turn back from his voyage. At that time, many believed the world was nearly flat. Some crew members were afraid that the ships would come to the edge of the world and fall off. People living on the earth cannot see that the earth's surface is curved. However, you have seen pictures taken from space that clearly show the earth as round. See Fig. 2–6. But how could you prove the earth to be round from your own direct observations?

Fig. 2–6 The earth as seen from outer space.

SIZE AND SHAPE OF THE EARTH

The ancient Greeks knew that the earth was round. About 300 B.C. Aristotle gave two reasons for believing in a round earth. He observed that when the earth's shadow was cast on the moon, it had a curved shape. Also, people moving north from Greece reported that stars moved higher in the sky. The stars seem to move lower in the sky as the people moved south. Aristotle said that this could only be explained if the earth was round. See Fig. 2–7.

Fig. 2-7 Evidence that the earth is round can be seen in the way that the North Star is seen higher in the sky when moving north or lower when moving south.

The ancient Greeks also were able to measure the size of the earth. A Greek named Eratosthenes (air-uh-**tahs**-thuh-nees) made this measurement about two thousand years ago. He observed that on June 22, at noon, the sun's rays reached the bottom of a deep well in Syene, a city in Egypt. He also observed that the buildings did not cast shadows. This meant that the sun was directly above Syene. On the same date at noon, Eratosthenes knew that in Alexandria, a city to the north, buildings did cast shadows. He asked how buildings in Alexandria could cast shadows and at the same moment buildings in Syene cast no shadows? The only possible answer was that the earth's surface was curved. See Fig. 2-8. The sun's rays did not strike all places on the earth's curved surface at the same angle.

Fig. 2-8 (a) On a flat surface, shadows cast by two different buildings of the same size would be the same length. (b) On a curved surface, shadows cast by two different buildings of the same size would be different.

Eratosthenes also saw that he could measure the distance around the earth by using the shadows. When the sun was not directly overhead, shadows would be cast. The sunlight would be striking everything at an angle. Eratosthenes was able to measure the angle of the sun's rays in Alexandria. He found the angle to be 7°. A complete circle is 360°. Seven degrees is about one-fiftieth of a complete circle. So the distance between Alexandria and Syene was one-fiftieth of the distance around the earth. See Fig. 2-9. Eratosthenes also knew that the distance between Alexandria and Syene was about 800 km. So 800 km was one-fiftieth of the distance around the earth. He then multiplied 800 km by 50. His result was 40,000 km for the distance around the earth. This was very close to the modern measurement of 39,989 km.

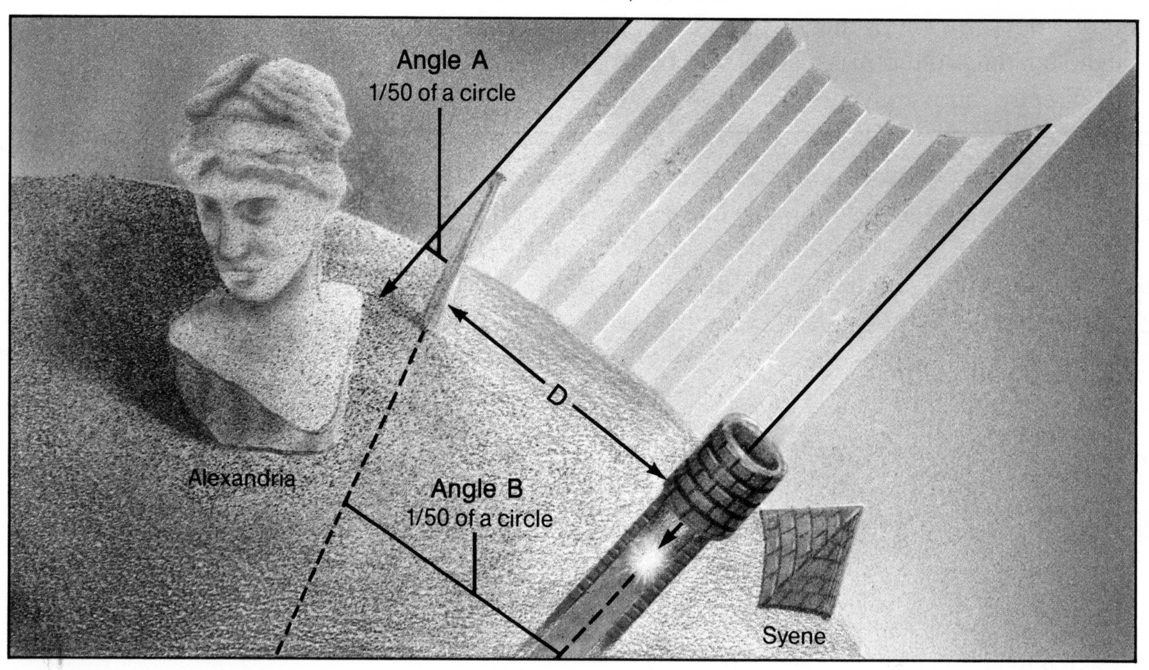

Fig. 2-9 Eratosthenes' method depended on measuring angle A. It was equal to angle B, which had the same relationship to a full circle that distance D did to a full circle.

Modern measurements have also shown that the earth is a little larger around the equator than around the poles. This means that it is not perfectly round. However, the earth is so close to being exactly round that its shape is said to be round.

MODELS OF THE EARTH

When we want to study the large features of the earth's surface, such as continents and oceans, some kind of earth model must be used. An example of an earth model is a globe. See Fig.

2–10. In many ways, a globe is the best model of the earth. Like the earth, the globe is round. But globes are often not easy to use. They are usually too small to show the earth's surface with much detail. A globe also cannot be carried easily. Because of this, maps often are used as models of the earth. Maps can show the earth's surface on a flat sheet of paper.

The earth's surface shown on a flat sheet of paper is always out of shape. Imagine trying to flatten a large piece of an orange peel. The curved shape of the peel will stretch and tear. In the same way, the curved surface of the earth, when shown on a flat surface, is stretched out of its true shape.

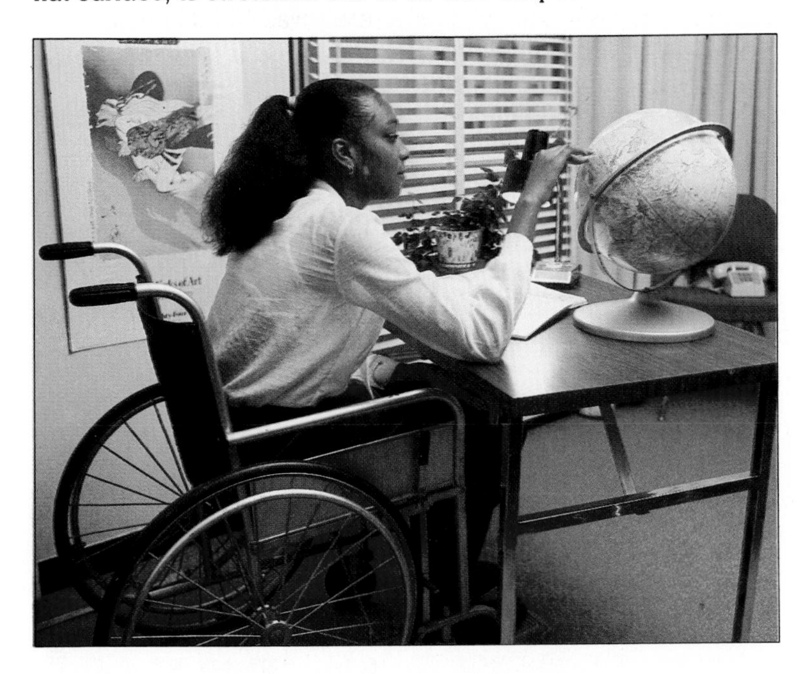

Fig. 2–10 A globe is a model of the earth.

MAKING MAPS

Making the curved surface of the earth into a flat map is done by using a **map projection.** There are different ways of projecting the curved surface of the earth onto a flat surface. No one way is perfect. Some projections distort the size of the continents, others distort the shape. You can imagine how a *map projection* is made by thinking of a globe made of glass. If you put a light inside the glass globe, the globe's surface could be projected on a piece of paper held next to the globe. One of the most common kinds of map projections is called a Mercator projection. It is made as if a sheet of paper were wrapped

Map projection The curved surface of the earth shown on the flat surface of a map.

around a globe, making a cylinder. See Fig. 2–11. This is the kind of map projection most often used in textbooks. Only the area of the earth's surface near the equator is shown accurately on this kind of map. Regions near the poles appear much larger than their true sizes. See Fig. 2–12.

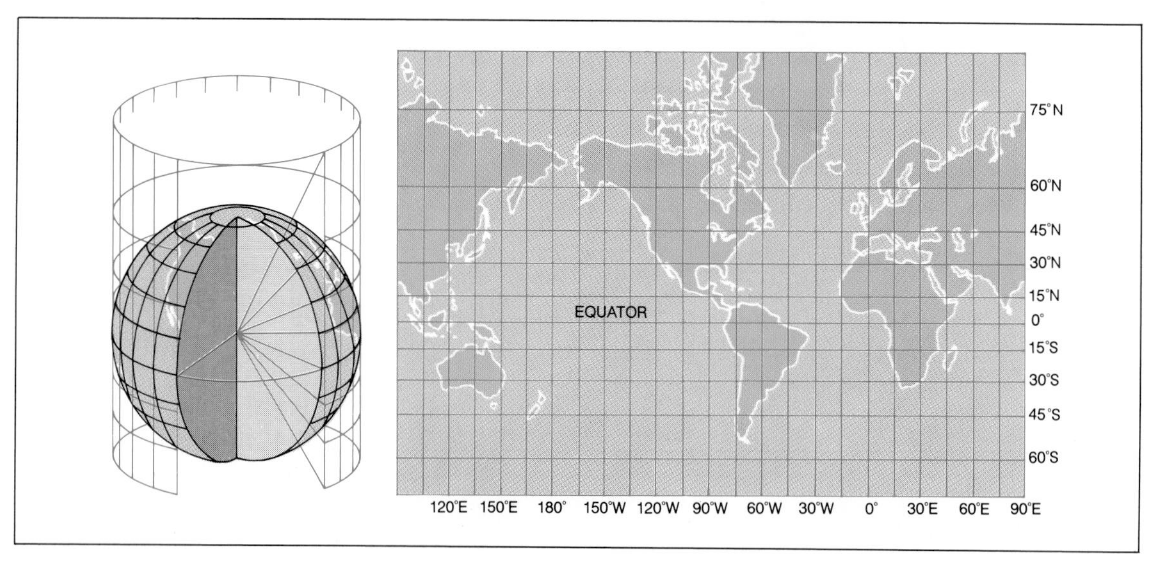

Fig. 2–11 (above) A Mercator projection shows only the areas of the earth's surface near the earth's equator accurately.

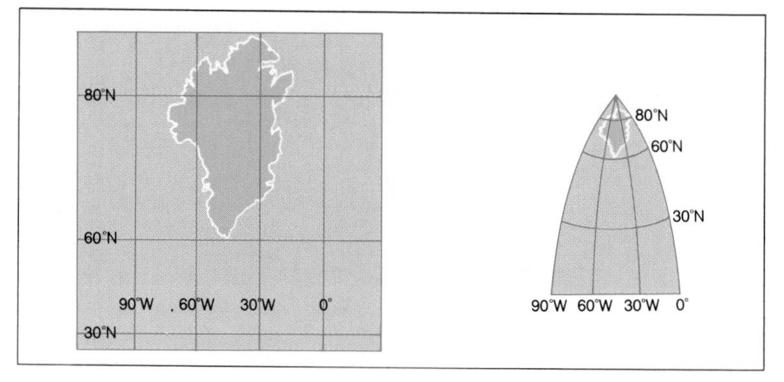

Fig. 2–12 (right) On the left is Greenland as it appears on maps using a Mercator projection. On the right is Greenland in its true size in relation to other masses.

Another type of map projection uses a cone instead of a cylinder. This is called a conic projection. See Fig. 2–13. Maps showing small parts of the earth's surface, such as road maps, use this kind of map projection. Suppose that a flat sheet of paper is held so that it touches a globe at a single point such as the North Pole. A map then can be projected up onto the paper. This kind of map is called a polar projection. See Fig. 2–14. A polar projection only shows the correct size of regions near the poles. As you can see, all maps are not the same. Each shows only part of the earth accurately and other parts out of shape.

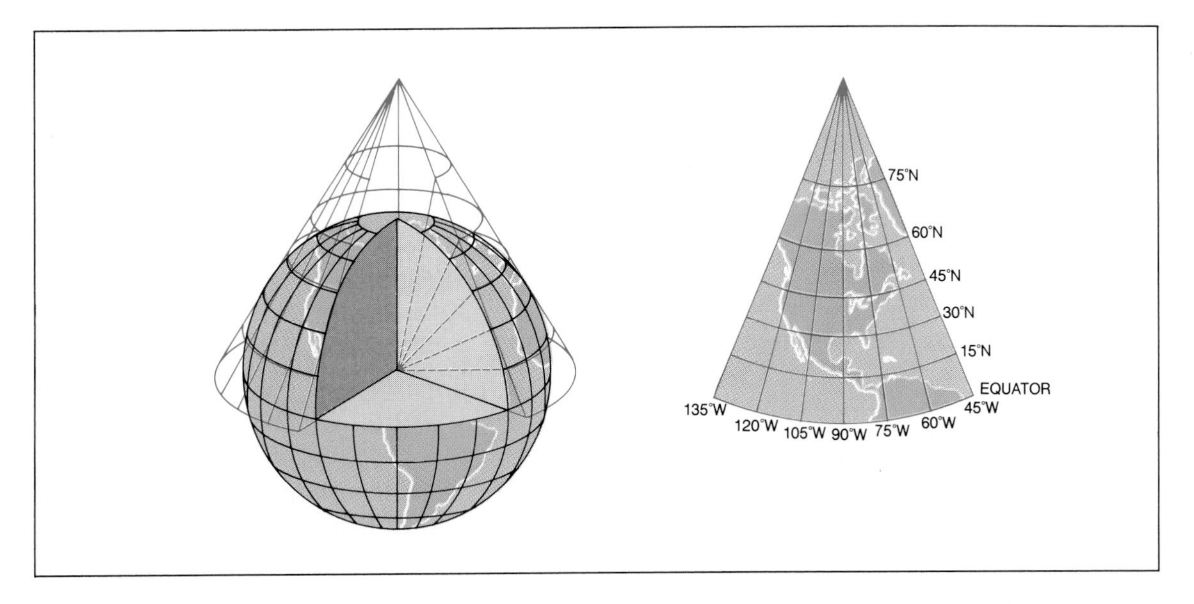

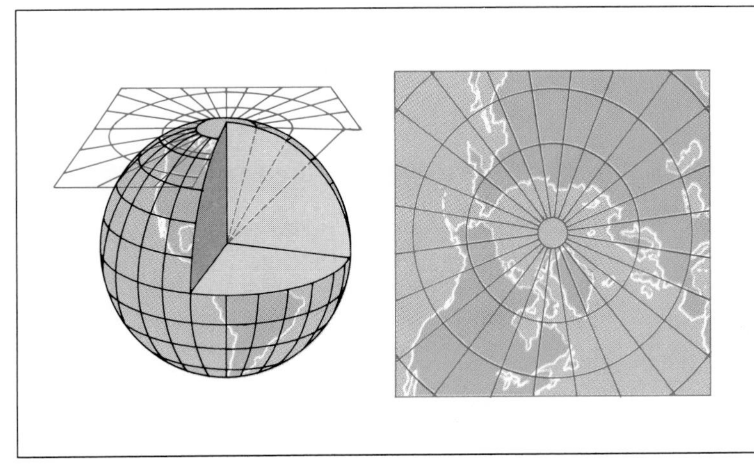

Fig. 2-13 (above) A map projection based on a cone. Maps made this way can show the true size of small areas of the earth's surface.

Fig. 2-14 (left) A polar projection shows the regions near the poles without distortion.

READING MAPS

A map can be used to provide different kinds of information about the earth's surface. For example, a *political map* shows boundaries between countries or states. These are human divisions of the earth's surface. Cities and towns are also shown. *Relief maps* show the land surface by using different colors for mountains and valleys. Political maps and relief maps are often combined. Weather can also be shown on maps.

Suppose that you want to use a map to guide you on a trip. The first thing you need to know is the direction you will be going. Common types of maps are made so that north is at the top and east is at the right when the map is held upright.

Scale The relationship between the distance on a map and the actual distance on the earth's surface.

You can also use a map to measure how far you will be going. This can be done by using the **scale** of the map. All maps have a *scale*. The scale shows the relationship between the distance on a map, or the size of a model, and the actual measurement that it represents. For example, a model airplane might have a scale of 1 to 1,000. This means that the model would be one-thousandth the size of the real airplane. Scales on maps show a relationship between distances. For example, 1 cm on a map might represent 1 km on the ground. The scale of the map may be shown as a bar graph. Or the scale may be written as 1/10,000. The numerator, 1, gives the map distance. The denominator represents the actual distance on the earth's surface. See Fig. 2–15. A map scale may also be written as 1:10,000. The first number is the distance on the map. The second number is the actual distance on the earth's surface. Thus, 1 cm on the map would be equal to 10,000 cm, or 100 m, on the ground. Most maps show the scale in a lower corner of the sheet.

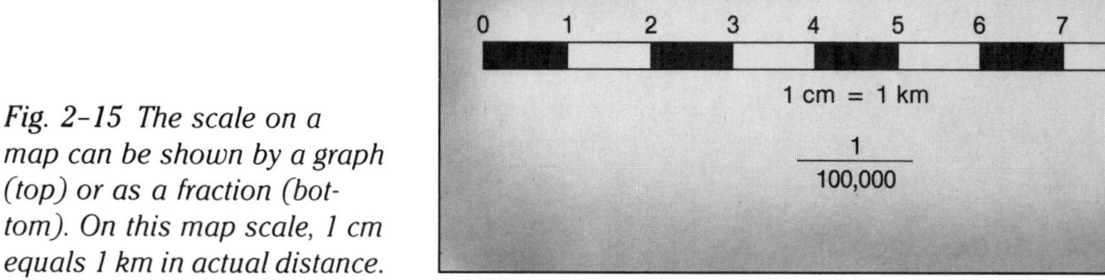

Fig. 2–15 The scale on a map can be shown by a graph (top) or as a fraction (bottom). On this map scale, 1 cm equals 1 km in actual distance.

Legend Information on a map that explains the meaning of each symbol used on the map.

Another kind of information found in the lower corners of maps are symbols. Symbols are used to show features such as roads, buildings, and railroad tracks. Some symbols look like the features they represent. To show what these and other symbols mean, maps have a **legend.** A *legend* explains the meaning of each symbol used on a map.

TOPOGRAPHIC MAPS

Topographic map A map that shows the shape of the land surface.

Contour lines Lines drawn on a topographic map joining points having the same elevation.

One type of map that is very useful in earth science is called a **topographic map.** See Fig. 2–16. This is because *topographic maps* show the surface features of the earth. This is done by using **contour lines.** Lines on a map that connect points having the same elevation, or height, are called *contour lines.* Contour means shape. Contour lines are used to show the shape of the land. Each contour line is labeled with the elevation of all the

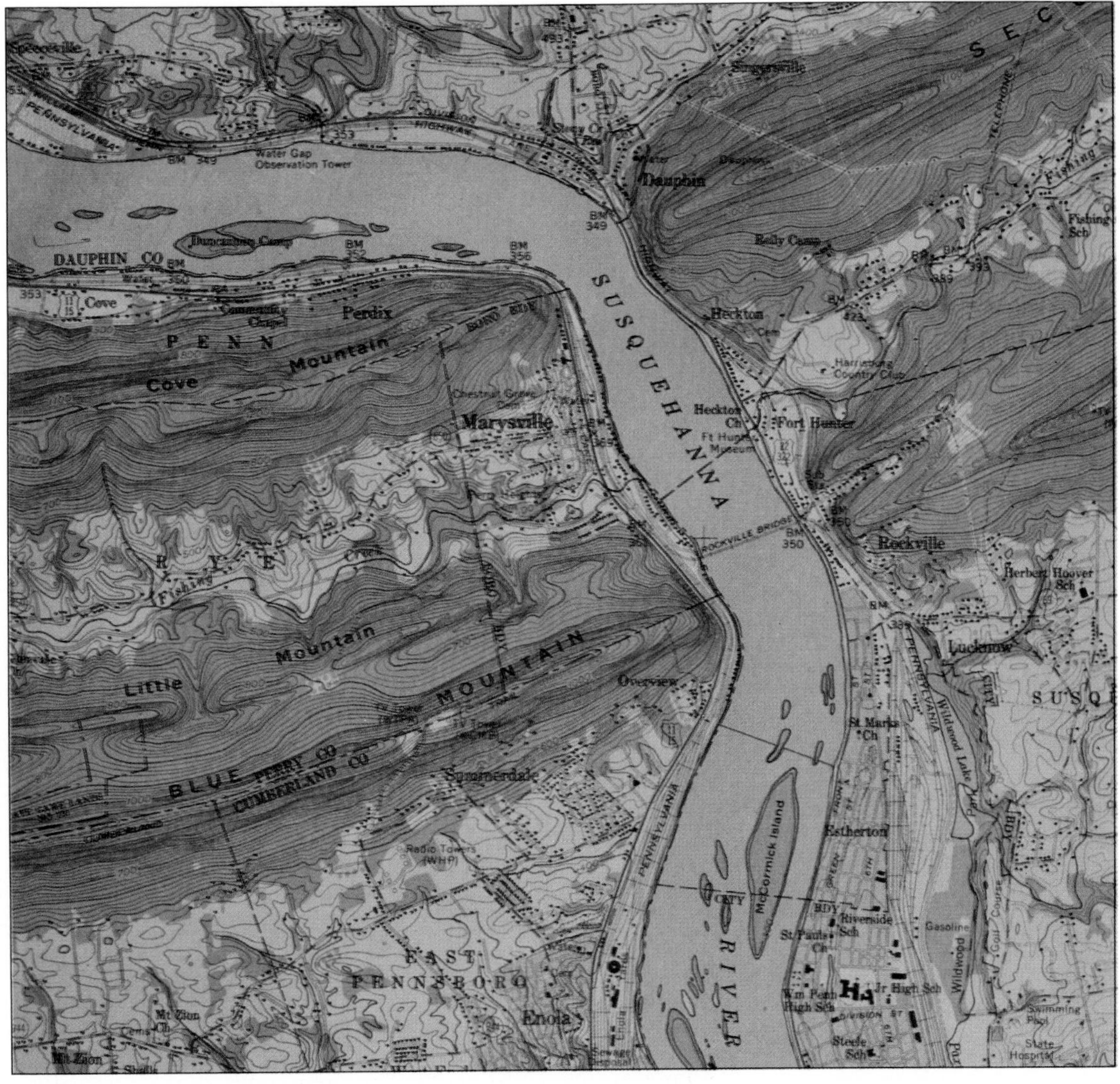

Fig. 2-16 A topographic map shows the surface features of the earth.

points found on the line. Fig. 2–17 shows how contour lines are drawn to show mountains and valleys. The difference in elevation between the contour lines is called the *contour interval.* The contour interval on different topographic maps is not always the same. The contour interval is chosen according to the size of the map and the elevations that must be shown. For example, the contour interval shown in Fig. 2–17 is 100 m. A smaller contour interval would cause the lines to be closer together and harder to see. Other symbols commonly used on topographic maps are shown in Fig. 2–18.

Fig. 2-17 (right) Contour lines show the changes in the elevation on the earth's surface.

Fig. 2-18 (below) Common symbols found on a topographic map.

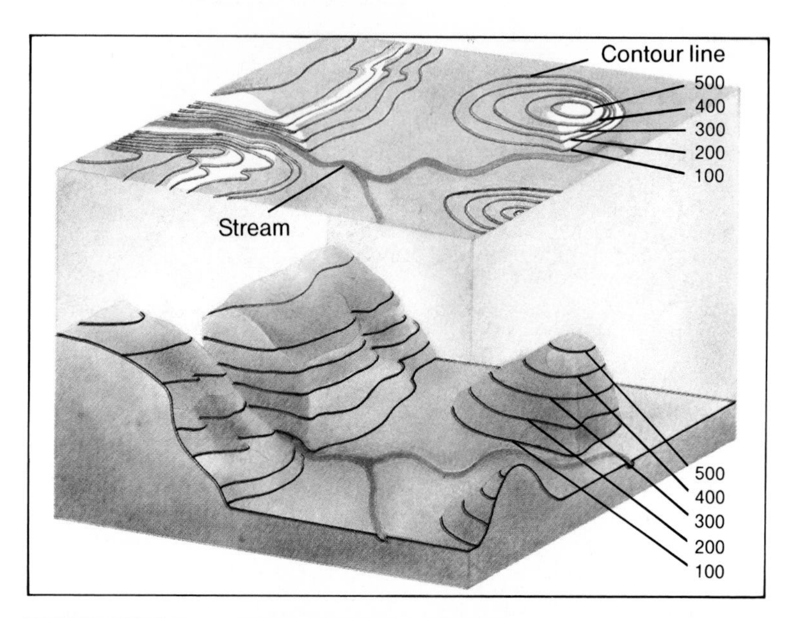

Symbol	Color	Meaning	Symbol	Color	Meaning
	Black	Buildings		Blue	Stream
				Blue	Stream that does not flow constantly
	Black	school and Church		Blue	Lake or pond
	Red or black	Road or highway		Brown	Depression
	Black	Railroad		Blue or green	Marsh or swamp

SUMMARY

The earth's size and shape can be found from observations. Maps are made by projecting the curved surface of the globe onto a flat surface. Maps give us information about the earth's surface.

QUESTIONS

Use complete sentences to write your answers.

1. Explain how the ancient Greeks were able to tell the earth's size and shape.
2. Name three kinds of map projections.
3. List three kinds of information that maps provide.
4. Describe the type of map that is very useful in earth science.

INVESTIGATION

CONTOUR MAPS

PURPOSE: To draw a contour map.

MATERIALS:

4 pipe cleaners	masking tape
soup plate	grease pencil
mixing bowl	metric ruler
plastic cup	2 sheets of paper

PROCEDURE:

A. Turn the plate, bowl, and cup upside down on a flat surface. Place the bowl on top of the plate and the cup on top of the bowl to form a "hill." Tape the three objects together. See Fig. 2–19.

B. Tape two sheets of paper together. Place the objects on the paper. Draw guide lines on the paper and on the plate so that you can align them when you remove the plate. Draw a line around the edge of the plate.

C. Hold the ruler straight up on the table. At the 5-cm mark, use the grease pencil to mark a line on the bowl. Make this measurement at several places around the bowl.

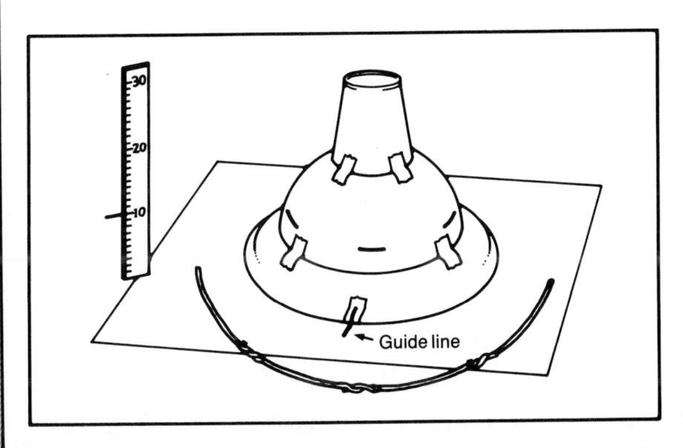

Fig. 2–19

D. Twist the ends of the pipe cleaners to form one long piece. Wrap it around the bowl at the 5-cm mark. Have a partner hold the pipe cleaner while you remove the taped objects. Place the pipe cleaner on the paper and trace the circle it makes.

E. Do the same for elevations of 10 cm and 15 cm.

1. How many circles have you drawn?
2. What are each of their elevations measured in centimeters? Label them.
3. If 1 cm = 2 m on your model, what contour interval would that represent?

F. Use the ruler to approximate the highest point of the hill.

4. What is the highest elevation in meters?
5. If there were a school at that elevation, how would you show it? Draw it in.
6. Many cups have an indentation on the bottom. How would you show this on your contour map?
7. How would you show a marsh between 10 m and 20 m?

CONCLUSIONS:

1. How could you change your model to show that one side is steeper than the other?
2. If you did that, what would the contour lines look like?
3. What would the contour lines look like if a stream ran down the hill?

Fig. 2-20 The location of any place on earth can be found. What two pieces of information are given here?

2-4. Position on the Earth's Surface

At the end of this section you will be able to:

- ☐ Describe a location by using latitude and longitude.
- ☐ Find a place on the earth's surface by using latitude and longitude.

If someone wants to find where you live, you can give the person an address. Your address probably is made up of two things. It has a street name and a number. In the same way, the "address" of any place on the earth's surface can be found. See Fig. 2-20. Even a place in the middle of the ocean can be found if it can be described as the point at which two lines cross.

LATITUDE

The equator is an imaginary line that divides the earth into two equal parts. It is drawn around the globe midway between the North Pole and South Pole. One of the kinds of lines used to find places is based on the equator.

Latitude The distance in degrees north or south of the equator.

Suppose that you want to find an island in one of the world's oceans. First you must know if the island is in the northern or southern part of the earth. Once you know which part, you need to know how far north or south of the equator the island is. This can be done by using other lines running parallel to the equator. Each of these lines is called a parallel of **latitude**. A parallel of *latitude* is measured by using the equator as a starting point of 0°. Then it is possible to find how many degrees north or south of the equator the parallel of latitude is. Such a description is called the latitude. Latitude tells the distance in degrees north or south of the equator. For example, the latitude of the North Pole is 90° north or 90°N. In the same way, the latitude of the South Pole is 90°S. A place halfway between the North Pole and the equator would have a latitude of 45°N. See Fig. 2-21. Every place on the earth's surface can be described by a latitude. Can you see why the latitude of a place can never be greater than 90°?

Latitude can be found by using the North Star. One way to do this is by measuring how far the North Star is above the horizon at night. The latitude can be found by measuring the number of degrees the North Star is above the horizon.

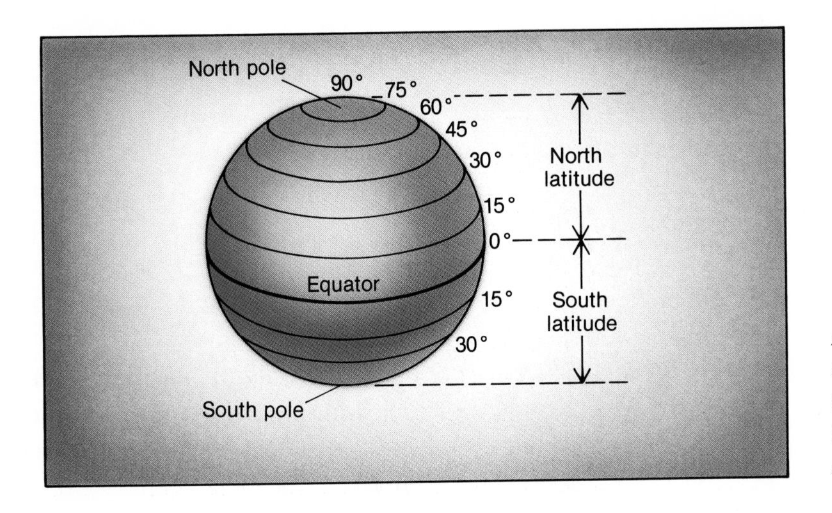

Fig. 2-21 Find the parallel of latitude that would pass through a place halfway between the equator and the North Pole.

LONGITUDE

The latitude of a place only describes how far north or south of the equator a place is. But you also need to know its east–west location. Another set of lines, called *meridians*, is used to show east–west locations. These lines are drawn from pole to pole. The prime meridian is used as the starting point of 0°. It runs through Greenwich (**Gren**-ich), England. The location of a place east or west of the prime meridian is called its **longitude.** *Longitude* shows the distance east or west of the prime meridian in degrees. For example, a place might have a longitude of 30° west or 30°W. It would be found on the meridian that is 30° west of the prime meridian. See Fig. 2-22. The east and west longitude lines meet at the 180° meridian on the other side of the earth.

Longitude The distance in degrees east or west of the prime meridian.

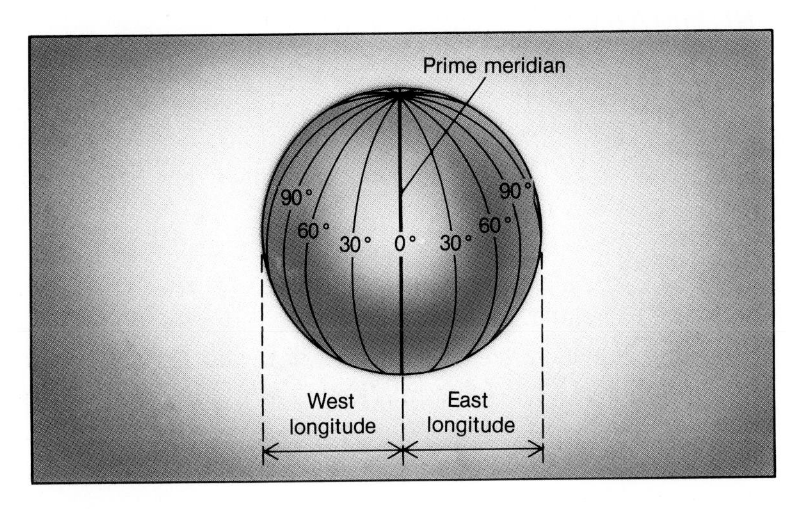

Fig. 2-22 Longitude measures the distance east or west of the prime meridian.

FINDING LOCATION BY LATITUDE AND LONGITUDE

By using both latitude and longitude, any place on the earth's surface can be found. For example, Washington, D.C., has a latitude of 39°N and a longitude of 77°W. Therefore, Washington can be found on the parallel of latitude that is 39° north of the equator. Washington is also on the meridian that is 77° west of the prime meridian. Washington, D.C., is found where these two lines meet. See Fig. 2-23. Thus the latitude and longitude of a place are like an address, showing where it is on the earth.

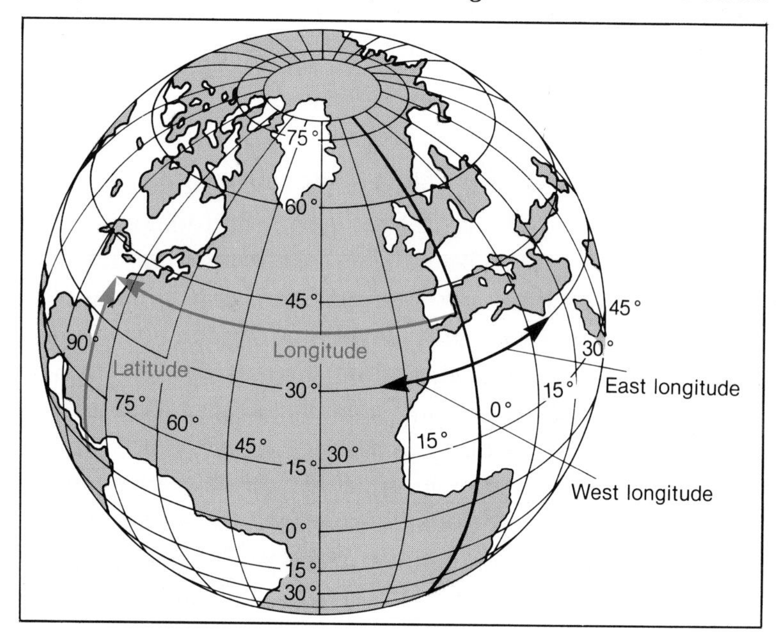

Fig. 2-23 Washington, D.C., is found at 39° N and 77° W.

SUMMARY

To find a place on the earth's surface, lines of latitude and longitude are used. Latitude is measured from the equator and longitude is measured from the prime meridian. Both latitude and longitude can be measured by using the position of the North Star and the sun.

QUESTIONS

Use complete sentences to write your answers.

1. Describe how you would find an island located near 15°S and 45°E.
2. What state is found at 60°N and 150°W?

INVESTIGATION

READING LATITUDE AND LONGITUDE

PURPOSE: To find places on a map using latitude and longitude.

MATERIALS:
pencil paper

PROCEDURE:

A. Study the map of the world in Fig. 2–24. Answer the following questions.
   1. At about what latitude and longitude is your school?
   2. At about what latitude and longitude is the capital of the United States?
   3. At about what latitude and longitude is the center of Africa?
   4. At about what latitude and longitude is the United Kingdom (England)?
   5. What longitude does not seem to pass through any land masses?
   6. What continent is found at 25°S and 135°E?
   7. What state of the United States is found at 20°N and 155°W?
   8. What country is found at 40°N and 140°E?
   9. What country is found at 25°N and 105°W?

CONCLUSIONS:

1. Can the location of a city be found by using only its latitude or only its longitude? Why?
2. Explain how the location of any place on the earth's surface could be described.

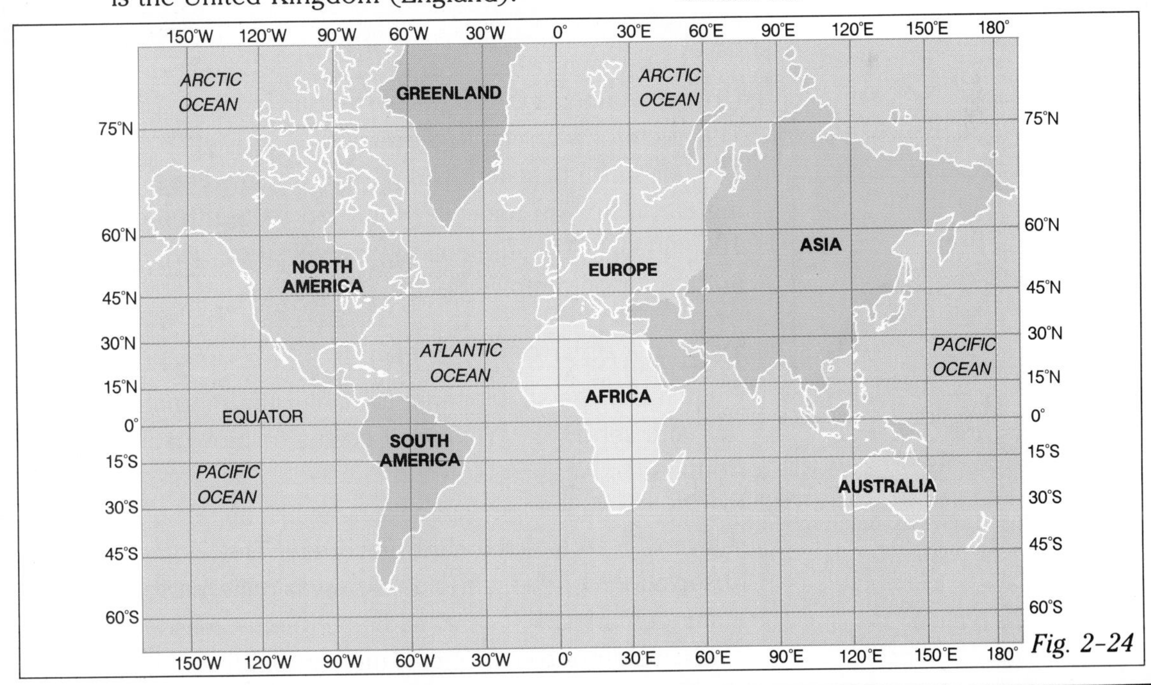

Fig. 2–24

2–5. Motions of the Earth

At the end of this section you will be able to:

- Explain the effects of the earth's rotation.
- Describe the earth's motion around the sun.
- Explain the relationship between the tilt of the earth's axis and the seasons.

Fig. 2–25 This photo shows the sun's position in the sky at the same time of day on different dates throughout the year. The figure-eight shape is caused by the sun's changing position in the sky.

How do you think the picture of the sun in Fig. 2–25 was made? The picture was made by someone moving at a speed of 106,000 km/h in space. The person was traveling on the moving earth. So you and everything on the earth are moving at 106,000 km/h. In this section you will study evidence that you are a rider on a moving earth.

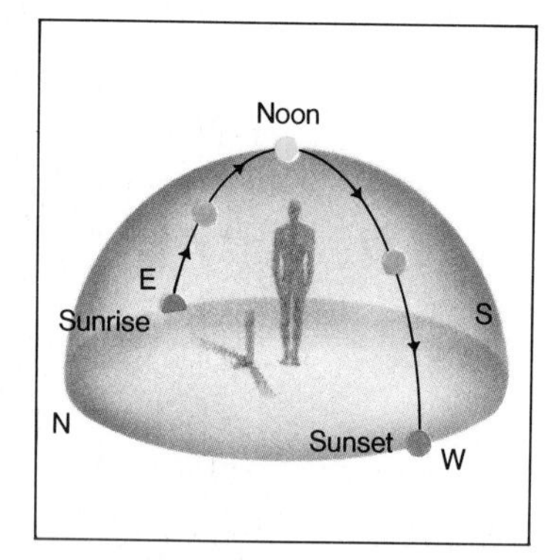

Fig. 2–26 The motion of the sun across the sky each day can be shown by the moving shadow cast by a stick.

THE EARTH'S ROTATION

Each day the sun appears to rise in the morning, cross the sky, and set in the evening. It seems to move across the sky from east to west, as shown in Fig. 2–26. There are two ways to explain the movements of the sun. First, it could be said that the sun moves around the earth each day. This idea explains the observations of the sun. This idea was believed to be true for many centuries. A Greek named Ptolemy reasoned that the earth could not be moving. If the earth moved, he said, a bird on a tree branch would be swept off. The idea that the sun moved around the earth was accepted for 1,500 years.

But in the 1500's a Polish scientist, Copernicus (koe-**pur**-nih-kus), proposed that the sun does not move around the earth. Why, then, does the sun appear to move across the sky? The correct explanation is that the earth is spinning like a bicycle wheel. This spinning motion is called *rotation*. Just as a wheel turns on an axle, the earth turns on its **axis**. The earth's *axis* is an imaginary line running through the center of the earth from pole to pole. The direction of earth's rotation is west to east. For example, if you put your finger on a globe and push toward the east direction on the globe, you will cause it to rotate west to east, the same way as the earth.

Axis The imaginary line through the center of the earth on which the earth rotates.

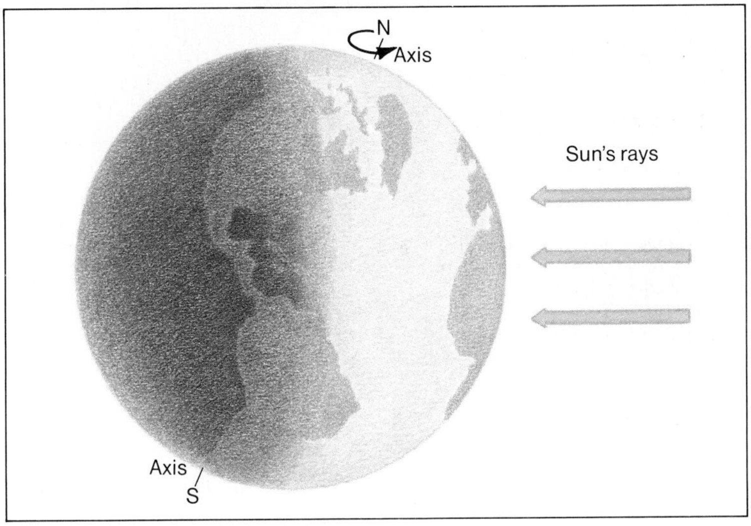

Fig. 2-27 The rotating earth produces day and night.

For everyone on earth, the most observable effects of the earth's rotation are day and night. As the earth rotates, every place passes through sunlight and darkness. See Fig. 2–27. A shadow made by a stick during a sunny day can be used to follow the path of the sun across the sky as the earth rotates. The shadow stick could be made into a sundial and used to tell time. When the sun is highest in the sky, the shadow's position on the ground could be marked twelve noon. Numbers could be put on each side of this mark to show daylight hours before and after noon. You would have made a kind of sundial. See Fig. 2–28. However, since the sun moves, people a short distance east or west of each other would have different times on the sundial. This would happen when people set their clocks

Fig. 2-28 A sundial can be used to measure the passage of time during the day.

for noon when the sun was highest in the sky. Because the sun moves from east to west, the sundials in the west would show noon later than those in the east. For example, Philadelphia is a short distance west of New York City. The sun is overhead in Philadelphia about four minutes later than in New York. Cities near each other would have different times. Thus the use of sundials is not the best way to tell time.

To prevent this problem, the earth has been divided into 24 equal north–south strips, or time zones. Each time zone covers 15° of longitude. The boundaries between time zones have been made so that they do not pass through cities. Within each zone, all places have the same time. This is called **standard time.** All clocks in a time zone are set to the *standard time* for that time zone. See Fig. 2–29. When you travel east and into the next time zone, you must set your watch ahead one hour. When traveling west and into the next time zone, you must set your watch back one hour.

Standard time The time based on one meridian and used for a given time zone.

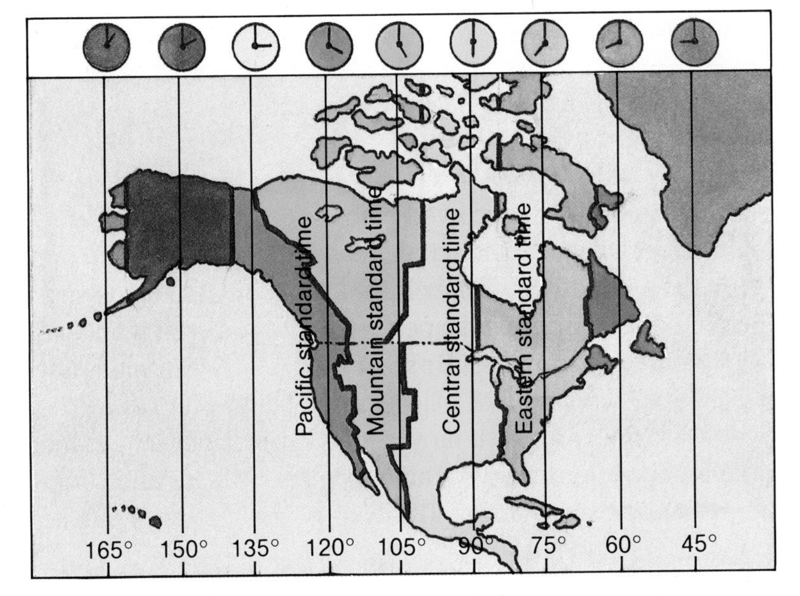

Fig. 2-29 This map shows time zones for North America and Hawaii. The boundaries between time zones are arranged so that all of any state is in the same zone. Clocks in Hawaii and most of Alaska are set two hours earlier than Pacific standard time.

If you traveled all the way around the world, you would cross 24 time zones. Each time zone differs from the one next to it by one hour. This means that a trip around the world would cause you to gain or lose 24 hours, or one day. To adjust for this, an international date line has been placed between two of these zones. See Fig. 2-30. It was placed through the Pacific Ocean at the 180° meridian in order to avoid crossing any land masses. When crossing the international date line going west, the day

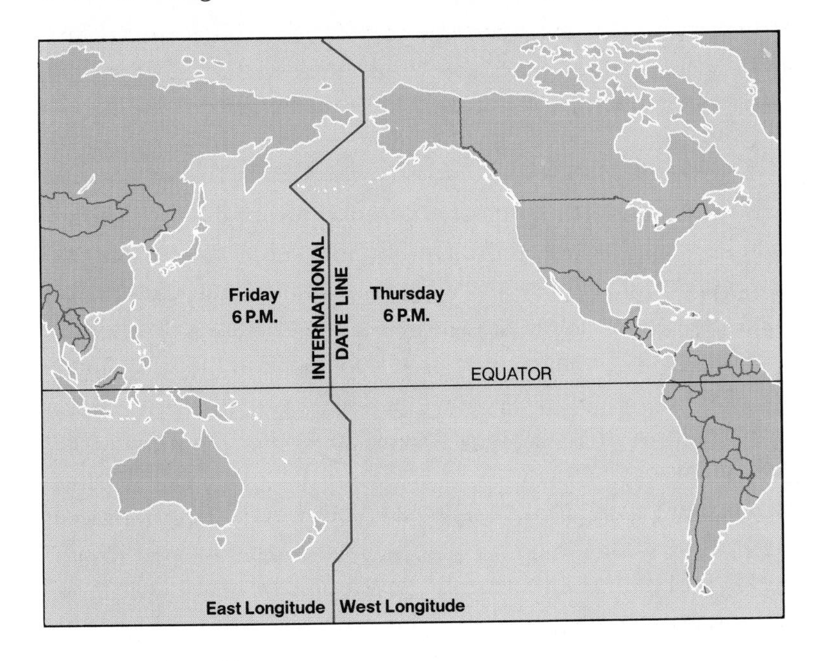

Fig. 2-30 The international date line is located in the Pacific Ocean to avoid crossing land where people live.

changes to the next day. When going east, the day changes to the day before. For example, when it is Sunday west of the international date line, it is Saturday east of it.

MOTION AROUND THE SUN

Each planet follows a curved path, or orbit, around the sun. There is a gravitational attraction between each planet and the sun. This attraction keeps the planet in its orbit around the sun. The orbit of the earth is not a perfect circle. The orbit is an oval shape called an *ellipse* (eh-**lips**). See Fig. 2–31. The sun is not located at the center of the ellipse but is slightly off to one side. As a result, the earth, as it moves in its orbit, is not always the same distance from the sun. During the year, the distance from the earth to the sun changes. The average distance between the earth and the sun is 150 million kilometers.

Fig. 2–31 The earth is not always the same distance from the sun because its orbit is elliptical. The stars seen in the summer sky are different from those seen in the winter sky.

Observation of the stars at different times of the year shows that the earth follows an orbit around the sun. If you look at the sky on a clear night, you will see stars and some patterns of stars. Six months later, you will see different stars and different patterns. As the earth moves in its orbit around the sun, different stars come into view. See Fig. 2–31.

If you think of a planet as moving in its orbit on a plane, its axis will not be straight up, or at 90° to that plane. The earth's axis is tilted 23½° to the plane of its orbit. As the earth moves around the sun, the North Pole will be tilted toward the sun during part of the earth's orbit. The axis will be tilted away from the sun during another part of the orbit.

SEASONS

The tilt of the earth's axis and the earth's movement in its orbit around the sun cause the seasons. In the Northern Hemisphere, the North Pole is tilted most toward the sun about June 22. This is the first day of the summer season. This day is called the **summer solstice** (**sol**-stus). The word "solstice" means "sun stops." This is because the sun's path on this day will reach its highest position in the sky for the year. The sun's rays strike the earth's surface more directly on the *summer solstice* than on any other day of the year. See Fig. 2–32b. Thus the Northern Hemisphere receives more solar energy about June 22 than at any other time of the year. This day also has the longest time of light and shortest time of darkness in the year. Within the area of the North Pole, there are 24 hours of daylight. See Fig. 2–33, position A. In some time zones, clocks are set ahead one hour because of the longer days. This system is called *daylight-saving time.* More daylight and the large amount of solar energy cause the warmer weather in the summer months.

Summer solstice The time of the year, about June 22, when the earth's axis is tilted most toward the sun.

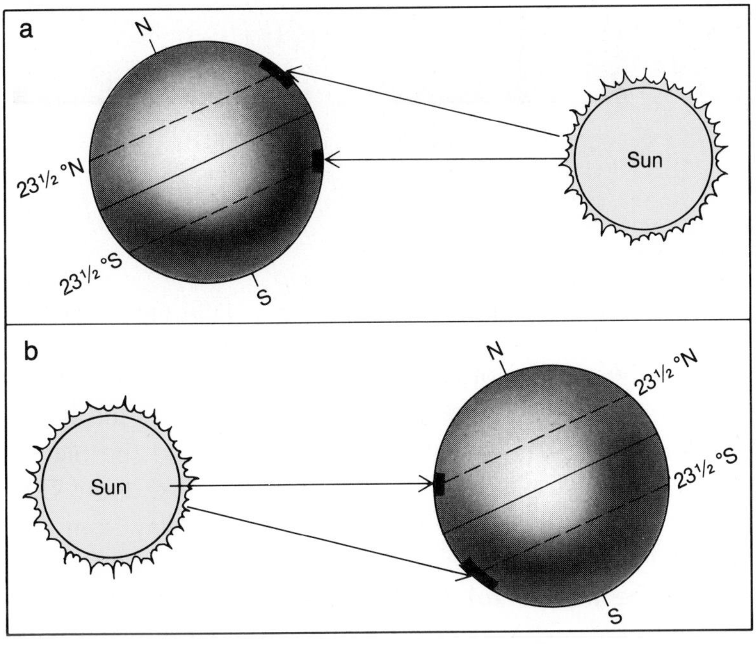

Fig. 2-32 The sun's rays deliver more energy when they strike directly than when they strike at an angle.

As the earth moves in its orbit, the sun's path across the sky each day becomes lower. The days become shorter and the nights longer. Three months after the summer solstice, about September 22, the length of day becomes equal to the length of

Autumnal equinox The time of the year, about September 22, when the earth's axis is tilted neither toward nor away from the sun.

night. The Northern Hemisphere is tilted neither toward nor away from the sun. See Fig. 2–33, position B. This is the first day of autumn, or fall. This day is called the **autumnal equinox** (**ee-**kwuh-noks). *Equinox* means "equal night."

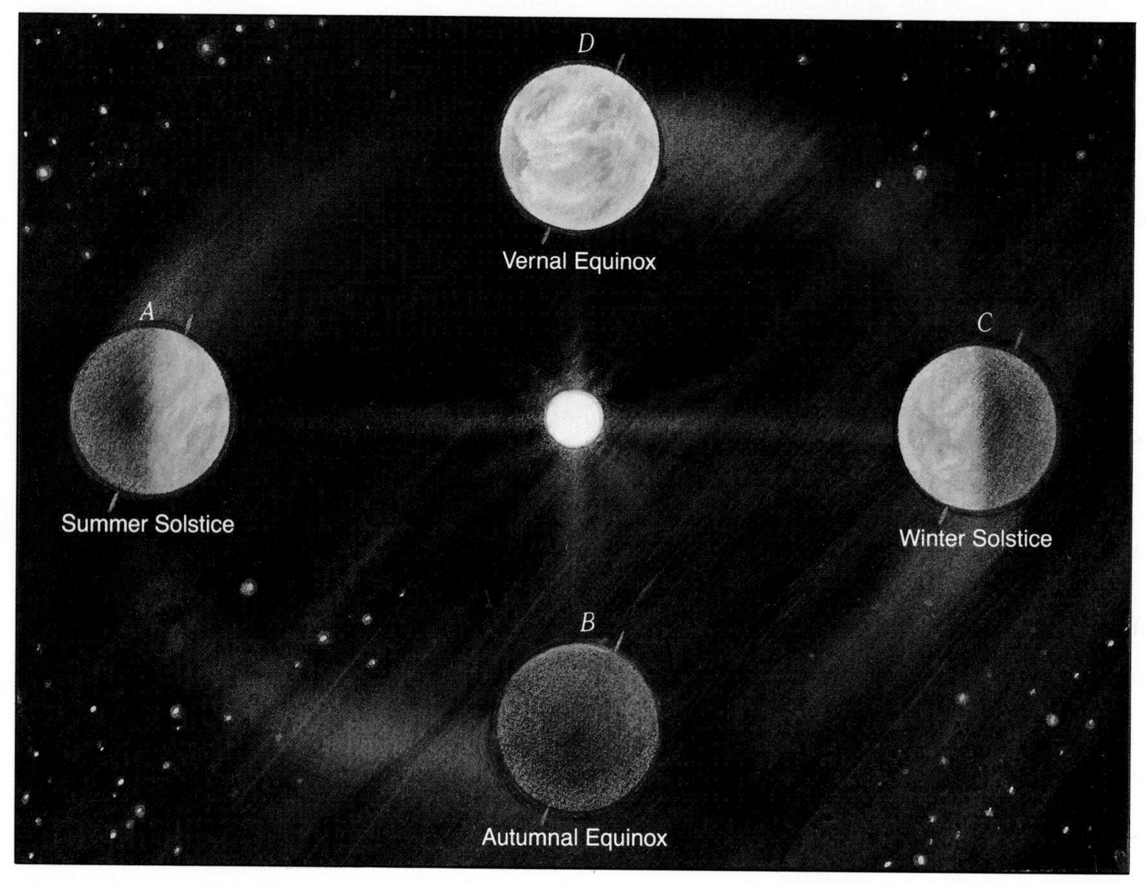

Fig. 2–33 The tilt of the earth's axis remains almost the same over very long periods of time. Position A: summer solstice; Position B: autumnal equinox; Position C: winter solstice; Position D: vernal equinox.

During the autumn, or fall season, the path of the sun across the sky continues to lower as each day passes. As the earth moves in its orbit, the days become shorter and the nights longer. Three months after the autumnal equinox, about December 22, the Northern Hemisphere is tilted away from the sun. See Fig. 2–33, position C. This is the first day of winter and is called the **winter solstice**. The sun's path has reached its lowest position in the sky for the year. The sun's rays make their smallest angle with the earth's surface. The Northern Hemisphere receives the smallest amount of solar energy of the year and has the shortest day and longest night. In the area of the North Pole, there are 24 hours of darkness. The small amount

Winter solstice The time of the year, about December 22, when the earth's axis is tilted most away from the sun.

of solar energy and the short days cause the cool winter weather during the winter season.

After the *winter solstice*, the days begin to grow longer and the nights shorter. The sun's path in the sky moves higher as each day passes. Three months after the winter solstice, about March 21, the length of the day becomes equal to the length of the night. The earth is tilted neither toward the sun nor away from it. See Fig. 2-33, position D. This day is called the **vernal equinox.** This is the first day of spring.

Vernal equinox The time of the year, about March 22, when the earth's axis is tilted neither toward nor away from the sun.

After the *vernal equinox*, the earth moves for another three months toward its summer solstice position. The days grow longer and the nights shorter. The days are now longer than the nights. The sun's path across the sky moves higher each day. In June the summer solstice will occur again. Then the cycle of seasons will begin again.

It takes about one year to complete the cycle of seasons. A year is defined as the length of time it takes the earth to complete one revolution in its orbit. One way of measuring the year would be to measure the amount of time between two vernal equinoxes. However, the time between two vernal equinoxes is about 20 minutes shorter than one year. This is due to the effect of the moon's gravity on the earth. It causes the earth's axis to move. In 13,000 years, the earth's axis will point toward the sun on about December 22. In other words, summer in the Northern Hemisphere will take place in December. After 13,000 more years, the earth's axis will return to its present position.

SUMMARY

Day and night are the most observable effects of the earth's rotation. The earth revolves around the sun in an elliptical orbit. The tilt of the earth's axis and the earth's revolution cause the seasons.

QUESTIONS

Use complete sentences to write your answers.

1. What causes day and night?
2. Describe the earth's path around the sun.
3. Why do we need standard time zones and the international date line?
4. Describe the tilt of the earth's axis during each season.

INVESTIGATION

STAR MOVEMENT

PURPOSE: Why do different stars appear in the sky during different times of the night and the year?

MATERIALS:
pencil paper

PROCEDURE:

A. The only stars that are visible at midnight are those that are above the horizon, as you can see in Fig. 2–34.
   1. What letters in Fig. 2–34 show stars visible at midnight?
   2. What letters in Fig. 2–34 show stars not visible at midnight?

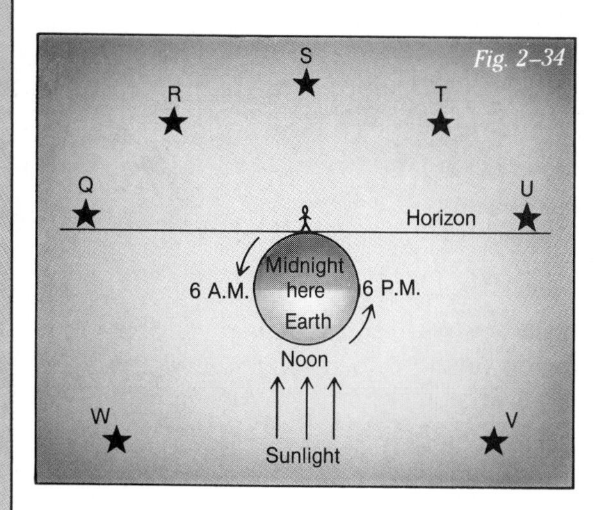

Fig. 2–34

B. The earth moves around the sun. It takes the earth one year to complete its orbit around the sun. Fig. 2–35 shows the earth at three different positions on its year-long orbit around the sun. They are shown in the diagram as X, Y, and Z. Stars are shown in positions A to P.

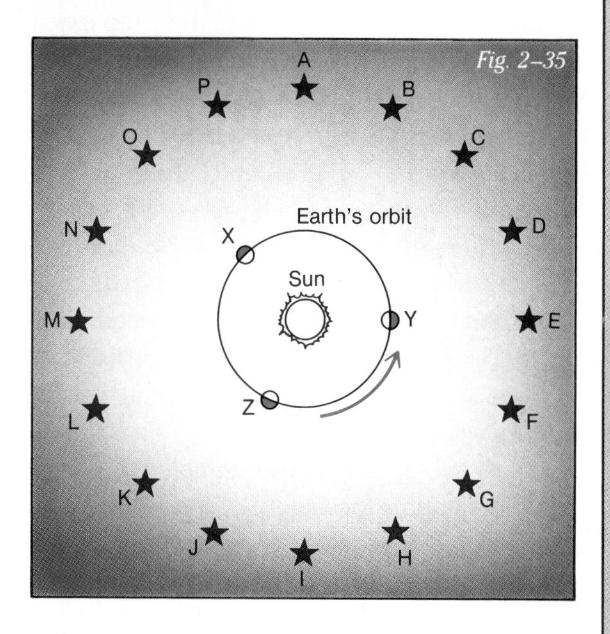

Fig. 2–35

C. Look carefully at Fig. 2–35. At earth position X, put a piece of paper on the diagram so that the edge of the paper represents the horizon at midnight.
   3. Which stars will be visible (above the horizon) at midnight when the earth is in position X?

D. Do the same thing for earth positions Y and Z.
   4. Which stars will be visible at midnight when the earth is in position Y?
   5. Which stars will be visible at midnight when the earth is in position Z?

CONCLUSIONS:

1. Why aren't all stars seen at night?
2. Explain why different stars are visible at midnight at different times of the year.
3. Explain why certain stars, such as the North Star, can be seen all year round.

CAREERS IN SCIENCE

CARTOGRAPHER

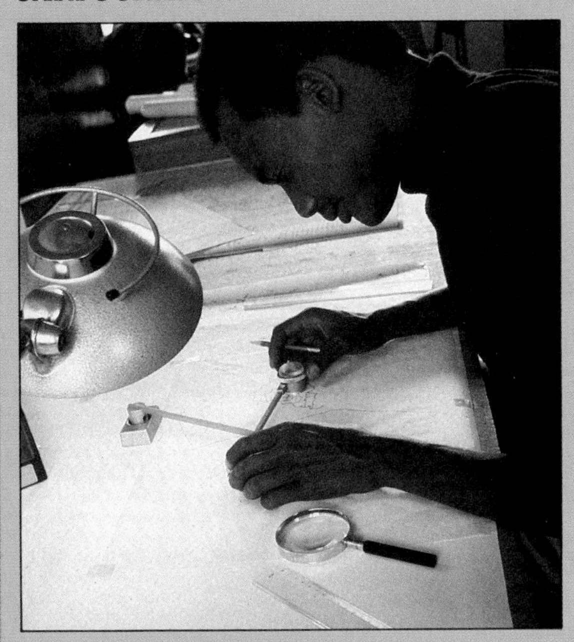

Since ancient times, maps have shown where places are and, often, how to reach them. Today, cartographers continue to produce a variety of maps. Road maps guide us on our trips. Topographic maps show the earth's land features. Resource maps show the distribution of our planet's natural resources. Star maps describe the universe.

Cartographers must present data accurately and according to scale. Their tools include measuring devices, rich colors, and even computers and images sent from satellites. In fact, some cartographers are involved in mapping what we know of other planets.

To work in the field, cartographers need a college degree and must have completed courses in geography, economics, drawing, design, and computer science. Jobs are available in publishing companies, the government, and universities. For further information, write: Association of American Geographers, 1710 16 Street N.W., Washington, DC 20009.

SCIENCE EDITOR

Science editors work on textbooks, magazines, and journals; some even develop computer software. A science editor supervised the way this book is designed and the information that appears in it. An editor reviews the work written by the author and decides if anything needs to be changed, or added, or omitted. He or she must make sure that the information contained in the article or book is accurate. Science editors also work with the artists and photographers who provide illustrations for the written material, which is called copy.

To become a science editor, you need a college degree and strong writing skills. A knowledge of science and math is essential, and teaching experience is helpful. Jobs are available with a variety of publishing companies. For further information, write: National Association of Science Writers, P.O. Box 294, Greenlawn, NY 11740, or Association of Earth Science Editors, c/o American Geological Institute, 5205 Leesburg Pike, Falls Church, VA 22041.

CHAPTER REVIEW

VOCABULARY

On a separate piece of paper, match the number of each sentence with the term that best completes it. Use each term only once.

axis	crust	longitude	winter solstice	scale
contour line	equinox	orbit	standard time	mantle
core	latitude	solar system	summer solstice	topographic map

1. The sun, planets, and smaller bodies orbiting the sun are called the ______.
2. The ______ is the thin, solid layer covering the outside of the earth.
3. The ______ is a very thick layer just under the crust of the earth.
4. The very dense center of the earth is called the ______.
5. A(n) ______ shows the features of a small part of earth's surface.
6. Points having the same elevation are shown by a(n) ______ on some maps.
7. The ______ of a location is the distance north or south of the equator in degrees.
8. The distance east or west of the prime meridian is called the ______.
9. A(n) ______ shows size relationship between a map or model and actual size.
10. The ______ is an imaginary line on which the earth rotates.
11. The time all clocks are set to within a given time zone is called the ______.
12. The ______ is the day when the sun follows its highest path across the sky.
13. The day when the sun follows its lowest path across the sky is the ______.
14. The curved path a planet makes around the sun is its ______.
15. A time when day and night are of equal length is called a(n) ______.

QUESTIONS

Give brief but complete answers to each of the following questions. Unless otherwise indicated, use complete sentences to write your answers.

1. Explain why the planets closest to the sun are different from the planets farthest from the sun.
2. If the planets formed from the same material as the sun, why didn't the planets become stars?
3. How can rocks found on the moon give scientists information about how the earth began?

4. What is the Moho?
5. What evidence is there that the earth was once melted?
6. Compare the seasons in the Southern Hemisphere with the seasons in the Northern Hemisphere.
7. Why do the boundaries of the time zones go around cities?
8. What would be the effect of the earth's axis being tilted more than 23½°? Less than 23½°?
9. Why are the summer months warmer than the winter months in the Northern Hemisphere?
10. Why was the international date line drawn in the ocean?
11. Where will the earth's axis be pointing toward on about June 22 in approximately 13,000 years?

APPLYING SCIENCE

1. Suppose you go to the moon. The first thing you see is some footprints in the dusty surface. Write a hypothesis to explain the footprints. What changes, if any, will you make in your hypothesis if you then see vehicle tracks? Following the vehicle tracks, you discover the remains of a moon-lander. What changes in your hypothesis will you make? Now will you know the true origin of the footprints?
2. From a topographic map of your area, find the elevation of your home, a friend's house across town, and your school. Also find out what is the highest elevation in or near your town. Your local Department of Engineering and Development will have a topographic map of your area if you can't find one in the library.
3. What problems would exist if there were just one time zone for the entire earth?

BIBLIOGRAPHY

Jespersen, James, and Jane Fitz-Randolph. *From Sundials to Atomic Clocks.* New York: Dover, 1982.

Owen, Tobias. "The Evolution of Titan's Atmosphere." *The Planetary Report,* November/December 1983.

Pampe, W. R., ed. *Maps and Geological Publications of the United States: A Layman's Guide.* Falls Church, VA: American Geological Institute, 1978.

Waldrop, M. Mitchell. "Before the Beginning." *Science 84,* 1984.

CHAPTER

3

Bent rock layers are the work of great forces inside the earth. These once flat layers of rock were squeezed together by moving continents. When continents collide, they act like a giant vise.

PLATE TECTONICS

CHAPTER GOALS

1. Give evidence for the theories of continental drift and sea floor spreading.
2. Explain how continental drift and sea floor spreading led to the theory of plate tectonics.
3. Use the theory of plate tectonics to explain how the plates in the earth's crust move.
4. Explain how some mountains, island arcs, and sea floor trenches may form at the boundaries of crustal plates.
5. Using the theory of plate tectonics, describe how continental landmasses become larger.

3–1. Moving Continents

At the end of this section you will be able to:

- Explain the theory of continental drift.
- Use the theory of *plate tectonics* to explain how plates in the earth's crust move.
- Relate *convection* to movements of large parts of the earth's crust.

Once there were oceans in the midwestern United States, and part of Alaska was a tropical island. Ice used to cover much of the Sahara desert, while swamps existed in Antarctica. These are pieces of a giant puzzle. Put together, the pieces can show how the planet we live on works.

DRIFTING CONTINENTS

For a long time, scientists looked at the many surprising discoveries on the continents of the earth. They tried in different ways to explain each of the discoveries. Look at the pieces of the jigsaw puzzle in Fig. 3–1. It is hard to see what the whole picture looks like when the pieces are apart. The same is true for seeing how the earth works. For a long time, scientists did not see their discoveries as pieces of the same puzzle. Thus they could not see how the different parts of the earth's picture could be put together.

Fig. 3-1 It is difficult to see the whole picture when pieces of the jigsaw puzzle are apart.

Have you noticed that a world map looks a little like a jigsaw puzzle? For example, the continents of South America and Africa seem to fit together like pieces of a puzzle. See Fig. 3-2. Could these continents once have been connected?

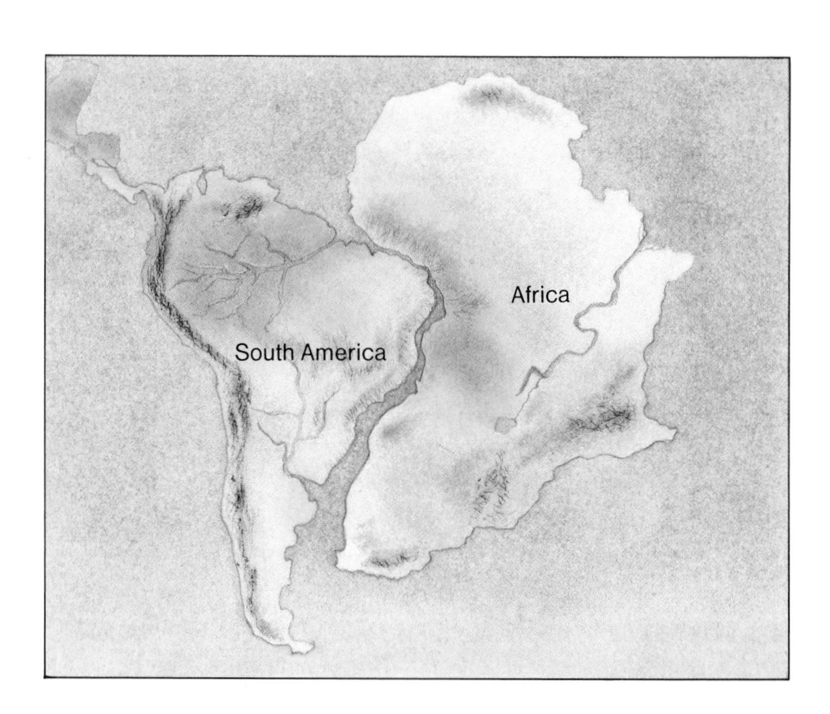

Fig. 3-2 The fit of South America and Africa led to the theory of continental drift.

The idea that the earth's continents were once a large, single landmass that moved apart is called the theory of *continental drift.* It was proposed as a scientific theory in 1912 by a German scientist named Alfred Wegener. Wegener called this landmass Pangaea, which means "all lands." Wegener pointed to the fit of South America and Africa, as well as other evidence. Remains of very similar plants and animals were found in Africa, South America, Australia, India, and Madagascar. See Fig. 3–3. He asked how plants and animals could be so similar if the lands were always separated by such wide oceans. Wegener also pointed to evidence of glaciers in warm lands, such as the continent of Australia. He reasoned that such continents once must have been in colder places on the earth and moved to where they are now. However, Wegener's theory of continental drift was not accepted by most scientists. They found it hard to believe that giant continents could push their way through the solid crust of the ocean floor. They said it would be as impossible as ships being able to move through a frozen sea. Some scientists tried to explain the similar plants and animals in another way. They said that the continents were once joined by a large landmass that was now below the ocean. There were still missing pieces to the puzzle. Thus the theory of continental drift did not become a part of the scientific view of the earth at that time.

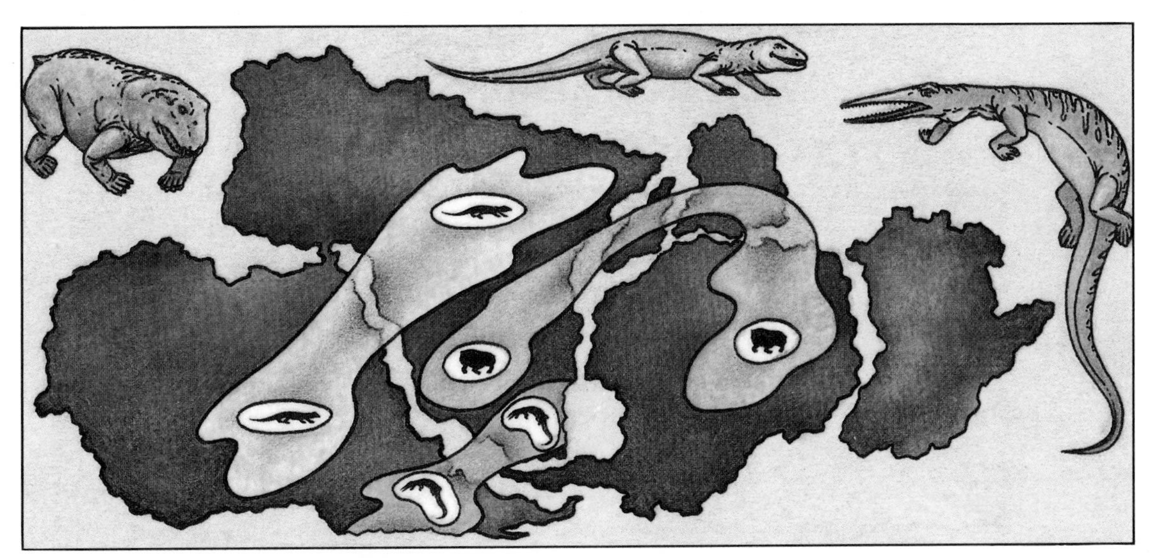

Fig. 3–3 The theory that the lands were once joined together can explain why the fossils of these ancient reptiles have been found across different continents.

SEA FLOOR SPREADING

Finding out how the continents move through the ocean floor was not easy. However, modern scientific instruments have been able to map the sea floor. One of the most surprising discoveries on the sea floor was a chain of mountains. These

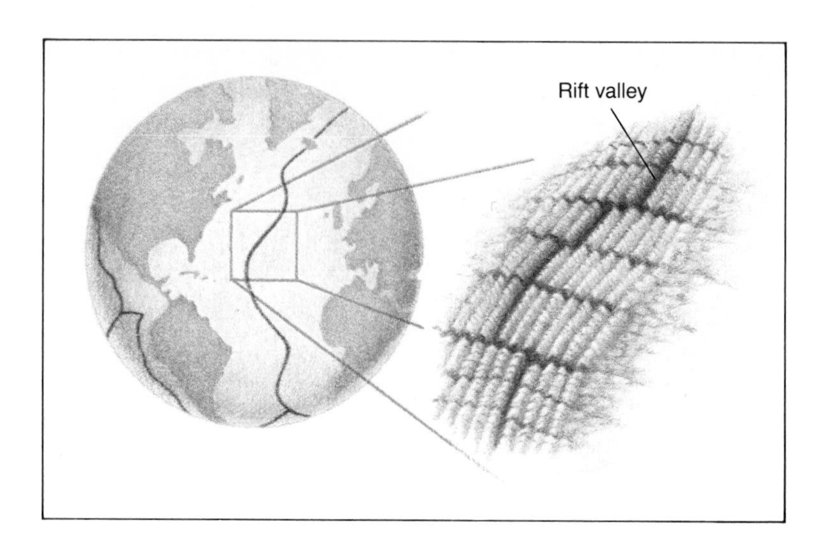

Fig. 3–4 The colored line on the globe shows the location of the mid-ocean ridge system. The enlargement shows the surface features of the mid-ocean ridge.

Mid-ocean ridge A chain of underwater mountains found on the floor of the oceans.

Rift valley A deep, narrow valley formed where the earth's crust separates.

mountains, almost completely covered by the oceans, were named the **mid-ocean ridge.** The *mid-ocean ridge* is the largest system of mountains in the world. It circles the earth like the endless seam of a baseball. See Fig. 3–4.

One of the best-known parts of this giant chain of underwater mountains is the mid-Atlantic ridge. In Fig. 3–4, you can see how those mountains would look if the water in the Atlantic Ocean could be drained away. Then the mid-Atlantic ridge would be seen as a wide range of mountains running down the middle of the floor of the Atlantic Ocean. In certain places, the ridge has been pushed out of line, forming other mountain ranges. Running through the middle of the ridge is a deep, narrow valley called a **rift valley.** A *rift valley* forms when the ocean crust separates along the mid-ocean ridge.

As scientists explored the mid-ocean ridge, they made two important discoveries. First, the layer of mud that covered most of the sea floor was missing from the mid-ocean ridge. Instead, rock samples taken from these mountains showed that the rocks were formed by volcanoes. *Pillow lava,* a kind of rock formed when molten lava is cooled quickly by water, was common. See Fig. 3–5. Second, scientists found that the center of the mid-ocean ridge was much warmer than any other part of the sea floor. Therefore, the center of the mid-ocean ridge must be a place where hot material from inside the earth is rising toward the surface. A new theory, called *sea floor spreading,* was proposed to explain these discoveries. The theory of sea floor spreading says that the mid-ocean ridge is a crack in the

Fig. 3-5 Pillow lava is common along the mid-ocean ridge.

earth's crust. Molten rock from the inside of the earth constantly rises from the center of this crack. Cooled by the sea water, the lava becomes solid. It is then pushed outward and away from the ridges by new molten rock. Thus new sea floor is always being formed. The new sea floor then spreads out everywhere along both sides of the mid-ocean ridge. See Fig. 3-6. Evidence for sea floor spreading is found in the different ages of parts of the sea floor. The youngest rocks are always found along the sides of the mid-ocean ridge. The oldest rocks are found farthest from the ridge.

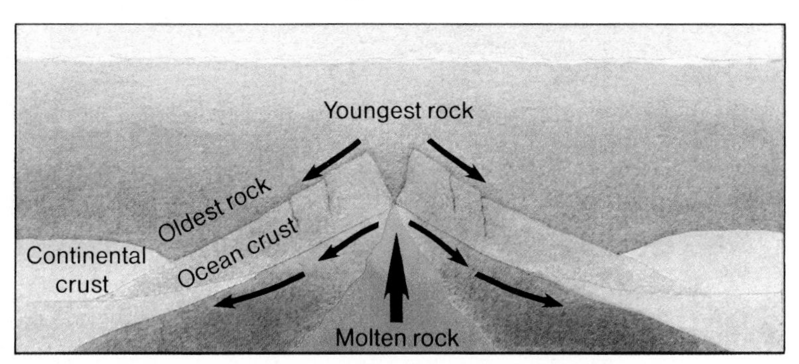

Fig. 3-6 Molten lava rises in the rift valley along the mid-ocean ridge. New sea floor is created and is moved outward along each side of the ridge.

THE THEORY OF PLATE TECTONICS

Some scientists saw sea floor spreading as the beginning of a new theory to explain how continents move. They proposed that the crust is broken up into large sections, or plates. These plates grow at the mid-ocean ridge and move away at a speed of 1 to 20 centimeters per year. Continents are carried along with some of the moving plates, the way logs frozen into blocks of ice are carried along a river. Continents are usually parts of larger plates, but they could also make up entire plates. Thus continents did not have to push their way through the solid crust of the earth to move. They could move as sections of the earth's crust moved. The new theory used the ideas of sea floor spreading and continental drift. This theory has been given the name **plate tectonics** (tectonics comes from the Greek word "tekton," meaning "builder").

Plate tectonics The theory that the earth's crust is made up of moving plates.

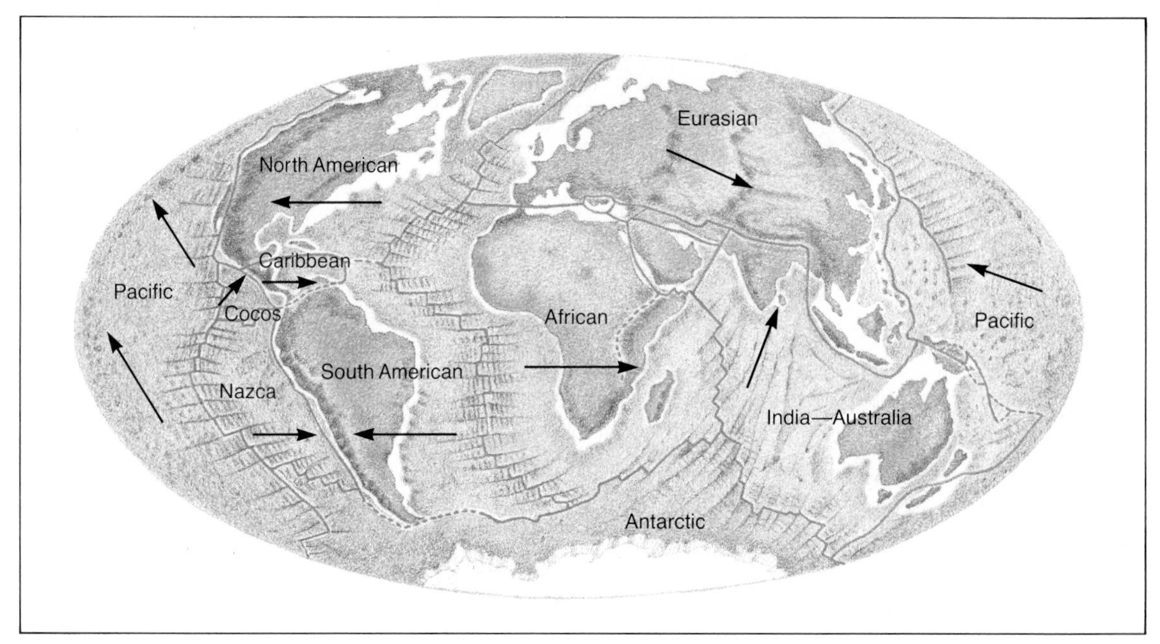

Fig. 3-7 The boundaries of the crustal plates are shown on this global map. Arrows show the direction each plate is moving.

The modern theory of *plate tectonics* says that the earth's crust is divided into seven major plates of different sizes. The largest plates have the same names as the continents they carry. See Fig. 3-7. As you can see, the size and shape of the continents are not exactly the same as the plates that carry them. Unlike the sea floor, which is always being formed along the mid-ocean ridge, the rocks that make up the continents are much older. They are also lighter than the rocks that make up most of the crustal plates. Thus the continents ride along on top of the moving crustal plates. The theory of plate tectonics has changed the way scientists think about the earth.

WHAT MOVES THE CRUSTAL PLATES?

Powerful forces are needed to move the gigantic crustal plates. Scientists believe that the forces start inside the earth.

The forces are caused by the heat inside the earth. They are similar to the forces produced when cold air is added to hot air. You probably have seen what happens to cold air when a freezer is opened. Since cold air is denser than hot air, it sinks. At the same time, the less dense warm air rises. A current is formed that moves and mixes the air. Movement of material caused by a difference in temperature is called **convection.** Scientists believe that there may be very slow-moving *convection* currents set up in the mantle as heat moves out toward the crust. See Fig. 3-8.

Convection Movement of material caused by a difference in temperature.

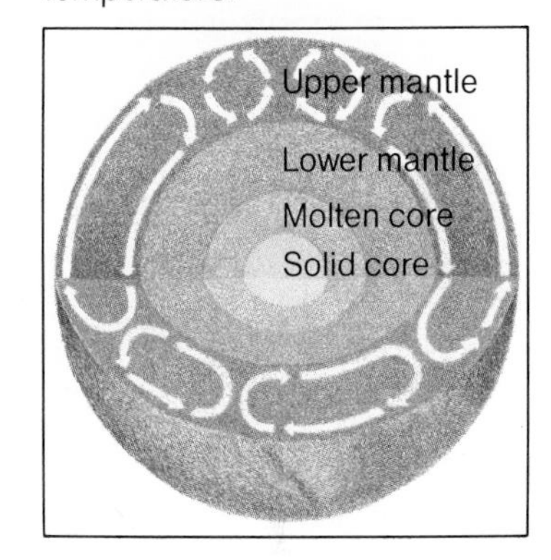

Fig. 3-8 Convection currents in the mantle are believed to cause the crustal plates to move.

Sea floor spreading could be caused by material being carried up by convection currents. The slow currents in the mantle cause material to be added to the crust along the mid-ocean ridge. At other places in the crust, currents in the mantle could cause one plate to be pulled down beneath the edge of another. This returns crustal material back to the inside of the earth.

Scientists do not have a complete answer to the question of what forces move the crustal plates. The question cannot be answered until there is a better understanding of what lies beneath the moving plates that make up the earth's surface.

SUMMARY

The theory of continental drift said that the earth's continents were once a single landmass. The discoveries of the mid-ocean ridge and sea floor spreading were used as evidence to form the theory of plate tectonics. This theory says that the earth's crust is divided into separate moving plates. Powerful currents in the mantle may be the driving force causing the plates to move.

QUESTIONS

Use complete sentences to write your answers.

1. List three observations Alfred Wegener used as evidence to support the theory of continental drift.
2. Explain what is happening along the mid-ocean ridge.
3. Describe how continents move, using the theory of plate tectonics.
4. How do convection currents help explain the movement of the crustal plates?

Fig. 3-9 A portion of the San Andreas fault in California.

3-2. Action at the Plate Edges

At the end of this section you will be able to:

- Compare three kinds of boundaries between the crustal plates.
- Describe how some mountains, island arcs, and sea floor trenches may be formed.
- Explain how continents grow.

Walk westward across the valley shown in Fig. 3-9 and you would leave northern California for what was once part of Mexico. Travel to southern Florida and you would be in a land that was once part of Africa. Or move eastward into Connecticut or Massachusetts. You would be leaving the continental United States for what was once part of Europe. Places such as these exist all over the earth.

PLATE BOUNDARIES

The earth's surface is cracked into many crustal plates. As these cracks open, crustal plates move. Each crustal plate is a strong, rigid piece of the crust with an average thickness of about 100 kilometers. The crustal plates "float" on the layer of the earth called the mantle. Studying the way earthquake waves move through the mantle shows that the mantle's upper layer is partly molten. Thus crustal plates are able to slide over the soft upper part. This hot, soft upper part of the mantle is called the **asthenosphere.** The *asthenosphere* goes down about 250 kilometers into the mantle.

Asthenosphere The partly molten upper layer of the mantle.

Since the crustal plates are solid, they move mostly along their edges, or boundaries. Boundaries are always marked by a kind of fault, or crack in the crust where plates meet. There are three kinds of plate boundaries. Each kind of boundary depends on how the separate plates move.

One kind of plate boundary is formed along the mid-ocean ridge. There the plates are being pushed apart as new crust is added. See Fig. 3-10 (a). If the plates move apart in one place, they must come together in another place. Thus not all boundaries between the plates are places where they are moving apart. Some plate boundaries are formed where two or three plates are coming together.

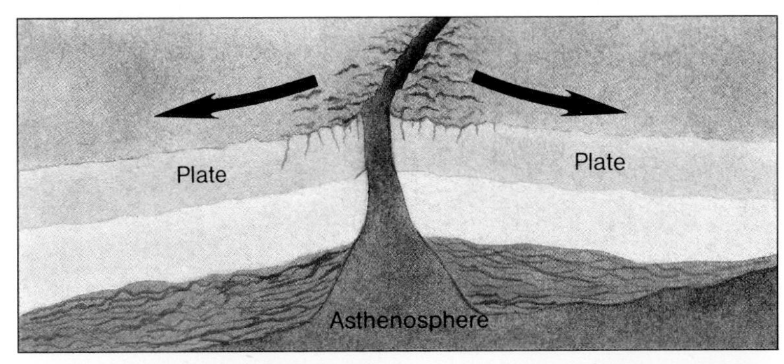

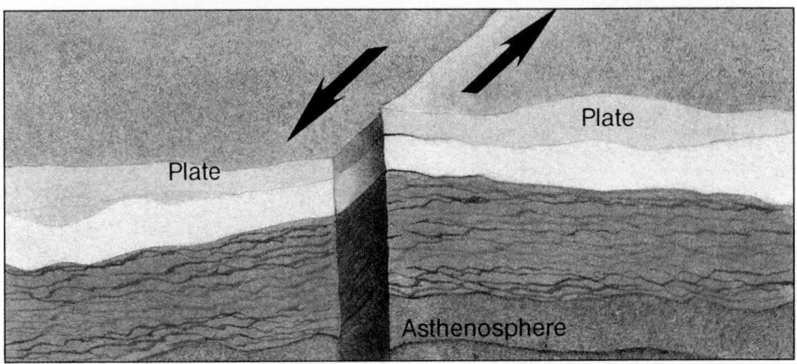

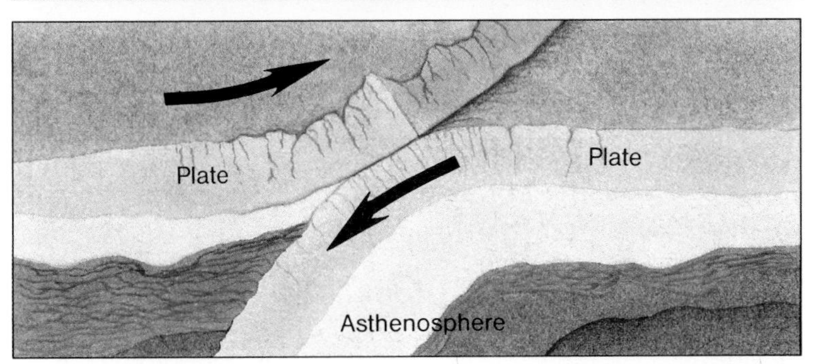

Fig. 3-10 (a) One kind of plate boundary is formed when molten rock rises up between two separating plates. (b) When two plates collide, they can slide past each other. (c) Sometimes when two plates collide, one may be pushed down while the other rides up.

A second kind of plate boundary is formed where the plates slide past one another. The motion of each plate is parallel to the plate boundary. See Fig. 3-10 (b). A third kind of plate boundary is formed where two plates move toward each other. The edge of the heavier plate is usually pushed down beneath the lighter plate. See Fig. 3-10 (c). This process is called **subduction.** When *subduction* takes place, one of the plates is forced down into the mantle. There the plate is destroyed by melting as its material becomes part of the mantle. As material is removed from the plate by subduction, new material is added to the plate along the mid-ocean ridge. Equal amounts of material are removed and added. Otherwise, the size of the earth would be changing all the time.

Subduction Where one plate is forced beneath another plate.

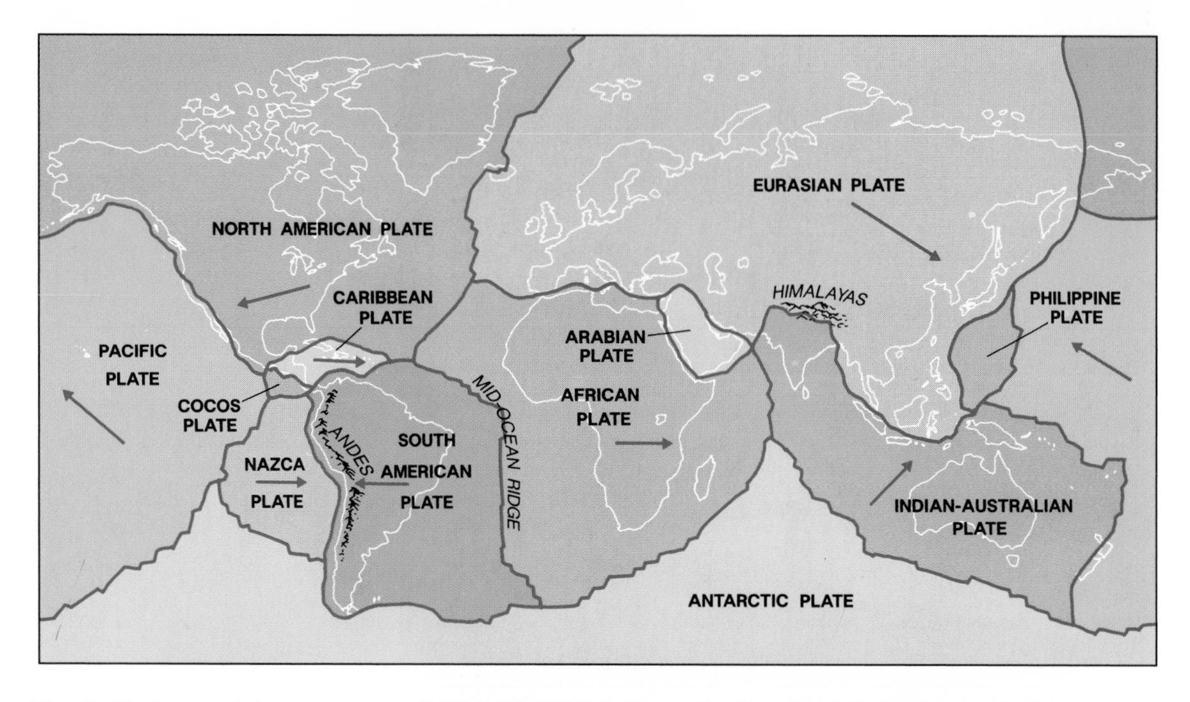

Fig. 3-11 Some of the world's largest mountain ranges occur along plate boundaries.

COLLISIONS THAT CHANGE THE CRUST

When you look at Fig. 3-11, you can see that the plate carrying the South American continent is moving westward. The Nazca plate on the Pacific Ocean floor, west of South America, is moving eastward toward the continent. These two plates meet along the western edge of South America. This collision of plates seems to be crumpling up the western edge of South America and forming the Andes (**an**-deez) Mountains at the boundaries of the crustal plates. Fig. 3-12 shows subduction occurs as South America rides up over the Nazca plate. It is believed that the melting material below the mantle is being added to the Andes Mountains in two ways. The rubbing of the plates adds material to the base, while some of the melted material rises to the surface and forms volcanoes.

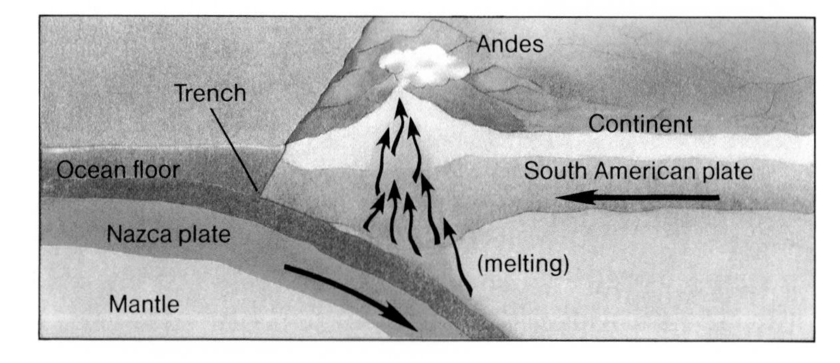

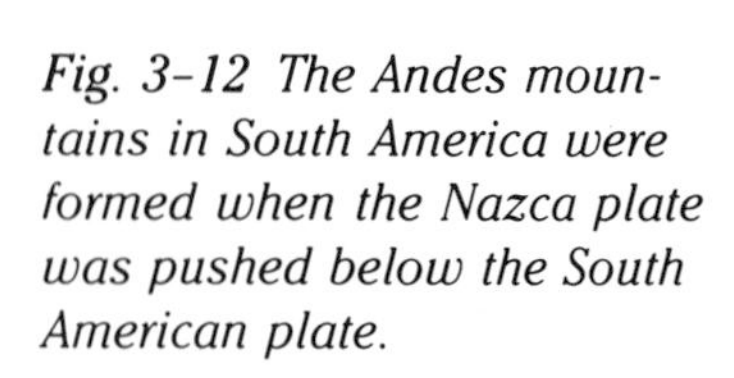

Fig. 3-12 The Andes mountains in South America were formed when the Nazca plate was pushed below the South American plate.

The rock that makes up the continents is lighter than the rock of the plates on which the continents are carried. Thus the continents "float" on the plates like gigantic rafts. When continents meet, they do not sink into the mantle. See Fig. 3-13. Instead, the two continents come together and crumple into mountain ranges. The two continents then become joined into a single landmass. For example, India became attached to Asia in this way. India was once a separate continent, carried on a northward-moving plate. The Indian plate collided with the Asian plate. The continents carried on the plates were crumpled and folded when this occurred. The Himalayas (him-ah-**lay**-ahs), a very thick layer of mountains, were formed. See Fig. 3-14. The Alps in Europe were also formed in a similar way. They were created when parts of the African plate collided with the Eurasian plate. Millions of years from now the Mediterranean Sea may disappear as Africa meets the Eurasian plate.

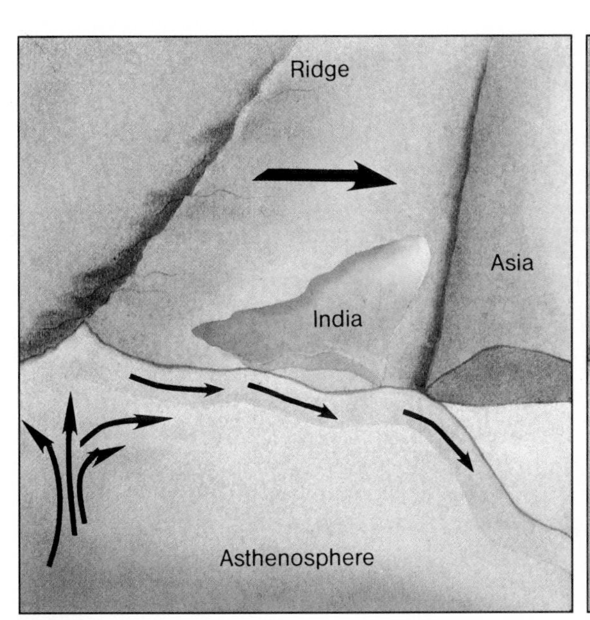

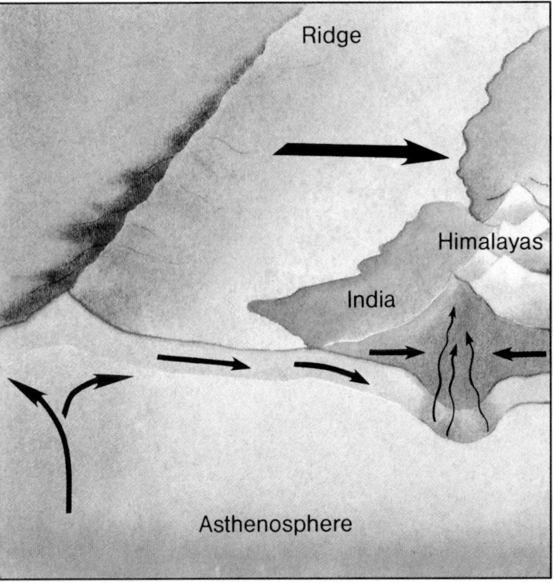

Fig. 3-13 India "crashed" into Asia about 45 million years ago, producing the Himalayas.

If you look at the world map in Fig. 3-7, a line of islands can be seen along the western side of the Pacific Ocean. These islands are caused by two ocean plates meeting. Scientists believe that the Pacific plate is slowly being pushed under Asia. As it sinks, it is melted by the high temperatures of the mantle. The light, melted material is forced up, forming strings of volcanoes. Because of the slow movements of these plates, the volcanoes are formed offshore under the water. The lava piles up

Fig. 3-14 The Himalayas.

until the volcanoes rise above the water and form islands. See Fig. 3-15. The volcanoes form a curved shape, called an *island arc.* The reason for the curved shape of island arcs can be seen by pushing down on one side of a tennis ball. The Aleutian Islands and Japan are examples of island arcs. As the plates continue to move, the island arcs may eventually get crushed against continents.

In some places in the ocean, one crustal plate plunges below another to form a *trench.* Some of the deepest parts of the ocean are at these places. Trenches are long, narrow valleys in the ocean floor. Their depth is usually greater than 6,000 meters. Most of the world's trenches are located near island arcs in the Pacific Ocean, as shown in Fig. 3-16.

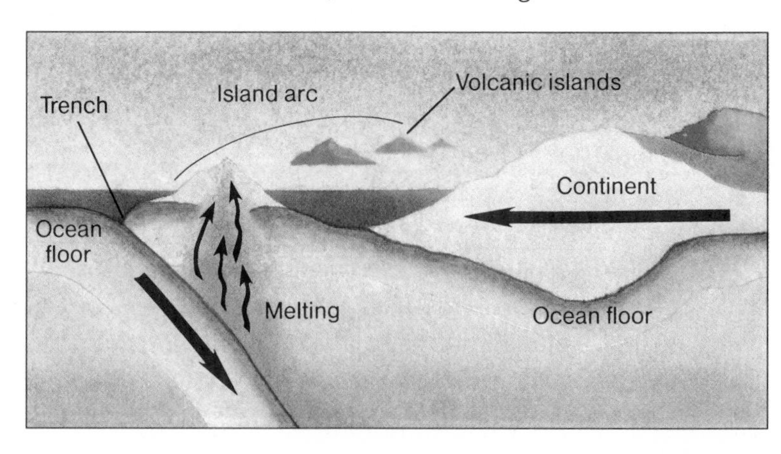

Fig. 3-15 Volcanic islands and sea floor trenches can be created when one crustal plate is pushed beneath another.

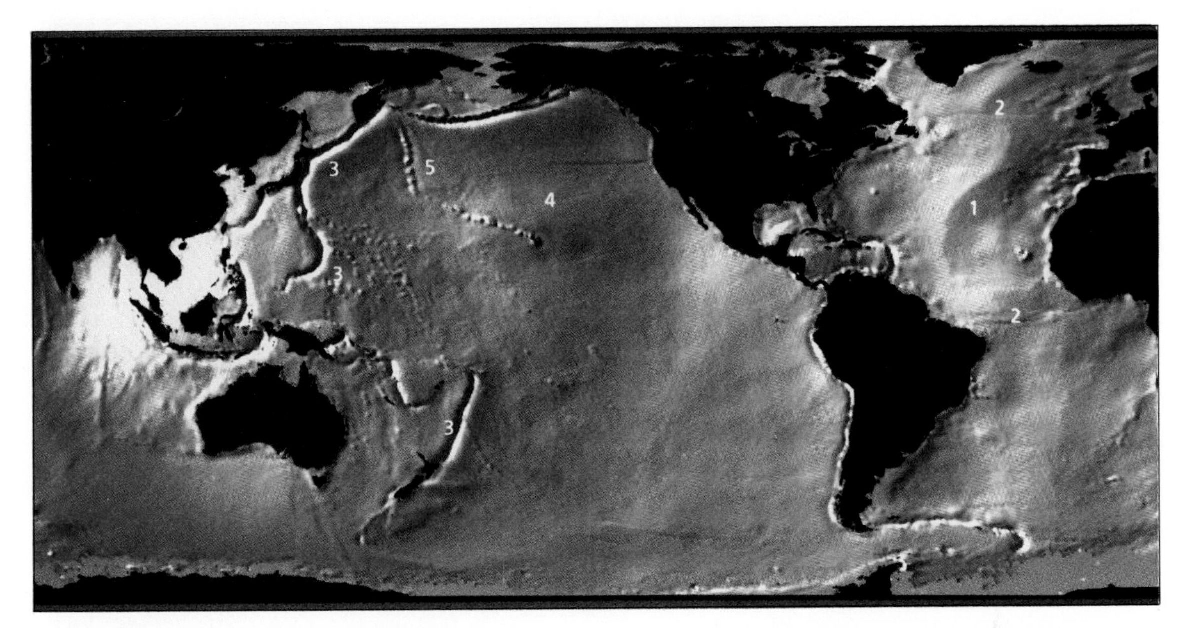

Fig. 3-16 *Many of the world's largest trenches on the sea floor are shown in this photo taken from space.*

GROWTH OF CONTINENTS

If you tried to measure the age of the North American continent, you would find that it isn't all one age. You would find that the continent is oldest at its center and becomes younger toward its coasts. Thus the outer parts of the continent must have been formed after the center formed. It seems as if the continent grew. Many scientists believe that continents do grow. Throughout the world, small pieces of continental crust are scattered over the ocean floor. These submerged *micro-continents* are usually several hundred kilometers across. They are believed to be parts of continents that somehow became separated. A part of the ocean floor, carrying a micro-continent, may collide with a continental plate. If the ocean floor is subducted, the small piece of continental crust can become attached to the edge of another continent. A continent can then grow by adding small pieces to its edges. Thus the original continents were much smaller than they are today. Much of the west coast of North America and almost all of Alaska seem to have been added in this way. See Fig. 3-17.

Continents also may have grown when large island arcs or pieces of sea floor crust were pushed against and over the continents, forming a "thin skin" over the older continental rock. There is evidence of a thin layer of rock in the Appalachians. The crust may have wrinkled to form these mountains.

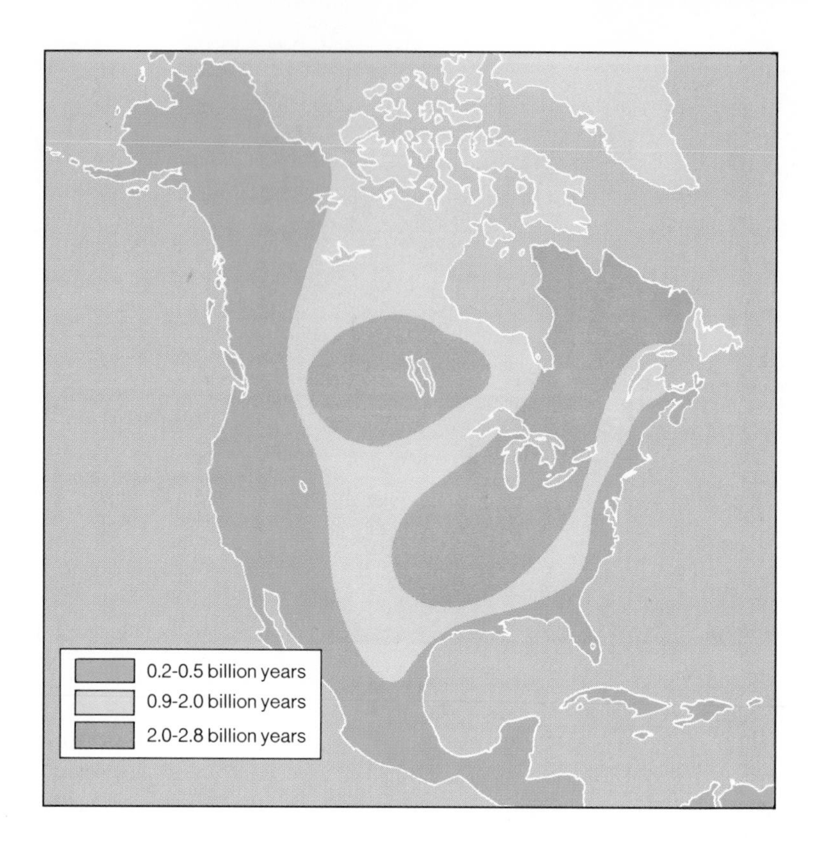

Fig. 3-17 The edges of the North American continent are younger in age than its center.

SUMMARY

Along their boundaries, crustal plates may be moving apart, past each other, or together. The world's great mountain ranges, volcanoes near these mountains, deep ocean trenches, and island arcs, are caused by plates coming together. As a plate moves, some material along its edge may be destroyed. In other places, new material is added to the plate's edge. Continents may grow when small pieces of continental crust are added to their edges.

QUESTIONS

Use complete sentences to write your answers.

1. Describe the three kinds of boundaries that exist between crustal plates.
2. How does plate tectonics explain the formation of mountain ranges, trenches, and island arcs
3. Explain how small pieces of continental crust may be added to a continent.

TECHNOLOGY

SEISMIC REFLECTION PROFILING

Scientists today consider the study of the earth's crust to be one of the great frontiers of modern earth science and a key to understanding the nature and evolution of our planet. For a long time, earth scientists concentrated their study on the uppermost surface of the earth, relying on general theories to explain the crustal layers beneath it. The most common theory held was that, under a thin layer of sedimentary rock at the surface, lay a thick band of granite with a thick layer of volcanic rock below that.

Today, new techniques, such as deep drilling and measurement of tremors that pass through the earth, have allowed scientists to explore the geology of the thick crust of the continents. The deepest hole drilled for scientific purposes is more than 10,900 meters (about 11 kilometers) beneath the surface, and that record will soon be broken. As a deep hole is drilled, a long cylinder, or core, of rock is cut and brought to the surface for scientists to study.

The COCORP Project, based at Cornell University, has used a second technique, called seismic reflection profiling, to explore deep beneath the earth's surface. The principle behind seismic reflection profiling is similar to that used in echo-sounding from a boat. Two crews using truck-mounted vibrators are needed for this technique: one to send tremors, or seismic waves, through the earth and another to receive and measure them. Crews move the source and receiver slowly over the earth's surface like a giant caterpillar. Huge quantities of data are recorded and must be processed by a computer. The final product is very detailed information about the structure and other physical properties of the rock along the path taken by the source and receiver.

Many deep holes and seismic measurements are needed to build a reliable base of information, and the cost of these information-gathering techniques has often limited scientists. Nonetheless, based on the information now available, many scientists have revised their theories. They found that, while it is true that the crust is generally divided into zones or bands of different types of rocks, considerable variation in the bands exists.

Earth scientists have long believed that the rocks of the crust, and thus the continents themselves, are shaped by the up-and-down movement or shifting of rocks over time. The variation in the bands of rock uncovered by these new techniques has strengthened the evidence for that theory. It also has provided evidence that makes many believe the masses of rock shift not only up and down but sideways as well.

The next few hundred years will be an exciting time for earth scientists, as the whole crust—not just its surface—is explored and charted. The results will be of practical use in the search for petroleum and the prediction of earthquakes. This exploration will also give earth scientists the tools to learn still more about how continents were formed and will change over time.

INVESTIGATION

MAKING A CRUSTAL PLATE MODEL

PURPOSE: To assemble a model of the seven major crustal plates, identify each by name, and locate some earth features related to crustal plate boundaries.

MATERIALS:

scissors	3 color pencils
glue	tracing paper

PROCEDURE:

A. Fig. 3–18, page 81, shows the pieces of a jigsaw puzzle. Each piece represents a map of one crustal plate. Carefully trace the edge of each piece onto a plain sheet of paper. Next, trace the continent outlines and the numbers shown.

B. Cut out the pieces you have drawn. Be careful not to cut off the numbers.

C. Assemble the plates as you would a jigsaw puzzle. The pieces should fit very closely. Glue the assembled map onto a piece of paper. You will use this map in several more activities, so save it and treat it with care.

D. The crustal plates are named according to continents or other major features on them. The names of the seven plates you traced are: African, Antarctic, Australian-Indian, North American, Pacific, South American, and Eurasian. On your map the Pacific plate is in two pieces. Outline each crustal plate on your new map by tracing around it.

E. Label the plates on your assembled map with their correct names.

1. Which plate does not bear the name of a continent?
2. List the names of the plates that bear the name of a single continent.
3. List the names of the plates that bear the names of more than one country or continent.

F. Fig. 3–4 on page 68 shows the location of the mid-ocean ridge. Draw the ridge on your map, using the color pencil listed in your key.

G. Refer to Fig. 3–11 on page 74, which shows some mountain ranges that are related to crustal plate boundaries. Make marks similar to those found in Fig. 3–11 on your crustal plate map, using the color listed in the key.

4. List the names of the mountain ranges shown on your map.
5. Explain how these are thought to be due to crustal plate activity.

H. See Fig. 3–16 on page 77, which shows where some major ocean trenches are. Using the proper color from your key, mark the locations of those shown.

CONCLUSIONS

1. List the names of the seven major crustal plates. Describe why each was so named.
2. Name three earth features that are found along crustal plate boundaries.
3. Explain how the mid-ocean ridge is related to crustal plate activity.

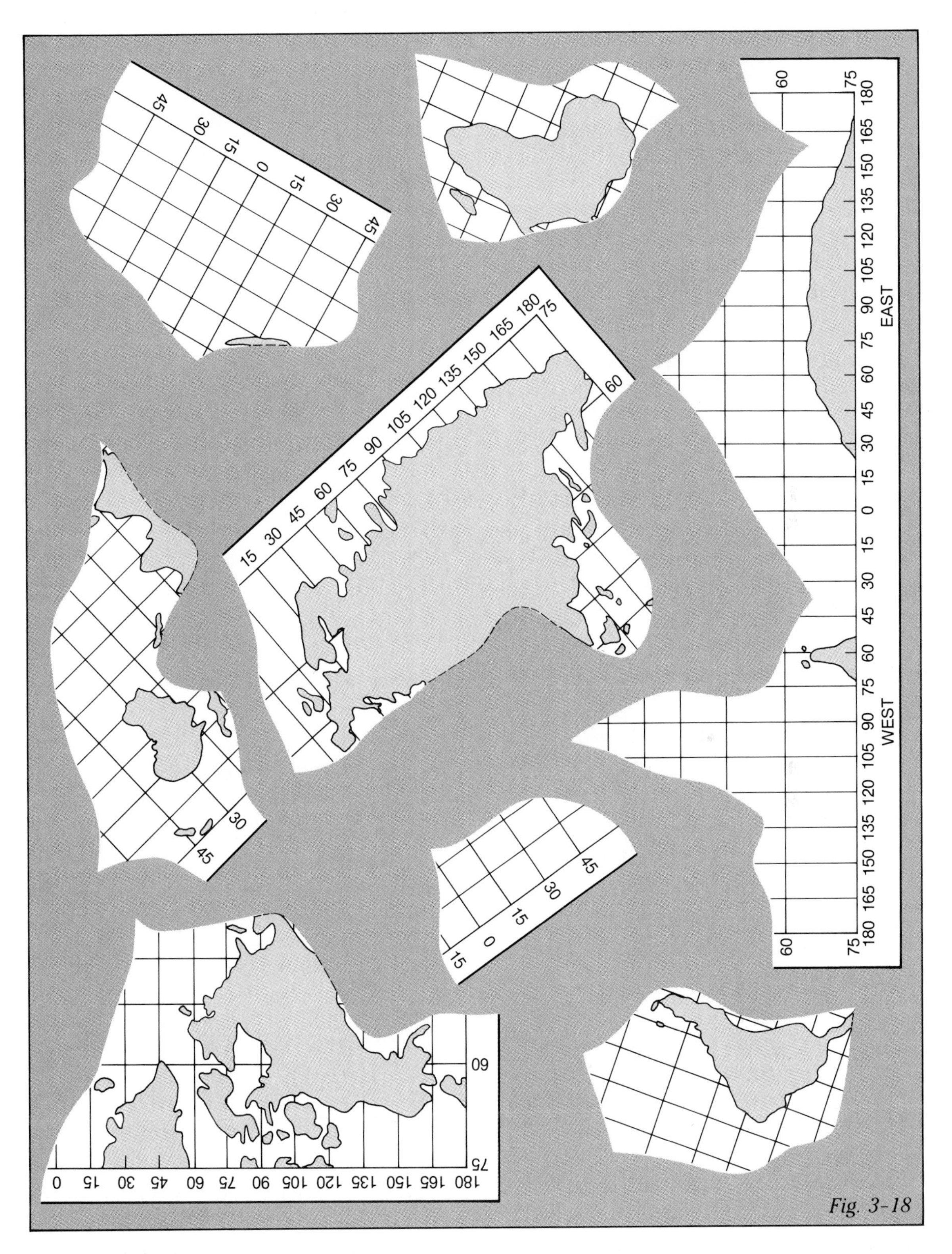

Fig. 3-18

¡COMPUTE!

SCIENCE INPUT

The study of crustal plates has produced many changes in the way earth scientists think about the beginnings of the earth's history. The theory of plate tectonics, or shifting plates, has been helpful in explaining how ridges and other landforms have been created. In addition, knowing the direction of plate shifting is useful in dating certain features of the earth.

Before you can use a theory, however, it is important to be familiar with all its parts. Understanding plate tectonics requires that you learn many new words and concepts.

COMPUTER INPUT

The computer can be used as a learning tool in a number of ways. One way is to create a game. Instead of asking straightforward questions, a game format can permit some hinting. In Program Plate Game, a game format is used. Your answer will not be input as a complete word, but as separate letters until the entire correct word is printed on the screen. There is, in this way, a continual cycle of feedback.

Feedback is information about your answers; it helps you to learn more efficiently. Your teachers' comments and grades on your assignments are feedback. They give you information you can use to judge the quality of your work. The computer's feedback helps in this way, too. It acts as a private tutor and has the advantage of giving you an immediate response.

In Program Plate Game you will use the computer to learn the words and concepts of plate tectonics explained in this chapter. The computer will give you feedback and, at the end of the program, will tell you how many times you have given the wrong answers.

WHAT TO DO

Make a data chart that lists brief definitions for the new words and concepts in this chapter. A sample chart appears below. Make your definitions short and clear.

Enter Program Plate Game using the information from your data chart to write data statements 311 to 319. Data statement 310 is printed in the program as an example.The program gives you the definition and asks you to enter the word or concept being defined, letter by letter. If the letter you choose is in the word, it will be printed on the screen in its proper place. If the letter is not in the word, it will count as a wrong choice.

SAMPLE DATA CHART

Term	Definition
Core	Center part of the earth
Mantle	Next to crust / largest layer
Crust	Thin solid outer layer of the earth
Mid-ocean ridge	Underwater mountains

PROGRAM NOTES

If you changed the nature of the data statements, this program could be applied in a variety of ways. You could use words and concepts from other chapters, or from other subjects. For example, you could change the word to the name of an historical figure, and change the definition to an event or achievement associated with that person. Then the program would print out the achievement and you would need to supply the person's name.

TECTONIC TUTOR: PROGRAM PLATE GAME

GLOSSARY

HARDWARE — The machinery itself, including the CPU (central processing unit), keyboard, and monitor (screen).

SOFTWARE — The programs used in the computer. Usually, these are available on a floppy disk or cassette tape.

REM — Short for *remark*, a command in BASIC, written for the user, not the computer. REM statements help the user understand the program, but they are not read by the computer.

PROGRAM*

```
100  REM PLATE GAME
110  FOR X = 1 TO 20: PRINT : NEXT
120  READ W$,H$: IF W$ = "0" THEN
     END
130  T$ = "":Q = 0
140  FOR X = 1 TO LEN (W$):T$ = T$ +
     "-":NEXT
150  PRINT : PRINT : PRINT : PRINT H$"-
     "T$: PRINT : INPUT "PICK A
     LETTER —> ";L$
160  G$ = T$:T$ = "":Q = Q + 1:Q1 =
     0
170  FOR X = 1 TO LEN (W$)
180  IF MID$ (W$,X,1) = L$ AND Q1 = 0
     THEN Q = Q - 1:Q1 = 1
190  IF MID$ (W$,X,1) = L$ THEN T$ =
     T$ + L$: GOTO 210
200  T$ = T$ + MID$ (G$,X,1)
210  NEXT
220  FOR X = 1 TO LEN (W$)
230  IF MID$ (T$,X,1) = "-" THEN GOTO
     150
240  NEXT
250  PRINT W$;"- ";Q;" WRONG
     LETTER(S)": PRINT : PRINT
260  FOR X = 1 TO 15: PRINT : NEXT :
     GOTO 120
300  REM DATA STATEMENTS
310  DATA  CORE, CENTER PART OF THE
     EARTH
311  DATA 0,0
320  END
```

BITS OF INFORMATION

Feedback is an important function of computer learning systems. It is a term that was first applied to both computers and humans by Norbert Wiener, a pioneer in computer electronics. He explained feedback as a process of communication involving a response to an original statement or input that helps a person or machine make a necessary adjustment. Wiener's most well-known book is called Cybernetics, *which is a term used to define the science of control and communication.*

Because they can manipulate data quickly, computers can give feedback almost instantaneously, making many pieces of modern technology self-adjusting. For example, as the space shuttle is operating, computers on the spaceship and on earth are constantly monitoring the performance of the many pieces of equipment on board. If a problem is spotted by the computer, many adjustments are made immediately and automatically to assure the success of these space missions.

* All the programs in this book are written for the Apple IIc.

CHAPTER REVIEW

VOCABULARY

On a separate piece of paper, match each term with the number of the description that best explains it. Use each term only once.

plate tectonics	island arcs	asthenosphere	rift valley
trenches	micro-continents	sea floor spreading	subduction
convection	pillow lava	continental drift	mid-ocean ridge

1. Describes the mid-ocean ridge as a place where hot material is rising.
2. A valley with steep sides.
3. The Aleutian Islands and Japan are examples of.
4. Differences in temperature causing movement in material.
5. The idea that the earth's continents are moving.
6. A chain of underwater mountains.
7. The theory that the earth's crust is made of moving plates.
8. Where one plate is forced beneath another.
9. A kind of rock formed when lava is cooled by water.
10. Long, narrow depressions in the ocean floor.
11. Pieces of continental crust scattered over the ocean floor.
12. The partly molten upper layer of the mantle.

QUESTIONS

Give brief but complete answers to each of the following questions. Unless otherwise indicated, use complete sentences to write your answers.

1. Why was the idea of continental drift not accepted by most scientists?
2. Describe the theory of continental drift.
3. What two important discoveries about the mid-ocean ridge led to the sea floor spreading theory?
4. How does sea floor spreading explain why the youngest rocks are found near the mid-ocean ridge?
5. Use the theory of plate tectonics to explain how the continents move.
6. What causes convection currents in the earth's mantle?
7. What did scientists find out about the ages of rocks on the sea floor?

8. Where are crustal plates moving apart?
9. Give an example of where two plates are sliding past one another.
10. Give an example of where crustal plates are coming together.
11. Name some mountain ranges that have been formed by plate collisions.
12. What is an ocean trench and where are most of them found?
13. Give two examples of island arcs.
14. Why do ocean plates sink under continental plates when they meet?
15. Why do continents crumple when pushed together?

APPLYING SCIENCE

1. A crustal plate moves an average of 5 cm a year. How far will the crustal plate move in (a) 5 years? (b) 50 years? (c) 1,000 years? How does this explain why it is difficult to see that landmasses move?
2. According to the theory of plate tectonics, will California ever fall into the Pacific Ocean? Explain.
3. If the sea floor is spreading at the mid-ocean ridge, does this mean that the earth is getting larger? Explain.
4. Why don't continents fit together exactly, the way a jigsaw puzzle does?
5. Using Fig. 3–7 on page 70, describe where your home will be on the earth millions of years from now.

BIBLIOGRAPHY

Alexander, George. "Going, Going, Gone." *Science 81*, May 1981.

Bingham, Roger. "Explorers of the Earth Within." *Science 80*, September/October 1980.

Harrington, John W. *Dance of the Continents: Adventures with Rocks and Time.* Boston: J. P. Tarcher (Houghton Mifflin), 1983.

Miller, Russel. *Continents in Collision* (Planet Earth Series). Chicago: Time-Life Books, 1984.

Siever, Richard. *The Dynamic Earth.* San Francisco: W.H. Freeman, 1983.

Time-Life Editors. *Floor of the Sea* (Planet Earth Series). Chicago: Time-Life Books, 1984.

CHAPTER

4

Life begins after only a few months on a hardened lava flow. Much of the earth's crust was formed from sheets of lava. Every day there is new crust added to the earth.

BUILDING THE CRUST

CHAPTER GOALS

1. Identify the ways that the earth's crust responds to strain caused by movement of the crustal plates.
2. Describe how the different types of volcanoes form.
3. Explain why earthquakes, volcanoes, and mountains occur in the same areas.
4. Describe the geologic features produced by magma.

4–1. Earthquakes

At the end of this section you will be able to:

- Explain why most earthquakes occur.
- Define the term "fault" and describe three ways that rocks can move along faults.
- Name and describe three kinds of earthquake waves.
- Describe safety procedures to follow in case of an earthquake.

On the afternoon of Good Friday, March 27, 1964, most of the people of Anchorage, Alaska, were at home. Schools were closed and offices let out early. The streets were quiet and peaceful. Suddenly, at 5:36 P.M., the ground began to shake. The people of Anchorage, however, were not worried. They were used to the ground shaking. But this would be the beginning of one of the largest earthquakes ever to take place in this century.

For the thousands of people who lived in Alaska, that day would be remembered for the widespread destruction that took place. Thousands of homes and most of the industry were destroyed. For scientists, however, this was an opportunity to study the forces at work inside the earth. They made many observations and measurements of the ground. From this information they began to understand what happens when an earthquake strikes. But even more important was that scientists began to understand what causes an earthquake and where earthquakes are most likely to take place. With this new information, scientists hope to be able to predict future earthquakes. In this way, lives may be saved.

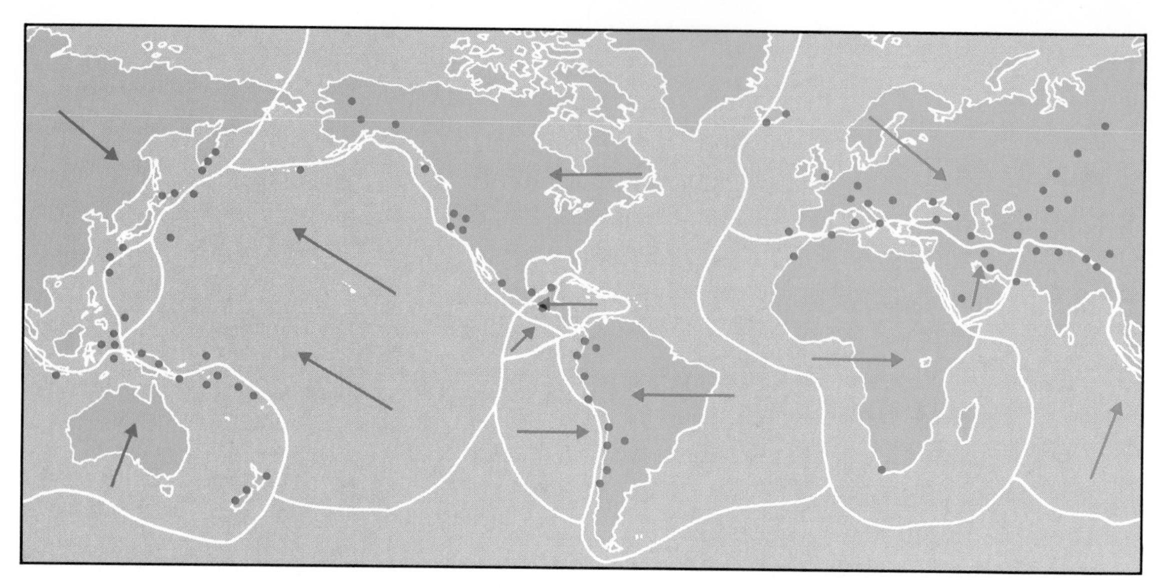

Fig. 4-1 The colored dots on the map show the location of large earthquakes during recent years.

WHAT IS AN EARTHQUAKE?

Fig. 4-1 shows the places where large earthquakes took place during a six-year period. This map shows that most serious earthquakes occur where the crustal plates come together. The boundaries between the plates are actually huge cracks in the crust. As the plates slowly move, their edges grind together. The plates move against each other along these cracks in crustal rocks. A place where the rock has moved on one or both sides of a crack is called a **fault.**

Fault The result of movement of rock along either side of a crack in the earth's crust.

The rocks along a *fault* may move in three ways. First, the rocks may be pulled apart. This can happen where the plates are separating, as they are along the mid-ocean ridge. Then the rocks along one side of the fault slide down against the rock on the other side. This is called a normal fault. See Fig. 4-2 (a). Second, the rocks may be pushed together, as they are when two plates meet head-on. One side of the fault usually moves up. This is called a reverse fault. See Fig. 4-2 (b). A third kind of fault takes place when the blocks of rock on each side move in opposite directions. This is called a strike-slip fault. This kind of fault occurs where two plates are sliding past each other. See Fig. 4-2 (c).

Faults are also found away from the edges of crustal plates. Many scientists believe, however, that almost all faults are caused by moving plates.

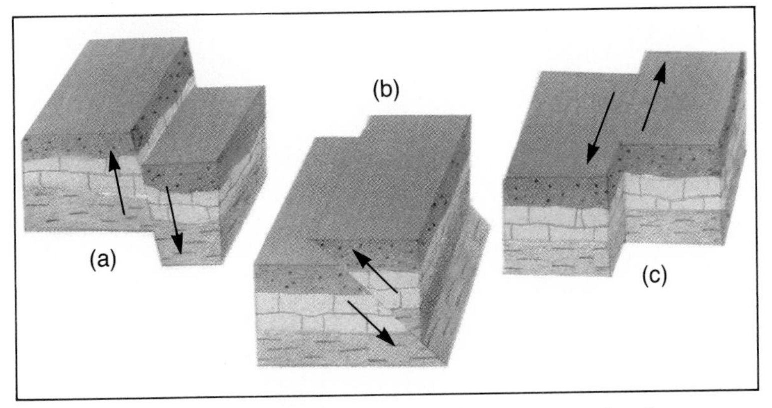

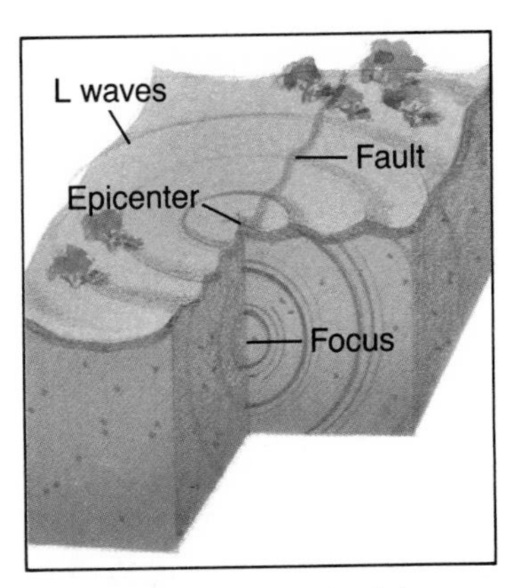

Fig. 4-2 (above left) *(a) Movement along a fault may be caused by plates moving vertically; (b) two plates can be pushed together along a fault where one plate moves up relative to the other; (c) plates can slide past each other.*

Fig. 4-3 (above right) *The focus is the exact place where movement occurs along a fault. The epicenter is the place on the surface above the focus.*

The plates that make up the earth's crust are in slow, constant motion. As they move, the rocks that make up the plates bend, stretch, or get squeezed. When rocks can no longer change shape, they break at their weakest point. These cracks, or fractures, are also faults.

Almost all large earthquakes take place when rocks suddenly snap along faults. If the rocks along a fault were constantly moving, their surfaces would soon be worn smooth. The rocks would slide gently like a well-oiled drawer. However, the rock surfaces are jagged. The sides of a fault area usually are locked together like a sticky drawer. This causes stress. When the stress becomes too great, the rocks suddenly snap and slip along the fault. This movement causes a sudden release of the built-up stress, which causes an earthquake.

The point beneath the earth's surface where the rocks move along the fault is called the **focus** See Fig. 4-3. Directly above the *focus* on the surface is the **epicenter.** The *epicenter* is the place where the earthquake is usually felt most strongly.

Focus The point of origin of an earthquake.

Epicenter The point on the earth's surface directly above the focus of an earthquake.

EARTHQUAKE WAVES

Sudden movement along a fault causes vibrations in the form of waves. These waves move out from the focus in all directions. The waves are felt on the surface as an earthquake. Earthquake waves are called **seismic (size-**mik**) waves.** *Seismic waves* are a form of energy. They can travel through solid rock, loose sand, and gravel. Seismic waves are felt most often on loose, water-soaked earth. During an earthquake this loose earth shakes like jelly in a bowl. In general, the buildings that are damaged most during earthquakes are those built on loose rock. See Fig. 4-4.

Seismic waves The vibrations caused by movement along a fault.

Fig. 4-4 This damage occurred when a large block of loose sand, gravel, and clay slid down toward the shore.

There are three kinds of seismic waves, *primary* (P) *waves*, *secondary* (S) *waves*, and *surface* (L) *waves*. Primary waves are caused by back-and-forth vibrations in the rock. They travel through the earth faster than the other waves. Secondary waves are caused by the up-and-down motions of the rock. They do not travel as fast as the primary waves. Surface waves cause the surface of the land to move in much the same way as ripples move on water. They cause most of the destruction on land. The three types of seismic waves are shown in Fig. 4-5.

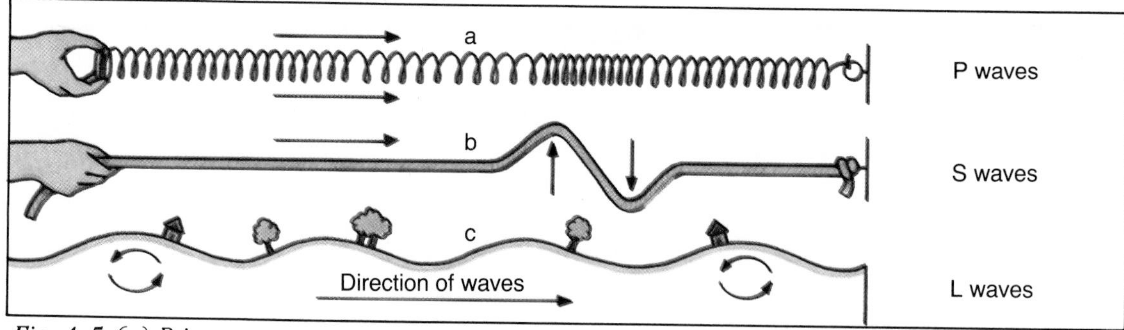

Fig. 4-5 (a) Primary waves are like the back and forth wave moving along a stretched spring; (b) secondary waves resemble the up and down wave moving along a rope; (c) surface waves are like waves on water.

Seismic waves travel out from an earthquake focus and pass through the earth. Each kind of seismic wave moves in its own way through different parts of the earth. For example, secondary waves cannot travel through liquids. The fact that secondary waves do not travel through the earth's core helped scientists to understand that the earth's core is not completely solid. With this information, scientists can study the inside of the earth.

MEASURING EARTHQUAKES

Seismic waves can be recorded by an instrument called a **seismograph** (**size**-muh-graf). Fig. 4-6 shows how one kind of *seismograph* works. In 1935, Charles Richter (**rik**-tur) used seismograph records to measure the energy given off by an earthquake. The *Richter scale* uses a scale of numbers to show the amount of energy released by an earthquake. Each number on the scale represents ten times as much energy as the next lower number. For example, an earthquake measuring 5 on the Richter scale would be ten times larger than one with a measurement of 4, while an earthquake measuring 6 would be a hundred times larger than one measuring 4. Almost all earthquakes are very small. That is, they have Richter measurements of 4 or less. Large earthquakes with values of 8 or more usually take place only once every five to ten years. However, any earthquake with a rating of 6 or more on the Richter scale can cause damage to buildings. Fig. 4-7 shows where the largest earthquakes have taken place in the continental United States.

Seismograph An instrument that records seismic waves.

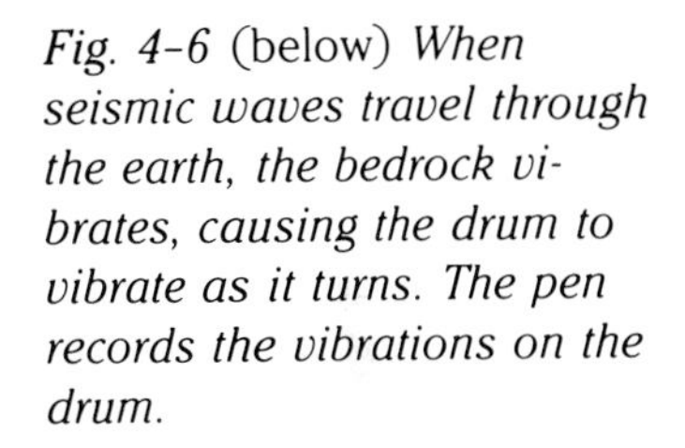

Fig. 4-6 (below) *When seismic waves travel through the earth, the bedrock vibrates, causing the drum to vibrate as it turns. The pen records the vibrations on the drum.*

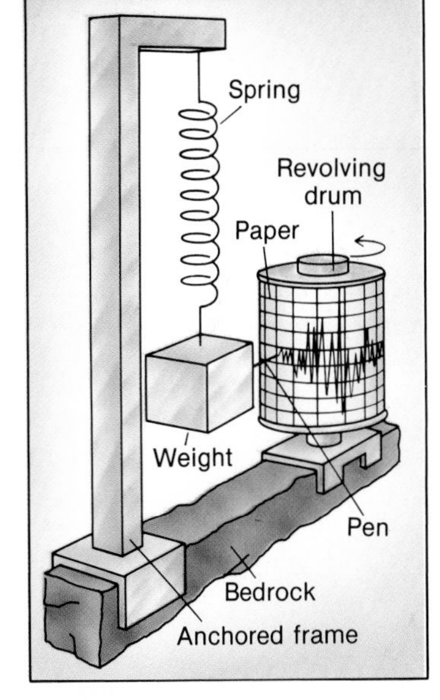

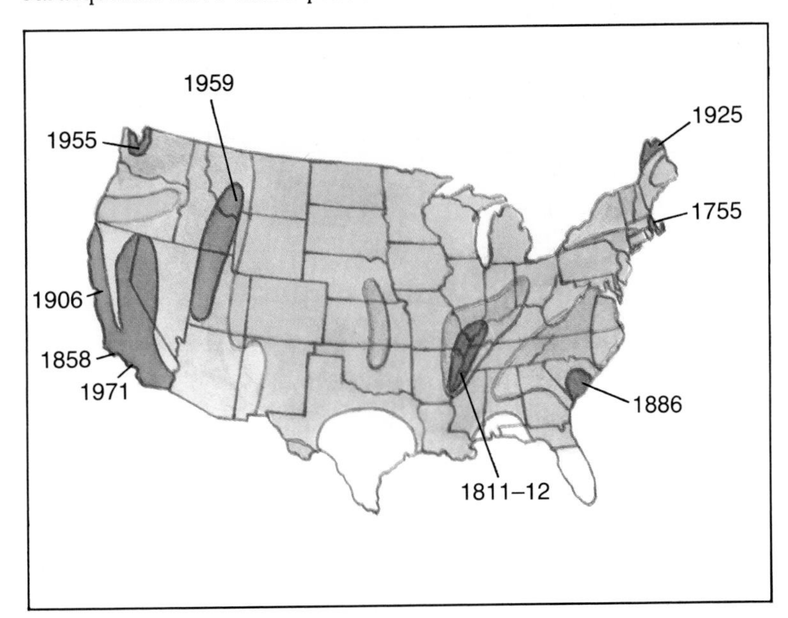

Fig. 4-7 (left) *The darkest color shows the areas where the largest earthquakes have struck in the past. Dates mark the location of very large earthquakes.*

Large earthquakes release more energy than hundreds of thousands of small ones together. This means that small earthquakes do not release large amounts of energy a little at a time. The total energy released by small earthquakes is not enough to prevent the buildup of stress in the rocks. Large, dangerous earthquakes give off most of the stored energy.

EARTHQUAKE SAFETY

It is estimated that throughout history 74 million people have lost their lives in earthquakes. See Fig. 4-8. Some places in the world have many more earthquakes than other places. However, no place is completely safe from earthquake damage. Some faults are known to be the source of earthquakes. For example, the city of San Francisco is located very near a fault that caused a major earthquake in 1906. Buildings in cities near such faults should be built in ways that will help to prevent them from falling during an earthquake. Dams should not be built near active faults. Failure of a dam during an earthquake can cause a very sudden and dangerous flood. Nuclear power plants should also be built in places away from faults. An earthquake could cause the release of radioactive materials from these plants.

Fig. 4-8 Earthquakes can cause damage to homes and property.

Earthquakes usually take place without any warning. If one takes place where you are, the actual shaking probably will last less than a minute. The entire earthquake may be over before you have time to protect yourself. But if there is time, a few simple rules can be followed to reduce the danger. Many schools and buildings already have plans in case of an earthquake.

Move into an open area to get away from buildings that might collapse. However, if you are in a large building, stay inside. Do not run outside where you might be struck by falling objects. Try to stand in a doorway. Get under a desk or table that will protect you from falling parts of the building. Stay away from chimneys, brick walls, and large glass areas. These are likely to fall during an earthquake.

After the earthquake, help any injured persons. Stay away from fallen electric wires or power poles. Do not move around the neighborhood to check the damage. You probably will be getting in the way of emergency crews. Pay attention to instructions you may be given by officials such as police officers and firefighters.

We cannot stop the movement of the great crustal plates. However, scientists soon may be able to predict when and where an earthquake will happen. They have found that the earth often gives signals before an earthquake. For example, the speed of primary waves seems to slow down before an earthquake. This change in speed can be observed from a small earthquake or explosion. Measurement of the change in speed may make it possible to predict larger earthquakes. Research also has shown that certain gases are released into well water just before an earthquake. The presence of these gases in well water could be a signal of a coming earthquake. Also, the earth bulges in places where major earthquakes have occurred. Animals seem to become very nervous and excited a short time before an earthquake takes place. Research is being carried on to try to find other signs that usually occur before an earthquake happens.

Experiments are now under way to try to measure the buildup of stress along faults. In the future, earthquake predictions may become as common as today's advance warning of bad weather. Being able to predict earthquakes will make it possible to prevent some loss of life and property.

SUMMARY

Earthquakes occur when rocks suddenly move along a fault. Earthquakes send out P, S, and L waves, which can be recorded on a seismograph. Knowing what to do during and after an earthquake can decrease the chance of injury.

QUESTIONS

Use complete sentences to write your answers.

1. What is a fault?
2. Describe three ways that rocks may move along faults.
3. Describe the three kinds of seismic waves.
4. Explain what you should do during and after an earthquake.

INVESTIGATION

PLOTTING EARTHQUAKES

PURPOSE: To plot a number of earthquake positions and to look for a pattern in these positions.

MATERIALS:
crustal plate map paper
pencil

PROCEDURE:

A. The following table lists 15 locations, in the degrees of latitute and longitude, where earthquakes have occurred. This list is only a small sample of all the earthquakes that occur each year.

Earthquake	Latitude	Longitude
1	60°N	152°W
2	45°N	125°W
3	35°N	35°E
4	30°N	115°W
5	30°N	60°E
6	20°N	75°W
7	50°N	158°E
8	40°N	145°E
9	15°N	100°E
10	15°N	105°W
11	10°S	105°E
12	5°S	150°E
13	0°	80°W
14	25°S	75°W
15	50°S	75°W

B. On the crustal plate map you made in the Investigation section on page 80, locate the positions where each earthquake occurred and place its number at that position. For example, earthquake number 1 occurred on the coast of southern Alaska at 60°N latitude and 152°W longitude. Place a small number 1 at this location.

C. After locating all 15 earthquakes, look at your map to answer the following questions. Use complete sentences to write your answers.
   1. Describe where the earthquakes are located.
   2. Is there a relationship between the earthquake positions and the shape of the continents?
   3. Describe how the locations of the earthquakes are related to crustal plate boundaries.

CONCLUSIONS:

1. Describe the pattern of the locations of the 15 earthquakes you plotted.
2. Copy and complete the following hypothesis: The pattern found by plotting a number of representative earthquake locations on the world map suggests that many future earthquakes may be expected to occur ______.
3. From the information on your map, predict if an earthquake will take place where you live.
4. Which is easier to predict, where or when earthquakes will happen? How can earthquake predictions be improved?

4–2. Volcanoes

At the end of this section you will be able to:

- Explain what causes volcanoes.
- Point out the locations on the earth's surface where most volcanoes are found.
- Describe how volcanoes erupt.

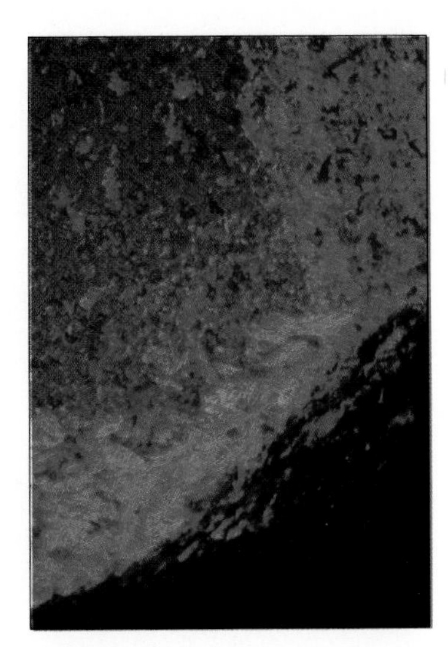

Fig. 4-9 Lava is magma that appears on the earth's surface.

On August 24, A.D. 79, the scene in Pompeii was peaceful and normal. Though the city was located near Mount Vesuvius, a large volcano, the people were not worried. The volcano had not erupted for hundreds of years. Suddenly, Mount Vesuvius erupted with an explosion ten times more powerful than the eruption of Mount St. Helens. Volcanic ash buried the city and its people. Scientists believe that there are 33 sites in the western states, Alaska, and Hawaii where an eruption could take place at any time.

MAGMA AND LAVA

In certain places in the earth's mantle there are pockets of red-hot, molten rock called **magma.** About 95 to 320 km beneath the surface is a zone of rock where seismic waves are slowed. The rock in this region is very close to its melting point. *Magma* is formed in this zone. For example, the action around the plate edges can cause deep pockets of magma to be formed.

Magma Molten rock beneath the earth's surface.

Magma is always red-hot. Its temperature is usually between 500°C and 1,200°C. The liquid magma is less dense than the solid rock around it. Being less dense, magma tends to work its way up toward the surface like a hot-air balloon rising in the air. Sometimes the magma cools before reaching the surface. It then hardens into solid rock while still beneath the surface. Often magma does reach the surface as a liquid. When magma comes out onto the earth's surface, it is called **lava** (**lahv**-uh). See Fig. 4–9. A **volcano** begins to form at the place where *lava* appears on the surface.

Lava Magma that has flowed out onto the earth's surface.

Volcano A place where lava reaches the surface and builds a cone or other surface feature.

THE RING OF FIRE

The world's most active *volcanoes* on land are found mainly in a narrow belt that circles the Pacific Ocean. This chain of volcanoes is often called the "Ring of Fire." See Fig. 4–10. Other

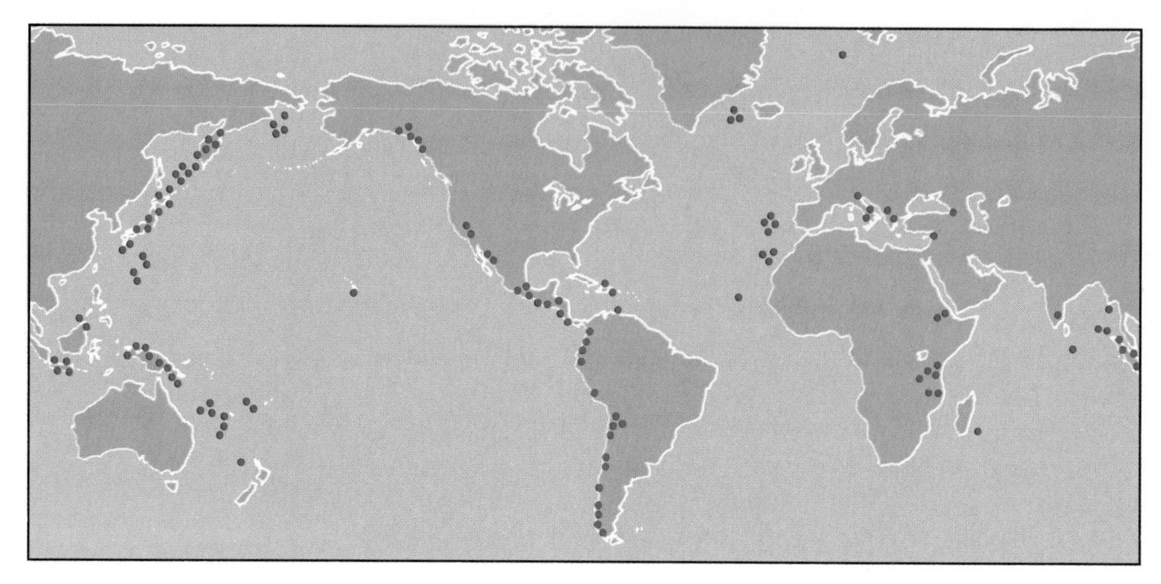

Fig. 4-10 The colored dots on the map show the locations of many of the earth's volcanoes.

centers of volcanic activity are located along the mid-ocean ridge and in the Mediterranean region. These areas are the active parts of the earth's crust along the boundaries of the crustal plates. These active zones are also the location of most of the world's earthquakes.

Volcanoes are found in greatest numbers along two kinds of plate boundaries. First are the zones of subduction. Here, one plate is sinking beneath the other and producing magma. For example, volcanoes such as Mount St. Helens in the Pacific Northwest are caused by subduction. See Fig. 4-11. There the Pacific plate is sinking beneath the North American plate.

Another example of volcanoes formed by subduction is on the other side of the Pacific near Japan. There the Pacific plate is being pushed beneath the Asian plate. See Fig. 4-11. The re-

Fig. 4-11 Many volcanoes are formed from the collision of two plates that make up a part of the sea floor.

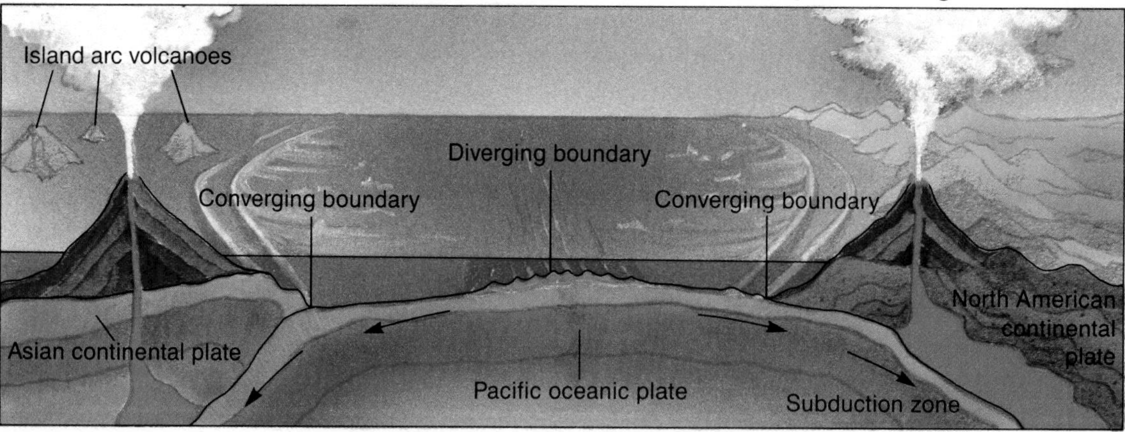

Fig. 4-12 Surtsey, a volcanic island, was the result of an underwater volcano. Surtsey rose from the sea in 1963.

sulting volcanoes form a chain of island arcs. These are the volcanic islands of the Aleutians, Japan, and the southeastern Pacific.

A second zone of volcanic activity is found along the boundaries where plates move apart. Along the mid-ocean ridge, for example, constant eruptions take place beneath the sea. In the North Atlantic, part of the mid-Atlantic ridge rises above the water. This is the island of Iceland. The volcanoes found in Iceland are caused by magma rising where plates are spreading apart. See Fig. 4-12. The volcanoes found in East Africa are also caused by plate spreading. There the continental plate is being pulled in two. A place where a continent is breaking apart is called a *rift.* Scientists believe that rifts in continents may become oceans millions of years from now.

Most volcanoes are located along plate boundaries. However, some volcanoes, like the Hawaiian Islands, are found in the middle of plates. Scientists believe that these volcanoes are caused by a huge column of rising magma. This hot spot remains in place while the moving plate is carried over it. A chain of volcanoes can result. The Hawaiian Islands are part of such a chain stretching out across the Pacific Ocean. See Fig. 4-13.

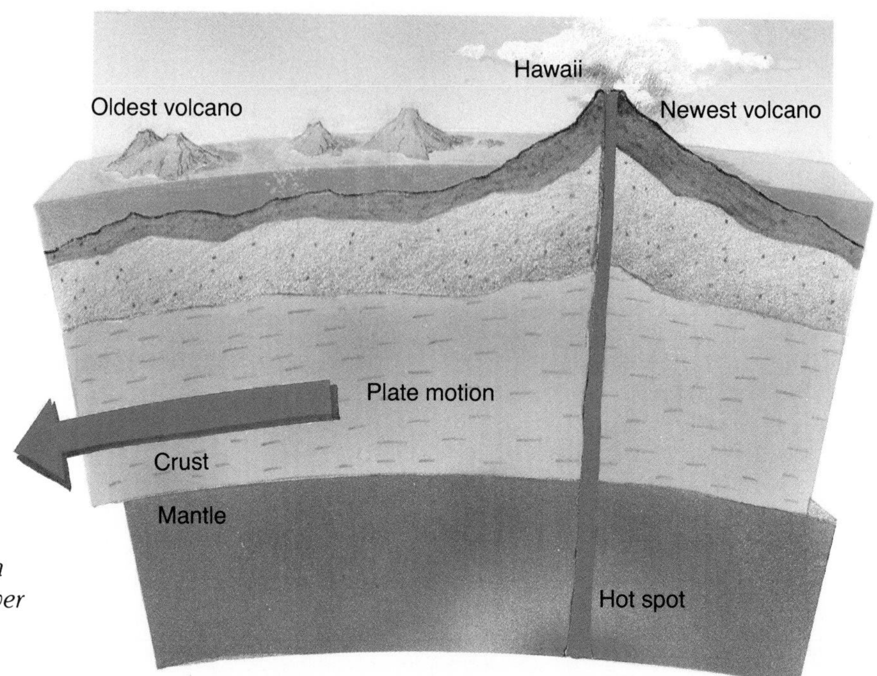

Fig. 4–13 The Hawaiian Islands were formed when the Pacific plate moved over a hot spot.

VOLCANIC ERUPTIONS

The opening through which magma reaches the surface is called a *vent.* Around a vent there is often a bowl-shaped feature called a *crater.* A single volcano often has several vents through which lava reaches the surface.

Volcanoes are not all alike. They differ in the way they erupt, or throw out materials. Also, magma is not always the same. It may be heavy, dark, and thick or it may be light and thin. Some magma contains steam and other gases that bubble up like the gas in a bottle of soda pop.

How a volcano erupts depends mainly on the kind of magma and what is inside the vent. Gentle eruptions usually happen when the vent is not blocked and the magma is thin. Then the lava flows out in a steady stream. If the vent is sealed by old, hardened lava or other materials, an explosion can occur. Trapped gases build up pressure in the magma. An explosion can be set off suddenly by any change. For example, Mount St. Helens erupted when part of the mountain broke away along a fault. See Fig. 4–14. An eruption along the side and near the top of a vent, as with Mount St. Helens, sends the blast to the side as well as upward. Many volcanoes may erupt quietly at one time and with explosions at another time.

Fig. 4–14 (opposite page) *(a) A landslide, caused by an earthquake, begins; (b) large amounts of rock and ice slide down the side of the mountain; (c) steam and volcanic ash are released; (d) Mt. St. Helens explodes, giving off rock and ash.*

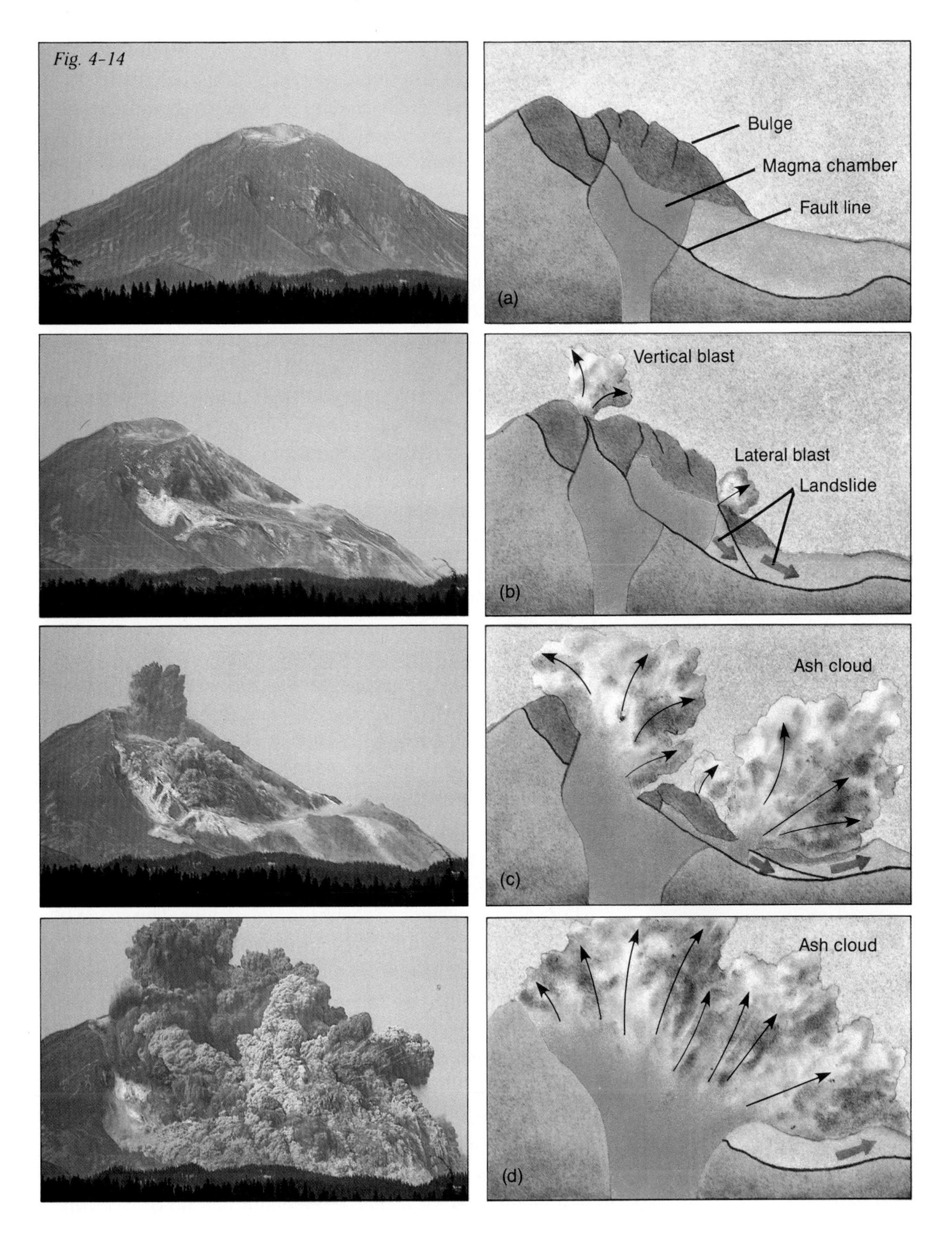
Fig. 4-14
Bulge
Magma chamber
Fault line
(a)
Vertical blast
Lateral blast
Landslide
(b)
Ash cloud
(c)
Ash cloud
(d)

When a volcano erupts, living things are destroyed by the high temperatures. The temperature of lava is about 1,000°C. Living things also may be affected by poisonous gases given off by the volcano. The volcano also gives off clouds of ash which can travel around the world. The ash can block out part of the sunlight and cause a cooling of the atmosphere. Ash deposits can clog air filters in cars, cause power failures, and make breathing difficult. But volcanoes can provide benefits as well. When they erupt, valuable materials are brought to the earth's surface. Volcanic deposits can contain metals, such as gold, silver, iron, and copper. Lava rocks supply materials for fertile farmland.

Today, some scientists are looking for ways to tap the heat given off by erupting volcanoes. Currently, some countries, such as Iceland, use heat from magma in the earth. This heat is called *geothermal energy*. Geothermal energy supplies much of Iceland's heat and hot water.

An eruption can end in three ways. First, the supply of gas in the magma may drop. Then there is not enough pressure to force the magma out of the vent. Second, the magma chamber may be emptied enough to bring an end to the eruption. Third, the vent may become blocked when magma hardens.

Fig. 4-15 Volcanic bombs are large pieces of hardened lava thrown out by volcanoes. They often make a whistling sound like a falling bomb as they fall.

During explosive eruptions, lava is blown out. Particles of lava harden as they fly through the air. The smallest of these particles are called *volcanic dust* and *ash*. Larger particles called *cinders*, which are several centimeters across, are thrown out as well. There may also be large pieces, up to one meter across. These are called *bombs*. Bombs are often soft enough to become shaped like loaves of bread as they fly through the air. See Fig. 4-15.

There are three kinds of shapes for volcanoes. The shape of a volcano is caused by the type of eruption. First, the ash and cinders may build up around a vent. The result is a tall, narrow *cinder cone* like the one shown in Fig. 4-16 (a). Second, a volcano may begin with an explosion followed by quiet lava flows. Eruptions of this kind usually build up a large, cone-shaped mountain. This is called a *strato*, or *composite*, *volcano*. The mountain consists of alternating layers of lava and pieces of rocks thrown out during the violent periods. See Fig. 4-16 (b). A third kind of volcano forms when there is almost never an explosive eruption. It forms a mountain built up of layer upon

Fig. 4-16 (a) A cinder cone, (b) a composite volcano, (c) a shield volcano.

layer of lava. This is called a *shield cone.* It has a broad, flat shape with gently sloping sides. See Fig. 4–16 (c). The Hawaiian Islands are a group of shield cones built up over a hot spot on the Pacific Ocean floor.

SUMMARY

Magma is present in certain places in the crust and upper mantle. Magma may reach the surface to form a volcano. Volcanoes form mainly along plate boundaries. The type of eruption determines the kind of volcanic cone that is formed.

QUESTIONS

Use complete sentences to write your answers.

1. How does a volcano begin?
2. Where are most of the active volcanoes found?
3. What two kinds of crustal plate boundaries produce the greatest numbers of volcanoes?
4. Describe three ways volcanoes may behave and the kind of volcanic cone formed in each case.

INVESTIGATION

PLOTTING VOLCANOES

PURPOSE: To plot the positions of a number of volcanoes.

MATERIALS:
world map
paper
color pencil

PROCEDURE:

A. The following table shows 17 locations, in degrees of latitude and longitude, where volcanoes have formed.

Volcano	Latitude	Longitude
1	60°N	150°W
2	45°N	120°W
3	20°N	105°W
4	0°	75°W
5	65°N	15°W
6	40°N	30°W
7	17°N	25°W
8	45°N	15°E
9	30°N	60°E
10	55°N	160°E
11	40°N	145°E
12	5°S	155°E
13	10°S	120°E
14	5°S	105°E
15	15°S	60°E
16	30°S	70°W
17	55°S	25°W

B. Use a color pencil to plot each volcano position on the map of the world. For example, plot the position for volcano number 1 by going across the 60°N latitude line to where it crosses the 150°W longitude line. Write a small 1 at this point. This point should be on the southern coast of Alaska. Continue to plot the remaining volcanoes. These volcanoes are only some of the many that have formed in these regions.

C. Look at your finished map and answer the following questions.
1. Which ocean has volcanoes that stretch to the north and south near its midpoint?
2. Which ocean has a ring of volcanoes on the land surrounding it?
3. Is there a general pattern in the location of volcanoes?

D. Look at the map on which you plotted the earthquakes. Compare the volcano locations to the earthquake locations.
4. How are they related?
5. What is the relationship between the location of volcanoes and crustal plate boundaries?

CONCLUSIONS:
1. In general, where on earth are volcanoes most likely to occur?
2. Explain why volcanoes and earthquakes may be caused by the same forces.
3. Earthquakes very often happen before volcanoes erupt. Explain how eruptions of volcanoes may be predicted using earthquakes.

4–3. Building the Land

At the end of this section you will be able to:

- Explain two ways that the earth's crust may be strained
- Describe three ways in which parts of the crust may be raised to form mountains.
- Compare the way that plateaus form with the way that mountains form.

Seashells can be found at the tops of some mountains. Old beaches can be found hundreds of meters above sea level. See Fig. 4–17. Could these locations once have been underwater and then risen to their present height?

Fig. 4–17 How can you tell that this area was once covered by water?

FORCES THAT CHANGE THE CRUST

The rocks that make up the earth's crust are always being put under stress and strain. Forces in the earth cause rocks to be squeezed together, melted, pulled apart, or twisted. Since rocks do not change shape easily, the rocks become strained. It is as if giant hands are always working to push, pull, and bend a rigid crust. When the solid rock of the crust is moved, it is called **diastrophism.** Huge forces must act on the rocks to produce *diastrophism.* You already have seen how the crust is made up of great moving plates. Much of the force needed to bend and twist the rocks in the plates comes from the moving plates. As they grind together, they create forces that are felt all through the rocks.

Diastrophism Movement of solid rock in the crust.

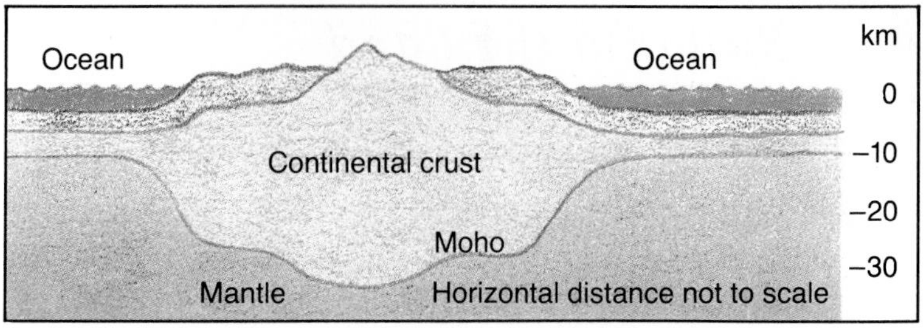

Fig. 4–18 The crust is thin under the ocean floor, thicker under the continents, and thickest under mountains.

The rock that makes up the mantle is more dense than the rock that makes up the crust. Because of the difference in density, the continents and sea floor are floating on the mantle. Suppose you drop a piece of wood in water. Part of the wood will sink into the water. The wood will then float because it is less dense than the water. The same is true for the crust floating on the mantle. The crust floating on the mantle is like different-sized logs floating in water. Large, thick logs float higher in the water than small logs. Mountains, being the thicker parts of the crust, float higher than other crustal parts. They also have roots that reach deeper into the mantle. See Fig. 4–18. Each part of the crust is in a state of balance as it rides on the mantle. This state of balance between different parts of the lower-density crust floating on the higher-density mantle is called **isostasy** (ie-**sahs**-tuh-see).

Isostasy The state of balance between different parts of the lightweight crust as it floats on the dense mantle.

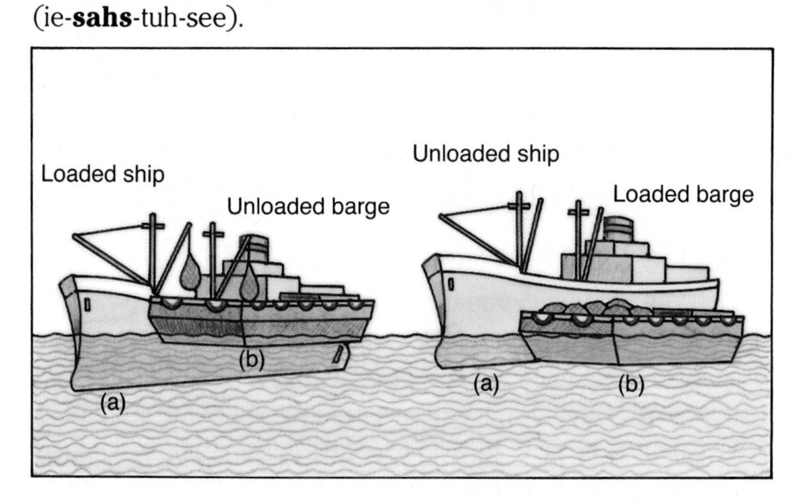

Fig. 4–19 Ships floating in water act like the crust floating on the mantle.

Observing what happens to floating ships shows how *isostasy* works. See Fig. 4–19. Cargo is moved from ship A to barge B. Ship A rises in the water and barge B sinks. The crust, floating on the mantle, acts in a similar way. For example, a heavy rock can fall from a mountain to the land below. The weight of the rock moves from one area of the crust to another. Because of

Fig. 4-20 Folds are caused when rocks are put under strain for a long period of time.

isostasy, the area that loses the weight rises. The area that gains the material sinks. Thus the rise and fall of different parts of the crust create strain in the surrounding rocks.

The rise and fall of parts of the crust take place very slowly. Strain builds up in the rocks over a long period of time. Sometimes the rocks crack. If there is movement along either side of such a crack, it is called a fault. If there is no movement along such a crack, it is called a **joint**.

Joint A crack in solid rock with no movement along the sides of the fracture.

Forces that squeeze rocks together do not always cause joints or faults. Instead, a slow, steady pressure can make some rocks bend without breaking. When rocks bend in this way, it is called **folding.** *Folds* in rocks may be small wrinkles. They also can be large enough to form mountains. See Fig. 4-20.

Folding Bending of rocks under steady pressure without breaking.

SHAPING THE LAND

Mountains are caused by diastrophism. There are three main ways that a part of the crust can be raised enough to form mountains. First, large blocks of rock can be separated by faults. Then the blocks may be tilted, like a row of books that have fallen over. Mountains made in this way are called **fault-block mountains.** The Grand Tetons in Wyoming are examples of *fault-block mountains*. See Fig. 4-21 (a).

Fault-block mountains Mountains formed when large blocks of rock are tilted over.

A second type of mountain range is formed by giant folds in the rocks. The Appalachian Mountains, for example, contain the remains of many great folds. During their long history the Appalachians have been worn down. Today we see long, parallel ridges and valleys, which were formed when the rock layers

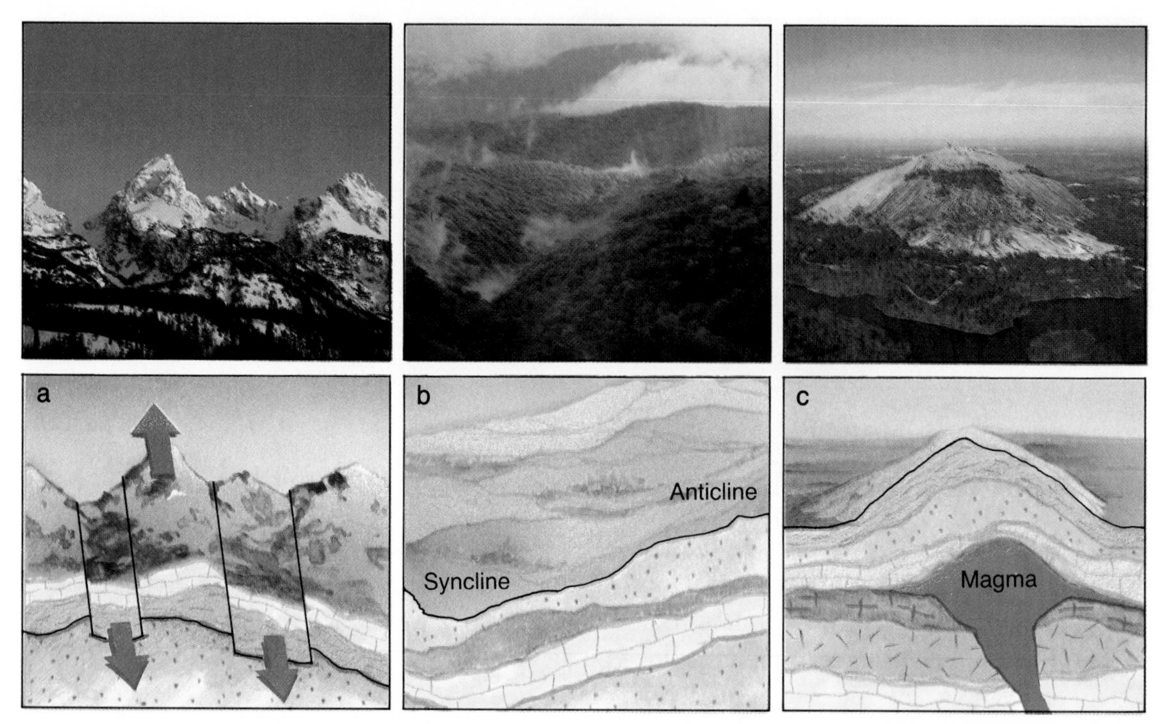

Anticline Rock layers folded upward.

Syncline Rock layers folded downward.

were bent into folds. See Fig. 4–21(b). The raised part of the fold that forms the ridges is called an **anticline (ant**-ih-kline). The ridges in folded mountains like the Appalachians are *anticlines.* The lowered section of a fold is called a **syncline** (**sin**-kline). *Synclines* form the valleys between the ridges. Look at Fig. 4–21b. again.

Dome mountains Mountains formed when a part of the crust is lifted by rising magma.

A third type of mountain is made when a large body of magma rises close to the surface. The crust is pushed up and lifted by the rising magma. The part of the crust that is raised makes **dome mountains.** The Black Hills of South Dakota are *dome mountains.* See Fig. 4–21(c).

PLATEAUS

Plateau Large raised area of level land.

Scattered all over the earth are very large regions of land that are lifted above the level of the surrounding crust. A large region of raised land is called a **plateau** (pla-**toe**). *Plateaus* can be caused by the same forces that build mountains. In the case of plateaus, large areas of the crust are bent upward. The Colorado Plateau, where the Grand Canyon is found, was formed this way. The Colorado River cut down into the land as the plateau was slowly raised. This formed the Grand Canyon. See Fig. 4–22 (a). Plateaus also can be formed when lava pours out and

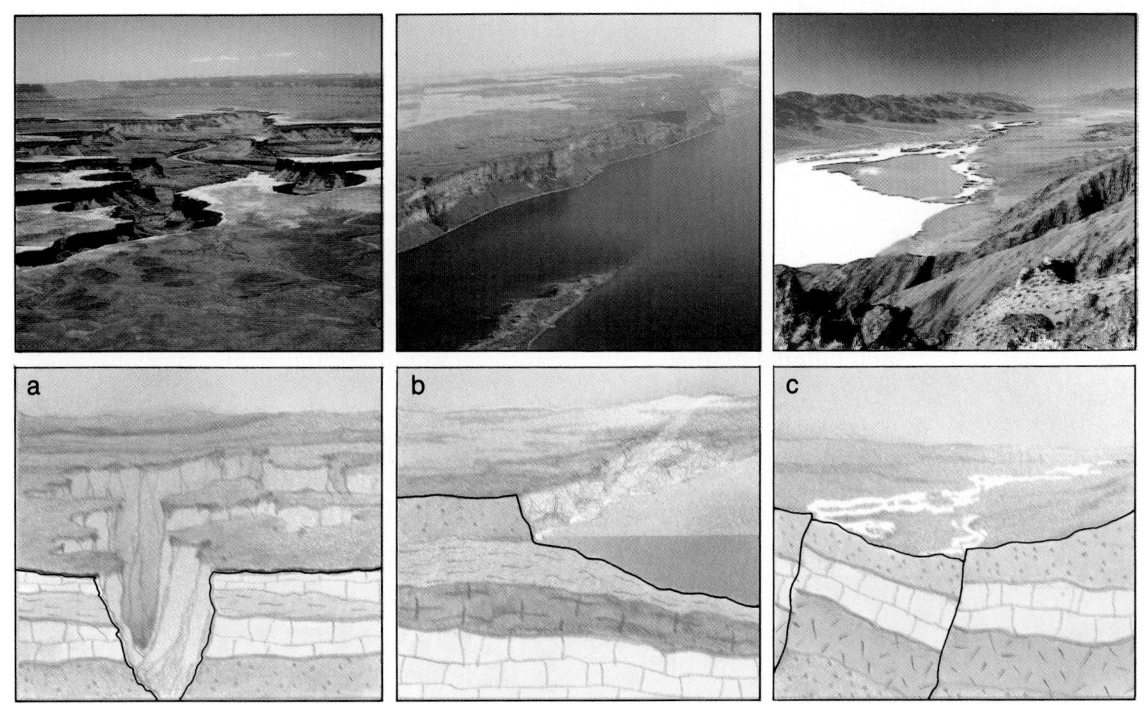

covers a large part of the land surface. The Columbia Plateau in the state of Washington is a lava plateau. See Fig. 4-22 (b).

While large areas of land are being uplifted into plateaus, some small areas are sinking. Death Valley in California is an example of such a place. The large block that makes up the floor of Death Valley is slowly tilting as one end sinks. See Fig. 4-22 (c).

Fig. 4-21 (opposite page) *(a) Large blocks of crust are pushed up to make fault-block mountains; (b) layers of rock bend to form folded mountains; (c) magma pushing up the crust forms dome mountains.*

Fig. 4-22 (above) *(a) The Grand Canyon, a plateau made by a river cutting into slowly rising land; (b) the Columbia Plateau, formed from lava; (c) Death Valley, formed when an area of land sank.*

SUMMARY

The solid rock is not as quiet as it seems. Huge forces are at work to strain the rock. Sometimes the rock moves to form mountains. Three types of mountains are folded, faulted, and domed. Parts of the crust can be lifted to form plateaus.

QUESTIONS

Use complete sentences to write your answers.

1. Describe two ways the crust may be strained.
2. Compare how folded, faulted, and domed mountains are formed.
3. Describe two ways in which a plateau may be formed. Give an example of each.

INVESTIGATION

DENSITY AND ISOSTASY

PURPOSE: To find the density of some common rocks and relate these to isostasy.

MATERIALS:

shale rock	balance
granite rock	graduated cylinder
basalt rock	water

PROCEDURE:

A. Copy the following table:

Rock	Mass (g)	Volume (cm^3)	Density (g/cm^3)
shale			
granite			
basalt			

B. Obtain samples of shale, granite, and basalt. Make certain each rock will fit into your graduated cylinder. Measure the mass of each rock and record the information in your data table.

C. Using the graduated cylinder and water, find the volume of each rock. Remember to tilt the graduated cylinder and add the rock slowly in order to avoid breaking the bottom of the cylinder. Record your results in the data table.

D. Find the density of each rock sample by dividing the mass by the volume. Record these measurements in the data table.
   1. Which rock has the greatest density?
   2. Which rock has the least density?

E. The three most common rocks found on the earth's surface are shale, granite, and basalt. Find the average density of your three rocks by adding their densities and dividing the total by three.
   3. What is the average density? This average is very close to the average density of all the rocks on earth's surface.
   4. Which rock has nearly this density?

F. Scientists have found the density of the earth by dividing the earth's mass by its volume. This gives an average density for the earth of 5.5 g/cm^3.
   5. How does the average density of the three rocks compare to the average density of the earth?
   6. In isostasy, the rocks of earth's crust are said to float on the mantle. Which of the rocks you measured would float highest if resting on the mantle?
   7. Which rock would float deepest in the mantle?

CONCLUSIONS:

1. Which has the higher density, the entire earth or the rocks found on the earth's surface?
2. Explain how the results of this activity support the idea that the earth has a lower-density crust floating on a higher-density mantle.
3. Explain how isostasy causes an uneven surface of the crust.
4. Explain why the largest mountains sink deeper into the crust.
5. Explain why mountains rise as sediment is worn away from the mountain's surface.

CAREERS IN SCIENCE

GEOLOGICAL TECHNICIAN

Geological technicians assist geologists and spend much of their time working outdoors. A technician's tools may be as simple as a hammer and chisel or they may involve heavy equipment. Technicians assisting geological researchers take samples of the earth's crust and determine which rocks are contained there. Other technicians are involved in studying earthquakes and their causes, and must be familiar with some of the computerized equipment in the field. Another group of geological technicians is engaged in the search for deposits of oil and natural gas, which are vital to our country's energy future. All of these technicians carry out the work designed by a geologist, who usually heads the team.

Most technicians have completed at least a two-year program at a technical school. Jobs are available with mining and petroleum companies and the government. For further information, write: American Geological Institute, 5205 Leesburg Pike, Falls Church, VA 22041.

SURVEYOR

Surveying is a job for someone who likes to work outdoors. Surveyors measure the shape and contour of the land and its waterways, and they help to set its boundaries. With this information, detailed maps and charts can be prepared. These are vitally important for engineers who are planning a highway, a pipeline, a mining project, or some other type of construction, such as a shopping center or a railroad.

Surveyors usually work in groups called "field parties." For example, one person may operate the theodolite (a device to measure angles), while another holds a rod used to measure the height and rise of the land at a particular location. High school courses in math, physics, and mechanical drawing provide good preparation for this kind of work. On-the-job training is available, but advancement usually requires college courses leading to a degree. For more information, write: American Congress on Surveying and Mapping, 210 Little Falls Street, Falls Church, VA 22046.

CHAPTER REVIEW

VOCABULARY

On a separate piece of paper, match each term with the number of the statement that best explains it. Use each term only once.

seismic wave	fault	lava	plateau	focus
diastrophism	volcano	seismograph	epicenter	magma

1. A place where the rock on either or both sides of a crack has moved.
2. The location along a fault where movement causes an earthquake.
3. The location on the surface directly above the focus of an earthquake.
4. Records seismic waves.
5. Vibrations caused by movement along a fault.
6. Molten rock below the earth's surface.
7. Place where lava is produced.
8. Magma that has reached the earth's surface.
9. The movement of the solid rock of the crust.
10. A large area of raised land.

QUESTIONS

Give brief but complete answers to each of the following questions. Unless otherwise indicated, use complete sentences to write your answers.

1. How are earthquakes related to crustal plates?
2. Describe three ways that rock may move along a fault.
3. What is the difference between the focus and the epicenter of an earthquake?
4. Describe the three kinds of earthquake waves. Explain how they differ.
5. What is the Richter scale? Give an example of how it is used.
6. Tell what you should do if an earthquake occurs when (a) you are outside a building or (b) you are inside a building.
7. What is the difference between magma and lava?
8. Where are earthquakes and volcanoes most likely to occur?
9. Describe two sources of strain which cause the rock of the earth's crust to become bent and twisted.
10. Compare and contrast the structure of cones produced by explosive, composite, and nonexplosive volcanoes.

11. What is the difference in the way the Aleutian Islands formed and the way Iceland formed?
12. What is the difference between a joint and a fault?
13. How much more energy is released by an earthquake that measures 7 on the Richter scale than one that measures 4?
14. Why are the ridges of the Appalachian Mountains called anticlines and the valleys called synclines?
15. How could you tell the difference between a folded mountain and a fault-block mountain by their layers of rock?

APPLYING SCIENCE

1. Imagine that a large earthquake occurred in your town or city. (a) List some of the dangers that could occur. (b) Would hospitals and police and fire departments be able to handle such an emergency? Explain your answer. (c) Describe an earthquake emergency plan that would reduce injuries to people and damage to property.
2. Erupting volcanoes can both damage and benefit humans. Make a list of the damages and a list of the benefits. Do you think erupting volcanoes have been harmful or beneficial to humans? Explain your answer.
3. The largest mountain systems in the United States, such as the Appalachians and Rockies, are both folded and faulted. Explain how a mountain range could develop both folds and faults.
4. Plot the locations of earthquakes and volcanic eruptions during the school year on the crustal plate map used in the activity section on page 80. Use a color pencil different from the one used in the activity.

BIBLIOGRAPHY

Courtillot, V., and G. E. Vink. "How Continents Break Up." *Scientific American*, July 1983.

Decker, Robert and Barbara Decker. *Volcanoes*. San Francisco: Freeman, 1981.

Francis, Peter, and Stephen Self. "The Eruption of Krakatoa." *Scientific American*, November 1983.

Kimball, Virginia. *Earthquake Ready*. Culver City, CA: Peace Press, 1981.

Walker, Bryce. *Earthquake* (Planet Earth Series). Chicago: Time-Life Books, 1982.

CHAPTER

5

Nothing on the earth escapes the effects of the atmosphere. This photograph shows a natural arch. The natural arch began as a solid wall of rock. But not even rock lasts forever.

WEATHERING AND EROSION

CHAPTER GOALS

1. Explain how the rocks on the earth's surface are always being changed.
2. Understand how water changes the earth's surface.
3. Describe how water occurs beneath the earth's surface.

5–1. Weathering and Erosion

At the end of this section you will be able to:

- ☐ Name the two processes that work together to wear down the earth's surface.
- ☐ Describe the two processes that break rock into smaller pieces.
- ☐ Explain how soil is formed.
- ☐ List three ways in which gravity can cause rocks and soil to move.

The earth's surface is always changing. Sometimes the changes are sudden, like the earth movement that destroyed the highway shown in Fig. 5–1 on page 114. More often, the changes are slow and take place in small steps.

ROCKS WEAR AWAY

No matter where you look on the surface of the earth's continents, you see the results of a constant battle. The battle is between two kinds of forces that cause the land to be pushed up and then worn down again. On one side of the battle are the forces beneath the surface. These powerful forces cause the crust to be faulted, folded, and lifted. On the other side of the battle are the processes of **weathering** and **erosion** (ih-**roe**-zhun). The term *weathering* means all the processes that break rock into small pieces. You have seen examples of weathering at work on sidewalks. A new sidewalk is smooth and free of cracks. Later the surface becomes rough and cracked. The top layer is worn off. Large cracks appear if roots from nearby trees grow underneath the sidewalk. These processes

Weathering All the processes that break rock into smaller pieces.

Erosion All the processes that cause pieces of weathered rock to be carried away.

Fig. 5-1 Landslides can be caused when heavy rains or earthquakes loosen material on a steep slope.

are similar to the way rocks weather at the earth's surface. See Fig. 5-2. Once rock has been broken up by weathering, small pieces can be moved by gravity, water, ice, or wind. Everything that happens to cause pieces of rock to be carried away is called *erosion.*

Weathering and erosion, along with folding and faulting, shape the mountains, hills, plateaus, and plains that make up the earth's surface. If there were no weathering and erosion, the earth would look very different. It might appear as rough as the surface of the moon. On the other hand, weathering and erosion alone could flatten mountains on the continents within 25 million years. The battle between the forces that build the

Fig. 5-2 The solid granite rock on this hillside is being reduced to fragments by the processes of weathering.

earth's crust and the forces that wear it down goes on all the time.

HOW WEATHERING HAPPENS

Rock material is usually formed below the surface of the crust. It is not exposed to the weather. As long as it remains buried, no weathering takes place. Only rocks that are formed on the surface, or are uncovered, are exposed to the weather. One of the main forces in weather is the water in the air. In general, weathering happens where the solid part of the earth (the *lithosphere*) comes into contact with the liquid water of the earth (the *hydrosphere*) or air (the atmosphere). These three parts of the earth meet in zones called **interfaces**. Weathering takes place in *interfaces*.

Interface A zone where any of the three parts of the earth (the lithosphere, hydrosphere, and atmosphere) come together.

Physical and **chemical weathering** are two processes of weathering. In *physical weathering*, rocks are broken into smaller pieces. In *chemical weathering*, substances in the rock are changed into different ones. Weathering is often a combination of both physical and chemical breakdown.

Physical weathering The ways rock breaks up into smaller pieces without any change in the materials in the rock.

Chemical weathering The ways rock breaks down by changing some of the materials in the rock.

PHYSICAL WEATHERING

1. Frost. The effect of *frost* action on a rock is an example of physical weathering. Frost action happens when water enters the cracks in rock. If the temperature drops below 0°C, the trapped water freezes. Water expands when it freezes. Freezing water produces a very large force as it expands. You may have seen water pipes that broke when the water inside froze. When water freezes inside a crack in a rock, the crack is made larger. The water in the crack freezing, melting, and freezing again repeatedly can break the rock apart. Frost action is an important agent of physical weathering in cold climates.

2. Exfoliation. Another kind of physical weathering begins with rock that is buried deep underground. A rock deep beneath the surface is under great pressure from the weight of all the rock above. This buried rock expands when it is exposed by erosion. Its outer layers may be loosened and peel off like the layers of an onion. This is one example of a process called **exfoliation**. See Fig. 5-3. Weathering by *exfoliation* causes rocks to take on a rounded shape as the layers peel off. Large rounded knobs of rock can be formed by exfoliation. See Fig. 5-4.

Exfoliation The peeling off of scales or flakes of rock as a result of weathering.

Fig. 5-3 How was this rock weathered?

3. Heating. Heating also can cause rock to break apart. Solid materials, such as the substances that make up rock, expand when they are heated. Most kinds of rock are made up of many different substances called *minerals.* A rock can be identified by the minerals it contains. The different minerals in rocks may expand by different amounts when the rock is heated. This can cause the rock to crack. Over a long period of time, the heating during the day and cooling at night can make the outer layers of the rock break into small pieces.

4. Plants. Plants can be agents of physical weathering. Plant roots work their way through small cracks in the rock. As the roots grow, they expand and break apart the rock.

Fig. 5-4 Half Dome in Yosemite National Park is an example of a large dome of rock produced by exfoliation.

CHEMICAL WEATHERING

1. Water. When *chemical weathering* takes place, the minerals in the rock are changed. Water is one agent of chemical weathering. Some materials in rock can be dissolved in water. These dissolved minerals are picked up by water running over rocks in streams or from rain. The rock is broken up by loss of some of its material to the water. Some minerals that make up rock are able to absorb water the way a sponge does. This absorbed water increases the size of those parts and strains the rock. The weakened rock then is more easily broken apart.

Fig. 5-5 Rust is caused by oxygen and water acting on iron.

2. Oxygen. The air contains the gas *oxygen*. Oxygen is another agent of chemical weathering. A small amount of oxygen from the air can join with some minerals in rock when they are wet. The minerals are then changed into new substances. The combining of minerals or other materials with oxygen is called *oxidation*. You have seen pieces of rusting iron. See Fig. 5-5. Oxygen can cause the oxidation of wet iron to form rust. Rocks often contain iron that can rust just like an old chain or car bumper. The red, yellow, and orange colors often seen in weathered rock result from a similar process.

3. Carbon dioxide. The air also contains *carbon dioxide* gas. Carbon dioxide joins with water to form a weak acid called carbonic acid. This acid is able to dissolve some of the minerals in rock.

4. Plants. Chemical weathering can also be caused by the acids from some plants. The simple plant lichen (**like**-en) is an example. Lichens grow on the surface of rock. See Fig. 5-6. An acid is produced that dissolves some of the materials in the

Fig. 5-6 Lichens produce acids that break down rock surfaces.

rock. Lichen is often called a "pioneer" plant because it is one of the first plants to grow in a rocky area.

SOIL

A gardener may think of soil as material in which to grow plants. A farmer knows that crops could not be grown easily without soil. The earth scientist is interested in soil as a substance formed by the physical and chemical weathering of rock and the decay of living matter. Looking at a handful of soil will show that it is made up of particles. These particles could be the remains of a mountain. The particles can be divided into two groups. One group is made up of **organic** material. The other group is made up of pieces of weathered rock. *Organic* means coming from living things. Most of the organic material in soil comes from decaying plants. Some organic material also comes from animal remains. Many living things too small to be seen also are present in soil. In time, all the organic material is changed into **humus** by bacteria. *Humus* is a very important part of the soil. It is one of the main sources of the food and minerals needed for the growth of new plants. Without a supply of the materials supplied by humus, soil is not fertile. Soils differ in the amount of humus and other organic matter they contain. For example, the soil in swamps may be made almost entirely of organic matter. In deserts, on the other hand, the soil usually has only a small amount of humus.

Organic Coming from any living thing.

Humus Material in soil that is from the remains of plants and animals.

Soil also contains pieces of weathered rock. These particles are not all the same size. The larger pieces of rock are called gravel. The smaller pieces, from grain size up to two millimeters, are called sand. The smallest pieces are called silt and clay. These particles are as fine as flour or dust.

Making soil is a very slow process. Weathering must first break down the solid rock. As time passes, and weathering continues, soil begins to form separate layers. The layers in soil are called **horizons** (huh-**rie**-zuns). Each *horizon* has its own properties and is different from the other layers. Humus in the top layer gives it a dark color. This dark-colored top layer is called the *A horizon.* The A horizon is also called topsoil. It is the best layer in which to grow plants.

Horizon Any separate layer of soil.

In young soil there are only two horizons. These two layers are the A horizon and the *C horizon.* The C horizon is made of partly weathered rock.

If erosion does not remove the topsoil, a third layer will develop in time. Water moving through the A horizon dissolves some of the minerals. The finest soil particles, the size of silt or clay, are carried through the soil by the water. These grains and dissolved minerals are collected in a layer called the *B horizon*, or subsoil. It may take about 10,000 years for this layer to form. The B horizon is a hard soil. It is difficult to plow or dig. Only large, strong plant roots can enter it. Water can not pass quickly through the subsoil.

A soil with well-developed A, B, and C horizons is called *mature*. The steps in the development of a mature soil are shown in Fig. 5-7. Mature soils are not formed in dry regions. The moisture needed for the chemical weathering processes is absent in dry areas.

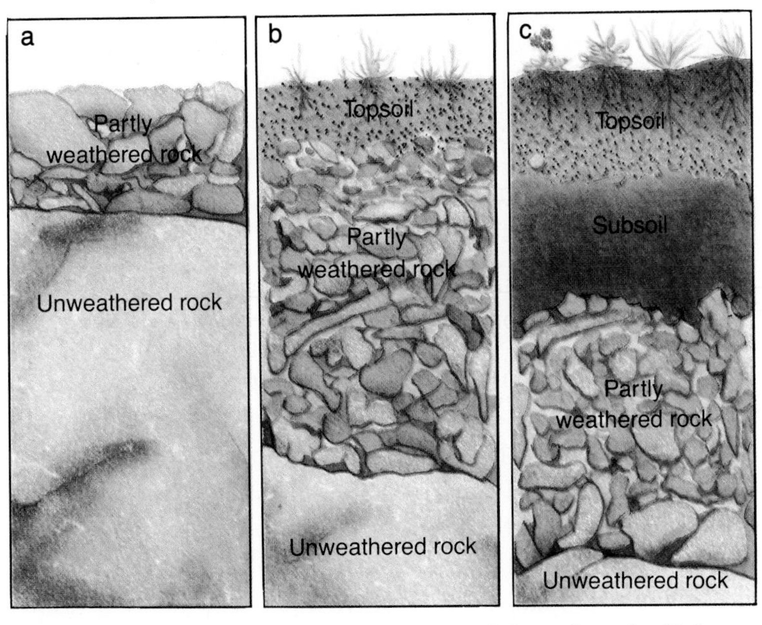

Fig. 5-7 (a) unweathered rock; (b) immature soil has an A horizon and a C horizon; (c) mature soil has an A horizon, a B horizon, and a C horizon. The photo shows mature soil.

Plants growing in soil remove some of the minerals. If these minerals are not replaced, the soil becomes less able to support the growth of new plants. If the plants are not removed, they will return the minerals to the soil when they decay. Soil that is used to grow crops must have the minerals replaced in the form of fertilizers.

EROSION CAUSED BY GRAVITY

The effects of weathering are best seen where the ground is not level. Broken rock pieces can fall from a cliff or slide down

Fig. 5–8 A pile of talus has collected at the bottom.

Talus A pile of broken rock that gathers at the bottom of a cliff or steep slope.

Landslide A sudden movement of large amounts of loose material down a slope.

Creep A slow downhill movement of loose rock or soil.

a slope. A pile of broken material at the bottom of a cliff or steep slope is called **talus**. Fig. 5–8 shows *talus* on the side of a canyon wall. Gravity is an agent of erosion because it pulls pieces of rock down to the bottom of a slope.

Gravity can also move large amounts of loose, weathered material. The sudden movement of large amounts of loose rock or soil is called a **landslide**. *Landslides* may be made up of part of a slope or the side of a mountain. Some landslides are very dangerous. These are the ones in which a large amount of rock material suddenly tumbles down a mountain slope. Such a landslide can take place where rock layers are tilted. If a weak layer beneath the surface gives way, the upper layers rush down the slope. Millions of kilograms of rock and loose material may suddenly move. Earthquakes often can be the cause of landslides. Landslides all over the world have cost many lives. Entire towns and villages have been wiped out. Sudden floods have been caused when large landslides block rivers or raise the level of lakes so that water spills over dams.

The movement of weathered rock by gravity is not always sudden. Sometimes the movement is only about one to two centimeters a year. Slow downhill movement of loose rock or soil is called **creep**. It is often hard to tell that *creep* is taking place. There is usually no change in the way the slope looks. However, careful observation of trees, telephone poles, and fence posts can show evidence of creep. See Fig. 5–9. Creep is the most common kind of erosion caused by gravity. It is the

Fig. 5-9 (left) *Soil creep.*

Fig. 5-10 (right) *Slump usually leaves a curved surface.*

main way weathered rock moves downhill. Sometimes a part of a slope may slide down more or less in one piece. This is called a **slump**. A *slump* usually takes place where the slope of the land has been changed due to erosion. For example, the banks of streams often slump into the eroded stream channel. See Fig. 5-10. Slumping is also seen along cliffs at the shore where waves have cut into the base of the cliff.

Slump The sudden downward movement of a block of rock or soil.

SUMMARY

The earth's land surfaces are the result of two opposing forces. Movements of the crust raise the land to form features such as mountains. At the same time, the agents of weathering and erosion work to wear down the land features. Rocks are made into smaller pieces by physical and chemical weathering. Soil is formed from pieces of weathered rock and the remains of plants and animals. Gravity causes rocks to move.

QUESTIONS

Use complete sentences to write your answers.

1. Describe the two processes of weathering and erosion.
2. Name two kinds of weathering and explain how they are different.
3. Explain how the three layers of soil are formed.
4. What causes slumping, landslides, and soil creep?

INVESTIGATION

WEATHERING

PURPOSE: To investigate some examples of chemical and physical weathering.

MATERIALS:
test tube
test tube rack
limestone rock
soda water
glass container

PROCEDURE:

A. Half-fill a test tube with soda water. Rest it carefully in the test tube rack.

B. Drop a small piece of limestone into the soda water.

   1. Describe any reaction you see.

C. Without moving or shaking the container, look closely at the limestone in the soda water.

   2. What forms on the rock itself?

D. Gently shake the container so that the rock moves.

   3. What happens to the bubbles on the rock?

E. Soda water is a solution of carbon dioxide gas in water, making the water acidic. Rain also dissolves some carbon dioxide gas from the air.

   4. Would you expect rain to have a similar effect on limestone rock?

F. Limestone is dissolved by an acid. In so doing, carbon dioxide gas is made. Look again at the limestone in the soda water. Do not shake the container.

   5. Are there gas bubbles rising from the surface of the rock?

G. Let the soda water and limestone sit for a few minutes. Look again at the limestone rock.

   6. What evidence do you see that the rock is dissolving?
   7. What kind of weathering is shown by the limestone in soda water?

H. Half-fill the glass container with water. Place a small piece of limestone in the water. Shake the rock around in the water until no more air bubbles rise from it.

I. Remove the rock from the water and place it in the waste container.

J. Look at the bottom of the water in the container.

   8. Are there any pieces of rock in the bottom of the container?
   9. What kind of weathering is shown here?

CONCLUSIONS:

1. Which was an example of: (a) chemical weathering? (b) physical weathering? Explain what happened in each example.
2. Compare how weathering took place in your investigation to how weathering occurs in nature.
3. What would be the effect of shaking the rock in soda water as compared to plain water?
4. Is it possible for rocks in nature to undergo both chemical and physical weathering? Explain your answer.

5-2. Water on the Land

At the end of this section you will be able to:

- ☐ Describe the parts of the *water cycle.*
- ☐ Name the parts of the earth's *water budget.*
- ☐ List three ways in which running water erodes the land.
- ☐ Compare the features of a young river and an old river.

A flood can make a peaceful river or stream become a killer. See Fig. 5-11. Swiftly moving water can tear houses from the ground. While floods destroy property and take lives, they are a natural part of the constant movement of the earth's water.

Fig. 5-11 Most floods are caused by heavy rainfall.

THE WATER CYCLE

About 97 percent of the earth's water supply is in the oceans. Smaller amounts, about 2 percent, are in glaciers and ice, below the ground, in lakes or rivers, or in the air. See Fig. 5-12.

The earth's water moves from one storage place to another. Water that was once in the Pacific Ocean may now be in your faucet. It may also be in the clouds in the sky. Water moves from oceans to air and land and back to the oceans again. This movement occurs all over the earth and is called the **water cycle.** A cycle is something that has no beginning or end. The *water cycle* consists of steps that are repeated over and over.

Water cycle The processes by which water leaves the oceans, is spread through the atmosphere, falls over the land, and runs back to the sea.

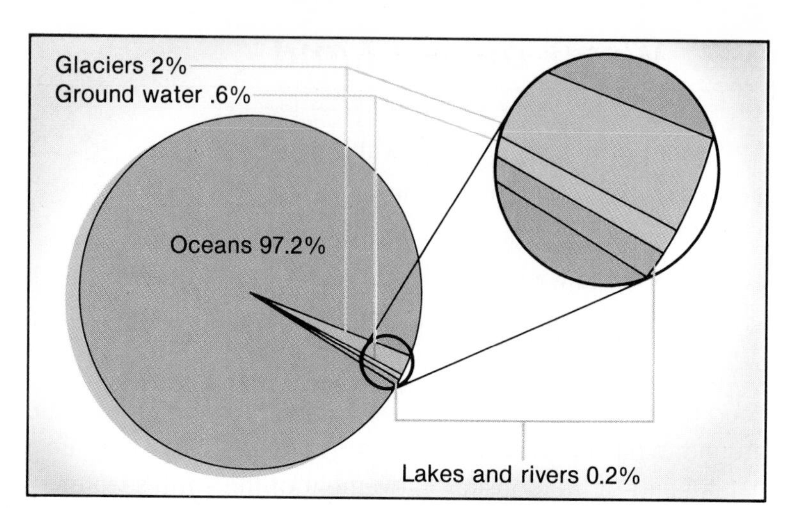

Fig. 5-12 This graph shows where the earth's water supply is found.

Evaporation The process of water changing from a liquid to a gas at normal temperatures.

The first part of the water cycle happens when sunlight falls on a body of water. Heat from sunlight causes some of the water to **evaporate** (ih-**vap**-uh-rate). *Evaporation* causes liquid water in the sea to change into a gas called water vapor. The water vapor then enters the atmosphere. Winds spread the water vapor through the atmosphere.

Precipitation Water that falls from clouds, usually as rain or snow.

The next part of the water cycle causes the water to fall out of the atmosphere. Water vapor in the air becomes rain or snow and falls as **precipitation** (prih-sip-uh-**tay**-shun). Large amounts of *precipitation* fall on the land. Several things can happen to this water that falls on the ground. It can sink into the ground or it can form rivers or streams. Rivers and streams move quickly over the land surface. The water in rivers and streams flows back into the sea within a few weeks. If the water did not return to the sea, evaporation would cause the sea level to drop by one meter each year.

Ground water The water that soaks into the ground, filling the openings between rocks and soil particles.

Transpiration The process by which plants take up water through their roots and release water vapor from their leaves.

Rainwater that soaks into the ground becomes **ground water.** Some *ground water* is returned to the air by plants. This process is called **transpiration** (trans-puh-**ray**-shun). Plants take in ground water through their roots. They then return water vapor to the air through openings in their leaves. Only a small amount of ground water is taken up by plant roots and *transpired* back into the air. Instead, most of the water sinks deep into the crust. It fills the spaces between cracks and other openings in the rock. Almost all ground water becomes part of an underground water system. The water in this system moves very slowly. As it seeks lower levels, it eventually flows back into the ocean. The complete water cycle is shown in Fig. 5-13.

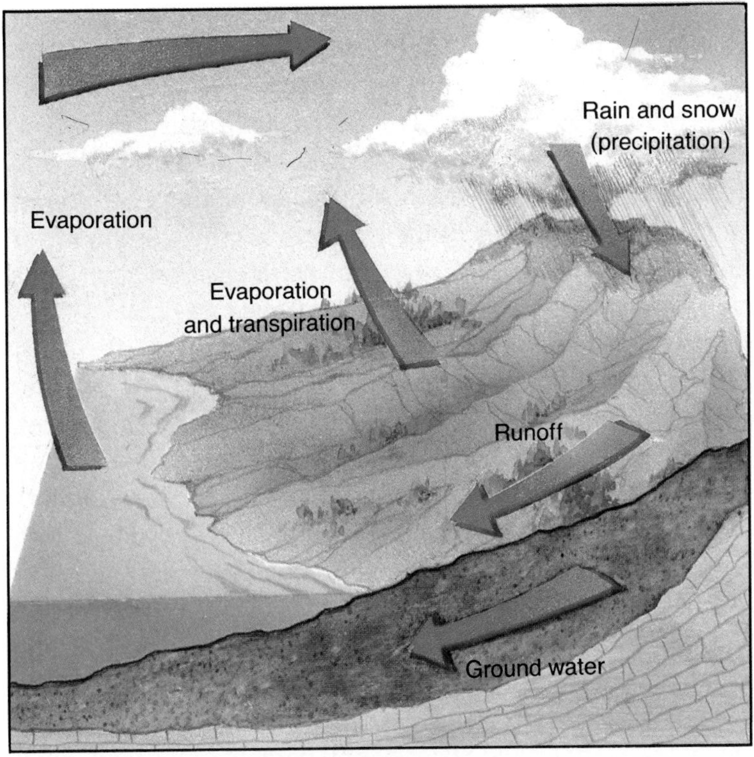

Fig. 5-13 The parts of the water cycle are shown by the arrows of this diagram.

A single drop of water may spend a few days in the air, or 50,000 years frozen in the ice of a glacier. It may be stored in a lake for a hundred years, or it can be deep underground for thousands of years. Given enough time, a drop of water will complete the cycle.

THE EARTH'S WATER SUPPLY

The earth is on a *water budget.* A budget is a record of income and outgo. Precipitation such as rain or snow is the income of the water budget. Evaporation and transpiration are the outgo. At any time, an area may have more water coming in than going out. Some water enters the ground. But ice melting in the spring or a heavy rainstorm can add extra water to an area. Thus the water budget may be out of balance for a while. The opposite may also happen. In the summer, evaporation and transpiration may increase. The weather may be very dry and the water supply may drop. One place may have too little or too much water. However, the water budget for the entire earth is balanced. The amount of water entering the air is equal to the amount of water falling as precipitation.

Modern civilization depends on a ready supply of water. All living things need water. For example, a person can live only a few days without water to drink. Water also serves in many other ways to support our lives. Crops that supply food are affected by the amount of water available. The oceans, lakes, and rivers of the world are used as highways to transport goods. Most of the great cities of the world have grown around river valleys or seaports. The cities, towns, industries, and farms have kept growing until many have reached the limits of their supply of water.

In the United States, each person uses about 7,000 L of water each day. This amount of water could fill a bathtub more than 100 times. Only about 400 L are actually needed for each person. The remaining 6,600 L are used for farm irrigation and industry. This water is "borrowed" from the water cycle. It must be collected, stored, and sent out to homes, fields, and factories. Only a tenth of the precipitation falling on the United States is used. The rest returns to the oceans. Altogether, the water cycle provides more water than is needed. However, the water is not always in the right place at the right time. For example, the western United States has 60 percent of the land. But it receives only 25 percent of the total water falling as precipitation. Many areas have dry periods, such as California in 1976–1977 and Florida in 1981.

We cannot change the water budget in a particular place. We must be careful how our valuable water is used. *Water conservation* must become a part of our lives. All the ways of saving water and making it available for use are included in water conservation. The way water conservation can be carried out is discussed in Chapter 10.

RUNNING WATER

Suppose the photograph shown in Fig. 5–14 could have been made about nine million years ago. It would not show the exposed layers of rock. Scientists believe that the San Juan River carved this great valley. Another example of erosion by running water is the Grand Canyon, which is about 1.6 km deep and 14 to 29 km wide. It shows the power of running water to cause erosion on the earth's surface.

Water moving over the earth's surface moves soil, rock, and even boulders. It can shape the land. Erosion by running water

begins with raindrops that come down on exposed soil. Raindrops break up lumps of soil and stir up the soil. Some rain sinks into the soil. Some moves over the surface. Water that moves over the surface of the land is called **runoff**.

Fig. 5-14 (left) *The San Juan River has created Gooseneck Canyon, Utah.*

Fig. 5-15 (right) *All the streams in an area join to form larger ones.*

Runoff may come from melting snow or ice as well as from rain. Runoff begins as a shallow layer, or sheet, of water flowing downhill. Pebbles and rocks on the surface quickly break the sheet into tiny streams. These tiny streams form shallow **channels.** *Channels* meet and are widened and deepened by the water flowing through them. Small streams come together forming larger ones. A group of streams collects runoff from a large area. All the streams from an area flow into one large stream or river to form a river system. See Fig. 5-15. Streams that flow into rivers are called **tributaries** (**trib**-yuh-ter-ees). The large area that sends runoff into a river is called the *drainage basin,* or *watershed.*

Runoff Water running downhill over the land surfaces.

Channel The path of a stream of water.

Tributary A stream that flows into a river.

EROSION BY STREAMS

Running water erodes the land surface in three ways. One way is when soil and rock materials dissolve in water. Over many years, running water can dissolve away some of the rock. But water itself is not able to change the rocks very much. Most erosion by streams is done in a second way. This is when particles of rock are carried along by the water. The rocks that are small enough to be picked up by the water wear away the channel, just like sandpaper wears away wood. A third way is when pieces of rock too large to be carried by the water are

Fig. 5-16 Contour plowing follows the shape of the land surface to help prevent rapid runoff with heavy erosion.

rolled and bounced along the channel. They chip and wear away the rock of the channel. Thus streams are able to move loose materials and, at the same time, cut down into solid rock.

Erosion takes place rapidly where the soil is loosely packed or where the slope is steep. Heavy rainfall washes away soil more rapidly than light showers do. This is because the soil cannot absorb all of the rain as it falls. A rainstorm may cause a huge flow of mud to move down a slope or valley suddenly. These sudden mud flows may destroy houses or other buildings in their paths.

Careless use of the land also speeds the rate of erosion. Plants protect the soil from washing away. When the natural plant cover is removed, the soil is exposed to runoff and is eroded. It takes a long time to replace the lost soil. Nature needs about 1,000 years to make 2.5 cm of soil. Grass and trees planted along roadways and overpasses help hold the soil in place. Farmers use contour plowing to reduce erosion in their fields. See Fig. 5-16.

THE LIFE HISTORY OF A RIVER

River systems are formed where streams of running water come together. The bottom of a river channel is called the *bed.* Its sides are called the *banks.* The place at which the river flows into a larger body of water is called the *mouth.* Rivers are often described as young or old. These terms refer to the amount of erosion of the river channel, not to the age of the river. The amount of erosion can be seen by the shape of the channel.

1. Young rivers. A young river is still cutting through the channel. Thus there is a large difference in elevation between its watershed area and its mouth. As a result, the water flows swiftly through the channel. The soil and rock, carried or rolled by the water, scrape the bed. Large amounts of rock and soil from the bed are carried off. The channel is cut deeper. Erosion of the river bed gives most young rivers a V-shaped valley. See Fig. 5–17. Because the bed is rocky, a young river is likely to have waterfalls and rapids.

Fig. 5–17 How can you tell if this river is young or old?

2. Mature rivers. As a river reaches maturity, the cutting action becomes very slow, or stops completely. The difference in height between its source and mouth becomes less. The water flows through the channel more slowly. Erosion speeds up along the river banks. The river bed widens. Waterfalls and rapids are smoothed over. The valley of the river channel becomes wider and may take on a U-shape.

3. Old rivers. Finally, the slope of the river channel becomes so small that the water moves slowly. The slow-moving water is able to carry only the finest material. A bend in an older river may become a wide curve. This is because the water always moves faster along the outer bank of a bend. The faster-moving water has more energy to erode rock and soil. Water flows slower along the inner bank of a bend and tends to drop its load of soil and rock. See Fig. 5–18. These two actions cause

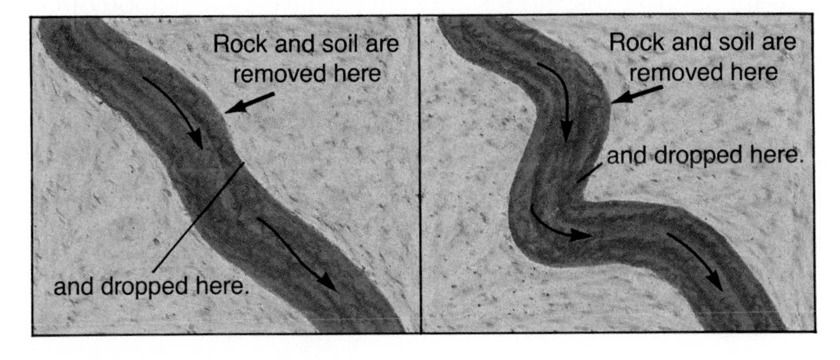

Fig. 5–18 The actions that cause the bend in a river to become wider are shown in this diagram.

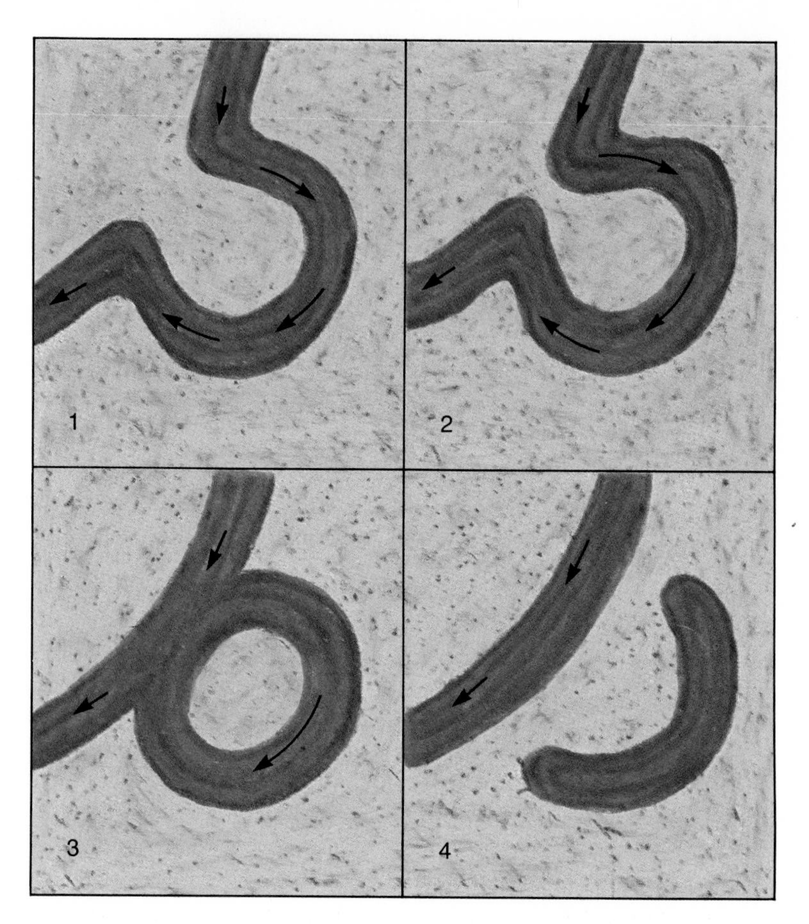

Fig. 5-19 Steps in the formation of an oxbow lake are: (1) a meander develops, (2) erosion cuts across the base of the meander, (3) the river bypasses the old meander, (4) a separate curved lake is left.

Meander A wide curve in the channel of an old river.

the curve to grow into more of a loop. In time, the bend can become a horseshoe-shaped loop called a **meander** (mee-**an**-dur). See Fig. 5–14. Further erosion may cut the base of the *meander*. An *oxbow lake* is formed in this way. See Fig. 5–19.

The river system may grow until it is large enough to drain the average runoff from its watershed. When the amount of runoff is greater than usual, the river is not large enough to handle the extra water. A flood results. Heavy rain or fast-melting snow in the spring can cause floods. Most rivers have a small flood once in a while. Major floods occur less often. Unfortunately, these floods often cause disasters. Lives may be lost. Buildings, roads, and crops can be destroyed.

Flood water deposits soil and rock on both sides of a river channel. Repeated flooding builds up layer on layer of this river soil. Such an area is called a *flood plain*. Older rivers often have wide, flat flood plains. The soil on a flood plain is fertile and makes good farm land.

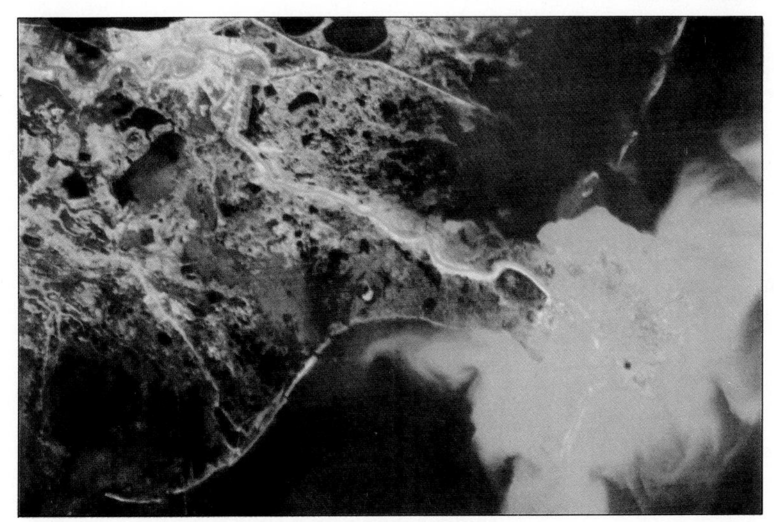

Fig. 5-20 The delta of the Mississippi River covers a large area at its mouth.

Rivers empty into a lake or the sea. Much of the soil and rock material carried by the river is dropped at the mouth. A deposit of these particles is built up in the larger body of water, causing a **delta** to be formed at the mouth of the river. The *delta* formed by the Mississippi River reaches far into the Gulf of Mexico. See Fig. 5-20.

Delta A deposit formed at the mouth of a river or stream.

SUMMARY

Almost all water on the earth's surface takes part in an endless cycle. Water moves from the oceans to the atmosphere, falls on the land, and runs back to the sea. Every area has a water budget. It shows the income from precipitation and the outgo through evaporation and transpiration. While running across the land, water erodes mainly by carrying or rolling rock and soil. All runoff water enters a river system. A river ages as the speed of water moving through its channel becomes slower.

QUESTIONS

Use complete sentences to write your answers.

1. Diagram and describe the processes in the water cycle.
2. Explain how the earth's water budget at a particular time and place could be out of balance.
3. Give an example of three ways in which running water erodes the land.
4. Compare the features of a young river with the features of an old river.

INVESTIGATION

WATER BUDGET

PURPOSE: To graph and study a typical water budget for Minneapolis, Minnesota.

MATERIALS:
graph paper
2 different color pencils

PROCEDURE:

A. Copy the following table.

MONTHLY PRECIPITATION AND EVAPORATION (in mm) FOR MINNEAPOLIS, MINNESOTA, DURING A ONE-YEAR PERIOD													
	Jan.	Feb.	Mar.	Apr.	May	June	July	Aug.	Sept.	Oct.	Nov.	Dec.	Total
Precip.	20	20	35	50	85	105	90	85	75	50	35	25	
Evap.	0	0	0	35	90	125	150	125	80	40	0	0	
P–E													

B. Plot a line graph of the monthly precipitation for Minneapolis. Make two axes. The vertical axis will represent the precipitation. The horizontal axis will represent the months. Use a color pencil to draw the line.
   1. Which month shows the greatest amount of precipitation?

C. On the same axes, plot a line showing the evaporation using a different color pencil or a dashed line.
   2. Which month shows the greatest amount of evaporation?

D. Shade in the area with one color where the precipitation is greater than the evaporation. Mark this area "surplus."
   3. Which months show a surplus?

E. Shade in the area with the second color where the evaporation is greater than the precipitation. Mark this area "deficit."
   4. Which months show a deficit?

F. Using your table, find the surplus or deficit for each month. Do this by subtracting the evaporation from the surplus. Record this in your table in the horizontal row headed "P–E." A positive value shows a surplus, a negative value is a deficit.

G. Fill in the last column of your table by finding the total precipitation and the total evaporation. Subtract the total evaporation from the total precipitation.
   5. How much is the surplus or deficit for the year?

CONCLUSIONS:

1. Under what conditions does a surplus occur?
2. How does a deficit occur?
3. At the end of the year, how can you tell if there was a surplus or deficit?
4. Why does the amount of evaporation during the months change?

5-3. Water Beneath the Land

At the end of this section you will be able to:

- Explain how water enters the ground.
- Relate the water table to springs and wells.
- Describe how limestone *caverns* are formed.

One day in 1981, in Winter Park, Florida, the hole shown in Fig. 5-21 suddenly appeared. Within a few hours it had swallowed buildings, cars, and most of a swimming pool. Could this also happen where you live? You will find out how to answer that question in this section.

Fig. 5-21 Homes in Winter Park, Florida, fell into a huge sinkhole when the ground collapsed.

HOW DOES WATER ENTER THE GROUND?

Three things can happen to water that falls on the ground. Some of it may be returned directly to the atmosphere by evaporation. Some of the water may form runoff. Most of the time, some also sinks into the ground. Any water that sinks below the surface is called *ground water.* Water can sink into the soil because there are open spaces between the soil particles. The ground water fills these spaces. The rock below the soil may contain many joints and cracks. Water fills these empty spaces too. Water also enters most rock that seems to be solid. This is because almost all rocks are made up of grains of different minerals. This causes spaces between the different grains in most solid rock. Water can fill these tiny spaces. The spaces are called *pores*. The amount of empty space in soil or rock is called its **porosity** (pour-**ross**-ih-tee). The *porosity* of some kinds of rock is 25 percent. This means that one-quarter of the rock is empty space! All rock and soil has a certain porosity. The porosity controls how much water rock or soil can hold. See Fig. 5-22.

Porosity The amount of open spaces or pores in soil or rock.

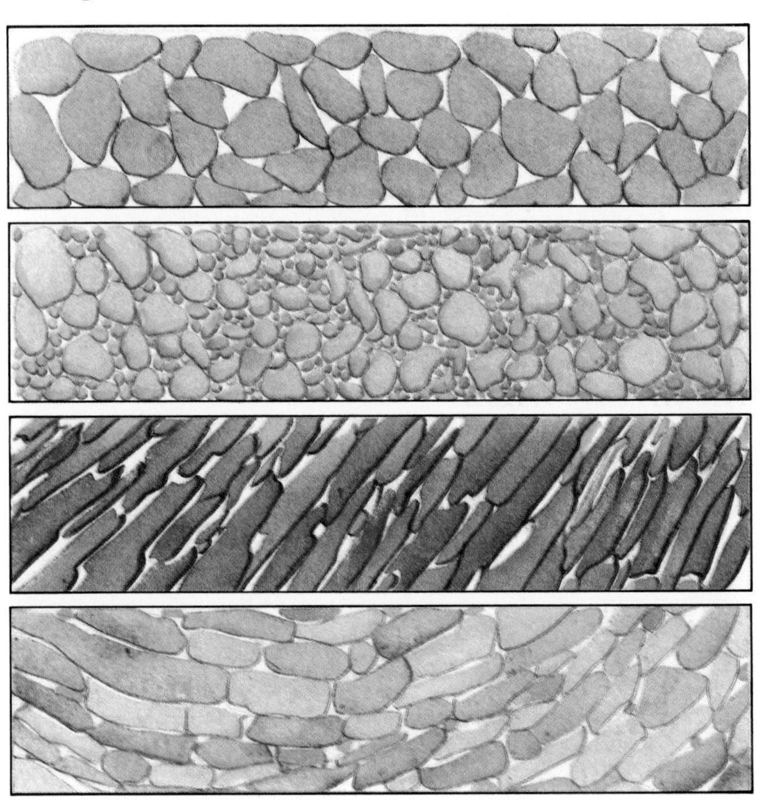

Fig. 5-22 There are open spaces or pores in soil or rock.

Another important property of soil and rock is **permeability** (per-me-uh-**bill**-ih-tee). *Permeability* describes how rapidly water can move between the pore spaces or cracks. In rock with many large pore spaces (high porosity) or with many cracks, water can usually move quickly. Therefore, the rock has high permeability. On the other hand, water cannot move quickly in rock that has very small pore spaces or is without cracks. Therefore, this type of rock has low permeability.

Permeability A measure of the speed with which water can move between the pore spaces in a rock.

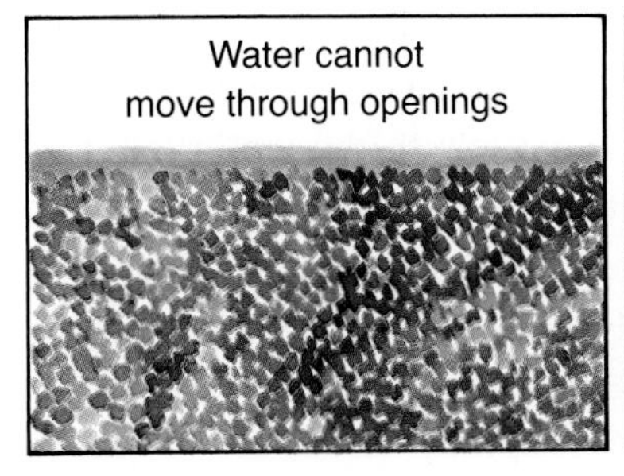

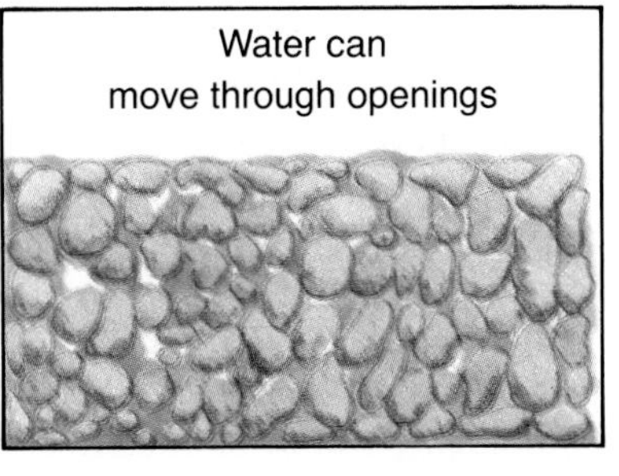

Fig. 5-23 Soils made up of fine particles have low permeability.

Soils, such as clay, with a large amount of very small particles have low permeability. The pore spaces between the tiny clay particles are very small. See Fig. 5-23. Once such small pore spaces are filled with water, the water is held in place. This means that more water cannot enter. Layers of clay or rock with low porosity stop the movement of ground water.

How far down into the crust can ground water go? The answer to this question is not known. It is known that water passes through joints, cracks, and pores in rock below the soil. There is probably a level, several kilometers deep, where the water can sink no further. Rock there is most likely under pressures great enough to close up the cracks and pore spaces.

THE WATER TABLE

Each time it rains some water sinks into the soil. The water moves downward through the soil and rock until it reaches a level where the pore spaces, cracks, and joints are already filled with water. The top of this level is called the **water table.** In general, the *water table* reflects the shape of the land surface. It rises under hills and falls beneath low places. In some places the land surface falls below the water table. Lakes, swamps, and rivers are found at these places. See Fig. 5-24.

Water table The level below the ground surface where all open spaces are filled with water.

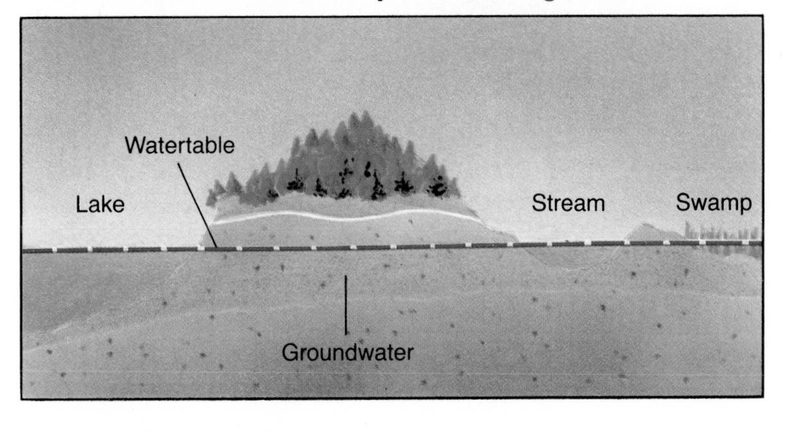

Fig. 5-24 In some low places the surface of the land falls below the water table.

The water table in a region may stay at the same level for a time. However, changes in the water budget may cause it to rise or fall. For example, heavy rains can cause the water table to rise. During dry periods, the depth of the water table usually falls.

Springs are found where the water table comes to the surface without a low place where the water can flow. Many springs are found on hillsides. Most springs disappear when the water table drops during a dry period. One kind of spring flows all the time. This kind of spring is supplied by ground water sitting on top of a layer of low porosity. The temperature of ground water is normally near the average temperature of the region. Thus spring water feels cool in summer and warm in winter. In some places, however, ground water is heated before it comes to the surface in a spring. Such places are called *hot springs.* The water of hot springs is heated by a body of magma or by hot gases escaping from magma. The water in hot springs often has a large amount of dissolved minerals. Some of these minerals may be deposited around the hot spring. See Fig. 5-25. *Geysers* are hot springs that shoot hot water and steam high into the air. Steam created underground forces the water out in a geyser. See Fig. 5-26.

Digging to a level below the water table creates a *well.* Most modern wells are made by drilling a hole to a depth below the water table. A pipe is usually put in to keep the well open.

Fig. 5-25 Where do the minerals around the hot spring come from?

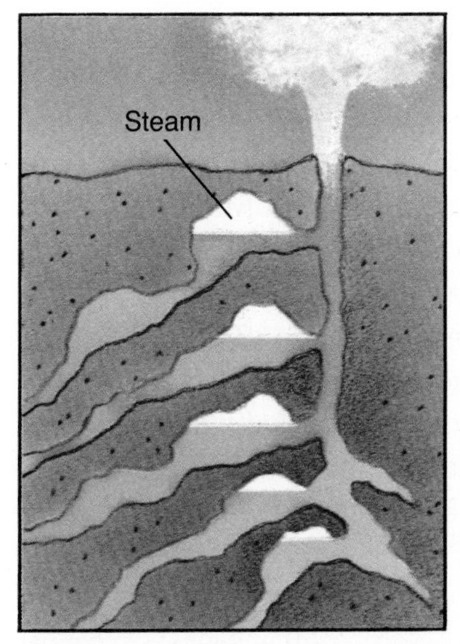

Fig. 5-26 A geyser is produced by steam pressure building up when water boils in underground chambers.

Openings in the bottom of the pipe allow water to flow into the pipe. Then the water must be pumped or lifted to the surface. A well provides water as long as the water table does not drop. Often a well will exhaust the water from the part of the water table close by. In time, the water flows back and the well again has water. If the well is drilled into rocks with low permeability, the water will flow into the well very slowly. Removing the water too fast from such a well will cause it to dry up until more water is able to flow in.

The water in some wells may flow out of the top of the well. These wells are called *artesian* (are-**tee**-zhun) wells. In artesian wells, the water comes from a porous rock layer. This layer is sandwiched between two nonporous layers. The porous layer is called the *aquifer* (**ak**-wuh-fer). The aquifer slants downward from both the place where it picks up water and from the surface. This downhill movement of water in an aquifer gives the pressure needed to force the water from an artesian well. See Fig. 5-27.

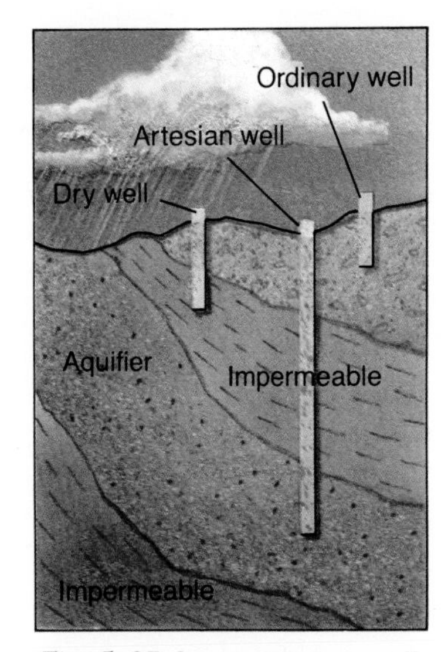

Fig. 5-27 In an artesian well, pressure is created by the downhill flow of water. This causes the water to rise to the surface.

THE WORK OF GROUND WATER

Mention a cave and almost everyone thinks of a dark and mysterious place. There is a mystery about caves, but it has nothing to do with dark secrets or hidden treasures. The real mystery is how a large cave can form in the rock beneath the surface. The answer to this mystery can be discovered by tracing the flow of ground water.

In some areas, thick layers of limestone are found. This rock lies beneath a thin covering of soil. Rain and ground water have a special effect on limestone. As the rain falls, it dissolves some carbon dioxide gas from the air. The rain becomes a very weak acid. Some types of limestone are easily dissolved by this acid. Water that contains dissolved limestone is called *hard water*. Hard water does not make lather with soap. As the acid ground water enters cracks in the limestone, some of the cracks are widened. With each new rainfall, more limestone is dissolved. The cracks widen further. In time, the ground water dissolves away a group of connected underground passages. A **cavern** (**kav**-ern) is formed. See Fig. 5–28. *Caverns* may run underground for several kilometers.

Cavern Any series of underground passages made when water dissolves limestone.

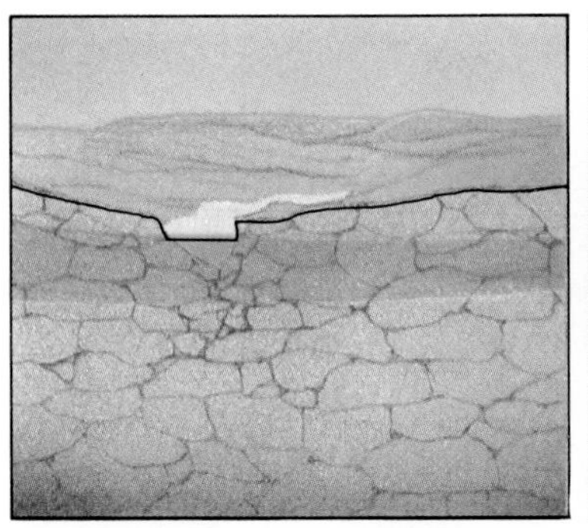

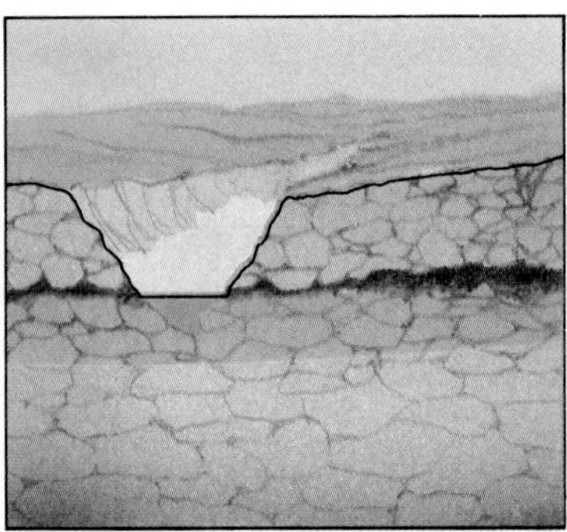

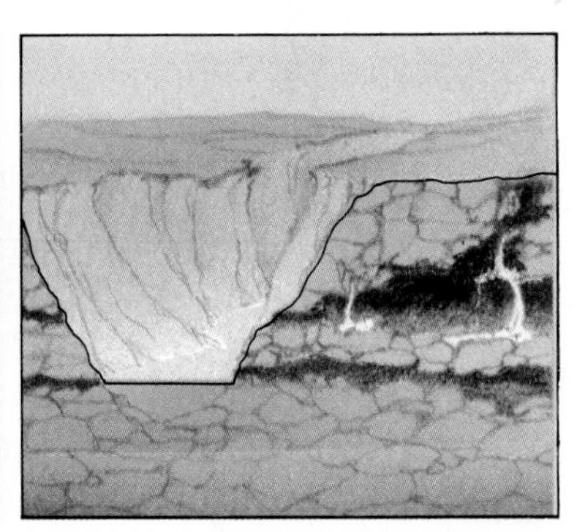

Fig. 5-28 A cavern is made when ground water dissolves limestone. The cavern first grows when the water table drops. As the river erodes the valley floor, the water table continues to drop until the cavern becomes empty.

There are many strange rock formations inside caverns. See Fig. 5–29. From the roof, long icicle-shaped stones hang down. These are called *stalactites* (stuh-**lack**-tights). The mounds of stone that seem to be growing up from the floor are called *stalagmites* (stuh-**lag**-mights). Stalactites and stalagmites feel cool and damp to the touch. They are formed when the ground water drips from the ceiling. The limestone, once dissolved in the water, is deposited. As each drop of water falls, it leaves behind a tiny amount of limestone. Limestone is also deposited on the floor where the drop lands. The stalactites slowly grow longer and the stalagmites grow taller. They may join as one solid stone column reaching from the cavern floor to ceiling.

Sometimes the roof of a cavern falls in. When this happens, a deep, round hole in the ground is formed. Caverns that have fallen in are called *sinkholes*. Fig. 5–21 shows a sinkhole. Sinkholes are likely to occur only in areas where there are also caverns. The sinkhole often fills with water to become a pond. In some regions where there are large deposits of limestone, there

Fig. 5-29 Unusual rock formations are found in caverns.

are many sinkholes. There are few surface streams in these regions because most runoff flows into the sinkholes. Surface water flowing into caverns can form underground streams. These streams, or small rivers, usually flow through the cavern until they reach an outlet such as a spring or river. The only place that ground water moves in underground streams is in caverns. Otherwise, ground water is found almost everywhere beneath the surface. It is spread through the pore spaces and cracks in soil and rock.

SUMMARY

Some of the water that falls on the earth's land surface sinks beneath the surface. The water moves through cracks or small openings in the soil and rock. Below the surface there are layers where all the spaces in the rock are filled with water. When the ground surface falls below this water-filled layer, the water flows out onto the surface, forming springs and artesian wells. Water moving through openings in certain rocks may dissolve the rock, creating caverns.

QUESTIONS

Use complete sentences to write your answers.

1. What two soil properties control the amount of water that sinks into the soil?
2. What relationship exists between the water table and a well?
3. How are springs produced?
4. Describe the conditions necessary to form a cavern.

SKILL-BUILDING ACTIVITY

COMPARING AND CONTRASTING THE PROPERTIES OF PERMEABILITY AND POROSITY OF SOIL SAMPLES

PURPOSE: To examine three soil samples, comparing their properties to find how they affect soil porosity and permeability.

MATERIALS:

3 beakers	100-mL graduated cylinder
gravel	
sand	water
soil	labeling tape

PROCEDURE:

A. Fill one beaker three-fourths full of gravel. Shake it a little so that the rocks settle. Label this beaker A.

B. Fill a second beaker to the same level with sand. Shake this beaker so that the sand settles. Label this beaker B.

C. Fill the third beaker to the same level with soil. Shake this beaker to settle the soil. Label this beaker C.

D. Look closely at the three beakers.

1. In which beaker does the amount of space between particles seem to be the greatest?
2. In which beaker does the amount of space between particles seem to be the smallest?
3. Which material do you think will hold the most water in the spaces between its particles?
4. Which material do you think water will flow through most easily?

E. Fill a graduated cylinder with 100 mL of water. Into beaker A, gently pour just enough water to cover the top of the particles. The idea is to fill the spaces between the particles of the material with water. Observe how quickly the material becomes soaked.

5. Subtract the amount of water in the graduated cylinder from 100 mL to find how much water was needed to fill beaker A to this level.

F. Repeat step E, using beaker B.

6. How much water was needed to fill beaker B to the top of the particles?

G. Repeat step E, using beaker C.

7. How much water was needed to fill beaker C to the top of the particles?
8. Which beaker held the most water in the material? The amount of water it held is a measure of its porosity.
9. Which beaker allowed the water to soak through the material the fastest? This is a measure of its permeability.

CONCLUSIONS:

1. Compare the porosity of the three soil samples.
2. How does the permeability of the soil samples differ?
3. What is necessary in order for a soil sample to have a great amount of porosity?
4. In what way is the porosity of a substance related to its permeability?

CAREERS IN SCIENCE

HYDROLOGIST

Hydrologists are scientists who study the earth's water supply. They probe beneath the surface to find out where underground aquifers and springs are located. (An aquifer is a layer of rock, gravel, or sand from which water can be obtained.) These underground sources can be especially important in dry areas that may be facing a water shortage.

Hydrologists also study surface waters. Some, for example, are involved with efforts to prevent flooding along our rivers. Others study irrigation methods and the effects of soil erosion on farmland. While hydrologists spend part of their time outdoors, they also conduct research in laboratories and in libraries.

To become a hydrologist, a college degree is required, with courses in geology, math, and computer science. Jobs are available in private industry and government agencies. For further information, write: American Geophysical Union, 1909 K Street N.W., Washington, DC 20006.

WASTEWATER-TREATMENT-PLANT OPERATOR

Keeping the country's lakes and rivers free of pollution is extremely important. A potential source of pollution is sewage from homes and work sites. This sewage and wastewater must first pass through treatment plants where it can be cleaned before entering our waterways. The machines in these plants are run by wastewater-treatment-plant operators. They monitor the antipollution devices and take water samples to make sure the water is clean. If breakdowns occur in pipes or valves, operators have the know-how to fix them. They also apply chemicals to water to purify it.

Operators learn through on-the-job training, and programs in wastewater technology are also available in two-year colleges. Employment opportunities in this field are expected to increase. Operators are employed in private industry and in community treatment plants. For more information, write: Water Pollution Control Federation, 2626 Pennsylvania Avenue N.W., Washington, DC 20037.

¡COMPUTE!

SCIENCE INPUT

Many landform features and events are created by the processes of geologic weathering and erosion. Slump, creep, landslides, caverns, and springs—landforms discussed in Chapters 5 and 6—all have been shaped through the forces of nature such as wind, waves, running water, glaciers, rain, and earthquakes. Studying how erosion occurs is important. In the practical sense, if the processes are understood, then damage may be prevented. Erosion is also a factor in the aging of landforms. Therefore, if you were a scientist of earth history, you would need to understand erosion to be able to estimate how old a landform is.

COMPUTER INPUT

Learning a new field requires memorizing a great many new terms and concepts. A modern computer can be used to make old-fashioned learning drills somewhat easier and even fun. The computer can be programmed to review material over and over again without ever getting tired and without ever getting angry at you for giving the wrong answers. The computer is a friendly machine when it comes to this type of activity.

Sometimes a computer, a computer language, or a word processor is described as being "user friendly." Usually this means that the terms for commands and responses, the labels on the keyboard, or the words and symbols used in operating the machine are more like everyday conversational words than like a new technical language.

Look at the words in Program Erosion. While you may not be familiar with the way the symbols are being used, the words and numbers are often used in the same way as in sentences or mathematical equations. These programs have been written in BASIC, a computer language that is not very technical and quite "user friendly."

WHAT TO DO

You will be programming your computer to ask you to make an association between a landform (a dune, for example) or a process (a landslide, for example) and the type of erosion that caused it. The computer will tell you whether or not you are right, and give you a score by telling you what percent of your answers are correct.

First, you must collect and record data on erosion for your data statements. In one column, make a list of all the landforms and processes discussed in Chapters 5 and 6. Next to each item write the type of erosion that caused it. When there is more than one kind of erosion at work, list each, separated by a slash (/). The program requires that you remember each part of your definition, or your answer will be considered incorrect. A sample data chart appears below. (The program will hold 100 data statements.)

After your data chart is complete, enter Program Erosion into your computer. Be careful to type your data statements as shown in the example statement 301, and to end with statement 401 as shown. Save the program on disk or tape and then run it.

GLOSSARY

DIM — Short for *dimension.* This instruction tells the computer how many spaces are needed in that part of its memory reserved for information or data. In this program, dimensions are given for information that will be used in the quiz questions. See line 110 below. There can be 100 questions and 100 answers.

PRINT — Instruction to the computer to print on the computer screen. There is a different instruction that starts the printer.

DRILL AND PRACTICE: PROGRAM EROSION

PROGRAM

```
100  REM EROSION DRILL AND
     PRACTICE
110  DIM Q$(100), A$(100)
115  REM READ IN QUESTIONS,
     ANSWERS
120  FOR X = 1 TO 100
130  READ Q$(X), A$(X)
140  IF Q$(X) = "0" THEN GOTO 160
150  NEXT
160  PRINT: PRINT
165  REM COMPUTER PICKS RANDOM
     QUESTION
170  N = INT(RND(1) * (X - 1) + 1)
180  PRINT "WHAT TYPE OF EROSION?"
190  PRINT Q$(N);: INPUT "-----> ";R$
200  PRINT
210  IF R$ = "F" THEN END
220  QC = QC + 1
225  REM COMPARE RESPONSE WITH
     ANSWER
230  IF R$ = A$(N) THEN PRINT
     "CORRECT" : C = C + 1: GOTO 250
235  REM ANSWER TO INCORRECT
     RESPONSE
240  PRINT "CORRECT ANSWER ---> ";
     A$(N)
245  REM SCORING
250  PRINT "OUT OF       "; QC; "    YOU
     WERE CORRECT ON =       ";
     C:PRINT "SCORE =       "; INT (C/QC
     * 100 + .5); "%"
260  GOTO 160
300  REM DATA STATEMENTS FOLLOW
301  DATA DUNES, WINDS
401  DATA 0,0
999  END
```

PROGRAM NOTES

This program can be used to help learn any kind of association. For example, your data chart could have listed football or baseball players and the league and club in which they play.

A computer that contains a large database can seem to "know" a lot. However, the computer only knows what has been put into it. If you put in the wrong definition, the wrong definition will come out. You also have to be quite precise. For example, if you enter "Rain/earthquakes" as agents of erosion for "landslide," and then, when the program asks for an answer, you type in "Earthquakes/rain," the computer will not recognize your answer as correct.

EROSION DATA CHART

Landform or Process	Agent of Erosion
Dunes	Winds
Sea cave	Waves
Mudflow	Rain
Landslide	Rain/earthquakes

BITS OF INFORMATION

Computer games are probably one of the most popular uses for personal computers. Complete with colorful graphics, there are adventure games, war games, mazes, and chess or checker games. For those who are absorbed with videogames and other uses of computers, there never seems to be enough software. You can exchange and buy software through the Young People's Logo Association (YPLA). YPLA lists programs in a number of computer languages. For membership information, write to: YPLA, 1208 Hillside Drive, Richardson, TX 75801.

CHAPTER REVIEW

VOCABULARY

On a separate piece of paper, match the number of each sentence with the term that best completes it. Use each term only once.

humus	meander	creep	precipitation
water cycle	delta	landslide	weathering
evaporation	chemical	interface	erosion
talus	slump	physical	organic

1. ______ is the breaking of rock into small particles.
2. The processes which cause rock particles to be carried away are called ______.
3. A(n) ______ is a zone where any of the three parts of the earth come together.
4. In ______ weathering, rock is broken up without any change in the material of the rock.
5. ______ weathering acts to change the materials of the rock.
6. The ______ is the material in soil that comes from the decay of plants and animals.
7. The pile of broken material at the bottom of a cliff is called ______.
8. A(n) ______ is the sudden movement of large amounts of loose rock or soil.
9. The slow downhill movement of loose rock or soil is called ______.
10. Downhill movement of part of a slope in one piece is called ______.
11. The ______ is the process by which the water leaves the oceans, is spread through the atmosphere, falls over the land, and runs back to the oceans.
12. ______ is the changing of a liquid into a gas at normal temperature.
13. A(n) ______ is a deposit formed at the mouth of a river.
14. The wide curve of an old river is known as a(n) ______.
15. Falling rain or snow is called ______.
16. ______ means "coming from living things."

QUESTIONS

Give brief but complete answers to each of the following questions. Unless otherwise indicated, use complete sentences to write your answers.

1. List four ways in which rocks can undergo physical weathering.
2. Name some agents of chemical weathering.

3. Give examples of how plants cause physical and chemical weathering.
4. Describe how the earth's rock surface becomes mature soil.
5. Why is gravity an agent of erosion?
6. What three things can happen to precipitation?
7. When does a water surplus occur?
8. Trace how runoff becomes a river.
9. What type of stream is most effective in cutting through bedrock?
10. Name the parts of a stream.
11. What is the relationship between soil porosity and permeability?
12. Why are lakes, swamps, and rivers found below the water table?
13. Compare the formation of a spring that disappears during a dry period to the formation of a spring that flows all the time.
14. Explain why ground water can produce a cavern.
15. Describe how each of the following form in caverns: (a) stalactites, (b) stalagmites, and (c) columns.

APPLYING SCIENCE

1. Look around your neighborhood for examples of weathering and erosion. List the examples you find in table form. Identify each example as either physical weathering, chemical weathering, or erosion. In the case of erosion, describe how the erosion occurred.
2. Visit a recent cut in a hillside or an excavation site. Take pictures of the soil layers that have been exposed. Determine whether the soil at the site is young or mature. Discuss your results and share the pictures with your class.

BIBLIOGRAPHY

Fodor, R. V. *Angry Waters: Floods and Their Control.* New York: Dodd, Mead, 1980.

Gardner, Robert. *Water, The Life-Sustaining Resource.* New York: Messner, 1982.

Gold, Michael. "Who Pulled the Plug on Lake Peigneur?" *Science '81,* November 1981.

Gunston, Bill. *Water.* Morristown, NJ: Silver Burdett, 1982.

Kitfield, James. "The Greening of Germany." *Omni,* April 1983.

CHAPTER

6

Nothing on the earth's surface can escape the effects of erosion. When ice, wind, and waves move over land, they pick up and move all kinds of material from the earth's surface.

ICE, WIND, AND WAVES

CHAPTER GOALS

1. Explain how ice, wind, and waves erode the land.
2. Give examples of ice, wind, and wave erosion.
3. Describe the deposits left by ice, wind, and waves.
4. Describe the areas most affected by ice, wind, and waves.

6–1. Glacial Ice

At the end of this section you will be able to:

- Define *valley glaciers* and *continental glaciers.*
- Explain how glaciers are able to cause erosion.
- Compare two types of deposits left by glaciers.
- Identify two types of theories that explain how *ice ages* may have been caused.

The greatest rivers on the earth are made of solid ice. Like huge plows, they move and shape large parts of the earth's surface. In some places huge amounts of ice reach the shore and flow out into the ocean. These slowly moving masses of ice are an important force in changing the surface of the earth.

GLACIERS

Most snow that falls on land melts to form runoff. The water returns to the sea. There are places on the earth, however, where it is too cold for all the snow to melt. As the snow falls, year after year, it piles up. The snowflakes are squeezed together and become ice. When the weight becomes great enough, the body of ice and snow begins to move. This large body of moving ice and snow is called a **glacier.**

Glacier A large body of moving ice and snow.

VALLEY GLACIERS

One kind of *glacier* on the earth today is a **valley glacier.** *Valley glaciers* are found in mountainous areas all over the world. See Fig. 6–1. They look like rivers of dirty ice filling mountain valleys. Often several valley glaciers join to form a large valley glacier. Valley glaciers are able to move because of the pull of gravity. Their great weight causes them to slide downhill. A very thin film of liquid water is formed under the

Valley glacier A river of ice in a mountain valley.

Fig. 6–1 (left) *A valley glacier.*

Fig. 6–2 (right) *The ice near the surface of a glacier breaks to form crevasses.*

glacier from melting ice. Heat caused by the friction of ice scraping against the mountainside usually causes this melting. The liquid water allows the ice to slide more easily over the rock. Glaciers also move because parts of the ice are under pressure for a long time. Each small crystal of ice is able to slip a tiny amount. Think of what would happen to a tall stack of cards if each card slipped a little. The sum of all the individual movements would make the whole stack move. In the same way, the ice in a glacier can creep slowly. As the glacier moves over uneven ground, the ice near its surface breaks and forms cracks that run across the glacier. Cracks also may appear near the edges when one part of the glacier moves faster than other parts. These great cracks are called *crevasses* (kruh-**vass**-ez). See Fig. 6–2. Since the ice near the surface is not under pressure, it does not flow. Instead, the ice cracks.

Some valley glaciers move down their valleys at the speed of about one meter each day. They are fed from snow that builds up on their upper ends. As they move down the valleys, their lower ends melt. This combination of downward movement and melting usually makes it seem as if the glaciers are not moving at all. Sometimes a glacier appears to retreat because its lower end melts faster than it moves forward. Other times the glacier may advance when it moves forward faster than its lower end melts.

CONTINENTAL GLACIERS

Continental glacier A sheet or mass of ice that covers a large area of land.

The other kind of glacier is a **continental glacier.** *Continental glaciers* are huge sheets of ice that may cover thousands

Fig. 6-3 *A continental glacier.*

of square kilometers. Two such glaciers, called icecaps, cover most of Antarctica and Greenland today. See Fig. 6-3. In places, the Antarctic icecap is over 4,500 meters thick. It covers great mountain ranges running across the continent. Ninety percent of the world's ice is in the Antarctic icecap. Antarctica also contains 75 percent of the world's freshwater.

Icecaps are shaped like wide domes. They are able to move because of their great weight. See Fig. 6-4. The weight of the ice causes outward movement in all directions from the thick center. Edges of icecaps break off when they reach the sea. These huge pieces of ice float away as *icebergs*. Valley glaciers that reach the ocean also form icebergs.

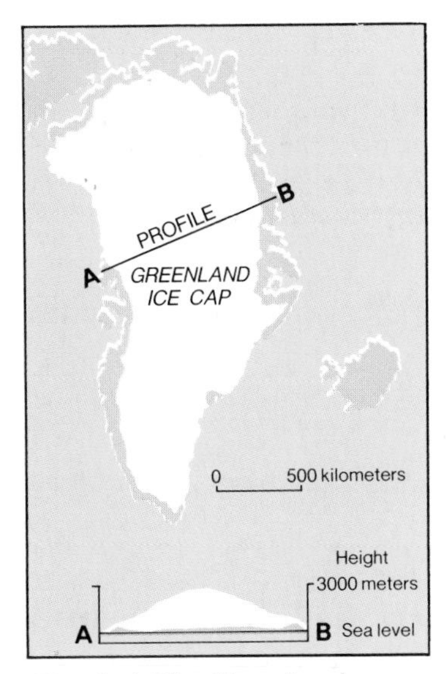

Fig. 6-4 The Greenland icecap is shaped like a dome. See profile A-B. Its weight at the center causes the ice to flow toward the shore.

THE WORK OF GLACIERS

Glaciers are responsible for the appearance of much of the earth's surface at high altitudes and latitudes. Sharp mountain peaks, wide valleys, level plains, lakes, and rivers have been made by glaciers. The movement of glaciers can erode and shape the land. Rocks buried in the ice at the bottom and sides of the glacier are like the teeth of a file. As glaciers move, the rocks scrape and dig at the land. Solid rock can be polished and often shows deep scratches where moving glaciers have left their marks. See Fig. 6-5 on page 150.

Glaciers that form in mountains change the shape of the mountains. The changes begin at the top of the valley where snow and ice collect. Frost action breaks rock from the valley walls. Rock is picked up from the valley floor as the glacier

Fig. 6–5 What evidence is found on these rocks to show that they were once covered by a glacier?

moves. The upper end of the valley is changed into a rounded, steep-walled basin called a *cirque* (**surk**). Many cirques can form close together in mountains. They cause the ridges between them to become very sharp and jagged. These ridges are called *arêtes* (uh-**retts**). Very steep peaks, called *horns*, may also be formed. As the glacier grows, it moves down the valley. The floor and walls of the valley are worn away. In time, the original V-shape of the valley is changed into a U-shape. The steps by which the glaciers cause mountains to become rugged are shown in Fig. 6–6. Unlike valley glaciers, which form only in mountains, continental glaciers cover large areas of the land surface except for the highest mountains. Continental glaciers also change the land differently from the way valley glaciers do. Instead of reshaping mountains, the moving icecap grinds down the land to a nearly level surface.

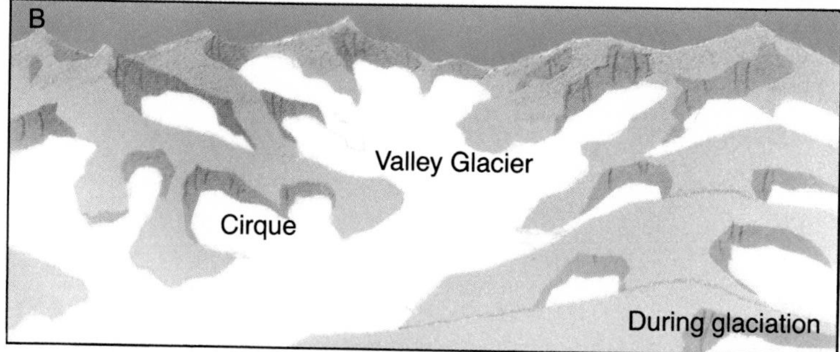

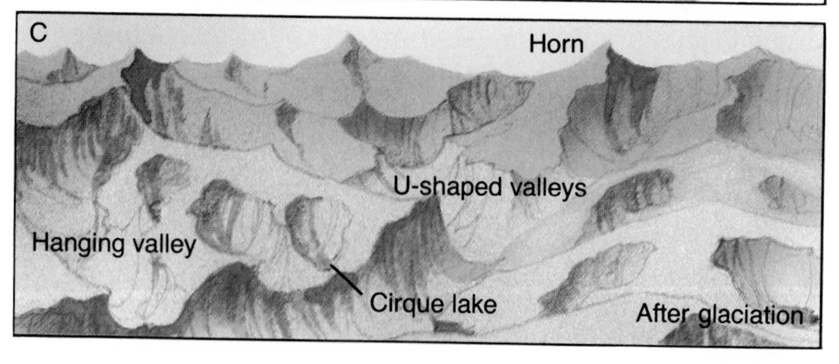

Fig. 6–6 (a) Mountains before they are changed by glaciers. (b) Glaciers growing in mountain valleys. (c) Appearance of mountains after the glaciers have disappeared.

GLACIAL DEPOSITS

A moving glacier can pick up rock material. When the ice melts, all of this material is deposited. There are two kinds of deposits left by glaciers. One kind consists of material deposited directly from the melting ice. This is called **till.** *Till* is made up of unsorted or broken rock of all sizes mixed together, from fine particles to boulders. Layers or ridges of till are called **moraines.** Ridges of till formed along the sides of glaciers are called *lateral moraines.* Lateral means "side." Till that is deposited in front of the glacier is called *terminal moraine.* Terminal means "end."

Till Material deposited directly from a glacier.

Moraine A layer of till.

A second kind of deposit is made when glacial ice melts and forms streams flowing out from the glacier. These are called outwash deposits. Streams flowing from the glacier's front drop sand and gravel, depositing the largest particles first, then the smaller particles. Unlike till, which consists of unsorted materials, outwash deposits are sorted. That is, outwash deposits are formed in separate layers of sand, gravel, and finer particles. If the streams flow out onto level ground, the material is deposited in wide, even sheets that form an outwash plain. Large blocks of ice may break off the glacier and become buried in material washed out of the melting glacier. When these blocks melt, a large water-filled hole called a *kettle lake* is left. See Fig. 6-7. Many of the lakes and ponds of Michigan, Wisconsin, and Minnesota were formed this way.

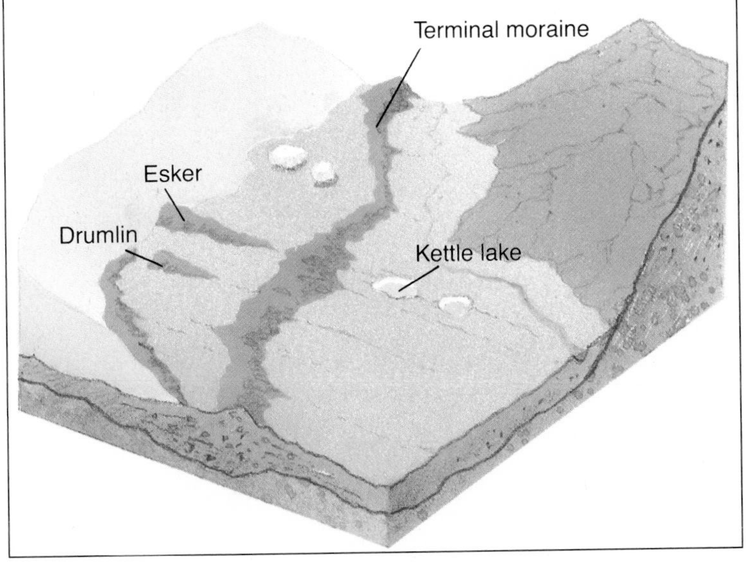

Fig. 6-7 The retreat of a glacier leaves a variety of glacial deposits.

Other deposits left by glaciers are called *drumlins* and *eskers*. Drumlins are long, low hills made of till. They range in height from about 15 to 60 meters. One side of the hill is usually steep and faces the direction the glacier moved. The Battle of Bunker Hill in the Revolutionary War was fought on a drumlin. Eskers are deposits made by streams flowing under the ice. Fig. 6–7 shows the kinds of deposits that can be made by glaciers.

ICE AGES

Glaciers covered much of the earth's surface ten thousand years ago. What is now New York City was then covered by a layer of ice about one kilometer thick. See Fig. 6–8. At that time, the sea level was much lower than it is today. This was partly a result of much of the earth's water supply being trapped in the ice sheets. The huge glaciers of that time eroded much of North America. They left features that now tell us where they were. For example, the Great Lakes were formed by glacial action. These huge lakes fill basins that were made when the advancing glaciers made river valleys wider and deeper. Water from the melting glaciers was trapped by dams built up from moraines. Other evidence of glacial erosion is seen in the plains of midwestern North America and all of the New England states. The climate of the earth has warmed since then. The last of the great ice sheets disappeared from North America about ten thousand years ago. Now glaciers are found only in high mountains and near the poles.

Scientists believe that several periods of glaciation have taken place during the earth's long history. Each cool period that allowed much of the earth's water to be frozen into ice was

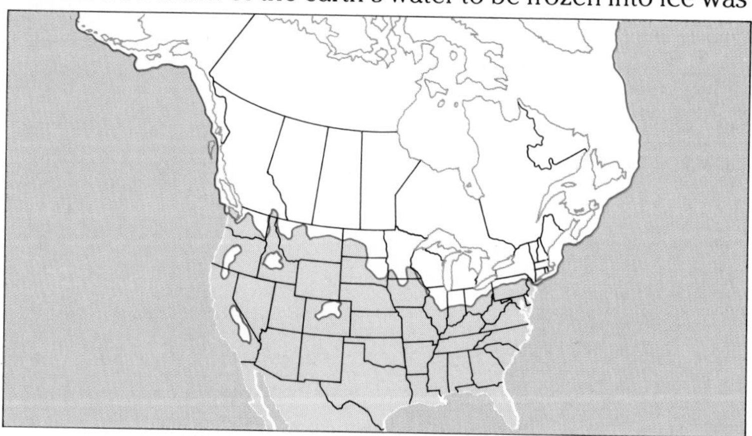

Fig. 6–8 About ten thousand years ago, glaciers covered many parts of North America.

an *ice age*. Each ice age was followed by a warm period. There is evidence that there have been at least four, and maybe as many as ten, ice ages during the last million years. During an ice age, glaciers grow and join together. In time, ice sheets several kilometers thick are formed. Because of their weight, these ice-caps spread out in all directions.

There is no one scientific theory that explains why the earth has had ice ages. Scientists look for reasons why the earth's climate should become cool.

1. Events on the earth. One type of theory uses events on the earth itself as reasons for cooling. For example, one theory is based on the drift of the continents. The positions of continents could prevent the movement of warm ocean waters toward the poles. This could cause regions near the equator to become very warm. Closer to the poles, the climate would be much cooler. Another theory is based on volcanic eruptions. An increase in volcanic activity might throw large amounts of dust into the atmosphere. This could prevent some of the sun's rays from reaching the earth's surface. The earth would then become cooler as it received less heat from the sun.

2. Events outside the earth. Another type of theory that explains the ice ages is based on reasons outside the earth. One theory suggested that at times the sun may produce less heat. The sun does go through short periods when it gives off slightly less heat. Whether this takes place over long periods of time is not known. A theory that is gaining support from many scientists is based on changes in the earth's orbit and the tilt of its axis. The elliptical orbit of the earth around the sun changes slightly. At times, this could have put the earth a little farther from the sun, producing cooler summers. In addition, changes in the tilt of the earth's axis can cause cooler summers. For example, the tilt of the earth's axis today is what causes the earth to have seasons. The combined changes in the earth's orbit and the tilt of its axis could have lowered the earth's temperature. This could have caused some snow to remain through the summer. If this occurred during many summers, snow would continue to collect until glaciers formed. Changes in the tilt of the earth's axis and orbit have their largest effect on the amount of heat received from the sun in the Northern Hemisphere. Those kinds of changes do seem to have taken place around the same time that the most recent ice ages have occurred.

In the future, scientists may be able to forecast when the earth will enter another ice age or a warm period. Presently, there are 25 million cubic kilometers of ice in the earth's glaciers. The melting of that ice would raise the sea level by about 65 meters. Many large cities of the world such as New York, London, and Tokyo would be flooded. This is a practical reason for trying to gain a scientific understanding of the reasons for the ice ages.

It is certain that the earth's climate will change. Even now the climate is changing. There is not yet enough scientific evidence to determine if the earth will enter another ice age. The world may be warming as a result of the burning of large amounts of fuels such as coal, oil, and natural gas. This has released large amounts of carbon dioxide into the atmosphere. Carbon dioxide can trap the heat the earth receives from the sun. This is similar to the way glass in a greenhouse traps heat from the sun. The trapped heat could melt the glaciers. Some scientists believe that this could raise the sea level enough to flood coastal cities such as New York.

SUMMARY

Large amounts of snow and ice may build up in mountain valleys and over areas near the poles. These masses of snow and ice that move become glaciers. Materials carried in the moving ice erode the land and are deposited when the glacier melts. The earth passes through periods called ice ages when its climate becomes cooler and huge glaciers form.

QUESTIONS

Use complete sentences to write your answers.

1. What are the differences between a valley glacier and a continental glacier?
2. Explain how the movement of a glacier can erode and shape the land.
3. Give an example of a land feature formed by till and one formed by outwash deposits. Tell how they differ.
4. Describe the two types of theories used to explain why ice ages occur.

SKILL-BUILDING ACTIVITY

EROSION BY RUNNING WATER: A STUDY IN CAUSE AND EFFECT

PURPOSE: To study the factors that determine the rate of erosion and amount of erosion caused by running water.

MATERIALS

stream table
water source
overflow container
soil
sand
gravel

PROCEDURE:

A. Water running off melting ice from a glacier or snow pack is one of the major causes of erosion.
   1. What conditions do you think affect the action of running water on soil, sand, and gravel?

B. One condition to study would be the effect of the slope of the land on erosion. Set up the stream table as shown in Fig. 6–9. Using a mixture of soil, sand, and gravel, let the water flow at the same rate for a given amount of time. Stop the water and observe the different-size particles that are moved by the running water.

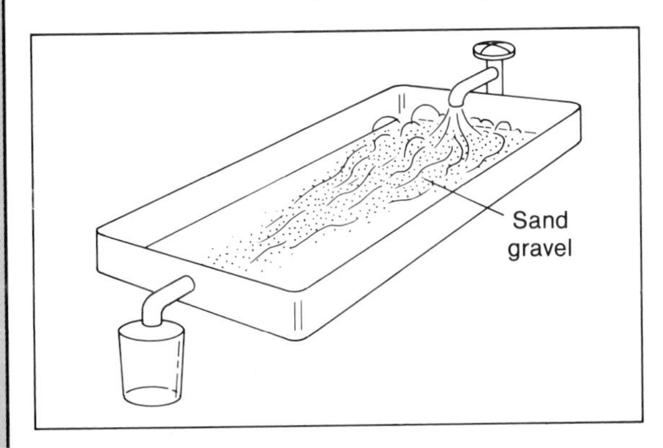

Fig. 6–9 A stream table.

   2. What size particles were moved by the water?

C. Increase the slope. This can be done by placing one book under the beginning of the stream table. If the speed of the water does not change, add one book at a time until you notice a change in speed.
   3. What effect did increasing the slope have on the speed of the flowing water?
   4. What effect does this have on the different-size particles moved?

D. Remove the books so that the stream table returns to its original height. Slowly increase the volume of water flowing. Observe the order in which the different-size particles move. Do this until particles of all sizes are moving.
   5. As the volume of water was increased, in what order did the different-size particles move?

CONCLUSIONS:

1. What effect would increasing the slope of a stream have on the amount of stream erosion?
2. What effect would increasing the volume of water flowing have on the amount of stream erosion?
3. What factors could change the slope or volume of water of a stream?
4. Imagine you were planning a house near a stream. From your observations in this Activity, how would your plans differ (a) if the stream were fast-moving? (b) slow-moving?

6-2. Wind and Waves

At the end of this section you will be able to:

- Explain how the wind can cause erosion as well as build deposits.
- Describe two types of wind deposits.
- Identify two ways in which waves wear away the land.
- Describe shoreline features caused by wave erosion.

Have you ever felt a strong wind? See Fig. 6-10. Sometimes the wind can blow you over unless you lean against its force or hold on to something. It is the energy in the wind that you feel. And it is this energy that is able to cause erosion and build deposits on the earth's surface.

Fig. 6-10 Some places in the world can have very strong winds.

WIND EROSION

Wind is moving air. Like running water, wind can move pieces of rock. But wind moves only small particles, such as dust and sand, through the air. Wind can only move large pieces of rock by rolling or bouncing them along the ground. Wind causes erosion only where many small rock particles are exposed. For example, in dry climates such as in deserts, the growth of plants that would cover the soil is prevented. Therefore, the wind is able to move the soil particles across the bare ground. Beaches, dry flood plains, and plowed ground are also exposed to wind erosion. Strong winds in these places carry away the sand and soil.

Besides moving small particles, wind has another effect. Particles of sand usually have sharp edges. Sand blown by the wind can **abrade** (uh-**brade**), or wear away, anything in its path. The ability of moving sand to *abrade* sometimes is used to clean the outside of stone or brick buildings. Particles of sand are blown against the dirty surfaces. As the sand is blown against the surfaces, the dirt is worn away. See Fig. 6–11 *(left)*. Wind can cause erosion of solid rock in the same way. Wind erosion does not have any effect on things higher than about one meter above the ground. The sand particles that shape rock by abrasion are not lifted very high by wind action. This means that most of the strange large rock formations often found in deserts were not carved by wind. Wind action only polishes and cleans structures near the ground. See Fig. 6–11 *(middle)*.

Abrade To wear away.

The ability of wind to carry dust or move sand depends on its speed. A strong wind is able to move large amounts of loose material. Most deserts have big areas of bare ground where all the sand and soil particles have been blown away. Only large particles such as gravel and stones are left behind. See Fig. 6–11 *(right)*. In dry regions, a strong wind may cause huge dust storms. During a large dust storm, a cubic kilometer of air may carry as much as 900,000 kilograms of dust. A storm covering a large area is able to move many millions of kilograms of dust. During periods of drought, the growth of plants and farm crops is prevented. Wind then becomes an important agent of soil erosion in farming areas.

Fig. 6–11 (left) *Sand is being blasted against this building to abrade and clean the surface.* (middle) *Wind causes unusual rock formations in deserts.* (right) *Everything except the heavier rocks has been blown away.*

WIND DEPOSITS

Windbreak A barrier that causes wind to move more slowly.

Dune A deposit of wind-carried materials, usually sand.

A barrier that causes the wind to lose speed is called a **windbreak.** For example, a bush or a large rock can cause the wind to slow down. When the wind slows down, it drops part of its load of sand. A mound of sand often is deposited around a *windbreak* such as a bush or a large rock. Once the sand deposit is started, it acts to block the wind further. In time, the mound grows and becomes a **dune.** *Dunes* are commonly found just inland from beaches. Wind carries sand from the beach and drops it around nearby plants or other windbreaks. Deserts contain much dry, loose material that is easily moved by the wind. Thus in many deserts there are dunes covering large areas.

Dunes can take many shapes. But all dunes have two things in common. Every dune has a gentle slope, called the *windward* side, which faces the wind. On the other side of the dune, away from the wind, the slope is much steeper. This is called the *leeward* side. Sand is blown up over the top of the dune. It then drops down on the other side. See Fig. 6–12. This causes the dune to move slowly in the direction the wind blows. Dunes can move over roads, railroad tracks, farmlands, and even buildings! Plants and fences are often used as windbreaks to stop the movement of dunes.

Not all the particles carried by the wind are deposited as dunes. Smaller and lighter particles are carried farther than the dune-building sand. This light dust eventually comes to rest. There is probably no place on the earth that does not have some wind-carried dust on it. In some places, the dust covers the land like a blanket. See Fig. 6–13. Thick layers of this wind-

Leeward
Wind
Windward
Sand grains

Fig. 6–12 (left) *A dune has a gentle slope facing the wind and a steeper side away from the wind.*

Fig. 6–13 (right) *A deposit of loess.*

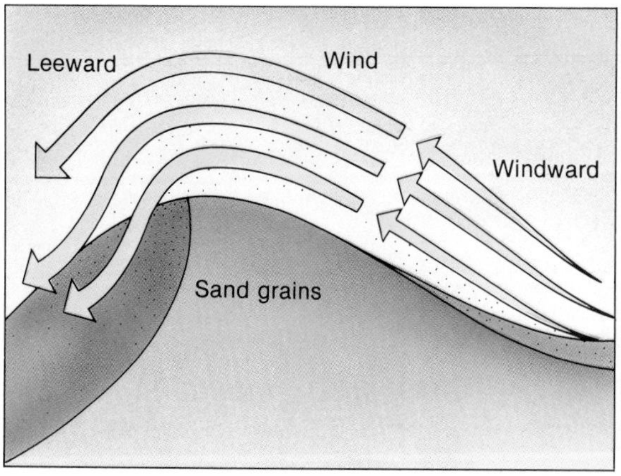

blown dust are called **loess (less)**. Deposits of *loess* are usually tan in color. They can be up to 30 meters thick, covering hills and valleys. Loess deposits are soft and easily eroded. A large area in northern China is covered with loess. The Chinese deposits came from the deserts of Asia. In the United States, loess is found along the eastern side of the Mississippi Valley. Deposits are also found in eastern parts of the Pacific Northwest. These deposits probably came from the beds of lakes and streams that dried after the glaciers retreated.

Loess A deposit of windblown dust.

WAVE EROSION

Wind is also able to move water and cause erosion. Winds blowing across oceans or lakes create waves. Waves travel across the water surface toward the **shoreline.** A *shoreline* is the place where land meets water. When waves reach a shoreline, they pound against it. During storms, large waves hit with great force. See Fig. 6-14. This constant pounding of the waves erodes the rocks of the shoreline. Waves break up rocks in two ways. First, waves force water and air into cracks in the rocks. This causes pressure to build up in the cracks. When the pressure is great enough, the cracks become larger. Large blocks of rock are broken up and moved by waves in this way. Second, pieces of broken rock cause additional erosion. This occurs when the rock pieces are smashed against the shore by the waves. Hitting the shore, they wear away more rock. As they are thrown around by the waves, these rock pieces also grind and rub against each other. This wears down the pieces. They become smaller and more round over a period of time.

Shoreline The place where the land meets water.

Fig. 6-14 Waves pound against the rocks along the shoreline.

Fig. 6-15 A magnificent sea cliff.

Sea cliffs The steep faces of rocks eroded by waves.

Sea cave A hollow space formed by waves at the base of a sea cliff.

Terrace A platform beneath the water at the base of a sea cliff.

Sea stack A column of rock that is left offshore as sea cliffs are eroded.

The effect of waves depends on the force of the waves and the type of rock that makes up the shoreline. Some rocks can be broken down at a faster rate than others. Fig. 6-15 shows a **sea cliff.** A *sea cliff* is a steep face of rock eroded by waves. The waves come straight into the shore and continually wear away the base of the sea cliff. The rocks at the top of the cliff eventually tumble down. Then they are ground up by wave action. Waves also cause **sea caves** to be formed. *Sea caves* occur when waves hollow out soft rock in a sea cliff. As the sea cliff is worn back by this process, a platform is formed at the base of the cliff. The platform is called a **terrace.** *Terraces* often slow down the erosion of the cliff. If the terrace is wide, waves lose most of their force before they reach the base of the cliff.

Unequal erosion also produces **sea stacks.** *Sea stacks* are columns of rock that stand a short distance offshore. They are formed when the softer rock is worn away. See Fig. 6-16.

Fig. 6-16 Sea stacks along a shoreline.

Fig. 6–17 (left) *A sandy beach.* (right) *A pebble beach.*

SHORE DEPOSITS

Rock particles produced by the grinding action of waves are usually pushed toward and along the shore. Many parts of the shoreline become covered by this ground-up material. These areas are called **beaches.** *Beaches* may be covered by fine sand or larger particles such as pebbles. See Fig. 6–17 *(left* and *right).* The kind of material found on beaches reflects the makeup of the coastline. Beaches also are made of materials carried to the shore by rivers and streams. The sand on a beach may be light or dark in color depending on its source. Much of the white sand on the coast of Florida came from erosion of the southern Appalachian Mountains.

Beaches Parts of the shoreline covered with sand or small rocks.

In the summer, waves that reach the beach are usually small. They carry small sand particles to the shore. The wide sandy beaches formed during the summer months can be changed during the winter months. Winter storms bring bigger waves to the beach. Large waves often carry sand away and deposit it offshore to form long underwater *sand bars.* See Fig. 6–18.

Fig. 6–18 A sandbar is formed offshore from material carried away from the shore.

Drift of sand along beach
Incoming waves
Path of typical grain of sand

Fig. 6–19 Sand is moved along a beach in shifting waves.

Waves almost always reach the beach at a slight angle. As a result, some of the water from the breaking wave moves parallel with the beach. Most of the water, however, flows straight back down into the surf. As a result of these two motions, sand grains are moved along a beach in a back-and-forth direction as shown in Fig. 6–19. Each day, this wave action can move sand and small beach pebbles many meters along the shore.

Longshore current Movement of water that is parallel with the shoreline.

Spit A long, narrow deposit of sand formed where a shoreline changes direction.

Waves striking the shore at an angle cause some water to move parallel to the shore. This motion is called a **longshore** current. *Longshore currents* also move large amounts of sand along the shoreline. This load of sand is often deposited when the shoreline changes direction. See Fig. 6–20 *(left)*. Sand is deposited at the point where the shoreline curves. This deposit is called a **spit.** *Spits* are long sand bars connected to the shoreline. They point in the direction that the longshore current moves. See Fig. 6–20 *(right)*.

The shape of a shoreline can also be caused by changes in the level of the sea. The shoreline, with its sea cliffs and terraces, seems to have been caused by a drop in sea level. It also

Fig. 6–20 (left) *Sand is deposited where a shoreline changes direction.* (right) *Toward the bottom of the photo, a spit has formed from sand moved alongside the beach.*

could have been caused by a lifting of the land. When the sea level drops or the land is lifted, new areas of shore are exposed to wave erosion. This causes many sea cliffs to be formed. On the other hand, the sea level may rise or the land may sink. Such a rise in sea level creates an irregular shoreline because stream valleys are flooded. The flooding of stream valleys can form bays and inlets along the coast.

Many things can change the shape and features of a shoreline. Careful study is needed to form some general conclusions about how a particular coastline was formed.

Fig. 6-21 Houses and other buildings are easily destroyed near the shoreline during a storm.

PROTECTING AND PRESERVING THE SHORES

Waves and currents make any shoreline a place of constant change. These changes are important because almost two-thirds of the world's population lives along the coasts. The coastal land is used for many human activities including seaports, fishing, and recreation. But many of the ways people use the shore can cause problems. For example, houses and other buildings can be destroyed if they are built too close to the water. See Fig. 6-21. There is no way to prevent the shoreline from changing. However, a scientific understanding of the way waves create beaches and other features will make it possible to predict and help control harmful changes.

SUMMARY

Wind can cause erosion in a dry region by picking up small particles of sand. These moving sand particles may strike against and wear away solid rock. Slowing down of the wind causes deposits to build up. Thick layers of windblown dust are found in some places. Waves wear away the land along the shore. Many features are formed along a shoreline by the work of waves and currents. The shape of a shoreline usually depends on the effect of the changing sea level.

QUESTIONS

Use complete sentences to write your answers.

1. How does wind cause erosion?
2. Describe two kinds of deposits caused by wind.
3. Explain how waves can erode a rocky shoreline.
4. Name and describe at least five shoreline features formed by wave erosion.

INVESTIGATION

WINDBLOWN DEPOSITS

PURPOSE: To study the nature of windblown deposits.

MATERIALS

newspaper
bottle of dry sand
box lid
spoon
goggles
water

PROCEDURE:

A. Spread out the newspaper to cover the desk top. Place the box lid near the center of the paper.

B. Remove the lid from the bottle of sand and place it in the middle of the box lid. Pour one spoonful of sand into the bottle lid. Your setup should look like Fig. 6–22.

Fig. 6–22 Your setup should look like this.

C. Put on goggles or other protective eye wear. With your face about 15 centimeters from the lid, blow gently down onto the sand. Increase the strength of your wind until sand is being thrown from the lid. Continue blowing like this for five to ten seconds.

D. Examine the material on the paper by rubbing your finger over it. Do the same thing to the material trapped in the box lid and the material left in the bottle lid. If no material is found on the newspaper, blow a little harder for five to ten seconds and retest.

1. Where are the finest grains found?
2. How did this process separate the sand into many different-sized particles?
3. Wind carries very fine particles for many miles before it slows down enough to settle and make a layer of dust. A layer of this material is called by what name?

E. Dump the material in the box lid and bottle lid into the waste container provided. DO NOT PLACE SAND IN SINK.

F. Repeat step B. Sprinkle this pile of sand with water until the surface is completely wet.

G. Repeat steps C and D.

4. What effect does wetting the sand have on the amount of material blown away?

H. Discard the wet sand in the container provided. NO SAND IN THE SINK.

CONCLUSIONS:

1. Describe the kind of material most likely to be blown some distance by wind.
2. Look at Fig. 6–11 *(right)* on page 157. Explain why large areas of desert have this appearance.

TECHNOLOGY

WIND FARMS

Sometimes a "new" technology is an old idea updated. That's what has happened to the windmill, a technology probably associated most closely with Holland. Versions of the windmill, however, have been in use in the United States for a long time, particularly in the rural areas of the country. They have been used primarily to pump water in irrigation systems, but they have also provided electricity. In fact, in the 1930's wind-generated electricity from the "windcharger" was common throughout the Midwest before utility lines covered the country. Now "wind farms" are being developed to "harvest the wind" in order to produce electricity on a large scale. The four giant wooden blades of the old storybook windmills have been replaced by a wide variety of designs. When used to generate electricity, they are called wind turbines. If they have moving blades that rotate around a stationary cylinder, you may see them referred to as rotors.

Many of the United States' wind farms are concentrated in the Altamonte Pass in California, shown in the photo. There the wind commonly blows for 15 to 20 days in a row without stopping, dies down for a day, and then starts all over again. This pattern is repeated from mid-March to mid-September. At one location in the Altamonte Pass, 100 wind turbines generate electricity day in and day out. Not far away, there are 50 more wind turbines designed and built by another manufacturer. At still a third location are the first five of several hundred wind turbines planned by a third developer. There are a few other sites of wind farms in Montana, Wyoming, and New Hampshire. Individual wind turbines are used to produce electricity on a smaller scale in both rural and urban areas.

Each type of turbine is slightly different in design. Ranging in height from 12 meters to 22 meters, they usually have two or three fiberglass blades, which may be 15 to 21 meters in diameter. Several of the newer "windmills" have particularly unusual designs. The Darrieus "egg beater" rotor has two or more curved aluminum blades that are attached, at the bottom and top of the blade, to a central power shaft. The Department of Agriculture has experimented with the Darrieus rotor as a water pumper to irrigate farms in Bushland, Texas. Engineers are also testing models of wind turbines with no moving parts at all, a design which creates far fewer mechanical problems.

A wind turbine is designed to be capable of generating a certain amount of electricity, but the exact amount it actually produces depends on the wind.

What if there isn't any wind? These latest wind turbine projects solve that problem. They are connected to an electrical system that stores the extra electrical energy produced when there is a great deal of wind. They can then be used when the air is calm and there is little or no wind. The biggest fault of windpower has always been that you couldn't depend on it as a constant source of energy. Backup energy sources have always been needed. The possibility now exists that wind could become a safe and reliable source of energy. This hope is shared in other countries where wind energy is being developed.

CHAPTER REVIEW

VOCABULARY

On a separate piece of paper, write TRUE next to the number of each statement that is true. Next to the number of each false statement, write FALSE, then make the statement true by writing the correct term in place of the underlined incorrect term.

1. A longshore current is a river of ice in a mountain valley.
2. Rock material deposited directly from a glacier is known as till.
3. A terminal moraine is a ridge of till.
4. Sand blown by wind can abrade anything in its path.
5. A barrier that causes wind to move more slowly is called a spit.
6. A place where water meets land is a windbreak.
7. A steep face of rock eroded by waves is known as a beach.
8. The underwater platform at the base of a sea cliff is called a terrace.
9. A sea cave is a hollowed-out space found at the base of a sea cliff.
10. A lone column of rock that stands a short distance offshore is known as a sea cliff.
11. A sea stack is a part of the shoreline covered by sand or small rocks.
12. The narrow deposit of sand formed where a shoreline changes direction is a dune.

QUESTIONS

Give brief but complete answers to each of the following questions. Unless otherwise indicated, use complete sentences to write your answers.

1. What conditions are necessary to cause a glacier to form?
2. Where on earth are continental glaciers found today?
3. Explain how glaciers are able to move.
4. Compare the two types of theories that explain the ice ages.
5. What is an ice age? When did the last one occur on earth?
6. Why is it important to understand what causes ice ages?
7. Where does wind erosion occur most often? Why?
8. Explain why wind erosion cannot account for most of the strange large rock formations found in the desert that appear to have been eroded.

9. How do plants help prevent erosion by wind?
10. What causes longshore currents?
11. How are sea caves, terraces, stacks, and sea cliffs related?
12. Contrast the kind of shoreline features that result from a lowering of the sea level with those that result from the raising of the sea level.
13. How can you tell if an area was eroded by (a) ice, (b) wind, or (c) waves?
14. Describe the deposits left by (a) ice, (b) wind, and (c) waves?
15. What types of areas are affected most by (a) ice, (b) wind, and (c) waves?

APPLYING SCIENCE

1. Find examples of erosion by ice, wind, or waves near where you live. Describe how the erosion is taking place. Explain what is being done about it.
2. Using an electric fan, some fine sand, a shallow box, and some paper to catch any escaping sand, set up the conditions necessary to study the movement of sand by wind. Study the use of windbreaks, the forming of dunes, the movement of dunes, etc. Take pictures or write a report of your results.
3. Refer to Fig. 6–20 *(right)* on page 162. Make a series of drawings to show what will happen in the future if the longshore current continues.
4. Make a poster for display in the classroom that shows the important features made by a valley glacier. Show and label such features as a cirque, arêtes, horns, U shaped valley, and the outwash plain formed by the glacier. Make the poster as drawings, cut-outs, or with pictures showing each of the features. Obtain permission before removing pictures from magazines, etc.

BIBLIOGRAPHY

Chasan, James Peter. "Columbia Glacier in Retreat." *Smithsonian*, January 1983.

Gallant, Roy. "Wind, Sand, and Space." *Science 81*, November 1981.

Radlauer, Ruth. *Grand Teton National Park*. Chicago: Children's Press, 1980.

Simon, C. "Columbia Glacier's Last Stand." *Science News*, January 1984.

Time-Life Editors. *Glacier* (Planet Earth Series). Chicago: Time-Life Books, 1984.

Time-Life Editors. *Ice Ages* (Planet Earth Series). Chicago: Time-Life Books, 1984.

CHAPTER

7

Most rocks can tell us something about the earth's past. Some rocks, like the one below, tell us that there were once fish living in the place where the rocks were found.

EARTH'S HISTORY

CHAPTER GOALS

1. Describe the principle of uniformitarianism and explain how it is used in the study of rock layers.
2. Discuss how fossils are formed and the kinds of information that can be obtained from studying them.
3. Describe two ways the relative age of rock layers can be found.
4. Explain how radioactivity can be used to find the age of rocks and fossils.
5. Name the four main parts of the earth's history and explain why the earth's history is divided into parts.

7–1. The Rock Record

At the end of this section you will be able to:

- Explain how the idea that "the present is the key to the past" helps to explain the earth's history.
- Compare the ages of different rock layers.
- List four kinds of information that may be learned from an *unconformity*.

The Grand Canyon is like a book that holds a part of the earth's history. The pages of the book are the layers of rock seen on the canyon walls. In the same way as printing on a book's pages carries information, the rocks contain the story of how they were formed. In this section, you will begin to understand how scientists can read some of the history recorded in the rocks.

TIME AND CHANGE

Almost all scientists believe that the earth was formed about 4.7 billion years ago. It is hard for anyone to imagine such a long period of time. The following comparison may help you to appreciate how long 4.7 billion years is. Suppose that an imaginary writer started to write a history when the earth was formed. This writer is very lazy and only types one letter every ten years. On each page there are 7,200 letters. One thousand pages make up one volume of the long history. Thus it takes the

Fig. 7-1 The rock layers of the Grand Canyon are like pages in a history book; each layer tells another story.

writer 72 million years to type each volume. During the 4.7 billion years since the earth was formed, the writer would have typed 65 volumes. The story would now be on page 278 of the 66th volume. See Fig. 7-2. The entire history of the United States would be recorded in the last line. Columbus' discovery of America would be found in the line above. About ten lines above would describe the first Egyptian pyramids. Cave paintings made 20,000 years ago would be on the same page. Most scientists believe that humans have been present on the earth for only a short part of its long history.

Each person can see only a very small part of the earth's long history. During a person's lifetime, it is hard to see that the earth's surface is always changing. If your grandfather climbed

Fig. 7-2 The "history of the earth" written at the rate of one letter every ten years and one volume every 72 million years. The formation of the oldest rocks known would begin in volume 14!

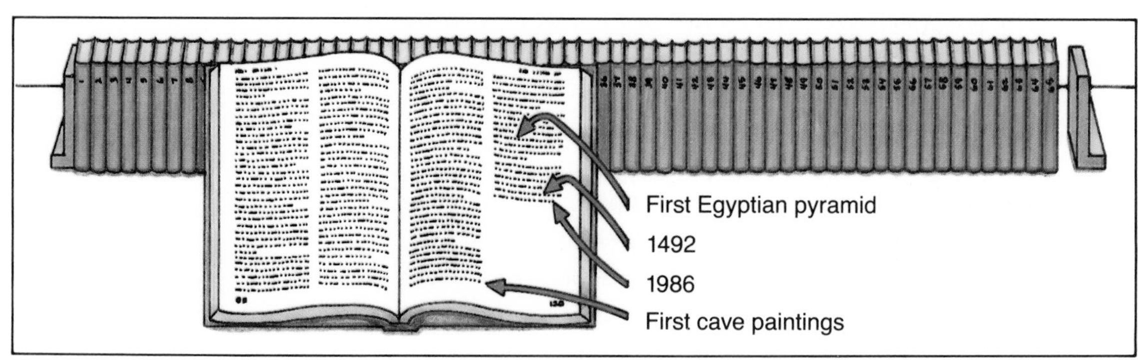

a mountain, you can climb that same mountain. But millions of years from now that mountain will probably be gone. In North America, erosion removes an average of six centimeters of the land every 1,000 years. During the billions of years of earth's history, even the tallest mountains have been eroded away. They have been replaced by other mountains raised by the forces that bend the earth's crust.

The changes on the earth's surface have been studied for many centuries. Writings about earthquakes date back to the Greeks more than 2,300 years ago. Aristotle believed that earthquakes occurred when air was absorbed by the ground. He said that when the air was heated by fires inside the earth, the ground exploded.

Then during the seventeenth and eighteenth centuries, many geologists believed that the surface features of the earth were formed by sudden events called *catastrophes*. For example, features such as mountains and canyons were explained to have been caused by sudden earthquakes. This idea said that the features on the earth were formed over a very short period of time. After these features were formed, they remained as we see them today. Thus the earth did not have to be very old.

But in 1795, a geologist named James Hutton proposed a different idea. He thought that the present is the key to the past. This idea is called the principle of **uniformitarianism.** The principle of *uniformitarianism* states that processes, such as mountain building, weathering, and erosion, take place very slowly and over long periods of time. It says that the geological processes now operating on the earth were active in the past.

Uniformitarianism The principle that the same processes acting on the earth today also acted on the earth in the past.

Geologists believe the processes that are working today have always existed. The forces that build mountains today were at work millions of years ago. The river now running in a canyon began wearing away the land thousands of years ago. Thus, by studying the present, we can learn about the processes in the past. For example, look at Fig. 7–3. According to the principle of uniformitarianism, the ripples in the ancient sandstone *(top)* were made by the same process that made the new ripples in the sand *(bottom)*.

We know the earth must be very old because its features take a long time to form or wear away. For example, rocks have been found with fossils of sea animals that lived more than 10 million years ago. Some of these rocks are now in mountains

Fig. 7-3 (top) *The ripples in ancient sandstone were formed by the same process, wind, that made the ripples in the sand* (bottom).

2,000 meters above sea level. These mountains have been building at a rate of only .2 millimeters per year for ten million years. Yet that is a short time compared to the billions of years the earth has existed. The evidence shows many cycles of mountain building and erosion. But the earth's features seem to change little in a human life span. Major changes can only be seen over spans of thousands or millions of years. Thus, the features we can see today are like a giant cover spread over the long history written in the rocks below. Beneath our modern geography is the record of ancient formations, many of which existed before human history began. Most modern scientists now believe that uniformitarianism is a useful guide to understanding the past. The forces that changed the earth in the past are thought to be similar to those causing changes today.

HOW ROCKS CAN TELL A STORY

Reading the rock record is like reading a story written in a mysterious code. The principle of uniformitarianism is one clue to the code. There are other clues.

In a group of rock layers, the bottom layer is generally the oldest. The top layer is the youngest. As an example, you can make a pile of books by placing one book on top of another. The book that was put down first will be in the pile for the longest time. It forms the oldest layer in the pile, and it is also the one at the bottom. This is an example of the **law of superposition.** The *law of superposition* is believed to work when rock layers are formed. Each new layer is laid over the one under it. Therefore, each layer is younger than the one beneath it. In the Grand Canyon, for example, the upper layers are seen near the top. Older layers are near the bottom of the canyon where the Colorado River is still eroding the land.

Law of superposition The principle that, in a group of rock layers, the top layer is generally the youngest and the bottom layer the oldest.

There are certain cases where the law of superposition does not apply. Forces are always creating folds and faults in the crust. Sometimes rock layers are bent or broken so much that old layers may be pushed on top of younger ones. As you can see in Fig. 7-4, it is sometimes difficult to tell how layers may have been arranged before they were folded or faulted.

Comparing the positions of rock layers can tell only which layer is older or younger. Older or younger refers to **relative age.** *Relative age* does not tell how many years have passed since a rock layer was formed. It only shows that one layer is older than another. Another principle can be used to tell the

Relative age A method of telling if one thing is older than another. It does not give the exact age.

Fig. 7-4 Can you tell which of these rock layers are the oldest from their position?

Crosscutting The principle that faults or magma flows cutting through other rocks must be younger than the rocks they cut.

relative age of rocks. It is called the principle of **crosscutting** relationships. For example, *crosscutting* can occur with a fault or a flow of magma within layers of rock. A fault is always younger than the layers it cuts across. A flow of magma is always younger than the rock layers around it.

ROCKS RECORD EARTH MOVEMENTS

Fig. 7–5 shows the arrangement of rock layers in a part of the Grand Canyon. You can see that the lower rock layers are tilted compared to those above. Layers of rock most often form in a level position. They generally lie flat with the other layers, like pages in a book. The lower layers in the photograph must have been tilted after they were formed. But how can the level layers above be explained? A reasonable explanation would be that the lower layers were formed, then tilted by crustal movements. Over a long period of time, erosion leveled the surface of the layers. Conditions then changed and the erosion stopped. New layers were deposited on the old eroded surface. The eroded surface forms a boundary between the old and new layers. Such a boundary is called an **unconformity.** An *unconformity* separates younger layers from older layers that were exposed to erosion. Unconformities may also exist between layers that have not been tilted. An unconformity usually shows that rock layers are missing. These are the layers that were lost to erosion. These missing layers are like pages torn from a book. They are gaps in the complete story that once existed in the rock record. With more information, these gaps may someday be closed.

Unconformity A boundary separating younger rock layers from older layers that were exposed to erosion.

Fig. 7–5 The boundary separating the tilted layers from the level layers is called an unconformity.

Fig. 7–6 shows four steps in the formation of an unconformity. First, layers are formed underwater. See Fig. 7–6 (a). Level layers such as these are made when material settles to the bottom of a body of water. Second, the land is lifted and tilted after the lower layers are formed. See Fig. 7–6 (b). Third, erosion levels these tilted layers. See Fig. 7–6 (c). Last, the eroded surface is submerged under the sea. Erosion stops and new deposits are formed. See Fig. 7–6 (d). Level rock layers on top are formed.

Unconformities give us clues about the history of a region. First, they help us to place events in order. For example, the rock layers below the unconformity must have been tilted before they were eroded. This is because all the layers have been

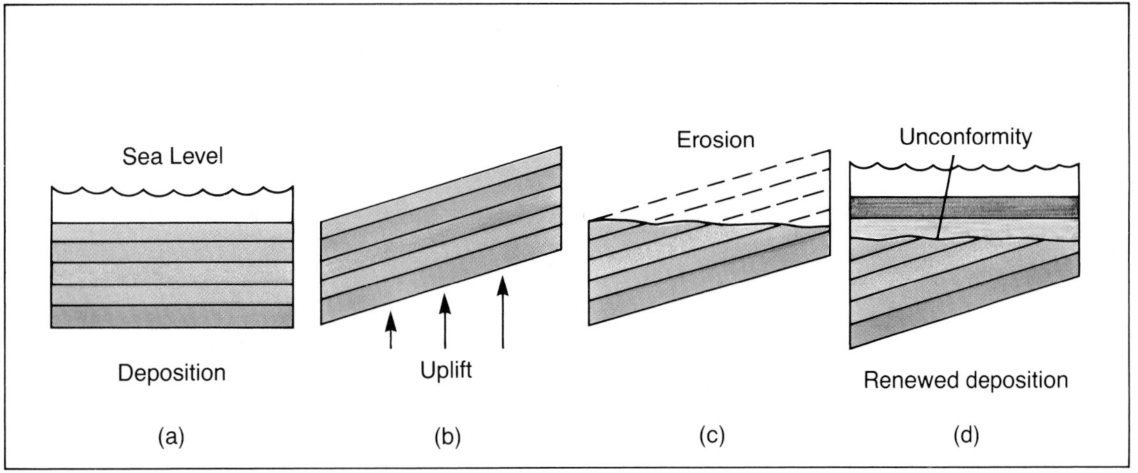

Fig. 7-6 These diagrams show the steps in the formation of an unconformity.

eroded. Otherwise, only the top layer would be eroded. Unconformities also tell us something about the location of the dry land and the sea in the past. The rocks below the unconformity were formed as level layers underwater. Their surfaces could be exposed to erosion only if they were raised above sea level. The rock layers above the unconformity were formed when the eroded surface submerged again. Thus unconformities are useful clues to the relative ages and history of layered rocks.

SUMMARY

Layers of rocks in the earth's crust contain a record of the earth's history. The same processes that shape the crust today also operated when those rock layers were formed. The law of superposition, the law of crosscutting, and the idea of relative age allow us to understand events in the past. Sometimes an unconformity is found in the rock record. Such a gap can also provide clues to the history of a part of the crust.

QUESTIONS

Use complete sentences to write your answers.

1. Use the principle of uniformitarianism to explain ripples found in some rock.
2. You are looking at the layers of rock made visible by a highway cut through a hill. Which layer is the oldest? Which layer is the youngest? Describe one case in which you cannot be sure of your answers.
3. List in proper order the four events that may have occurred where an unconformity is found.

INVESTIGATION

INTERPRETING ROCK LAYERS

PURPOSE: To use a diagram of a section of the Grand Canyon for reviewing some of its history.

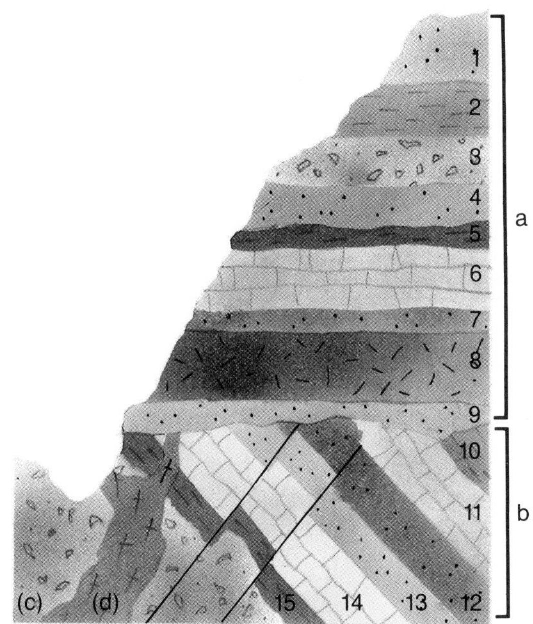

Fig. 7-7

MATERIALS:
pencil paper

PROCEDURE:

A. Study Fig. 7-7 and write the answers to the following questions:
   1. Which numbered layer is the oldest of the horizontal layers in group a?
   2. Which numbered layer is the oldest of the tilted layers in group b?
   3. Which is the older, group a or b?
   4. Was the tilted layer of group b level at one time in history?
   5. Did the movement along the fault in the group b layers take place before or after layer 9 was laid down?
   6. Which layer of rock was the first to form after the tilting of group b?
   7. Which layers of rock must have been laid down before the magma, labeled d, pushed its way in?

B. An unconformity represents a time during which erosion was taking place at a particular location. Material was being carried away to some other location. The layer being eroded lost its flat appearance. When new layers form over this uneven surface, the bottom of the first new layer is also uneven. Look at Fig. 7-7 again to answer the following questions.
   8. Which two layers have uneven bottoms and thus show periods of erosion? Identify the layer that shows the oldest unconformity.

C. Layer 3 and layer 7 are deposits of limestone. Layer 3 contains the remains of ancient sea life. Layer 7 contains no such remains and is more fine grained and dense than layer 3. Answer the following questions using this information.
   9. What can you say about the conditions that existed at the time layer 3 was laid down? What reasons can you give for layer 7 not having ancient remains?

CONCLUSION:

1. This activity reviewed four clues about interpreting rock layers. List the four in the order they were used.

7–2. The Record of Past Life

At the end of this section you will be able to:

- Define *fossil.*
- Describe six ways fossils can be formed.
- Point out several kinds of information that can be obtained from the study of fossils.

Fig. 7–8 tells a story. It shows human footprints made about 6,000 years ago. The rock in which the footprints are found was made from volcanic ash. Scientists believe that the footprints were made by a person running from an erupting volcano in Hawaii. What information do you think can be gained about this person from the footprints?

Fig. 7–8 Human footprints about 6,000 years old.

EVIDENCE OF ANCIENT LIFE

What could you learn by studying a person's footprint left in soft ground? Measuring the depth of the print could give an idea of the person's weight. The area covered by the impression gives a clue about the person's height and general size. Many of the kinds of plants and animals that have lived on the earth no longer exist. Scientists know about those ancient life forms only by their remains left in rocks. It is like describing a person from a footprint. Both remains and impressions are records of what happened in the past. See Fig. 7–9 (*top*).

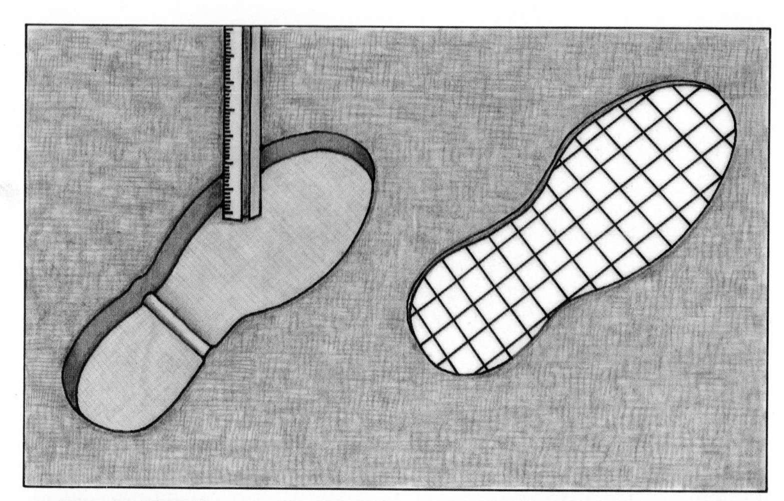

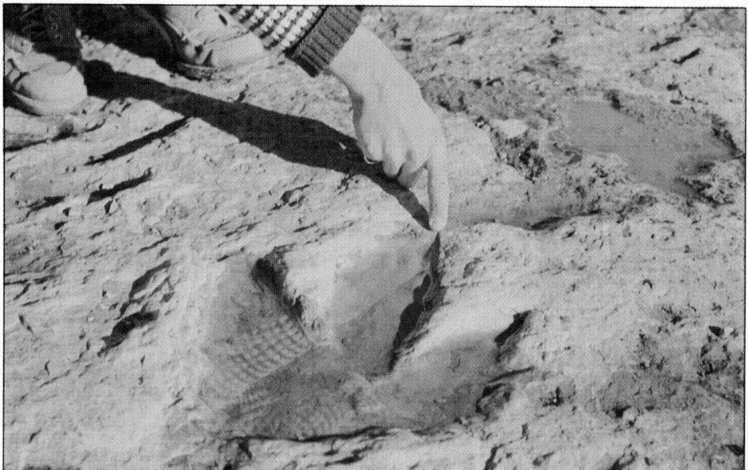

Fig. 7-9 (top) *Measurements are made of footprints in order to hypothesize about the animal that made them.* (bottom) *What can you tell about the animal that made this footprint?*

Fossil Any evidence of, or the remains of, a plant or animal that lived in the past.

Any naturally preserved part, trace, or entire remains of a plant or animal that lived in the past is called a **fossil.** A *fossil* can be a piece of wood, a tooth, or a bone that was part of a living thing. A footprint or an impression also can be a fossil. Although a footprint is not a part of an animal, it is evidence that the animal once lived. See Fig. 7-9 (*bottom*). A footprint is an example of a *trace fossil.* Any mark left by an animal walking, crawling, or burrowing is called a trace fossil.

HOW FOSSILS FORM

A fossil also can be a part of a living thing. Usually the remains of plants and animals are destroyed by the process of decay. Of all the different plants and animals that have lived, very few have become fossils. However, sometimes decay is prevented from occurring. For example, the remains of plants

Fig. 7-10 (left) *This young woolly mammoth was preserved in the permafrost in Siberia. It was discovered in 1977.* (right) *The woolly mammoth lived during the last ice age.*

and animals may contain hard parts, which are preserved better than the soft parts. Also, the plant or animal remains may be protected very quickly.

1. Ice. In a few cases, fossils have been protected by ice and preserved by freezing. Some frozen woolly mammoths have been found in the frozen areas of Siberia. See Fig. 7-10 (*left*). Woolly mammoths were ancestors of today's elephants. They lived 8,000 to 20,000 years ago during the last ice age. Unlike elephants, they were very hairy. See Fig. 7-10 (*right*).

2. Amber. Fossils of ancient insects are found in hardened tree sap *(amber)*. The insects became stuck in the sap. Later the sap became as hard as stone. Amber containing whole, perfectly preserved insects can now be found. See Fig. 7-11.

Fig. 7-11 The remains of this insect are implanted in a piece of amber.

3. Tar. In the center of the city of Los Angeles there are famous pools of tar. They are known as the La Brea Tar Pits. This sticky tar was formed by petroleum oozing out onto the surface. See Fig. 7–12. Tigers, wolves, horses, and other animals became trapped in the tar about 15,000 years ago. Many sank into the tar when they probably mistook the tar for a pool of water. Others attacked the trapped animals and were trapped themselves. The bones and teeth of these animals have been dug up from the ancient tar pits. These remains are very well preserved.

Fig. 7–12 The La Brea Tar Pits hold the remains of many prehistoric animals that were accidentally trapped in them.

4. Burial. Most organisms are not preserved as well as those in ice, amber, or tar. This is because decay begins quickly in most places when the remains are exposed to air. If the remains are covered by sand or soil, the decay process is slowed. Bones, teeth, shells, and other hard parts do not decay as fast as soft parts. When an animal or plant falls in a place where it becomes covered, its hard parts may be preserved. The remains of animals living in the sea and other bodies of water often sink to the bottom. There they are covered by mud and often become fossils. Most rocks that contain fossils were formed from the material on the bottom of bodies of water. Another way that fossils form is when ash falling from volcanoes has covered whole forests. These fossil forests are later found with the trees still in place.

5. Replacement by minerals. Although plant and animal remains may be covered over, decay can still continue. Other processes can also begin. The water that passes through the soil carries dissolved minerals. These minerals may slowly enter the remains of the dead animal or plant. The minerals take the place of some of the original material. The bones and shells become stronger with the addition of hard mineral material. When all of the original material has been washed away the fossil becomes completely mineral. Such fossils are said to be **petrified.** Fig. 7–13 shows *petrified* trees. These trees were once part of a forest. Most of the trees' living material was replaced by minerals.

Fig. 7–13 A part of a petrified tree.

6. Molds and casts. Some plant and animal remains are not replaced by minerals. They are completely dissolved. For example, a clam shell might become buried in soft mud. The mud would harden into rock. The shell would slowly dissolve and leave an impression of the shell in the rock. This type of impression is called a *mold.* Shellfish, fish, and plants leave molds in many rocks. A mold sometimes becomes filled with mud or mineral material. This results in the formation of a *cast.* The cast has the original shape of the living thing. See Fig. 7–14. Many kinds of soft-bodied plants and animals, such as jellyfish, leave fossils only in the form of molds or casts.

Petrify To replace the material in the remains of a living thing with hardened mineral matter.

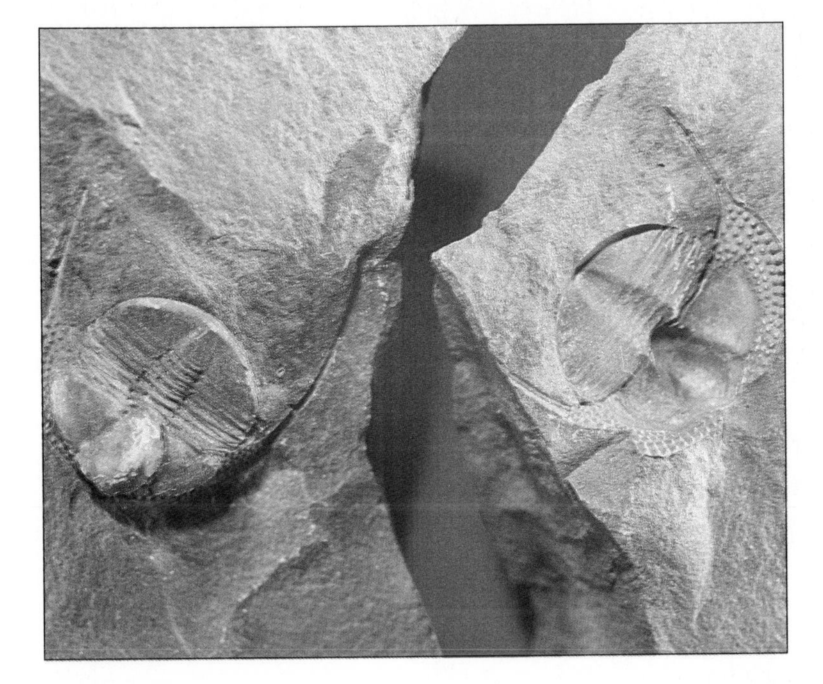

Fig. 7–14 (left) *A cast and* (right) *a mold fossil. The cast does not contain any part of the original animal. It was formed inside a mold fossil.*

Fig. 7-15 The dinosaur, Tyrannosaurus Rex.

WHAT CAN BE LEARNED FROM FOSSILS?

The study of fossils can provide us with important knowledge about the earth's past. Dinosaur bones, for example, can give scientists much information about the animal, such as its size and shape. The shape of the teeth gives clues about the kind of food the animal ate. Marks on bones show where muscles were attached. The shapes of the bones indicate how the different parts of the animal's body might have been used. See Fig. 7-15. Trace fossils such as footprints were made while the animals were still alive. These prints are records of where the animal lived. See Fig. 7-16.

Fossil plants provide clues about the conditions that existed in ancient times. We know, for example, that swamps are found only in warm climates. Ancient swamps, buried long ago, have developed into beds of coal. Today, coal is found in Antarctica. This suggests that the climate of this now-frozen land was once very warm. Also, fossil seashells are found on Antarctic mountain tops. This suggests that this land was once underwater.

Fig. 7-16 Fossil footprints of the same kind of dinosaur squatting (left) *and walking* (right). *No direct fossils of this dinosaur have been found. The drawing was made by using the trace fossils to estimate the animal's size and general appearance.*

One important theory coming from the study of fossils is the story of changing life. Many plants and animals that were common at one time are no longer in existence. Their stories are found in the layers of rock. A climb up the walls of the Grand Canyon is like a trip through the earth's history. The oldest layers at the bottom of the canyon contain no fossils. The next layers contain fossils of simple water plants. Climbing higher, layers with fossils of simple animals are found. Fish fossils are found in higher layers. Still farther up, fossils of land animals can be found. See Fig. 7-17.

Fig. 7-17 This shows the different forms of life in different geological times as they are represented by the fossil remains in the rock layers of the Grand Canyon.

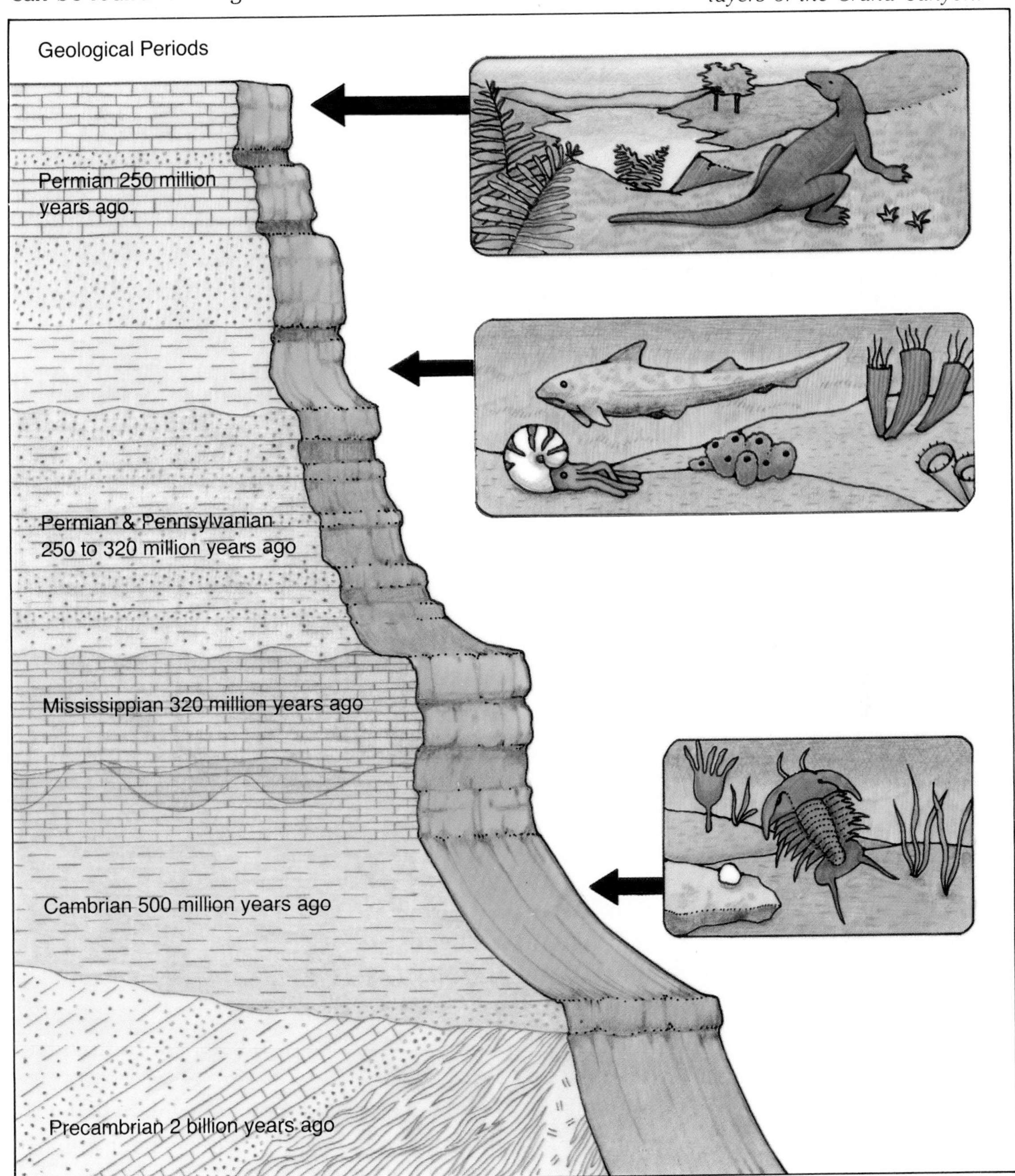

Fossils found in the Grand Canyon and in many other places have been studied. Based on the fossil evidence, most scientists agree with the following theories:

1. Life has probably existed on earth for at least three billion years.

2. Simple forms of life appeared first. Higher forms of life developed from simple forms.

3. Some forms of life no longer exist.

The fossil record is difficult to read. There is no evidence from fossils showing how the different forms of life may have developed. There are many missing parts in the fossil record. One scientist estimates that only 1 percent of all existing fossils have been discovered. If and when these missing parts are discovered, other theories may develop to explain the earth's history of life. Much work remains before the story of the earth and its life forms is nearly complete.

SUMMARY

Ancient forms of life have left evidence of their existence in the form of fossils. In a few cases, a living thing may be completely preserved. More often, the remains are changed in some way. A part or all of the original material may be replaced with mineral matter. Some fossils consist only of the impressions left by the remains or the activities of the living things. The study of fossils suggests that life has developed slowly during the earth's history, from simple forms to more complex forms. Fossils provide information about past environments and have helped us to learn about the earth's history.

QUESTIONS

Use complete sentences to write your answers.

1. What is a fossil?
2. Describe the ways fossils may be formed.
3. How does a fossil become petrified?
4. List some of the information that can be obtained from the study of dinosaur fossils.
5. Coal is found in Antarctica, and fossil seashells are found in Antarctic mountains. What does this tell us about Antarctica's past?

INVESTIGATION

MOLD AND CAST FOSSILS

PURPOSE: To make some molds and casts of common objects.

MATERIALS:

modeling clay
object for mold
plaster of paris
mixing container
mixing stick
water

PROCEDURE:

A. Roll the modeling clay into a smooth, flat layer.

B. Press a leaf, shell, coin, key, or similar object into the clay. Make sure there is a definite impression in the clay. Remove the object.

   1. How is the mold similar to the original object?
   2. How is the mold different from the original object?

C. Using additional clay, build a low wall that measures two or three millimeters high around the impression.

D. Using the mixing container, mix enough plaster of paris to fill the impression to the top of the wall. Mix enough water with the plaster of paris so that it looks like whipped cream.

E. Fill the mold to the top of the wall. You will have to wait about 30 minutes for the mold to harden. If time is not available, label your container with your name and set the mold aside to harden overnight.

F. After the mold has hardened, carefully pull off the clay.

G. Study the cast you have made.

   3. How is the cast like the original object?
   4. How is the cast different from the original object?
   5. How much of the cast would you need in order to identify the object that made it?

H. Exchange casts with another student. See if you can identify the object the other student's cast represents.

   6. What is the object?

I. Look back at Fig. 7–9 (*bottom*), page 178. Use it to answer the following questions.

   7. What kind of trace fossil is this?
   8. What kinds of information does this fossil provide?

CONCLUSIONS:

1. Explain the difference between a cast and a mold.
2. Which types of things about the original object cannot be identified from a cast?
3. Identify the kind of fossil in Fig. 7–18 below. Explain how the fossil probably formed.

Fig. 7–18

7-3. Age of Rocks

At the end of this section you will be able to:

- Explain how scientists can determine the relative ages of rocks.
- Describe how the age of rocks can be measured using *radioactive decay*.
- Name three *radioactive* materials used to find the ages of rocks and fossils.

How could you describe the ages of the people in the photograph shown in Fig. 7-19? Some could be described as teenagers. Others could be called adults or elderly. Another way is to measure their ages in years. A certain person might be 45 years old. Scientists express the ages of rocks in similar ways.

Fig. 7-19 How might you describe the ages of the people in this photo?

RELATIVE AGES OF ROCKS

Earlier in this chapter, you learned that the *relative age* of a rock layer could be found. This is done by studying the position of rock beds along with the presence of other features such as hardened magma and faults. Another way to compare the ages of rock layers is to look at the rock material itself.

Key bed A rock layer that is easily recognized and found over a large area.

Any rock layer that is found over a large area and can be identified easily is called a **key bed.** A layer made from volcanic material can be a *key bed.* A volcanic eruption may throw a large amount of ash into the air. The ash may settle over a large area, forming a single layer. For example, the eruption of Mount St. Helens spread a layer of ash over hundreds of thousands of square kilometers. In time, the ash will be buried and form a rock layer. Scientists studying the Mount St. Helens eruption have learned much about how to recognize rock layers formed by volcanoes.

Other events can create key beds. For example, a change in sea level produces layers that may serve as key beds. See Fig. 7-20. This layer can be traced among the other beds of rock. All layers below the key bed are older than the layers above it. Wherever the key bed is found, it can be used to find the relative ages of rock layers above and below. See Fig. 7-21.

Fig. 7-20 Can you see a layer that might be used as a key bed?

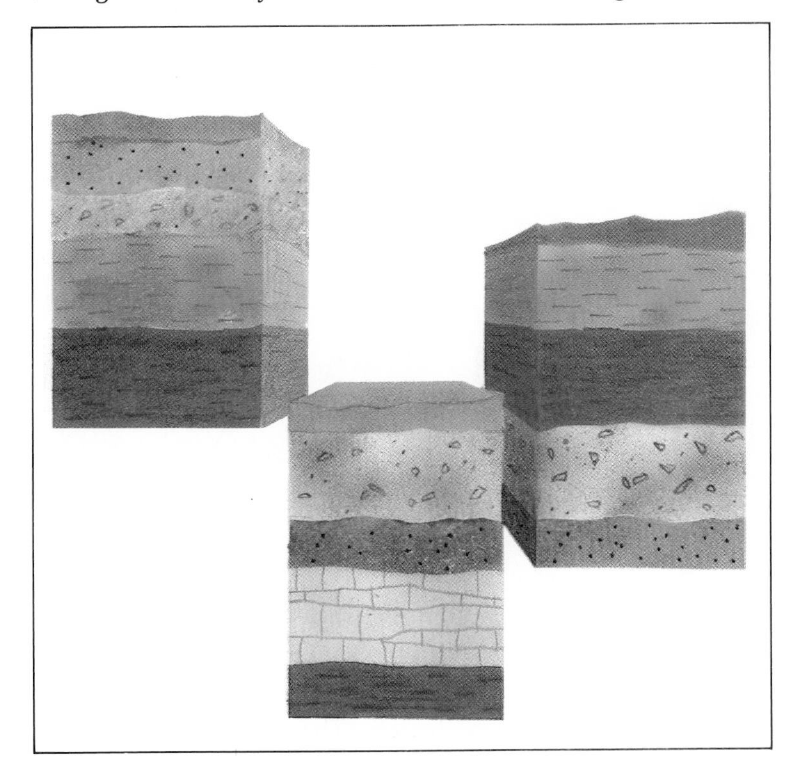

Fig. 7-21 Which of the rock layers could be used as a key bed?

INDEX FOSSILS

Fossils can be used to find the more accurate age of rocks. Such a fossil is called an **index fossil.** An *index fossil* is the remains of a plant or animal that lived during only a small part of the earth's history. Any rock containing an index fossil must have formed during the time the fossil lived. Thus the age of the rock can be determined from the period when the fossil was alive. Also, an index fossil must come from a form of life that

Index fossil The remains of a plant or animal that lived during only a small part of the earth's history.

lived in many parts of the world. An example of one kind of index fossil is the *trilobite.* See Fig. 7-22. The trilobite was a shell animal. It is believed to have lived about 500 to 600 million years ago. There were over a thousand kinds of trilobites living in oceans all over the world. Most were only a few centimeters long. A few hundred million years ago, all trilobites disappeared from the earth. Their shells are now commonly found as fossil molds and casts in certain rock layers all over the world. Trilobites lived for a short part of the earth's history and were very plentiful. Thus they make good index fossils. Any rocks that contain trilobite fossils probably formed between 500 and 600 million years ago. There are many other kinds of index fossils. Each kind comes from a certain part of the earth's history. The discovery of any kind of index fossil in a rock layer helps to find the age of that rock.

Fig. 7-22 Trilobites are good index fossils.

Suppose that the same kind of index fossil is found in rock layers at different places. Then the rocks in the layers must be about the same age. For example, trilobites are found in rocks on one side of the Grand Canyon. The same fossils are also found in rocks on the other side of the canyon. This means that the layers in each side of the canyon must have been formed at the same time. The layer was cut through, forming a canyon.

The same thing must be true for rock layers in all places on the earth containing a certain index fossil. No matter where those rocks were formed, they must be nearly the same age. For example, certain rocks in Europe contain the same kind of trilobites as found in the Grand Canyon. Thus the European rocks must be the same age as the Grand Canyon rocks. Scien-

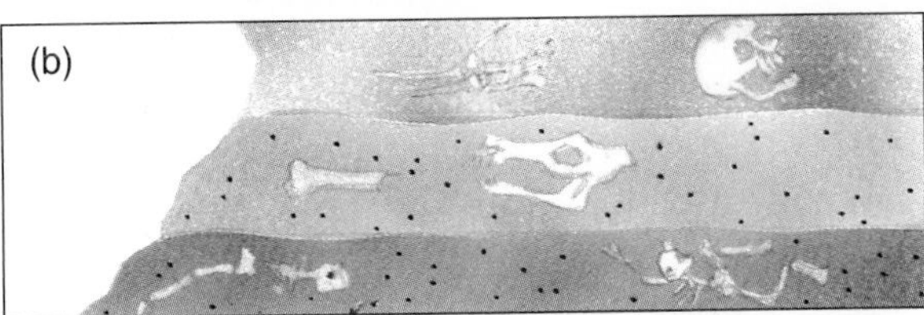

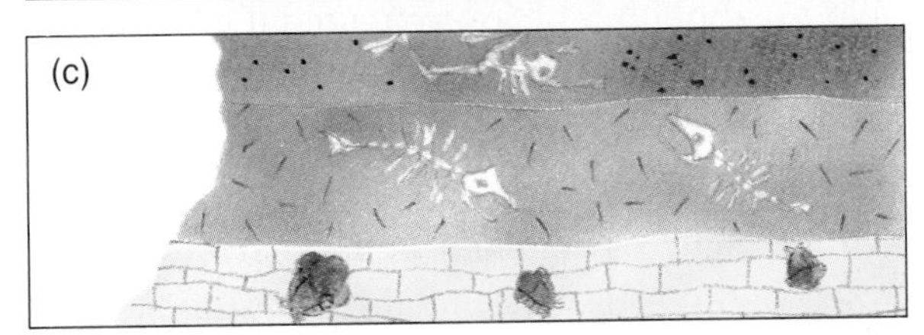

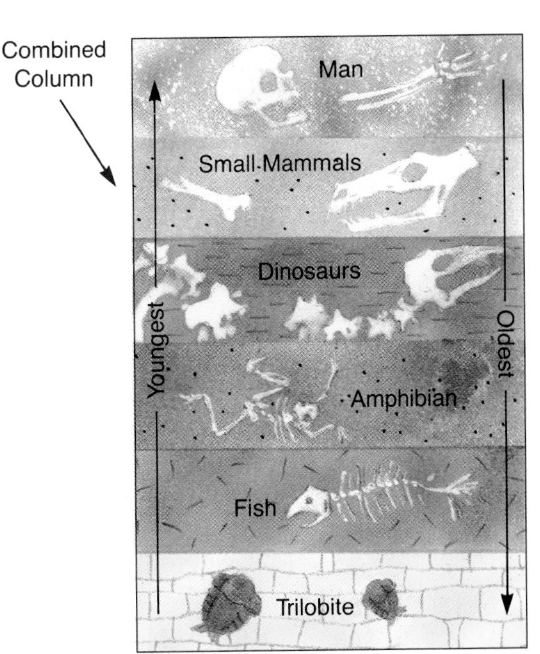

tists have used fossils to compare the ages of the rock at locations all over the world. They have combined all these observations to make a **geologic column.** Fig. 7-23 shows how a *geologic column* can be made. A geologic column has been made for the different rock layers found all over the world. The geologic column represents rock layers everywhere on the earth's surface. They are in the same order in which they were laid down. It is like an imaginary Grand Canyon where rock layers all over the earth could be seen at the same time. Scientists can compare any particular rock layer to its position in the geologic column. The relative age of any rock can then be found.

Fig. 7-23 (left) *Layers having the same color and fossils are the same geologic age.* (right) *The diagram shows a geologic column with each layer in order by its age.*

Geologic column An arrangement showing rock layers in the order in which they were formed.

FINDING THE EXACT AGE OF ROCKS

Key beds can be used to tell the relative age of rocks. Index fossils show that rocks were formed at a certain period of time. But scientists also can measure the ages of many rocks more exactly. They do this with the help of **radioactive** substances. A *radioactive* substance is one that changes over a period of time. All matter is made up of tiny units called atoms. A radioactive substance has atoms that change into other atoms. This is called *radioactive decay.* A radioactive atom decays into another kind of atom at a regular, steady rate. Thus radioactive decay can be used to measure the ages of certain rocks.

Radioactive Describes a substance that changes over a period of time into a completely new substance.

Half-life The length of time taken for half a given amount of a radioactive substance to change into a new substance.

1. Uranium. Uranium is a radioactive substance found in many rocks. It slowly decays into lead. Measurements show that it takes 4.5 billion years for one-half of the uranium in a rock to change into lead. See Fig. 7–24. This length of time is called a **half-life.** Because of this very long *half-life*, radioactive uranium is used to find the age of very old rocks. Scientists carefully measure the amount of uranium and lead in a rock. Using these measurements, the length of time that decay has been going on in the rock can be found. For example, equal amounts of uranium and lead in a rock would show that one-half of the original amount of uranium had changed into lead. Such a rock would be about 4.5 billion years old. One half-life would have passed since the rock was formed. A rock that old has not yet been found. The oldest rock known today is about 4 billion years old. It was found in western Greenland.

2. Potassium. Radioactive potassium is another substance found in many rocks. It too is used to find the age of rocks. Potassium decays into argon. The half-life of radioactive potassium is 1.3 billion years. Radioactive potassium has a shorter half-life than uranium. Thus it can be used to measure the age of younger rocks.

3. Carbon 14. A scientist can find the age of once-living materials such as a fossil shell or a piece of wood. All living things absorb from the atmosphere a radioactive form of carbon called carbon 14. Carbon 14 has a half-life of about 5,800 years. As long as the plant or animal is still alive, the amount of carbon 14 in its cells remains the same. This is because the decaying carbon 14 is constantly being replaced. When the plant or animal dies, the decaying carbon 14 is no longer replaced. The amount of carbon 14 in its cells starts to decrease. The fossil's age is found by measuring the amount of carbon 14 in the fossil. That amount is compared to the amount found in a modern living thing. For example, suppose the amount of carbon 14 in an ancient piece of wood is measured. It is found to be one-half the amount of carbon 14 found in wood from a tree still living. This would mean that one half-life of carbon 14 had passed since the old wood stopped taking in the radioactive carbon. See Fig. 7–25 (a) and (b). Thus the wood would be said to have an age equal to one half-life of carbon 14, or about 5,800 years. Can you tell the approximate ages of the wood shown in Fig. 7–25 (c) and (d)?

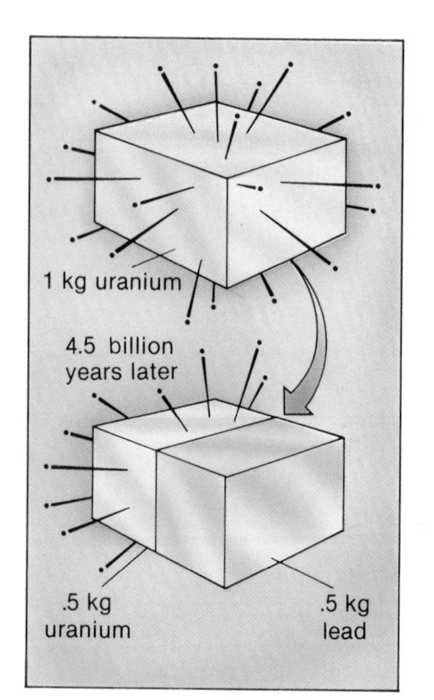

Fig. 7–24 After 4.5 billion years, half of all the uranium atoms will have changed into lead.

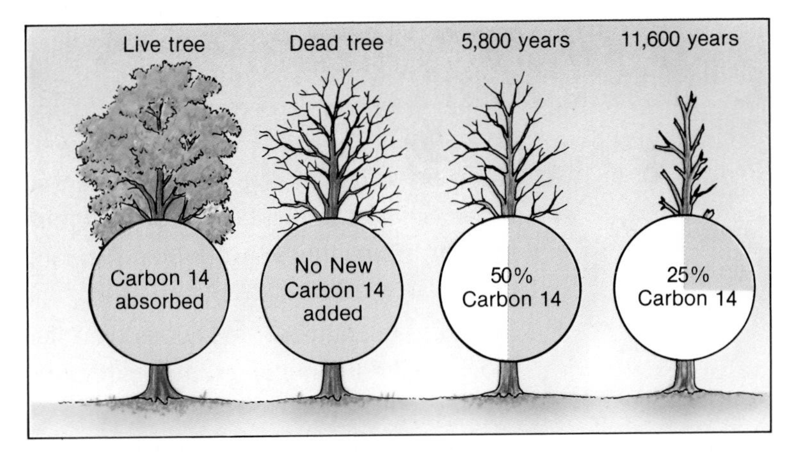

Fig. 7-25 The amount of radioactive carbon in a piece of wood can be used to find its age.

Carbon 14 is useful for finding the age of young fossils that are between 1,000 and 75,000 years old. Carbon 14 dating has provided a way of dating fairly recent events in the earth's history. For example, the age of fossil wood in Europe and North America has been measured. The results show that these areas were covered by ice as recently as 11,000 years ago.

SUMMARY

The relative ages of rocks can be found by using key beds. Index fossils can be useful in finding the period in the earth's history when the rock was formed. Measurement of the amounts of radioactive materials in rocks can be used to find the number of years since the rocks were formed. Radioactivity can also be used to measure the ages of the remains of once-living things.

QUESTIONS

Use complete sentences to write your answers.

1. Describe two ways that key beds may be formed.
2. Explain why index fossils are used to help find the age of a rock layer.
3. Why can radioactive decay be used to measure the age of rocks?
4. Explain how uranium can be used to tell the age of very old rocks.
5. Name a radioactive substance that can be used to tell the age of young rocks. Give its half-life and tell what substance it changes into.

SKILL-BUILDING ACTIVITY

RECORDING DATA

PURPOSE: To gather and record data using a model of radioactive decay and then to determine half-life based on the data obtained.

MATERIALS:

50 sugar cubes	pencil
shoe box	paper
felt-tip pen	graph paper
clock or watch	

PROCEDURE:

A. You will carry out instructions which will give data similar to radioactive decay. If possible, data should always be recorded in a table as the data is collected. Copy the table below. You may work in teams to perform steps B through F.

B. With a felt-tip pen, mark one side of each cube with an "X."

Shake Number	Number with Marked Side	Number Remaining
1		
2		
3		
Etc.		

C. Carefully place the 50 marked cubes into a shoebox. One member of the team should act as timekeeper. You will have two minutes to do steps D, E, and F. Work quickly and carefully. Read the directions before you begin.

D. Gently shake the shoebox so that the cubes roll around inside.

E. Open the box and remove all cubes that show their marked sides up. Count them. Do not return them to the box. Replace the lid on the box.

F. Record the number of cubes you removed. Find the number of cubes remaining and record.

G. Repeat steps D, E, and F nine more times. Add lines to your data table to record the data.

H. The half-life is the time it takes for one-half of a radioactive substance to change into a new substance. In this model, the cubes in the box represent the radioactive substance. The ones you removed represent the new substance.

1. How many shakes were needed before half of the cubes changed to a new substance?
2. If you followed the two-minute time schedule, how much time did it take for half the cubes to change?
3. What is the half-life for the sugar cubes in this case?

CONCLUSIONS:

1. Suppose the half-life of a radioactive substance is 5,000 years. After 10,000 years, how many atoms out of each 100 present at the start will have changed to a new substance?
2. Plot your data on graph paper. The horizontal axis will represent the shake number. The vertical axis will represent the number of cubes that showed their marked sides up. Describe the line plotted.

7–4. An Earth Calendar

At the end of this section you will be able to:

- Name four main *eras* of the earth's history.
- Describe the conditions that existed during each era of the earth's history.
- Compare the length of time occupied by each main division of the earth's past.

You can make a kind of calendar for your life by using important events. This calendar would mark off the time you live into separate divisions. One part of the calendar might be from the time you were born until you started to walk and talk. Another period might begin at the time you started going to school. Each change in the way you live could divide your life into parts. The record left in rocks shows that the earth also has a history that can be divided into a kind of calendar. One difference between the two calendars would be that, unlike your life history up to now, most of the earth's history is still unknown.

A TIME SCALE FOR THE EARTH'S HISTORY

Scientists believe that the oldest rocks found to date on the earth's surface have an age of about 3.8 billion years. This conclusion has been reached by measuring radioactive materials in many rocks.

The earth's history, as recorded in the rocks, begins that long ago. However, the earth itself must be older. The radioactive clocks in rocks begin ticking when the rock hardens from magma. Most scientists believe that the earth came into existence about 4.7 billion years ago. There must have been a period of about a billion years when the newly formed earth was developing the crust we find today. The calendar of the earth's history began when the solid crust formed.

What kind of picture of the earth's history comes from the study of rocks and their fossils? The record is not complete. Scientists do not agree on all the details. However, one fact seems clear. There have been great disturbances in the crust along with sudden changes in the forms of life on the earth. It is the evidence of these turning points that provides the reasons for dividing the earth's history into parts.

Era A time period in the earth's history. Each era is separated by major changes in the crust and changes in living things.

1. Eras. Scientists divide the history of the earth into long time periods called **eras.** The *era* in which we are now living began about 65 million years ago. It is called the *Cenozoic* (sen-uh-**zoe**-ik) *Era* (recent life). The era before the Cenozoic was the *Mesozoic* (mess-uh-**zoe**-ik) *Era* (middle life). It lasted about 160 million years. Before the Mesozoic was the *Paleozoic* (pay-lee-uh-**zoe**-ik) *Era.* The length of the Paleozoic Era was about 345 million years. As you see, the divisions that make the earth calendar are not of equal length.

The time covered by these three eras may seem long. However, they cover only 13 percent of the earth's history. The time before the Paleozoic makes up the remaining 87 percent of the earth's total history. The record of all the events before the Paleozoic is very difficult to read. Almost all fossils are found in Paleozoic and younger rocks. Since very few fossils are found before the Paleozoic Era, scientists cannot divide the time before the Paleozoic into separate eras. Thus it is simply lumped together into one very long time space called *Precambrian* (pree-**kam**-bree-un) time. See Fig. 7–26.

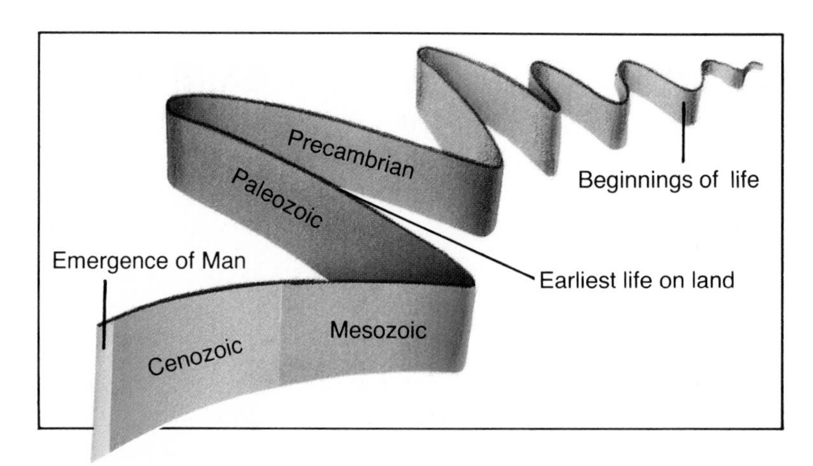

Fig. 7-26 In this chart, the length of Precambrian time is compared to the geologic eras.

Just as the yearly calendar we use is divided into months, weeks, and days, the earth calendar is divided into smaller and smaller parts. A very long division is an era. Study of the fossils found in the rocks of each era shows that the life forms changed during each era. For example, during the Mesozoic Era, dinosaurs appeared. They developed into many different forms. Some of the dinosaurs became large meat eaters. Others developed into huge plant eaters that lived in shallow lakes. Some even were able to fly. See Fig. 7–27.

Fig. 7-27 Many different kinds of dinosaurs lived during the Mesozoic era.

2. Periods. Study of fossil remains of life forms such as dinosaurs allows scientists to divide the eras into smaller divisions called *periods*. The Paleozoic Era has been divided into seven periods. The Mesozoic Era has been divided into three periods. The Cenozoic Era has two periods.

3. Epochs. Careful study of the fossils within each period also shows changes. Thus periods are further divided into smaller divisions called *epochs*. However, only the seven epochs that make up the two periods in the Cenozoic Era have been named. Epochs that make up periods in other eras have not been named. Table 7-1 lists the eras, periods, and epochs of the earth's history.

Era	Period	Epoch	Length in Years	How Long Ago It Ended (years)
Precambrian				570 million
Total Length of Precambrian Time			4 billion	570 million
Paleozoic	Cambrian Ordovician Silurian Devonian Mississippian Pennsylvanian Permian		70 million 70 million 35 million 50 million 25 million 40 million 55 million	500 million 430 million 395 million 345 million 320 million 280 million 225 million
Total Length of Paleozoic Era			345 million	225 million
Mesozoic	Triassic Jurassic Cretaceous		35 million 54 million 71 million	190 million 136 million 65 million
Total Length of Mesozoic Era			160 million	65 million
Cenozoic	Tertiary	Paleocene Eocene Oligocene Miocene Pliocene	12 million 16 million 11 million 14 million 10 million	15 million 37 million 26 million 12 million 2 million
	Total Length of Tertiary Period		63 million	2 million
	Quaternary	Pleistocene Present Time	2 million ______	______ ______
Total Length of Cenozoic Era			65 million	______

Table 7–1

PRECAMBRIAN TIME

According to the most widely accepted scientific theory, the solid earth was formed about 4.7 billion years ago. In time, an atmosphere and then oceans were produced. Most scientists agree that life probably began in the ancient seas. It is likely that at least one billion years passed before the first signs of life appeared in the oceans. At first there were probably only tiny bits of living matter. From these came the great variety of plants and animals that have lived on the earth. The actual record of life during Precambrian time is scarce. Rocks formed during this long part of the earth's history contain almost no fossils. This is probably because most very ancient forms of life did not have the hard parts that commonly form fossils. There is some evidence of life in Precambrian seas, such as impressions of soft-bodied animals. But these impressions are very rare. Microscopic examination of some Precambrian rocks shows traces of a simple kind of life. As Precambrian time came to an end, the stage was set for the appearance of many living forms.

THE PALEOZOIC ERA

The beginning of the Paleozoic Era, about 600 million years ago, marked the start of a time filled with important events. The earth's surface was very different from what it is today. The moving crustal plates caused continents to wander all over the face of the earth. At the beginning of the Paleozoic Era, there seems to have been one large and five smaller continents. They were widely separated from each other. All were located near the earth's equator. There was no land near the North or South poles. Later in the Paleozic Era, these continents moved together. By the end of the Paleozoic Era, all the separate continents had joined to form a single supercontinent called *Pangaea* (pan-**gee**-uh). An enormous single ocean covered the remainder of the earth's surface. See Fig. 7–28.

There were many changes in living things during this era. During the early Paleozoic Era, the level of the sea was high. As a result, shallow seas covered much of the land surfaces. These warm, shallow seas became filled with **invertebrates.** These are animals that do not have backbones. However, many had outside shells or plates. Some of the *invertebrates* living in the

Invertebrates Animals without backbones.

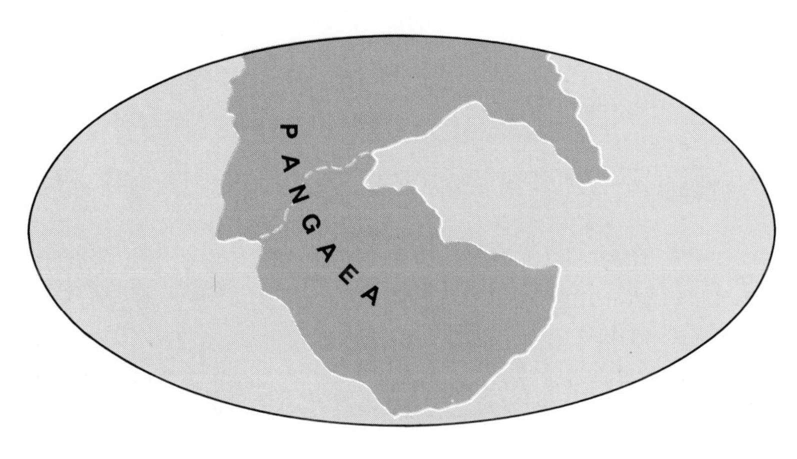

Fig. 7–28 During the Paleozoic era, separate continents came together to form the supercontinent called Pangaea.

early Paleozoic seas were trilobites, worms, sponges, snails, and starfish. As time went on, the first fish appeared in the Paleozoic seas. These ancient fish were the first of the **vertebrates.** *Vertebrates* are animals with backbones. Insects also first appeared during this time.

Vertebrates Animals that have backbones.

Forests of giant ferns and similar plants grew in the swampy areas that covered much of the land. See Fig. 7–29. Most of the world's supply of coal comes from these Paleozoic forests. The growth of land plants also provided shelter and food for the animals that were coming out of the sea, such as insects, scorpions, and land snails. Late Paleozoic rocks also have fossils of land vertebrates. *Amphibians* such as frogs and their relatives appeared along with the first reptiles.

Toward the end of the Paleozoic Era, the continents began to join and form larger landmasses. The collisions formed mountains. Thus the total area of the land grew smaller, just as a rug covers less floor space if it is crumpled up. The total size of the oceans grew as the continents took up less area. Since the same amount of water covered a larger area, the level of the sea fell. The shallow seas that covered much of the continents drained back into the oceans. The continental climates became drier. Living things on land had to change to fit the new conditions. These changes marked the end of the Paleozoic Era.

Fig. 7–29 The life in a Paleozoic swamp may have looked like this.

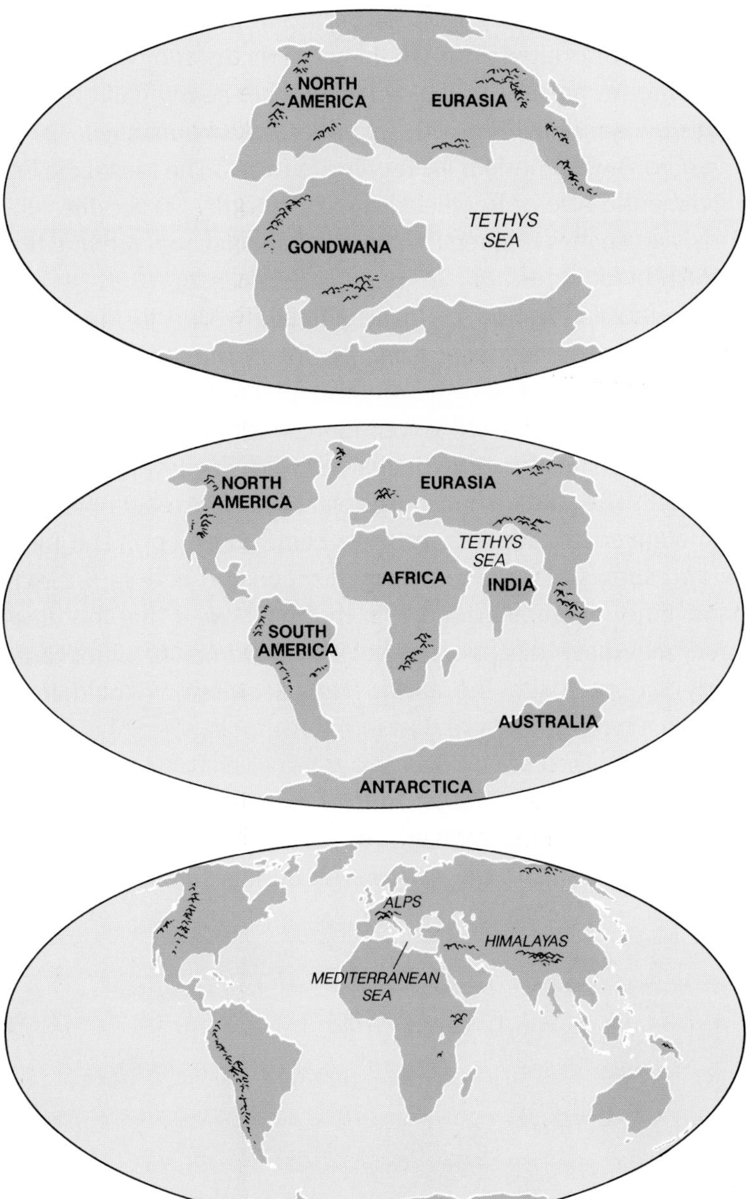

Fig. 7-30 During the Mesozoic era the present continents were formed. (a) First, the supercontinent Pangaea separated into North America and Eurasia and Gondwana. (b) Then, Gondwana divided into South America and Africa, India, Australia, and Antarctica (c) India moved north to collide with Eurasia.

THE MESOZOIC ERA

The Mesozoic Era began with the breakup of Pangaea. It first separated into three parts. See Fig. 7-30 (a). Later, during the Mesozoic Era, they separated into the continents as we know them today. See Fig. 7-30 (b) and (c). The changes in the earth's crust probably caused changes in the life forms. A rise in sea level may have been caused by sea floor spreading. When large amounts of magma pour out from the mid-ocean ridge, the sea floor is lifted. This would cause the level of water in the ocean to be higher. As the sea level rose, nearly half the land areas became covered with shallow seas and marshes. These conditions were perfect for reptile life forms. The Mesozoic Era became the Age of Reptiles. Dinosaurs, turtles, crocodiles, lizards, and snakes roamed the thick forests that covered much of the land.

The most common of these animals were the dinosaurs. These giant reptiles were some of the largest animals ever to live. See Fig. 7-31. But the dinosaurs, along with many other Mesozoic reptiles, were doomed. The moving crustal plates that opened the Mesozoic Era had slowed. The level of the seas dropped. The climate became cooler. The warm, wet condition that allowed the reptiles to thrive came to an end. The fossil record shows that the dinosaurs disappeared in a surprisingly short time. There is evidence in the rock layers that the dinosaurs may have disappeared because a comet or other large body collided with the earth. Such a collision would have thrown a huge amount of dust into the atmosphere. This would have caused the climate over the entire earth to become much cooler. Dinosaurs and other reptiles may have been wiped out because they could not change enough to meet the new conditions. The Mesozoic Era ended with the disappearance of the dinosaurs and many of their relatives.

Fig. 7-31 The outstanding form of life during the Mesozoic era was the dinosaur, of which there were many varieties.

Fig. 7-32 Some mammals that existed in the late Cenozoic era.

THE CENOZOIC ERA

The Cenozoic Era began about 65 million years ago. During this time the earth's surface has developed into the form we see today. The shallow seas that covered much of the land during the past are gone. However, the crustal plates continue to move. Their grinding collisions raise mountains and cause volcanic activity. The climates of the modern world range between the chill of the poles and the heat of the tropics. It is an era of great variety and change.

Mammals are one form of life that has been successful in living with the variety of climates in the Cenozoic Era. If the Mesozoic Era can be called the Age of Reptiles, then the Cenozoic Era can be called the Age of Mammals. Most of the ancestors of modern mammals appeared during the late Mesozoic. See Fig. 7-32. During the Cenozoic Era, many more kinds of mammals developed. They include hunting animals such as cats and wolves. There are also plant eaters such as horses and cattle. Animals able to run fast, such as deer and rabbits, have all developed to fit the conditions of the Cenozoic Era. Some mammals, such as whales and porpoises, have even gone back to the sea.

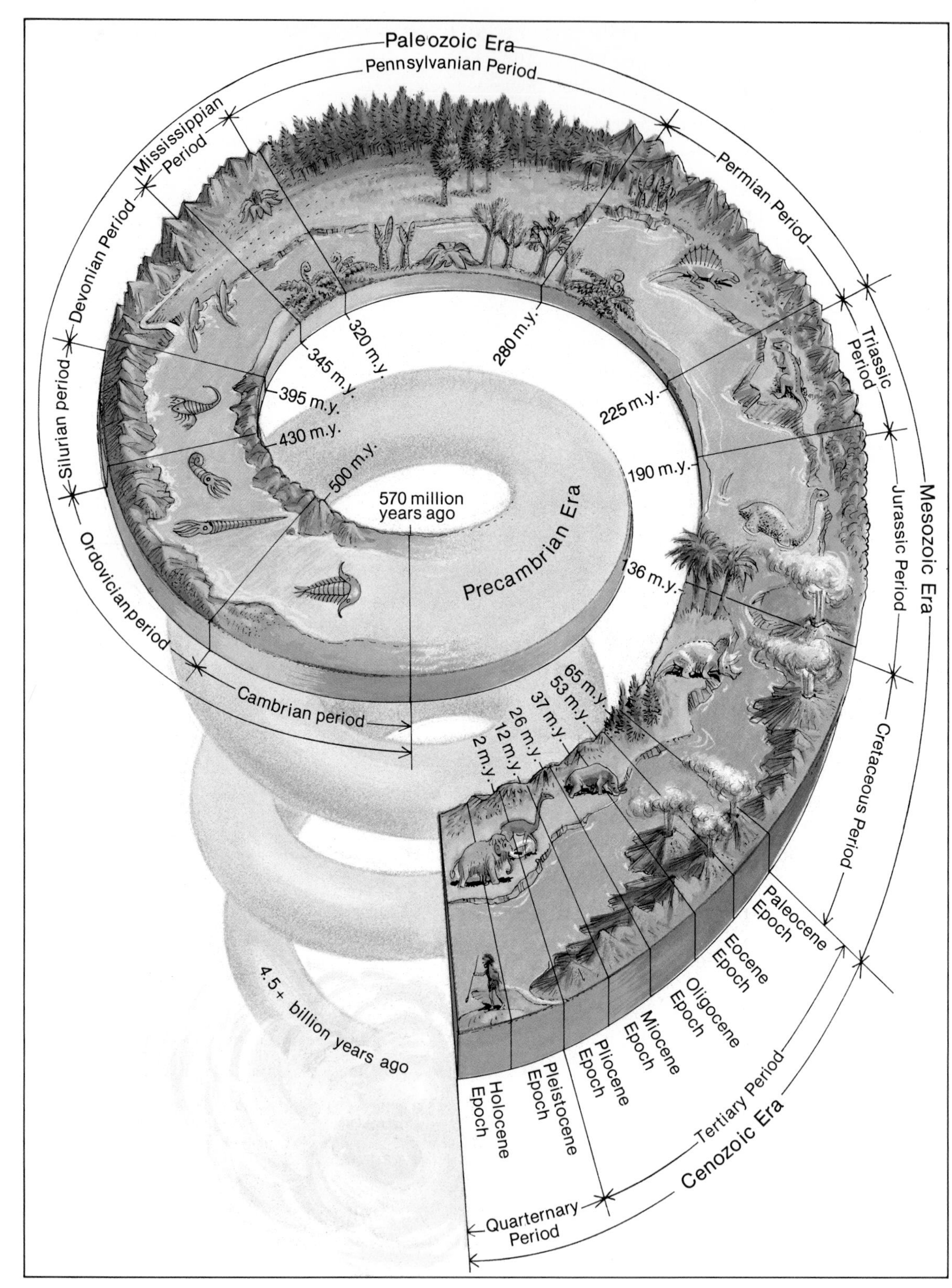
Paleozoic Era
Pennsylvanian Period
Mississippian Period
Devonian Period
Silurian period
Ordovicianperiod
Cambrian period
Permian Period
Triassic Period
Jurassic Period
Mesozoic Era
Cretaceous Period
Paleocene Epoch
Eocene Epoch
Oligocene Epoch
Miocene Epoch
Pliocene Epoch
Pleistocene Epoch
Holocene Epoch
Tertiary Period
Cenozoic Era
Quarternary Period
Precambrian Era
570 million years ago
4.5+ billion years ago
320 m.y.
345 m.y.
395 m.y.
430 m.y.
500 m.y.
280 m.y.
225 m.y.
190 m.y.
136 m.y.
65 m.y.
53 m.y.
37 m.y.
26 m.y.
12 m.y.
2 m.y.

There is reason to believe that the changes marking the Cenozoic Era will continue into the future. The modern world is only one stage in the earth's history. See Fig. 7-33. There is evidence that the Atlantic Ocean will become larger. It has been growing wider during the past 180 million years. Within the next 100 million years, a huge new mountain range will probably appear. It will grow along the Atlantic coast of North America. On the other side of North America, much of southern California and Baja California may become a small continental island. Australia will probably move north. There may be a collision between Australia and Japan or the eastern part of Asia. The resulting combination of the Indian, Antarctic, and Pacific oceans would result in the world's largest ocean.

Fig. 7-33 (opposite page) *This diagram shows the earth's history divided into the eras, periods, and epochs that most scientists use to describe the earth's past.*

SUMMARY

The earth's history can be divided into parts to make an earth calendar. Each major part of the earth calendar is separated from the next by some worldwide change. The record left in the rocks shows that there have been four main divisions, or eras, in the earth's past. Each era shows changes within its time space that allows it to be divided into smaller parts. Each era of the earth's history has been marked by certain conditions that favored the development of different life forms.

QUESTIONS

1. Name the era with which each of the following is identified: (a) Age of Reptiles, (b) Age of Mammals, (c) 87 percent of earth's history, (d) coal and oil formation, (e) vertebrates, and (f) invertebrates.
2. Make a chart like the following. List the four main eras of the earth's history in the first column and then complete the other columns.

Era	Conditions Existing	Organisms Found	Time Era Lasted
1.			
2.			
3.			
4.			

INVESTIGATION

GEOLOGIC TIME SCALE

PURPOSE: To prepare a model of the geologic time scale and relate to it some of the events that occurred during earth's history.

MATERIALS:

adding machine tape
meter stick
paper
pencil

PROCEDURE:

A. Copy the following table on a separate piece of paper.

Era	Length (years)	Length on Tape (cm)
Precambrian	4 billion	400
Paleozoic	345 million	
Mesozoic		
Cenozoic		

B. The left column in the table above shows the main divisions of the geologic time scale. Use the table on page 196 to fill in your table with the remaining number of years for all four eras.

C. Measure and cut off a strip of adding machine tape five meters long.

D. One meter's length of tape will represent one billion years. One billion is the equivalent of 1,000 million.

1. How many years would one millimeter represent if one meter represents one billion years?
2. How many years would one centimeter represent on this scale?

E. Change the number of years of each era to the number of centimeters that will represent it. Use the scale of one meter equals one billion years. Record these values in your table. Note that the first one is done for you.

3. Which time period is the longest?
4. Which time period is the shortest?

F. Draw a line across the tape very near one end. Label it "Begin Precambrian." Measure four meters from this line and draw another line. Label it "End Precambrian." Starting at this line, measure the distance on the tape that will represent the Paleozoic Era and draw a line. Label the beginning and end of this era. In the same manner, draw in the lines to represent the remaining two eras. Next, refer again to the table on page 196 to answer the following question.

5. In what era did the following exist or begin to exist? (a) invertebrates, (b) dinosaurs, (c) mammals.

G. For each of the other eras, mark on your tape information about the types of life that existed.

6. Which of the life forms in question 5 lived on the earth for (a) the longest period of time? (b) the shortest period of time?

CONCLUSION:

1. Compare the length of Precambrian time to the lengths of the different eras.
2. Give one way in which the geologic time scale is divided.

CAREERS IN SCIENCE

ARCHAEOLOGICAL ASSISTANT

Much of what we know about the past is due to archaeologists and their assistants. At a "dig," as an investigation site is called, they uncover the remains of earlier civilizations. Archaeological assistants dig through the rocks and soil. There they might find hunting tools, pottery, or cooking utensils used by the people who lived there. They might even discover part of a skeleton. Assistants also take part in marking and classifying these items for study. These artifacts, as they are called, will provide a picture of how people lived centuries ago.

Each year, digs are under way throughout the world. Students in high school or college may apply to participate in them. Courses in social studies, history, and geography provide a particularly useful background. It is also important to be able to do careful and detailed work. For more information, write: Archaeological Institute of America, 53 Park Place, New York, NY 10007, or contact your local museum or the anthropology department of a university near you.

GEOCHRONOLOGIST

How old is a newly discovered animal fossil? When was a particular mountain formed? Such questions can be answered by geochronologists. These scientists try to determine when events occurred in the earth's past by studying earth's formations and fossil remains. Geochronologists rely on a variety of dating methods. One of these involves the radioactive element carbon 14. Carbon is present in all living things, but the amount and type of carbon depends on how long it has been there. Because early humans used obsidian in many of their implements, this substance is useful for dating too. With this evidence, geochronologists try to piece together the earth's history and build a picture of what it might have looked like.

To work as a geochronologist, you will need at least a college degree, with courses in geology, chemistry, and paleontology (the study of fossils) including experience in field work. For more information, write: American Geological Institute, 5205 Leesburg Pike, Falls Church, VA 22041.

¡COMPUTE!

SCIENCE INPUT

Scientists estimate the dates of events in the earth's history in a number of ways. For example, they may study the types of rocks and sediment found in different layers of the earth, or they may look at evidence of plant and animal fossils in different layers. These events are dated relative to one another. In other words, a very specific time usually cannot be given. Instead, one event is said to come before or after another. In this way, the geological or biological events are arranged in an orderly fashion. Another way to state this is to say that they are placed in a sequence. Two sequences are listed below: the first is a sequence of geological eras; the second is a sequence for the evolution of animals on earth.

SEQUENCES

A	B
Geologic Eras	**Animal Evolution**
Precambrian	Bacteria
Paleozoic	Tribolites
Mesozoic	Dinosaurs
Cenozoic	Mammals

COMPUTER INPUT

The term "sequencing" refers to the logical arrangement, or order, of instructions placed in the computer. The computer follows each instruction in the numbered order, from lowest to highest. The programmer, through his or her understanding of (1) the purposes of the program and (2) the most efficient ways of using the computer's capacities, designs the order of instructions.

WHAT TO DO

Using data from this book, you will program the computer to print out several sequences. You will also learn how to add data into an existing program in order that it may be read in proper sequence. Written below is Program Order, which is designed to list the changes in continent formation. Enter the program and run it. Then, enter it again, using different numbers for each statement. For example, try the following two sets of numbers: (A) 204, 205, 206, 207, and (B) 100, 150, 160, 170.

Did either of these changes alter the order in which the sequence was printed out?

Suppose you want to add another event, the breakup of the supercontinent, to the process. In the program the statement would be written:

PRINT "BREAKUP OF SUPERCONTINENT"

But where would it go in the sequence? Which numbering system, example (A) or (B), would allow room to add a number? Give the statement a number and add it to the program. Run it. How did it change the order of events printed on the screen, or monitor?

Now, you can write your own sequencing program. On a separate piece of paper, rearrange the list of scrambled geologic time periods in proper sequence. Then, using Program Order as a model, write a program that will print the events on your screen correctly. By using multiples of ten for line numbers, leave space in the sequence for remark (REM) statements. Divide the periods by using the era in your rem statements.

To write your program, rearrange the following according to the information in Table 7-1, page 196.

Cretaceous Period
Precambrian Period

SEQUENCING: PROGRAM ORDER

Ordovician Period
Quaternary Period
Jurassic Period
Permian Period
Cambrian Period
Tertiary Period
Triassic Period
Pennsylvanian Period
Devonian Period
Mississippian Period
Silurian Period

Enter and run your new program. Each time you enter a program, remember first to save it on tape or disk.

GLOSSARY

RUN	The command in BASIC that starts the program you have entered.
PROGRAM	A list of step-by-step instructions that a computer reads and follows; also called software.
STORE	To save information in the computer's memory for future use.

PROGRAM

```
100  REM CHANGING CONTINENTS IN
     ORDER
110  PRINT "ONE SUPERCONTINENT"
120  PRINT "NORTHWARD DRIFTING OF
     SUPERCONTINENT"
130  PRINT "PRESENT CONTINENT
     SHAPES"
140  END
```

PROGRAM NOTES

The statements in programs are usually numbered in a way that will allow for additions without disturbing the sequence. However, instructions that should follow one after the other may be given consecutive numbers so that the necessary steps are not interrupted. If you would like to look at a number of programs to study their sequencing and numbering, check your library or local newsstand for one of the many computer magazines that are written for young adults. They contain a wide variety of programs.

BITS OF INFORMATION

Are you thinking of buying a computer? First, read about computers and do some window-shopping. You want a computer that will fit the family budget as well as perform the kinds of functions you require. You might also want one that is compatible with your school computer so you can practice at home what you learn in school. For a 50-page illustrated guide to home computers, write to: "How to Buy a Home Computer," Electronics Industries Association, P.O. Box 19100, Washington, DC 20036. Enclose a stamped, self-addressed envelope that is 6" × 9" or larger, with 54 cents postage.

CHAPTER REVIEW

VOCABULARY

On a separate piece of paper, match the number of each blank with the term that best completes each statement. Use each term only once.

eras	geologic column	relative age	uniformitarianism
fossils	invertebrates	key beds	vertebrates
half-life	petrified	superposition	
index fossils	radioactive decay	unconformity	

To understand the processes of the past, scientists use the principle of __1__, which states that they are the same as the processes acting in the present. By studying the positions of rock layers, and applying the law of __2__, the __3__ of the rock layer can be found. An eroded surface that sometimes can be seen between two rock layers is called a(n) __4__.

Remains of plants and animals, called __5__, can be found in rocks. One example is when the remains are replaced by minerals in __6__ wood.

Plant or animal remains that can be used to tell the age of rocks are called __7__. They can also be used to compare rock layers all over the world, particularly rock layers called __8__ that are easily recognized over a large area. The different rock layers can be combined to form the __9__. The most accurate method used today to measure the age of rocks is __10__. One example is uranium, which has a(n) __11__ of 4.5 billion years.

The earth's history can be divided into four main __12__. During the early Paleozoic Era, there were many animals without backbones. They are called __13__. As time went on, animals with backbones, called __14__, appeared.

QUESTIONS

Give brief but complete answers to each of the following questions. Unless otherwise indicated, use complete sentences to write your answers.

1. Why are older rock layers sometimes found on top of younger rock layers?
2. Give an example of a formation that demonstrates uniformitarianism.
3. Compare the fossils found in recent rock layers with the fossils found in old rock layers.
4. In what way can a fault or flow of magma in rock layers be used to tell the age of the rock layers?

5. What conditions are necessary for the preservation of a fossil?
6. Describe five ways in which ancient animals and plants can become fossils.
7. Why do only a few plants and animals become fossils?
8. Give three examples of how trace fossils are formed.
9. What can each of the following tell about the age of rock layers? (a) A key bed, (b) an index fossil, (c) uranium, (d) radioactive potassium, and (e) carbon 14.
10. What is the geologic column?
11. Why can't carbon 14 be used to tell the age of all rocks?
12. What evidence is used to divide the earth's history into eras?
13. What special events mark the beginning of each of the four eras?
14. Why did reptiles flourish during the Mesozoic Era?
15. Compare the general climate of the Mesozoic and Cenozoic eras.

APPLYING SCIENCE

1. Describe how a leaf falling from a tree could become a fossil.
2. Write a list of all the living things in your home. From this list, make another list of those living things that probably could become fossils. Answer the following: (a) Is your fossil list smaller than the list of living things in your home? Why? (b) How is your fossil list similar to the fossil list of living things one million years ago?

BIBLIOGRAPHY

Burgess, Robert F. *Man: 12,000 Years Under the Sea: A Story of Underwater Archaeology.* New York: Dodd, Mead, 1980.

Harrington, John W. *Dance of the Continents: Adventures with Rocks and Time.* Boston: J.P. Tarcher (Houghton Mifflin), 1983.

Mannetti, William. *Dinosaurs in Your Backyard.* New York: Atheneum, 1982.

Mossman, David, and William A. S. Sarjeant. "The Footprints of Extinct Animals." *Scientific American,* January 1983.

Stuart, Gene S. *Secrets From the Past.* Washington, D. C.: National Geographic Society, 1979.

Weitzman, David. *Traces of the Past: A Field Guide to Industrial Archaeology.* New York: Scribner's, 1980.

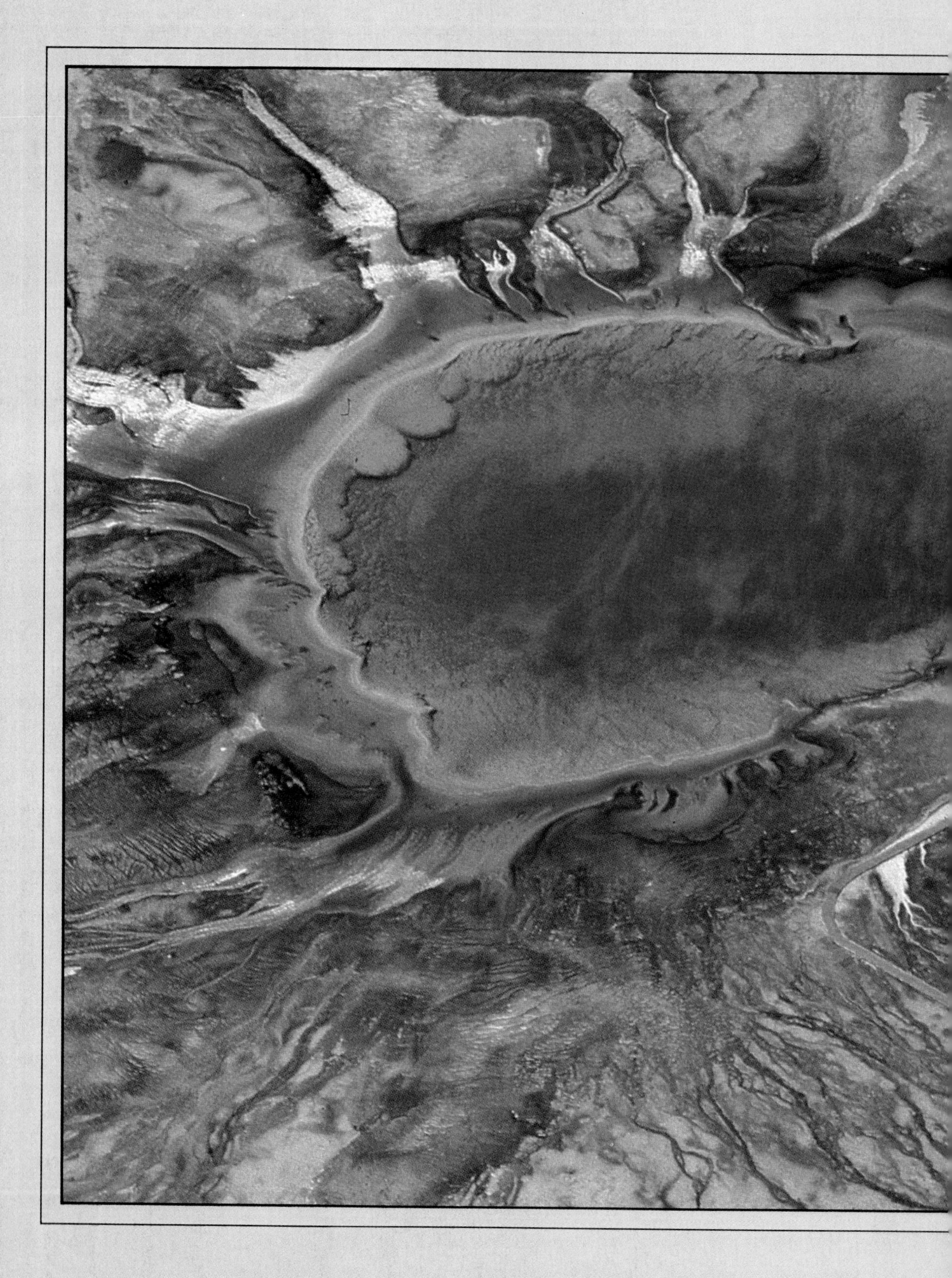

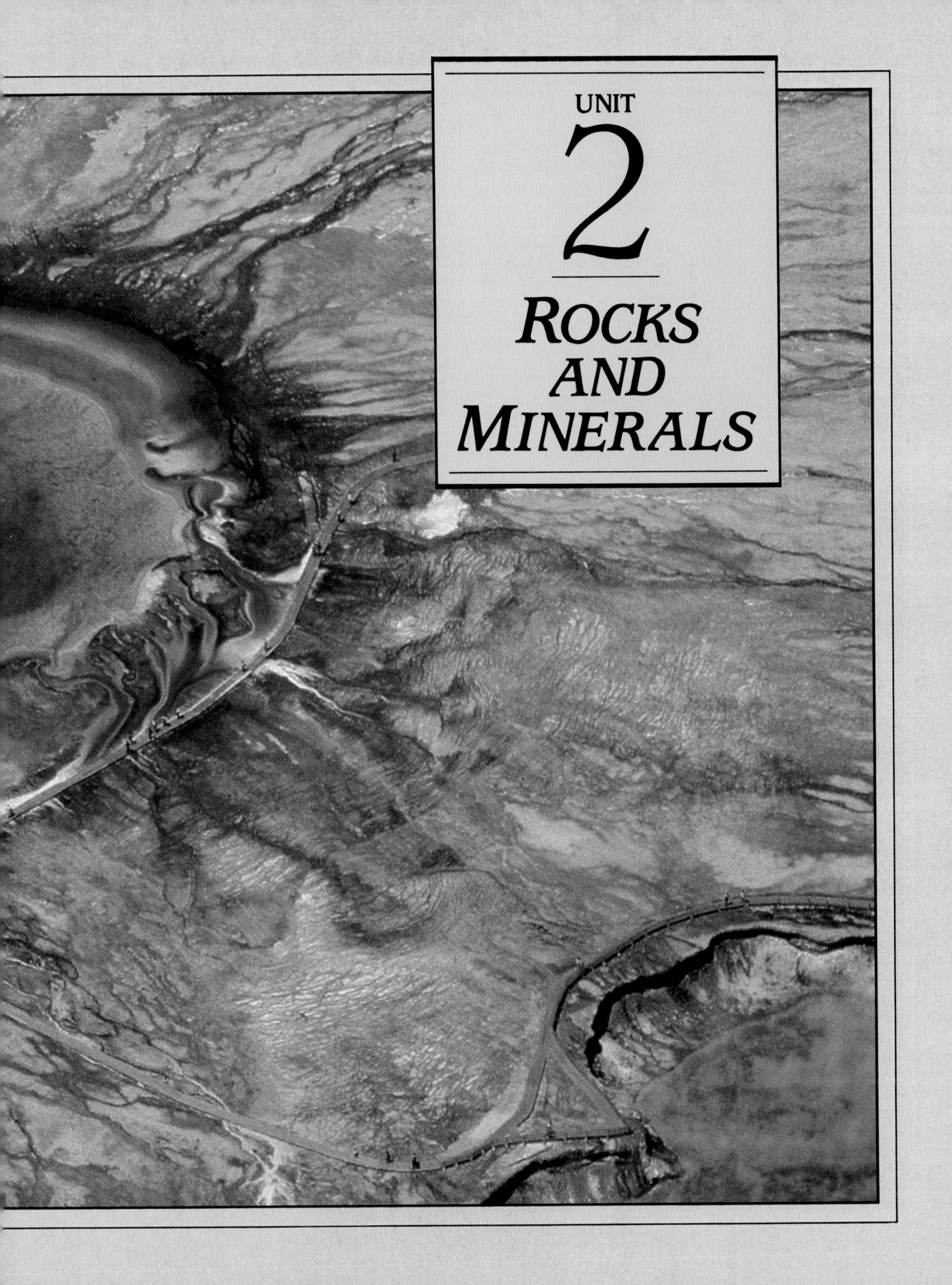

UNIT

2

ROCKS AND MINERALS

CHAPTER

8

There are materials in the earth's crust with properties that are the same all around the world. These materials are called minerals. Sulfur is one of about 2,000 minerals found on the earth.

MINERALS

CHAPTER GOALS

1. Describe an atom using the words *nucleus*, proton, neutron, and electron.
2. Explain how atoms may differ in structure and how these differences account for their ability to form compounds.
3. Describe three main divisions of the earth.
4. Explain the difference between a rock and a mineral.
5. Relate atoms, *elements, compounds*, and *minerals*.
6. List and describe several properties useful in identifying minerals.

8–1. Atoms, Elements, and Compounds

At the end of this section you will be able to:

- ☐ Describe how *atomic particles* make up an atom.
- ☐ Explain how one kind of atom is different from another.
- ☐ Show how atoms are able to join together to make a chemical *compound*.

The pieces of a jigsaw puzzle can fit together in only one way to make a complete picture. Atoms also can join together only in certain ways to make chemical compounds. The mineral quartz, for example, is a compound formed when silicon and oxygen atoms fit together in a particular way, like pieces in a puzzle. The joining together of atoms produces the many different kinds of materials that form the crust. In this section you will see why atoms behave as they do.

INSIDE THE ATOM

A single atom is so small that it is invisible even under the most powerful microscope. Yet atoms are made up of even smaller particles called **atomic particles.** Although *atomic particles* cannot be seen, scientists have been able to learn some important facts about them. For example, the particles with the greatest mass are *protons* (**proe**-tahnz) and *neutrons* (**noo**-trahns). Protons and neutrons are about the same size.

Atomic particles Small particles that make up atoms. Protons, neutrons, and electrons are three kinds of atomic particles.

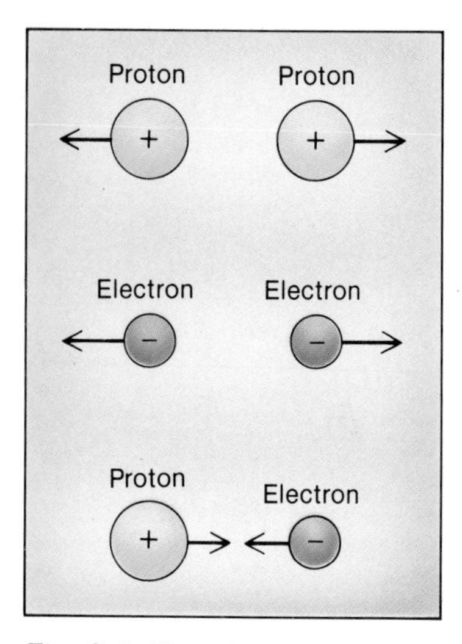

Fig. 8-1 Two protons or two electrons carry like charges that cause them to repel each other. A proton and an electron carry opposite charges and attract each other.

Nucleus The central part of an atom that contains the protons and neutrons.

One of the smallest particles is the *electron* (eh-**lek**-trahn). An electron is about two-thousandths (0.002) of the mass of a proton. These particles—protons, neutrons, and electrons are the parts that make up atoms.

Protons and electrons have electrical charges. Protons have a positive electrical charge. Electrons have a negative electrical charge. They behave the same way as magnets. If you have ever used magnets, you probably have seen what happens when the ends of two magnets are brought together. They may be pushed apart or pulled together. The forces that are active between protons and electrons are shown in Fig. 8–1. Atomic particles that carry like electrical charges push each other apart. On the other hand, different charged particles, such as a proton and an electron, pull together, or attract. The attraction between protons and electrons helps to explain how atoms are arranged. Protons and neutrons are found only in the center of the atom. Neutrons have no electrical charge; they are neutral. The protons and neutrons in the center of an atom make up the **nucleus** (**noo**-klee-us). Because the *nucleus* always contains protons, it has a positive electrical charge. Electrons are found around the nucleus. They move at a very high speed. In fact, they move so fast that they can make millions of complete trips around a nucleus in one second. The attraction between the positive nucleus and the negative electrons keeps the electrons around the nucleus.

CLIMBING THE ATOMIC LADDER

Table 8–1 lists 20 kinds of atoms, beginning with hydrogen. Hydrogen is the smallest atom. Look carefully at the number of protons and electrons as you go down this list. The table shows that there is always the same number of protons and electrons in an atom. Now look at the number of neutrons. As you can see, the number of neutrons can be equal to, greater than, or less than the number of protons or electrons.

The study of all known atoms, from smallest to largest, is like a trip up a ladder. At each step a proton and an electron are added. Look again at Table 8–1. Neutrons also may be added, but not always one at a time. At the top of the ladder is the largest atom, with more than 100 protons, more than 100 electrons, and more than 100 neutrons.

Name	Protons	Neutrons	Electrons	Symbol	Electron Shells 1 2 3 4
Hydrogen	1	0	1	H	1
Helium	2	2	2	He	2
Lithium	3	4	3	Li	2 1
Beryllium	4	5	4	Be	2 2
Boron	5	6	5	B	2 3
Carbon	6	6	6	C	2 4
Nitrogen	7	7	7	N	2 5
Oxygen	8	8	8	O	2 6
Fluorine	9	10	9	F	2 7
Neon	10	10	10	Ne	2 8
Sodium	11	12	11	Na	2 8 1
Magnesium	12	12	12	Mg	2 8 2
Aluminum	13	14	13	Al	2 8 3
Silicon	14	14	14	Si	2 8 4
Phosphorus	15	16	15	P	2 8 5
Sulfur	16	16	16	S	2 8 6
Chlorine	17	18	17	Cl	2 8 7
Argon	18	22	18	Ar	2 8 8
Potassium	19	20	19	K	2 8 8 1
Calcium	20	20	20	Ca	2 8 8 2

Table 8–1

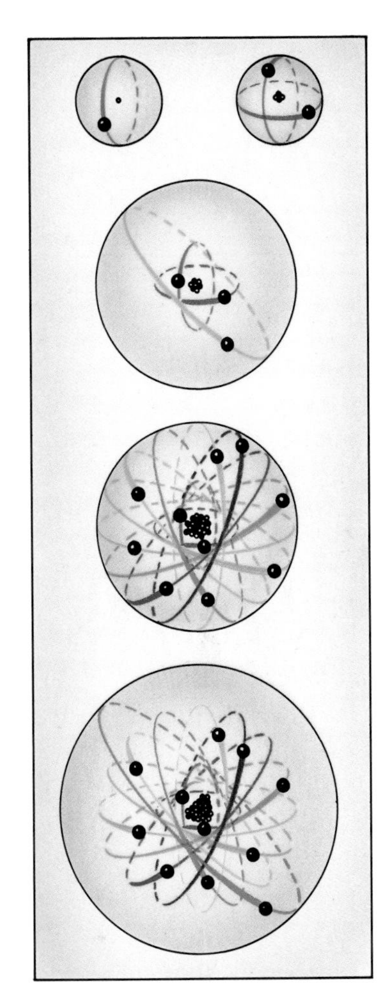

Fig. 8–2 Electrons move around an atomic nucleus within shells. Each kind of atom shown in the diagram has a different number of electrons and shells. Using Table 8–1, identify each kind of atom.

Also listed in the table are letter symbols for each kind of atom. These symbols are a short way of writing the names of the atoms. For example, the symbol for a hydrogen atom is H. The symbol for a helium atom is He, and for lithium, Li. Sometimes the symbol comes from Latin. For example, the symbol for iron is Fe. Fe comes from the Latin name for iron, *ferrum*.

Electrons in atoms move in a path, or an **electron shell.** Each *electron shell* is an area that is a definite distance from the nucleus. See Fig. 8–2. Electrons move within these shells. Each shell is a certain size and can hold a definite number of electrons. Electron shells for some atoms are shown in Table 8–1.

Electron shell An area where a certain number of electrons move at a definite distance from the nucleus.

ELEMENTS

Each different kind of atom is said to belong to a different chemical **element.** An *element* is a substance that cannot be broken down further into any other substance. An atom is the smallest complete part of an element. An element contains

Element A substance made up of only one kind of atom.

only one kind of atom. For example, the element silicon is made up of only silicon atoms. The element oxygen is made up of only oxygen atoms. Elements are the building blocks of all kinds of matter. There are more than a hundred different kinds of chemical elements known to exist. However, only eight elements make up most of the earth's crust. Oxygen makes up almost half of the weight of the earth's crust. The next most common element is silicon. Oxygen, silicon, and the six other elements shown in Table 8-2 make up about 99 percent of the weight of the earth's crust.

Element	Percent by Weight in Crust
Oxygen	46.60
Silicon	27.72
Aluminum	8.13
Iron	5.00
Calcium	3.63
Sodium	2.83
Potassium	2.59
Magnesium	2.09
Other elements	1.41

Table 8-2

HOW ATOMS COMBINE

Compound A substance made up of two or more kinds of atoms joined together.

1. Gaining and losing electrons. Very few atoms are found alone in nature. Usually two or more different atoms combine to form a chemical **compound.** A *compound* is a substance made up of two or more kinds of atoms combined together. Atoms combine because of their electron shells. Each shell can hold a definite number of electrons. For example, the first shell can hold only two electrons. The second shell can hold eight electrons. If an atom has an outer electron shell that is only partly filled, it is able to become part of a compound.

Look at chlorine in Table 8-1 on page 215. As you can see, a chlorine atom has seven electrons in its outer shell. This third shell can hold a total of eight electrons. Thus chlorine needs one electron to fill its outer shell. Now look at sodium in Table 8-1. It has only one electron in its outer shell. When one sodium atom transfers its one outer electron to a chlorine atom, sodium chloride, or table salt, is formed. This gives both atoms complete outer electron shells. However, the transfer of one electron changes the sodium and chlorine atoms. Each no longer has an equal number of negatively charged electrons and positively charged protons. The loss of one electron in sodium causes it to have one more proton than electron. This extra proton causes the sodium atom to be positively charged. The chlorine atom that gains the electron now has one more negative electron. The chlorine atom becomes negatively charged. An atom that takes on a positive or negative electrical charge caused by the gain or loss of an electron is called an **ion.** A sodium atom becomes a positive *ion* by the loss of an electron. A chlorine atom becomes a negative ion by gaining an electron. Since opposite electrical charges attract, the sodium ion and chloride ion join together to form sodium chloride. In the same way, many compounds are made up of ions held together by electrical charges.

Ion An atom that has become electrically charged as a result of the loss or gain of electrons.

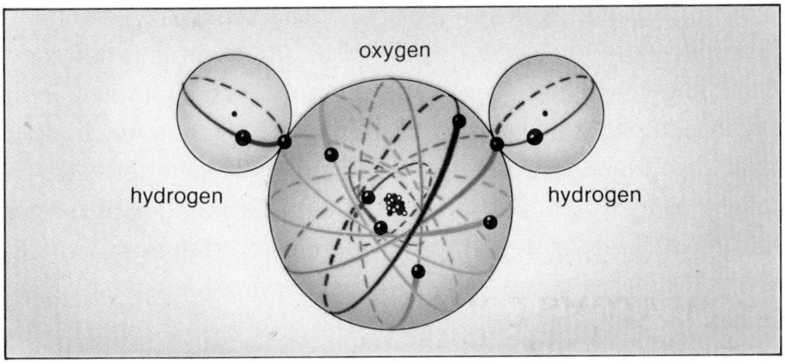

Fig. 8-3 Two hydrogen atoms and one oxygen atom can share their electrons and join together.

2. Sharing electrons. Not all atoms combine by gaining or losing electrons. Another way that atoms may combine is by sharing electrons. Look at the number of electrons in the hydrogen and oxygen atoms in Table 8-1. Oxygen has only six electrons in its outer shell. It can accept two more electrons to complete its outer shell. Two hydrogen atoms can give those two electrons to an oxygen atom. Each hydrogen atom can share its one electron with the oxygen atom. The oxygen atom then has a full electron shell of eight. In turn, the oxygen atom can share its outer electrons with the hydrogen atoms. One electron is shared with each hydrogen atom. Each hydrogen atom then has a full electron shell. See Fig. 8-3. When two hydrogen atoms share electrons combined with one oxygen atom, a **molecule** (**mol**-ih-kyool) of water forms. A *molecule* is the smallest part of a substance that still has the properties of the substance. The water molecule is represented by the chemical formula H_20. A formula tells how many of each kind of

Molecule The smallest part of a substance with the properties of that substance.

atom are in each molecule. The 2 in the formula means that there are two atoms of hydrogen for every atom of oxygen.

All the molecules of any one substance are alike. They all have the same properties. A molecule of a compound such as water, for example, always has the same kind and the same number of each type of atom. A water molecule always contains two hydrogen atoms combined with one oxygen atom. The properties of water come from the properties of these molecules. Every chemical compound has its own set of properties because it always contains the same number and kind of atoms. Molecules of different substances are not alike. For example, a molecule of water is different from a molecule of sugar. They differ in the number and kinds of atoms.

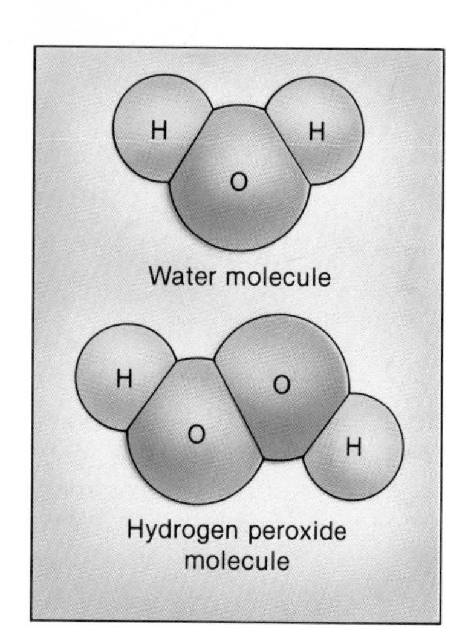

Fig. 8-4 A water molecule compared to a molecule of hydrogen peroxide.

Atoms also may combine in different ways. For example, two compounds can be formed from hydrogen and oxygen atoms. The two compounds are water, H_20, and hydrogen peroxide, H_2O_2. In water, there is one oxygen atom in each molecule. In hydrogen peroxide, there are two oxygen atoms in each molecule. See Fig. 8-4. Although water and hydrogen peroxide are made of hydrogen and oxygen atoms, the number of oxygen atoms is different. Thus their properties are different. You can drink water, whereas pure hydrogen peroxide would cause burns in your mouth and throat.

SUMMARY

Everything in the earth is made up of atoms. Atoms, in turn, are made up of smaller atomic particles. Each atom is made up of a nucleus with electrons moving around it. Each kind of atom contains a different number of protons, electrons, and neutrons. If their electron shells are incomplete, atoms can join together to form chemical compounds. Atoms can combine by gaining, losing, or sharing electrons.

QUESTIONS

Use complete sentences to write your answers.

1. List the particles found in the atom and give the location as well as the charge of each.
2. Describe the difference in the number of electrons, protons, and neutrons between one atom and the next larger atom.
3. Give an example of each of the two ways atoms may combine to make compounds.

INVESTIGATION

MODELS FOR ELEMENTS AND COMPOUNDS

PURPOSE: To make models of elements and compounds using paper clips as atoms.

MATERIALS:
9 paper clips (3 each of 3 different sizes)

PROCEDURE:

A. Separate the paper clips into piles according to size.
   1. If each size represents one type of atom, how many different elements would these represent?

B. In order to be seen, real atoms must combine in different ways. You will join paper clips together one way only, end to end, to make chains. Join together three paper clips of the same size.
   2. Which does this model represent, an element or a compound?

C. Use three of the remaining paper clips to form a chain that represents a compound.
   3. How does this model differ from the one made in B above?

D. Take apart your element and compound models and return the clips to their separate piles.

E. Using only two paper clips at a time, make as many different chain models that represent compounds as you can. Make a sketch of each model before you take it apart.
   4. How many different models could you make with just two clips?

F. Using only three paper clips at a time, make as many different models that represent compounds as you can. Remember, compounds are made when two or more kinds of atoms combine. You will need to take apart models to make new ones. Make a sketch of each model before you take it apart.
   5. How many different three-clip chain models did you make from the three sizes?

G. Using only four paper clips at a time, make as many different models that represent compounds as you can. Make a sketch of each model before you take it apart to make others. Use "s" for small, "m" for medium, and "l" for large to sketch your models.
   6. How many different models did you make with four clips?

CONCLUSIONS:

1. What effect did increasing the number of different atoms have on how many compounds could be formed?
2. There are a number of atoms that can combine with more than two other atoms each. What effect would this have on the number of compounds possible?
3. There are about 90 different kinds of natural elements. Most of these combine with other elements in several different ways. Explain why these elements could combine to make the millions of compounds known to exist.

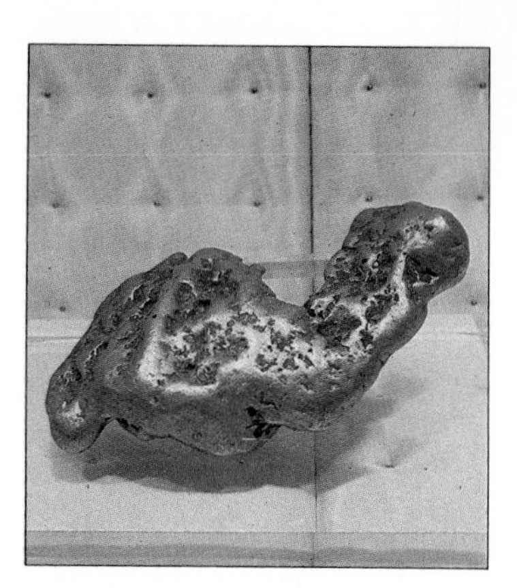

Fig. 8-5 A gold nugget found in a stream.

8-2. Earth Materials

At the end of this section you will be able to:

- ☐ Describe the *lithosphere* of the earth.
- ☐ Compare *minerals* and rocks.
- ☐ Name a mineral that is a compound and list the elements found in it.

Fig. 8-5 shows a gold nugget. Once a part of a larger rock, the nugget was separated from its parent rock by erosion. Its lucky finder picked it from a deposit made by the running water that once carried it. A lump of gold stands out from the other pieces of rock. It has its own special properties that make it easy to identify. Thus you would have no trouble picking a gold nugget from a handful of rocks.

MINERALS AND ROCKS

Atmosphere The layer of gases that make up the outside of the earth.

Hydrosphere All the bodies of water on the earth's surface.

Lithosphere The earth's outer layer averaging about 100 km in thickness.

The materials that make up the earth are often divided into three parts. The **atmosphere** (**at**-muhs-fear) is the layer of gases that surround the earth. All the bodies of water on the earth make up the **hydrosphere** (**hie**-druh-sfear). The outer layer of solid material that makes up the earth is called the **lithosphere** (**lith**-uh-sfear). Each of these three divisions is made up of materials that can be observed and described. The materials of the *atmosphere* and *hydrosphere* will be studied in other sections. The upper part of the *lithosphere*, commonly called the crust, will be studied in this section.

The solid continents and sea floors make up the earth's crust. Materials in the crust can be compared to materials in a kitchen. Some things are the same in all kitchens. Salt and ordinary sugar are good examples. Each substance has its own properties, such as taste and color. Even if you had different brands of table salt, you could not tell one from another by its taste or color. Since they are all the same substance, they have the same properties. Similarly, there are materials in the crust with properties that are the same no matter where they are found. These substances are called **minerals** (**min**-uh-rulz). A *mineral* is a solid, naturally forming substance that is found in the earth. Each mineral has its own definite set of properties.

Mineral A solid, naturally forming substance that is found in the earth and that always has the same properties.

Just as salt and sugar can be identified by their properties of taste, minerals can be identified by their properties.

On the other hand, some things found in a kitchen are not always the same substance. Soups or salads, for example, can be different combinations of foods. These combinations are not always the same. In the same way, different parts of the earth's crust are not always the same. If you pick up a piece of rock at one place, it may be made of materials that are different from a rock taken from another place. These two pieces of rock would not be the same because they are probably made up of different combinations of minerals. *Rock* is a solid part of the crust that is made up of one or more minerals. Most rocks contain more than one kind of mineral. See Fig. 8–6.

A mineral compares to a rock as a tree compares to a forest. A forest is made up of separate trees. Likewise, a rock is made up of separate mineral grains. Just as trees come in different shapes and sizes, the mineral grains in a rock are also different in size and shape. Most forests contain different kinds of trees

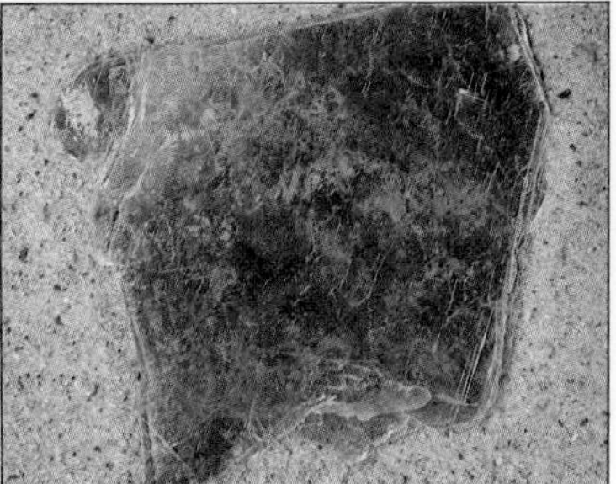

Fig. 8–6 (top) *Granite is a rock made up of various minerals.* (bottom) *Three minerals found in granite.*

just as rocks usually contain different minerals. However, in some forests there may be only one kind of tree. Some rocks also can be made up almost entirely of grains of a single type of mineral. Forests may be alike and be put into the same group even though they are never exactly alike. Rocks may also be alike and be put into the same group. But even rocks in the same group are never exactly the same.

MINERALS: A CLOSER LOOK

Suppose that you could take a piece of rock into a laboratory to find out exactly what it is made of. If you looked at a thin slice of the rock under a special microscope, it would be possible to see the mineral grains that make up the rock. See Fig. 8-7. But what would you see if you looked at a single mineral grain under a microscope with no limit to its power? The super microscope would show that the mineral grain was made up of atoms. It would take more than one million of those atoms to reach across the period at the end of this sentence.

Fig. 8-7 Individual mineral grains of gypsum magnified 30×.

Fig. 8-8 A sample of quartz crystals.

If you could see the atoms in the mineral, you would find that they are not all alike. For example, suppose that you were examining the mineral shown in Fig. 8-8. It is a common mineral called *quartz*. You would see that it contains two kinds of atoms, silicon and oxygen. Most minerals are made up of at least two kinds of atoms. That is, minerals are usually made up of two or more chemical elements. However, a few elements, such as gold, may be found in the pure form. A gold nugget is a lump of a single kind of mineral. It always has the same properties. Thus gold is easy to identify when found mixed with pieces of rock.

Each of the separate atoms in a mineral substance must be joined to one another. Otherwise the mineral grain would not exist as a solid piece of matter. In the mineral quartz, for example, the silicon and oxygen atoms must be joined together. The way in which these two kinds of atoms combine gives quartz special properties. Thus quartz can be identified as a separate mineral, just as gold can be identified by its properties. Quartz and almost all other kinds of minerals are examples of mineral compounds.

The atoms of a compound are not simply mixed like peanut butter and jelly in a sandwich. The atoms combine in such a way that a new substance, a compound, is formed. All mineral compounds, such as quartz, always contain the same kinds of atoms joined in the same way. See Fig. 8–9. Thus a mineral can always be identified by the special properties that come from the arrangement of its atoms.

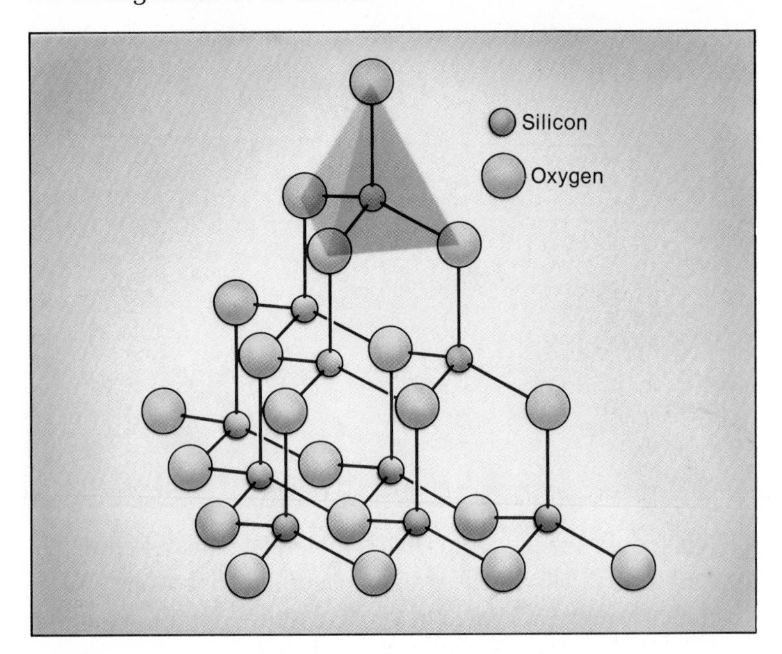

Fig. 8–9 A silicon atom and oxygen atoms are joined in this pattern. How does the number of oxygen atoms and the number of silicon atoms compare?

SUMMARY

The part of the lithosphere nearest the surface, the crust, is made of rock. If you looked closely at a piece of rock, you would probably see that it is made up of separate grains. Each of these individual grains is likely to be a mineral. Each mineral has its own set of properties. Most minerals are made up of two or more kinds of atoms. Atoms combine to produce chemical compounds. Each compound has its own properties because it is always made up of the same kinds of atoms.

QUESTIONS

Use complete sentences to write your answers.

1. Name and describe the three main divisions of the earth's materials.
2. What is the difference between a rock and a mineral?
3. List in order the three most abundant elements found in the minerals of the earth's crust.
4. What two elements combine to form the mineral quartz?

INVESTIGATION

MODELS OF MOLECULES

PURPOSE: To make models showing how atoms combine.

MATERIALS:
set of 12 marked cards

PROCEDURE:

A. The cards you will use represent atoms of hydrogen, oxygen, carbon, and nitrogen. Cards may be joined together only by placing the mark on the edge of one card next to a similar mark on another card. Select two cards marked "H," and one marked "O," and place them separately in front of you. To understand why the marks must be on the edges of the cards, answer the following questions.
   1. How many electrons does an atom of hydrogen have? (See Table 8-1, page 215.)
   2. The first electron shell of any atom is filled when it has two electrons. How many more electrons would a hydrogen atom need to fill its shell?
   3. Look at the H card. How many marks does it have on its edges? How many electrons does an atom of oxygen have in its second shell?
   4. The second shell of any atom is not filled until it has eight electrons. How many more electrons does an atom of oxygen need to fill its second shell? Look at the O card. How many marks does it have on its edges?

B. Join the three cards to make a model of a water molecule.
   5. What does the "2" in H_2O stand for?

C. Using two hydrogen (H) cards and two oxygen (O) cards, make a model of an H_2O_2 molecule. This is a model of hydrogen peroxide.
   6. Compare your model to Fig. 8-4, page 218. How are they alike?

D. Use the N card, which stands for nitrogen, and attach as many hydrogen cards to it as will fit.
   7. How many H cards fit on one N card?

E. Use the C card, which stands for carbon, and attach as many hydrogen cards to it as will fit.
   8. How many H's fit on one C? Writing the C first, then the H, write the symbols that would represent this molecule similar to the way that H_2O represents the water molecule.

F. Remove one H card from the C card. Attach in its place another C card.
   9. How many H cards are needed to satisfy the marks on this C card?

CONCLUSIONS:

1. Look at Table 8-1 and explain why there are four places marked on the edges of the C card for joining to other atoms.
2. Write the symbols for the nitrogen–hydrogen molecule you made in D.
3. Name and give the symbols for two kinds of molecules that contain only hydrogen and oxygen atoms.

Fig. 8-10 The British crown jewels contain many valuable minerals.

8-3. Mineral Properties

At the end of this section you will be able to:

- ☐ Describe how to identify a mineral such as salt.
- ☐ List and describe four properties of minerals that can be observed with the eye or hand alone.
- ☐ List and describe four properties of minerals that can be observed by using simple tools.

The large red jewel in the British crown shown in Fig. 8-10 is called the Black Prince's ruby. For centuries it has been part of the treasure that makes up the British crown jewels. Now it is known that this is not a ruby. It is a mineral called red spinel. Spinel is a form of the mineral corundum. It is much less valuable than a ruby. All minerals, such as spinel and ruby, have individual properties that can be used to identify them.

ORDERLY ATOMS

Like rubies and spinels, common table salt is also found as a mineral. When salt is found in its natural state in the earth, it is called *halite* (**hal**-ite). Halite is the mineral name for salt. Like most minerals, halite is a chemical compound. It is made up of the two elements sodium and chlorine. How can halite be identified? Tasting is not always a good method. Many mineral compounds are poisonous. You should never taste any substance that is not known to be a food. Also, many compounds other than halite have a salty taste. However, halite does have one characteristic that makes it easy to identify.

Crystal A mineral form with a definite shape that results from the arrangement of its atoms or similar units.

You can see this characteristic if you look at grains of ordinary salt with a magnifying glass. Many of the salt grains will be seen as tiny cubes. See Fig. 8-11. Each cube is a salt **crystal** (**krist**-'l). *Crystals* are formed because of the orderly way in which atoms or ions are arranged. For example, the compound sodium chloride, or salt, is made up of sodium and chloride ions. These ions can fit together only in the pattern shown in Fig. 8-12 (a). Thus salt and its mineral form, halite, are seen as crystals with the shape of a cube. See Fig. 8-12 (b).

Each mineral is made up of crystals with a definite shape. Fig. 8-8 on page 223 shows the crystal shape of the common mineral quartz. Usually the crystal shape of a mineral is not easily

Fig. 8-11 Many of the grains of salt from a salt shaker appear as cubes.

seen. Large and perfectly formed mineral crystals are not commonly found. This is because most of the time the crystals grow in a limited space. They are crowded together in the mineral. This results in mineral grains made up of very small crystals. The shapes of the individual crystals are not visible to the eye. However, the arrangement of the atoms or similar units in very small crystals can be found by using X-rays. For example, X-ray examination of the "ruby" in the British crown revealed the crystal pattern of spinel.

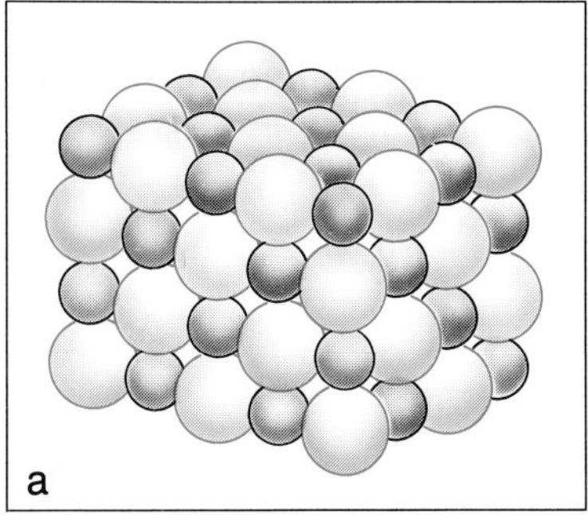

Fig. 8-12 (a) The shape of a crystal of halite is the result of the orderly arrangement of the sodium and chloride ions. (b) Pieces of the mineral halite.

OBSERVING THE PROPERTIES OF MINERALS

Since the crystal shape of a mineral usually is not easy to observe, other properties must be used. Most minerals have other properties that can be observed with the eyes and hands alone or with simple tools. Some of those properties are listed below.

1. Color. Minerals come in many colors. Quartz, for example, may be colorless. It can look clear. But it also can be milky white, brown, or pink. See Fig. 8–13 *(left)*. Ruby and red spinel are found in different shades of red. As you see, the same mineral, such as quartz, may have different colors. Or different minerals, such as ruby and spinel, may have the same color. For

Fig. 8–13 (left) *Quartz is found in many colors due to the dissolved mineral impurities.* (right) *The mineral pyrite has a metallic luster.*

these reasons, color by itself is not the best way to identify minerals. Color is helpful only when used with other properties.

2. Luster. The way light is reflected from the surface of a mineral is called luster. Some minerals reflect light in the same way that the shiny surface of metal does. See Fig. 8–13 *(right)*. Such minerals are said to have a metallic luster. Other minerals have a luster like glass, wax, or pearl. Some minerals have dull surfaces that hardly reflect light at all.

3. Feel. Rubbing your finger over the surface of a mineral can help you to identify it. Some minerals feel powdery. Some feel rough like sandpaper. Pencil lead is made from the mineral called graphite. This mineral is said to have a greasy texture.

4. Density. How heavy a piece of a mineral feels is determined by its density. You can tell the difference between salt and iron from their densities. A cupful of salt is lighter than a cupful of iron. In the same way, you can compare the weights of equal size pieces of different minerals. A piece of gold, for example, weighs eight times as much as the same size piece of halite. The density of gold is much greater than that of halite.

The four mineral properties described above can be identified by the use of eyes and hands alone. Four additional properties can be revealed by using simple tools. These tests are:

5. Streak. A powder mark may be left when a mineral is rubbed against a rough surface. This mark is called a *streak*. Streak color helps to identify a mineral. The color of the streak can be seen best on a white surface. The back of a bathroom tile can be used as a plate for making streak marks. The color of

Fig. 8–14 Making marks on a streak plate can help identify a mineral.

the streak is not always the same as the color of the mineral used to make it. See Fig. 8–14.

6. Fracture or cleavage. How can you quickly tell the difference between a piece of glass and a piece of clear plastic? One way would be to break each piece. Glass breaks in a completely different way than plastic. The difference is easy to see. Many kinds of minerals break in ways that help to identify them. Most minerals break unevenly. An uneven break is called a *fracture.* Fig. 8–15 shows a curved fracture. Not all minerals fracture. Some split along smooth surfaces. A smooth break is called *cleavage.* Minerals cleave in one, two, or three directions. See Fig. 8–15. Some minerals show their cleavage only slightly. The broken surface of a mineral sample must be examined carefully to discover how it breaks. Cleavage should not be confused with crystal shape. A mineral that shows cleavage will break in such a way that the broken pieces have the same general shape as the large original piece. A crystal, on the other hand, breaks into pieces that do not look like the original crystal.

Fig. 8–15 (left) *This sample of obsidian fractures when broken.* (right) *Iceland Spar, a type of calcite, exhibits cleavage.*

HARDNESS SCALE		
Hardness Number	Simple Test	Mineral
1	Fingernail scratches it easily.	Talc
2	Fingernail barely scratches it.	Gypsum
3	Copper penny just scratches it.	Calcite
4	Steel knife scratches it easily.	Fluorite
5	Steel knife barely scratches it.	Apatite
6	Scratches glass easily.	Feldspar
7	Scratches steel and glass easily.	Quartz
8	Scratches all other common minerals.	Topaz
9	Scratches topaz.	Corundum
10	Hardest mineral.	Diamond

Table 8–3

7. Hardness. Minerals differ in hardness. Hardness is a measure of a mineral's resistance to being scratched. Hardness is one of the best ways of identifying minerals. If mineral A can scratch mineral B, A is harder than B. If A can scratch B and B can scratch A, they both have the same hardness. Scientists have developed a scale to test the hardness of minerals. This scale is shown in Table 8–3. As you can see, it ranks minerals from one to ten. Ten is the hardest. The hardest mineral is diamond. Number one on the scale is the softest mineral. A mineral with this hardness, called talc, is found in talcum powder. A mineral with a high hardness number can scratch any mineral with a lower number. The hardness of an unknown mineral is tested by finding out which minerals on the scale it can scratch and which it cannot scratch. For example, gypsum is harder than talc or a fingernail. But gypsum is softer than calcite or a copper penny. Thus gypsum has a hardness between two and three on this scale.

Gem An attractive or rare mineral substance that is hard enough to be cut and polished.

Some minerals, such as diamond and quartz, are hard enough to be cut and polished into **gems.** To be considered a valuable *gem,* a mineral substance must also have an attractive appearance and be rare. See Fig. 8–16.

8. Magnetism. A few minerals are attracted to magnets. Perhaps you have put a magnet into a pile of sand. If so, you found that some small mineral particles, called magnetite, cling to the magnet. These magnetic properties are useful in helping to identify such minerals.

Fig. 8-16 Attractive, rare minerals, when cut and polished, can be very valuable.

9. Special properties. There are some properties that are shown by only a few minerals. For example, some minerals give off visible light when exposed to ultraviolet light or X-rays. This property is called *fluorescence*. Other minerals give off light when they are heated. Properties such as these are useful for identifying only a small number of the many different kinds of minerals.

SUMMARY

Atoms or other units that make up minerals are arranged in patterns that give each mineral a crystal form. Other properties of minerals can be observed with simple tests. Color, luster, feel, and general density can be determined without using anything other than eyes and hands. To test for the properties of streak, fracture or cleavage, hardness, and magnetism, simple tools must be used.

QUESTIONS

Use complete sentences to write your answers.

1. How can you tell that the mineral halite and table salt are the same substance?
2. List four mineral properties that can be tested using only the eyes and hands.
3. Describe four tests of mineral properties that require simple tools.
4. Why is the diamond called a gem?

SKILL-BUILDING ACTIVITY

SEQUENCING: MAKING AND USING A MINERAL-HARDNESS SCALE

PURPOSE: To make a sequence, or scale, of hardness for use with minerals.

MATERIALS:
chalk
penny
mineral sample
pencil

PROCEDURE:

A. Scratch the wood of a pencil with your fingernail.
   1. Did your fingernail scratch the wood or did the wood make a scratch on your fingernail?
   2. Which is harder, pencil wood or your fingernail?

B. Perform the same test to see which is harder, your fingernail or chalk.
   3. Did your fingernail scratch the chalk? (A mark that rubs off is not a scratch.)
   4. Which is harder, fingernail or chalk?

C. Look at Table 8–3, page 230.
   5. Name two minerals that are not as hard as your fingernail and give the hardness number for each.

D. Perform a test to see which is harder, your fingernail or pencil lead.
   6. Which is harder, pencil lead or your fingernail?

E. Perform a similar test to see if the wood of a pencil is harder than a piece of chalk.
   7. Which is harder, pencil wood or chalk?

F. Perform tests to find out which is harder, chalk or pencil lead. Do the same for pencil wood and pencil lead.
   8. Which is harder, chalk or pencil lead?
   9. Which is harder, pencil wood or pencil lead?

G. Perform a test to see if your fingernail is harder than a copper penny.
   10. Which is harder, fingernail or copper penny?
   11. List in order of hardness the five objects you tested, with the softest first and the hardest last. This is a sequence of objects according to their hardness.

H. Obtain a mineral sample from your teacher and test it for hardness, using only your fingernail and a copper penny.
   12. Where does the hardness of this sample fit in the sequence you made for the answer to item 11?

CONCLUSIONS:

1. Explain how a mineral-hardness test is performed, using an example.
2. Refer to Table 8–3 on page 230. What was the hardness of the mineral you tested in part H?
3. You are given two gems. Both look like diamonds but one is an imitation. Use what you have learned from this activity to tell how you could identify the real diamond.

TECHNOLOGY

A NEW GOLD RUSH: CLASH WITH THE ENVIRONMENT?

Not every new technological advance is welcomed with open arms. Some are considered a mixed blessing and raise as much anxiety as interest. For example, take the new "gold rush" planned to begin in 1985 in the Mother Lode area in Jamestown, California. (*Mother Lode* refers to the main vein of a deposit.) The last operating mines, the survivors of the gold rush of 1849, were closed in 1942. Gold mining, however, has continued in the United States in one form or other since those early gold rush days. Prospectors still pan for gold dust in streams or dig in the surface of rocks for "placer" gold, which consists of nuggets and grains of gold. While most individuals do not strike it rich, many still find gold mining profitable and exciting.

But the newest gold rush is not on a small scale. The Jamestown project is large-scale. It will operate 24 hours a day, using new technology to separate the gold from other minerals and the old technology of strip mining to pull the ore, a mixture of gold and quartz, from the hills. Instead of shafts being dug, mountains will be taken apart and millions of tons of rock will be moved. Conveyor belts miles long will carry ore directly from open pits to mills. There, the ore containing the gold will be crushed. The ore will then be washed in a mixture of water, pine oil, and other chemicals. This mixture is made to foam so that the gold and iron pyrites (fool's gold) will stick to the bubbles and be skimmed off the top. The iron pyrites is then separated from the gold by mixing it with a compound called thiourea. Thiourea is found in the urine of cattle and other animals. The iron pyrites–thiourea byproduct can be used as fertilizer, in glassmaking, or to produce sulfuric acid. This method is said to be environmentally safe. Since thiourea can be considered a fertilizer, a spill would not harm the environment.

Previous methods of recovering gold were borrowed from the techniques used in copper mining. In that method, the finely ground ore was mixed with cyanide in large open pits. The gold concentrate that resulted was then melted and formed into ingots. Environmentalists were concerned that the cyanide technique could create a potential for disaster because the cyanide might leak into and poison the ground water. The thiourea method was chosen instead.

Another environmental concern with the new project is the noise that will be generated by the strip-mining activities operating 24 hours a day, seven days a week. To protect the environment, the mining companies intend to pay the county to monitor both the noise and ground water pollution. The mining companies will also replant some areas they've mined and flood the mine pits to create reservoirs.

Newer techniques may be required if this 25-year project develops. The problems of strip mining in this region of California may grow. The population in this region is expected to increase at least five times during this period. Strip mines and housing developments do not usually become good neighbors. It will be interesting to follow this project in years to come to see how industrial and environmental interests can be managed.

¡COMPUTE!

SCIENCE INPUT

Geologists and geological engineers use their knowledge of the physical properties of a mineral to distinguish it from other minerals. Knowing a mineral's characteristics is essential to using that mineral. For example, if you are drilling through a mountain, you need to know how hard the rock will be and how it is likely to react when drilled or blown away with explosives. If you are building with rock, you have to be aware of the degree of softness of your material, or else the structure may fall. Knowing the properties of minerals is necessary for both their identification and use.

COMPUTER INPUT

It is possible to use the computer as an encyclopedia. Data (information) can be saved in the computer's memory and then recalled (brought back at a later date). One advantage of the computer is that you can manipulate the information in a computer more easily than the information in a book. You can choose to recall only part of the information, or you can rearrange it. These capacities exist because of the way in which computers store information. Computers do not actually store words and numbers as we normally use them. The computer works on the basis of a *binary number system. Binary* means "having 2 parts." A computer cannot actually read. It can only tell whether or not magnetic impulses are present. (Electrical flows create a magnetic field.) When the flow is on, a magnetic field is present; when the flow is off, the magnetic field is absent. These different states are represented by 1 (present) or 0 (absent). In other words, they are represented by a binary number system. This kind of number system allows the computer to rearrange the data more quickly than if it were using words and conventional numbers.

WHAT TO DO

On a separate piece of paper, make a chart of minerals and their properties. The presence or absence of a mineral property will be coded into a system of 0's and 1's. Use this book and other earth science reference books to research the properties of the different minerals.

In Program Search, the computer will help you to identify an unknown mineral by its properties. To provide the data for statements 301 to 400, code the characteristics of as many minerals as you wish, according to the chart code. Each mineral code should have 7 digits. For example, calcite would be 0000010. It has no streak color (0) and therefore belongs to streak color group 000. The mineral itself is a light color (0), has cleavage (1), and is not harder than 5 (0). The mineral codes will be your database. A database is a fund of information stored in the computer's memory. When you run the program, it will ask you to describe, according to the binary number system, the characteristics of the mineral to be identified. After you have input the information, the computer will review its memory and, if information about that mineral has been entered into the database, the computer will print out its name.

CHART CODE

1. Does the mineral have a streak color?
   1 = Yes; 0 = No
2. To which color group does the streak belong?
   000 = no streak color
   100 = black/grey
   010 = red/brown
   001 = some other color
3. Is the mineral itself a dark color (black, green, red, blue, or brown)? 1 = Yes; 0 = No
4. Does the mineral have cleavage?
   1 = Yes; 0 = No

5. On the hardness scale, is the mineral harder than 5?
   1 = Yes; 0 = No

PROGRAM

```
100  REM SEARCH
110  DIM M$(100),C$(100)
115  REM REPLACE THIS STATEMENT
     WITH CLEAR SCREEN STATEMENT
120  FOR X = 1 TO 100
130  READ M$(X),C$(X)
140  IF M$(X) = "NO" THEN GOTO 160
150  NEXT
160  M = X - 1
170  PRINT : INPUT "STREAK COLOR
     (1=YES/0=NO)? ";C1$
175  C2$ = "000": IF C1$ = "1" THEN
     PRINT : PRINT "100 = BLACK/
     GREY": PRINT "010 = RED/
     BROWN": PRINT "001 = OTHER":
     INPUT "WHICH COLOR? (100, 010,
     001)? ";C2$
180  PRINT : INPUT "DARK COLOR =
     BLACK, RED, GREEN, BROWN, BLUE
     (1=YES/0=NO)?";C3$
190  PRINT: INPUT "CLEAVAGE (1=YES/
     0=NO)?";C4$
200  PRINT: INPUT "HARDER THAN 5
     (1=YES/0=NO)?";C5$
210  M$ = C1$ + C2$ + C3$ + C4$ +
     C5$
230  FOR X = 1 TO M
240  IF M$ = C$(X) THEN PRINT M$(X)
250  NEXT
260  GOTO 170
300  REM PLACE DATA STATEMENTS
     BELOW
301  DATA CALCITE, 0000010
400  END
```

GLOSSARY

BIT — Short for *binary digit.* A bit is a piece of computer-coded information. The coding process is done by the computer hardware, not by the program. The hardware is designed to translate the data input into binary numbers, which can be stored electronically in the computer's memory.

BYTE — A collection of eight bits. The power of a computer's memory is often expressed as 8K or 16K. Each K represents 1,024 bytes of information. The more bytes in the memory, the greater its capacity.

PROGRAM NOTES

The longer the program, the more chances there are for error. Enter the program carefully. Besides typing it into the computer without errors, you must also research the data carefully. If the information in your data chart is incorrect, then, even if your program runs, you will be learning misinformation.

BITS OF INFORMATION

There is a computer program called "Prospector," which can analyze geological field data and identify different kinds of deposits. In 1982 Prospector, through the analysis of information provided by geologists, identified a $100 thousand mineral deposit in an area that had been mined for years with little success. Prospector's program was designed to imitate the thought processes of 20 top geologists.

CHAPTER REVIEW

VOCABULARY

On a separate piece of paper, match each term with the number of the statement that best explains it. Use each term only once.

atmosphere	lithosphere	compound	electron shell
atomic particles	nucleus	element	
crystal	mineral	ion	
hydrosphere	molecule	gem	

1. The part of an atom that contains protons.
2. The earth's crust.
3. The layer of gases that surrounds the earth.
4. A substance found in the earth that always has the same properties.
5. An area a definite distance from the nucleus where a number of electrons move.
6. An electrically charged atom.
7. An attractive mineral that is hard enough to be cut and polished.
8. Electrons, protons, and neutrons.
9. A mineral form having a definite shape.
10. A substance made up of only one kind of atom.
11. A substance composed of two or more kinds of atoms joined together.
12. All the bodies of water on the earth.
13. Two or more atoms joined together.

QUESTIONS

Give brief but complete answers to each of the following questions. Unless otherwise indicated, use complete sentences to write your answers.

1. Explain the relationship between (a) atoms and elements, and (b) compounds and minerals.
2. Explain the sentence, "A mineral compares to a rock as a tree compares to a forest."
3. Name one mineral and tell why it is a mineral and not a rock.
4. Name three atomic particles and tell where they are found in the atom.
5. Why does an electron attract a proton and repel another electron?
6. How does an oxygen atom differ from a carbon atom?
7. An atom has 45 protons. (a) In what part of the atom are they found? (b) How

many electrons would this atom have? (c) In what part of the atom are the electrons found?

8. Why does it take two atoms of hydrogen to join with one atom of oxygen to make a water molecule?
9. What is the difference in the way the atoms join to make a water molecule and the way the atoms join to make sodium chloride?
10. Give an example of how the same two kinds of atoms may combine in different ways.
11. Why is it not a good idea to taste a mineral to help you identify it?
12. Compare the test for streak with the test for hardness of minerals.
13. Compare fracture and cleavage of minerals.
14. List several minerals that can be scratched by feldspar. Why does feldspar scratch these minerals?

APPLYING SCIENCE

1. Find several minerals. Use a mineral identification key from a high school earth science or geology text to identify the minerals. Write a report of your results.
2. Grow crystals of some common compounds and bring in the crystals to show to the class. See *Crystals and Crystal Growing*, by Alan Holden and Phyllis Morrison, for information on procedures.
3. Alloys are mixtures of metallic elements. Find out what elements are in the following alloys: (a) stainless steel, (b) bronze, (c) sterling silver, (d) brass, (e) 14-K gold, (f) alnico.

BIBLIOGRAPHY

Dietrich, R. V. *Stones, Their Collection, Identification, and Uses*. San Francisco: Freeman, 1980.

Holden, Alan, and Phyllis Morrison. *Crystals and Crystal Growing*. Cambridge, MA: MIT Press, 1981.

MacFall, Russell P. *Rock Hunter's Guide: How to Find and Identify Collectible Rocks*. New York: T. Y. Crowell, 1980.

O'Donoghue, Michael. *Minerals and Gemstones*. New York: Van Nostrand Reinhold, 1982.

CHAPTER

9

The rocks of the earth come in many different colors, shapes and sizes. But each rock tells a story of how it was formed. In this chapter, you will learn to read some of this story.

ROCKS

CHAPTER GOALS

1. Explain how geologists classify rocks.
2. Discuss two ways that igneous rocks are formed, giving an example of a rock formed in each way.
3. Describe three kinds of sediments and how they form rocks, and name a rock formed from each.
4. Explain why sedimentary rocks are an important tool in the study of earth's history.
5. Tell how heat and pressure may change existing rock layers into new rock.
6. Describe the steps in the natural recycling of rocks on earth.

9–1. Magma and Igneous Rocks

At the end of this section you will be able to:

- ☐ Name the major groups of rocks and briefly describe how each is formed.
- ☐ Define *igneous* rock.
- ☐ Classify igneous rock into two major types.

You may know people who can't resist carrying home a rock or two every time they go on an outing in the country. Perhaps you have a rock or two in your room right now. Why collect rocks? Often it is simply a "different" or "colorful" rock that ends up in one's pocket. But John Young and Charlie Duke, shown in Fig. 9–1, page 240, were sent to the moon to gather rocks for a different reason. By gathering and studying rocks from a number of locations, scientists hope to gain a better understanding of the moon's history. In this section you will learn how rocks reveal their history.

THE THREE ROCK GROUPS

A **rock** is a piece of the earth's crust. *Rocks* are usually made up of two or more minerals mixed together. They come in all sizes, shapes, and colors. Some rocks are denser than others; some are harder than others. There seems to be a large variety of rocks found in the earth's crust. Careful study shows that

Rock A piece of the earth that usually is made up of two or more minerals mixed together.

Fig. 9-1 Studying rocks can tell us how they formed. Here, Apollo 16 astronauts are collecting samples of moon rocks.

there are some important differences among rocks. These differences result from the way rocks are formed.

Igneous rock Rock formed as molten material cools.

Geologists classify rock into three basic groups: *igneous* (**ig**-nee-us) *rock, sedimentary* (sed-ih-**ment**-uh-ree) *rock,* and *metamorphic* (met-uh-**mor**-fik) *rock.* **Igneous rock** is formed when molten material cools. Most sedimentary rock is created when small pieces of rock are cemented or squeezed together. Metamorphic rock is igneous or sedimentary rock that has been changed by great heat and/or pressure. In this section, you will learn about *igneous rock.*

IGNEOUS ROCK

Rock formed from molten material is called igneous rock. The word "igneous" means "coming from fire." Molten material reaching the earth's surface is called lava. Molten material beneath the earth's surface is magma. Igneous rock may be formed from either lava or magma.

1. Extrusive rock. Rock formed by lava coming to the surface is called *extrusive* (ek-**stroo**-siv) *igneous rock.* Extrusive means "pushed out." Lava that pours out onto the earth's surface begins to cool and harden as it comes into contact with the air. See Fig. 9-2. The appearance of igneous rock depends

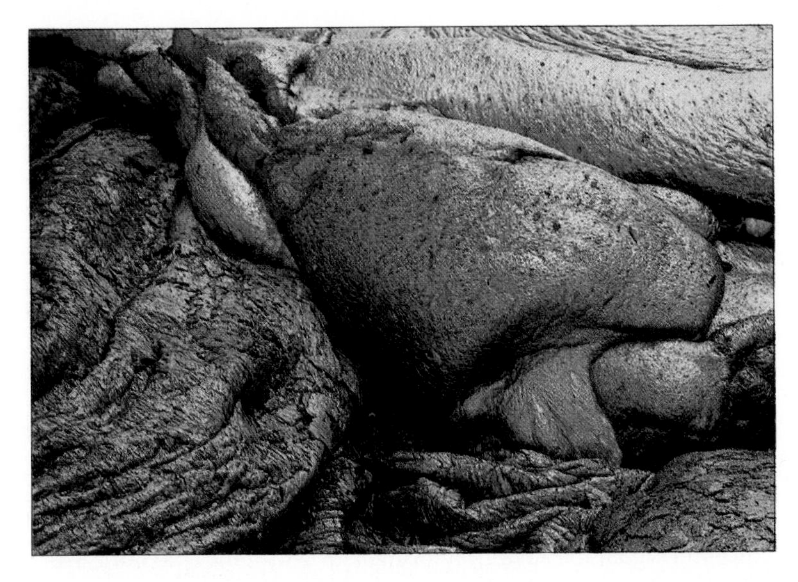

Fig. 9-2 Igneous rock forms as lava hardens.

largely on how fast the lava cools. When lava cools very quickly, there is little time for crystals to grow in the hardening rock. The rock formed has a glasslike texture and is called *volcanic glass*. The rock called *obsidian* is an example of volcanic glass. See Fig. 9-3. Obsidian was used to make arrowheads because its broken edges are as sharp as broken glass. However, lava usually cools slowly enough for small crystals to form in the rock. Thus most of the igneous rock that is formed from lava cooling at or near the earth's surface has small crystals. It is classified as fine textured. It does not look or break like glass.

Fig. 9-3 This deposit of obsidian (volcanic glass) formed when lava cooled very rapidly.

Fig. 9-4 Volcanic ash was deposited in a large area surrounding a volcano in Iceland.

Volcanic lava usually contains gases. If there is a small amount of gas, the lava pours out quietly. If there is a large amount of gas, the lava may come out with an explosion. Molten lava may be blown high into the air. Tiny pieces of hardened lava fall back to the ground as volcanic ash. See Fig. 9-4. As volcanic ash piles up on the ground, the pieces are squeezed together to form rock called *tuff. Pumice* (**puhm**-iss) is formed when gas bubbles are trapped in hardened lava. Pumice looks like a hard sponge and can float on water.

Not all lava comes to the surface of the crust as a result of volcanic eruptions. It can also come to the surface through many fractures. Then the lava spreads out in great sheets. Such an outpouring may cover the existing landscape and form a lava plateau.

Huge amounts of lava also reach the surface beneath the sea along the mid-ocean ridge. The cooling of this lava creates new sea floor. This slowly pushes apart the great crustal plates. The ocean floor is primarily made up of the igneous rock *basalt,* a dark-colored, fine-grained rock.

2. Intrusive rock. What about the huge amount of magma that never flows out onto the surface? This magma can also cool into rock. Igneous rock formed beneath the surface of the earth is called *intrusive* (in-**troo**-siv) *igneous rock.* Intrusive means "forced in." Fig. 9-5 shows intrusive rock called shiprock. Lava in the throat of a volcano first hardened. Then erosion wore away the mountain around the body of hardened lava, exposing it. Intrusive rocks cool slowly. There is time for large crystals to grow. Intrusive rocks are classified as coarse textured.

Fig. 9-5 Shiprock in New Mexico.

There are two types of igneous rock bodies formed deep in the earth. **Dikes** and **sills** are examples of one type. They form when magma is forced into cracks in rock and hardens. See Fig. 9-6. If erosion wears away the rock, *dikes* and *sills* are exposed at the surface. Dikes cut across the layers of rock. Sills run parallel to the layers.

Dikes Bodies of igneous rock formed underground in cracks that cut across layers of existing rock.

Sills Bodies of igneous rock formed underground in cracks that run parallel to layers of existing rock.

Fig. 9-6 A dike formed by intrusive igneous rock.

A second type of intrusive rock forms when very large bodies of magma harden within the crust. These are called **batholiths** (**bath**-uh-liths). *Batholiths* may stretch for hundreds of miles under the surface. Batholiths usually form the cores of mountain ranges. The upward movement of magma in the crust seems to be one of the steps in mountain building. The Sierra Nevada range is the result of a great batholith. See Fig. 9-7. A part of the batholith may be exposed as the mountains erode.

Batholiths Very large bodies of igneous rock that are formed underground.

Fig. 9-7 The Sierra Nevada batholith.

In general, intrusive igneous rock cools very slowly. Magma in batholiths probably takes many thousands of years to cool. Therefore, rock formed this way usually has large crystals. Granite, a kind of rock with large mineral crystals, is an example of the kind of rock found in batholiths.

SUMMARY

Every piece of rock has a story to tell. The origin of a rock is an important part of its history. One of the best ways to sort different kinds of rock is to classify them according to how they are formed. Igneous rock is formed directly from molten material. The cooling of the molten material can take place on the surface or beneath the ground. The size of the mineral crystals shows how fast the rock has cooled.

QUESTIONS

Use complete sentences to write your answers.

1. List the three major groups of rocks and describe how each is formed.
2. What does the word "igneous" mean?
3. Classify the following as either extrusive igneous rock or intrusive igneous rock: (a) obsidian, (b) a dike, (c) tuff, (d) pumice, (e) granite.

SKILL-BUILDING ACTIVITY

THE LABORATORY BURNER

PURPOSE: To become familiar with the operation of the laboratory burner.

MATERIALS:
burner goggles
matches

PROCEDURE:

A. There are several types of laboratory gas burners. All are alike in the way they work. The Bunsen burner is the most common kind. Look at Fig. 9–8. Locate on your burner the labeled parts of the burner.

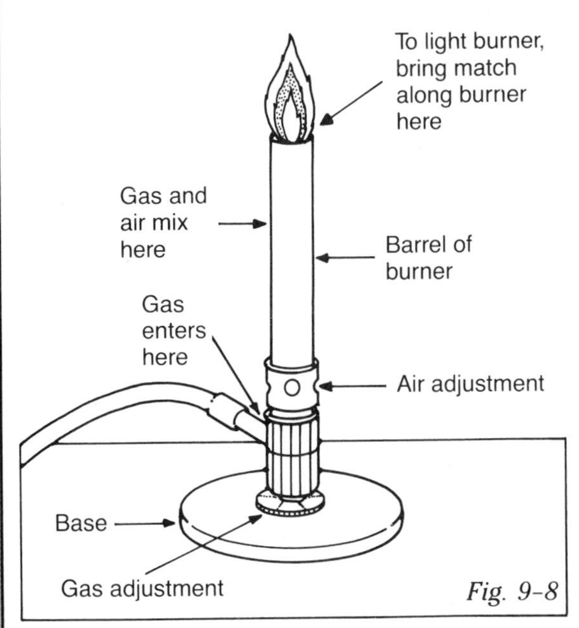

Fig. 9–8

B. Connect your burner to the gas source. CAUTION: WEAR GOGGLES ANYTIME YOU ARE LIGHTING AND USING THE LABORATORY BURNER.

C. To light a burner, bring a lighted match up alongside the top of the barrel, then turn on the gas source. If the lighted match is held directly over the barrel, the escaping gas may blow out the match. Fig. 9–9 shows what the flame should look like.

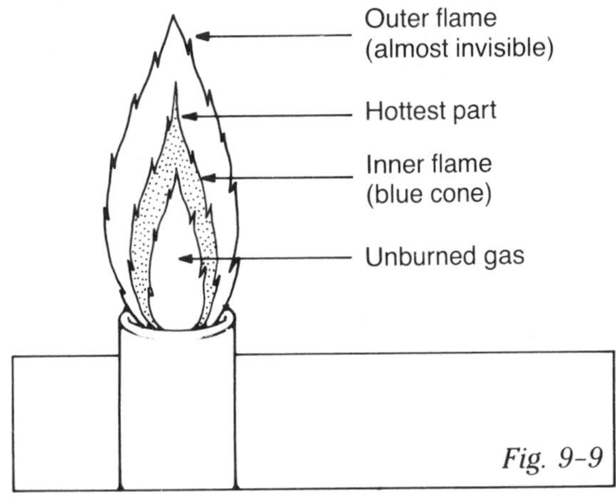

Fig. 9–9

D. If the flame is large and bright yellow, there is too much gas. Use the air adjustment to allow more air to enter until the flame is blue.

E. A gap between the flame and the top of the barrel means that too much air is being admitted. Use the air adjustment to decrease the amount of air entering at the base.

F. If the flame seems to be burning inside the barrel, turn off the gas. Cut down on the amount of air and light the burner again. CAUTION: THE BARREL MAY BE HOT.

CONCLUSIONS:

1. Make a drawing of your burner and label all parts.
2. Describe how to light the burner with a match and adjust it to have a blue flame.

INVESTIGATION

CRYSTAL FORMATION IN IGNEOUS ROCK

PURPOSE: To show how the rate of cooling affects crystal size.

MATERIALS:

test tube	test tube holder
stearic acid	index card
goggles	small plastic spoon
Bunsen burner	metric ruler

PROCEDURE:

A. Fill the test tube about one-third full of stearic acid. Because molten rock (lava) is extremely hot, you will use stearic acid to represent it. Stearic acid melts at about 80°C and forms crystals upon cooling. CAUTION: USE GOGGLES WHILE LIGHTING BURNER OR HEATING ANYTHING IN A TEST TUBE. ALWAYS POINT THE OPEN END OF THE TEST TUBE AWAY FROM YOURSELF AND OTHERS.

B. Light and adjust the burner to give a blue flame about six centimeters high.

C. Attach the test tube holder to the test tube near its open end.

D. Heat the test tube until the stearic acid is completely melted. Hold the test tube about four centimeters above the flame. Move the test tube around in the heat to allow all parts of the stearic acid to be heated. The test tube will blacken if the burner is not properly adjusted. If this happens, stop heating and wipe the test tube clean with a dry paper towel. Then adjust your burner to give a pale blue flame.

E. When the stearic acid is completely molten, pour about half of it onto an index card. Pour the remaining molten material into the small plastic spoon. Prop up the handle of the spoon so that the molten material will not spill.
   1. Which sample is cooling faster?

F. Look closely at the sample on the index card while the other sample continues to cool.
   2. Describe what you see.

G. Locate one of the longest crystals on the card. Measure its length, using a metric ruler.
   3. What is its length in millimeters?

H. After the stearic acid in the spoon has hardened, look closely at its crystals.
   4. Describe the shape of the crystals.
   5. In general, how do these crystals compare in size to the ones made on the index card?
   6. What is the length of the largest crystal you can find in the spoon sample?
   7. How does the rate at which stearic acid cools affect its crystal size?

CONCLUSION:

1. Based on your observations in this activity, how do you think the rate at which lava cools affects crystal size in igneous rock?
2. What size crystals would you expect to see in lava that has poured out across the earth's surface and hardened?
3. What size crystals would most likely occur when magma cools and hardens inside the vent of a volcano?

9–2. Sediments

At the end of this section you will be able to:

- ☐ Define *sediment.*
- ☐ Describe three ways that sediments are formed.
- ☐ Name one rock formed by each of the three kinds of sediments.
- ☐ Explain how rocks can give information about the earth's past.

The coal you see being mined in Fig. 9–10 was formed from plant material laid down long ago in ancient swamps. Dead plant material builds up in swamps because the lack of oxygen in the water prevents decay. The organic sediment has been pressed between layers of rocks until it has become a rock layer itself. The exact cause of the formation of the original sediments may never be known. However, one thing is certain. In the future, coal most likely will continue to be a major source of energy. It has been estimated that, within the continental United States, the earth's crust holds more energy in coal than the total known earth reserves of oil. We need only to learn how to use this source with minimum damage to our environment.

Fig. 9–10 Coal mining in the United States.

FORMATION OF SEDIMENT

Scientists believe there is evidence showing that about four billion years ago most of the earth's surface was covered with molten rock. The first rock of the crust was igneous rock formed as this molten material cooled. Igneous rock formed the surface of the young earth. Since its formation, the igneous rock on all the continents has been exposed to the agents of erosion.

The chief agent of erosion is running water. Water flowing over the surface can pick up small pieces of rock. Also, small amounts of many of the minerals in rock dissolve in the water. The small rock pieces and dissolved minerals carried by running water are called a *load*. The flooding stream in Fig. 9–11 is carrying a large load. Every stream, from the smallest trickle to the biggest river, carries a load. For example, near its mouth the Mississippi River moves a load of about 15,000 kilograms each second.

The amount of material carried by a stream depends on how fast the water is moving. Increasing the speed of the water in a stream increases the size of the load it can carry. When a stream flows into a larger body of water, such as a river, lake, or ocean, it slows down. When it slows down, the stream can carry less material. Part of the load falls to the bottom. The material that settles at the bottom of the water is called **sediment** (**sed**-ih-ment). *Sediment* is deposited where streams of running water slow down.

Sediment Any substance that settles at the bottom of water.

Fig. 9–11 This flooding stream is carrying a large amount of sediment.

The rock pieces in sediments are not all the same size. The smallest rock fragments make up what is commonly called *mud.* Mud is the most abundant of all sediments. It is a mixture of clay and silt particles. Larger pieces of broken rock make up *sand.* One-fourth of all sediment is sand. Coarse pieces of gravel are also found in sediments. Gravel sediments are abundant in some places, but usually make up only a small portion.

Sediments are usually deposited in layers when they settle at the bottom of a body of water. Fig. 9–12 shows an example of such layers. Generally, the larger rock particles are the first to fall to the bottom. Gravel settles before sand, and sand before silt and clay. Because the different particles settle at different rates, layers are formed.

Fig. 9–12 Sedimentary rock showing its many layers.

SEDIMENTARY ROCK

Most sediments collect very slowly. Many thousands of years may be needed to build a layer of sediment that is only a few centimeters thick. Each layer will grow in thickness until conditions change. For example, a layer of mud might settle in quiet water that is not moving. When the water begins to move, the layer of mud stops growing. When the water becomes quiet again, a new mud deposit is started.

Layers of sediment usually become covered by other layers. Particles are squeezed together by the weight of the layers above. Sometimes, minerals such as quartz or calcite fill the space between the particles. When this happens the sediments are cemented together. The rock made when a layer of sediment becomes solid is called a **sedimentary rock.**

Sedimentary rock A kind of rock made when a layer of sediment becomes solid.

TYPES OF SEDIMENTARY ROCK

1. Cementing and compaction. The most common of all *sedimentary rocks* is *shale*. See Fig. 9–13 *(top left)*. Shale is formed when clay particles are compacted, or squeezed together. When grains of sand become cemented together, *sandstone* is formed. See Fig. 9–13 *(top right)*. Gravel and large, round particles become cemented together to make *conglomerates* (kun-**glom**-uh-ruts). See Fig. 9–13 *(bottom)*.

Fig. 9–13 (top left) *Shale;* (top right) *sandstone;* (bottom) *conglomerate.*

2. Chemical. The second kind of sedimentary rock forms from *chemical sediments*. A body of water may contain a large amount of dissolved minerals. Chemical sediments form when the water evaporates and the dissolved mineral matter remains as a deposit. For example, if the water supply to a lake is shut off, the water in the lake eventually will disappear. As the water evaporates, the lake becomes a salt lake. When the lake is completely dry, a deposit of minerals remains. *Gypsum* (**jip**-sum) and *rock salt* are formed in this way. See Fig. 9-14.

Fig. 9-14 The white sand in White Sands National Monument is almost pure gypsum.

3. Organic. The third type of sediment is called *organic* (or-**gan**-ik) *sediment.* The word "organic" means "life." Organic sediment is made from the remains of living things. Clams, oysters, snails, coral, and some types of fish are among the animals that live in salt water. Each of these has a hard shell or a skeleton. When these animals die, their soft body parts decay. Their hard parts, however, fall to the bottom of the sea. Layers of such sediment become a sedimentary rock called *limestone.* Limestone also may contain mud and sand. See Fig. 9-15. Another organic sediment was formed from plants that lived in swamps millions of years ago. When these plants died, their remains eventually formed coal. Coquina (koh-**kee**-nuh) is also a sedimentary rock that formed from shells that hardened together.

Fig. 9-15 Limestone is formed from organic sediments.

Fig. 9-16 This sandstone was once an ancient desert. The layers of sediment are clearly seen.

HISTORY FROM SEDIMENTS

Sedimentary rocks are an important tool in the study of the earth's history. They are a record of the conditions in the earth's past. Coal is a sedimentary rock that comes from the remains of plants that lived in swamps. The coal found in Antarctica shows that the climate of that land was different in the past.

Sedimentary rocks containing molds of seashells tell of seas once spread over the present continents. Sedimentary rocks containing glacial deposits tell of former glaciers. Sandstone gives evidence of ancient deserts. See Fig. 9-16. Sedimentary rocks also tell about past life. The fossil remains of many plants and animals are preserved in the sedimentary layers.

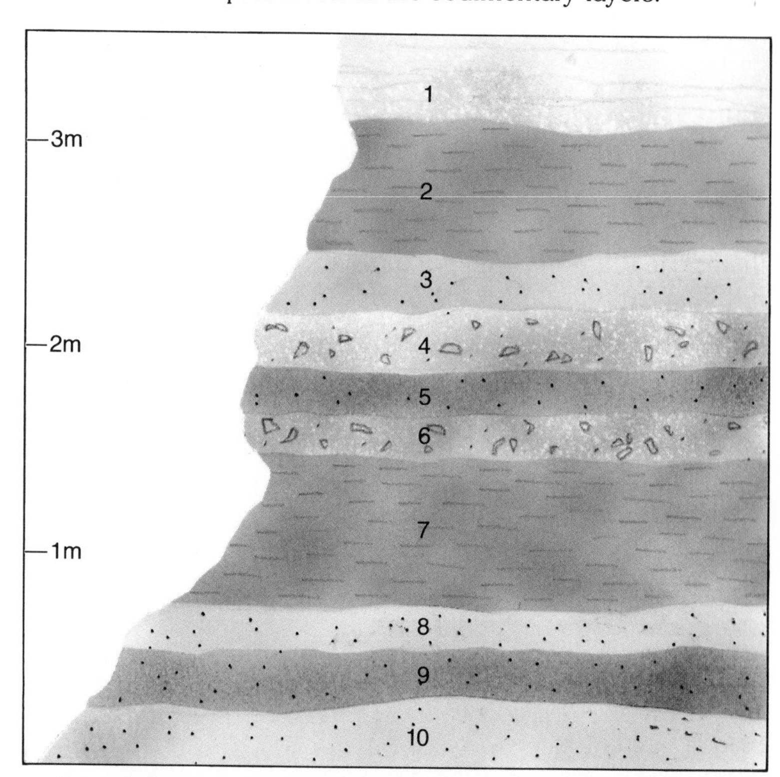

Fig. 9-17 Profile of a section of land. Layer 1: gypsum; layers 2,7: shale; layers 3,5,8,9,10: sandstone; layers 4,6: conglomerate.

One of the first things a student in geology learns to do is to study drawings that represent profiles of measured sections of land. Fig. 9-17 shows such a drawing. The layers of sedimentary rock differ in thickness and composition. Often samples of the rock from each layer are taken back to the laboratory to be studied. One property that is usually checked is the grain size. The grain size gives information about the movement of the water that deposited the sediment. For example, large grains would show the water to have been moving rapidly. Only the largest grains would have settled. The numbers in Fig. 9–18 give the grain size of the samples in order, from the finest grains found in shale samples to the coarsest grains found in conglomerates. Also shown is a graph of the grain size of each sample alongside the layer in which the sample was found.

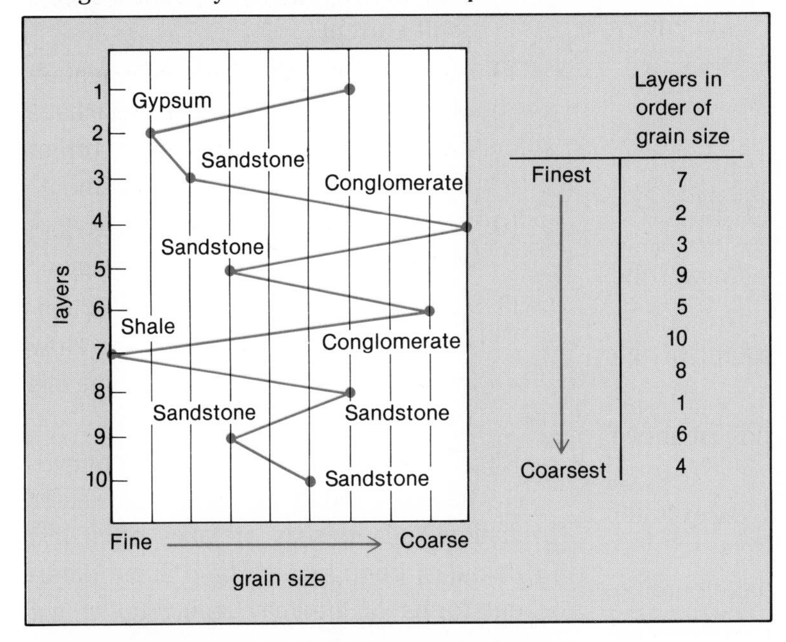

Fig. 9–18 Grain sizes of the samples.

SUMMARY

The earth's surface is no longer completely covered with a layer of igneous rock. There are processes at work on the surface that reduce igneous rock to sediments. Sediments are moved from one place and deposited in another. Layers of sediments can become changed into many different kinds of sedimentary rocks. These rocks contain a record of much of the earth's history.

QUESTIONS

Use complete sentences to write your answers.

1. What is meant by "sediment"?
2. Describe three ways that sediments may form and give an example of a rock made from each kind.
3. Give three examples of what sedimentary rocks can tell us about earth's history.

INVESTIGATION

STUDYING SEDIMENTARY ROCK

PURPOSE: To study a profile of sedimentary rock layers.

MATERIALS:
paper pencil

PROCEDURE:

A. Look at Fig. 9–17, page 252. Answer the following questions.
   1. What kind of rock is found at the 1-meter level?
   2. What kind of rock is found just above the 2-meter level?

B. Study Fig. 9–17 and Fig. 9–18. Answer the following questions.
   3. What would cause the shale found at the 1-meter level to be slightly finer grained than the shale located near the top of the 2.5-meter level?
   4. Which one of the sandstone layers has the largest grain size?
   5. Which kind of rock layer shown is a chemical sediment?
   6. Which rock layer was originally deposited as a very fine silt?

C. Much more can be learned about the history of a section of land. Remember that larger grains of sediment require faster-running water to carry them. Study Fig. 9–18 and refer to Fig. 9–17 to answer the following questions.
   7. What kind of rock was formed by sediments transported by the fastest-moving water?
   8. How many times during the history of this section did the water show such a swift current?
   9. As the grain size of the sedimentary rock became finer, what must have happened to the speed of the current of water that delivered the sediment?
   10. Which kind of rock layer was laid down when the water was very calm?
   11. Calm water with little current is typical of large, shallow inland seas. How many times did the water become this calm?
   12. How do you explain the layer of gypsum just above the last layer of shale?

D. Shallow inland seas or lakes often last decades, during which time many different forms of life may spring up in the water. When water no longer drains into the low area, the sea will evaporate.
   13. What kind of sediment would show that the water had completely evaporated from an inland sea or lake?
   14. How many times did this happen for the profile shown?
   15. What reason can you give that sea life may not have existed in this inland sea or lake?

CONCLUSIONS:

1. Starting with the oldest rock layer, list its number and name the sediment from which it formed.
2. Give the conditions that were present to cause each of the sediments.

9–3. Changed Rocks

At the end of this section you will be able to:

- Define *metamorphic.*
- Name two agents that produce metamorphic rock and explain how each may occur.
- Describe the *rock cycle.*

Can you identify the lump of material shown in the middle of the photograph in Fig. 9–19? It is the remains of a car. If you look carefully, you can see that the two headlights have been pushed together. They are now located next to each other. Applying pressure to the car certainly changed it! Rock in the earth's crust can be changed in a similar way. In this section you will learn about the ways in which rock is changed.

Fig. 9–19 Minerals in rocks can be crushed together the way these cars are crushed.

METAMORPHIC ROCK

Look again at the remains of one of the wrecked cars shown in Fig. 9–19. Is there anything in the lump of twisted metal that tells you what kind of car it was? Now look at Fig. 9–20. The rock, like the car, has been squeezed by great pressure. The rock was once a sedimentary rock. The sediments have been flattened and pressed tightly together. The original sediments can still be seen. The rock is a **metamorphic rock.** Metamorphic means "changed in form." Great pressure has changed a sedimentary rock into a *metamorphic rock.*

Metamorphic rock Rock that is changed by the action of heat or pressure without melting.

Fig. 9-20 Great pressure has squeezed these rocks together.

MAKING METAMORPHIC ROCK

Both igneous and sedimentary rocks can become metamorphic rocks. This can occur when heat, or pressure, or both act on the rock. Many igneous rocks and all sedimentary rocks are formed on or near the earth's surface. Then how are they put under pressure great enough to change them? There are several ways this can be done. For example, rocks on the surface may be covered by layers of sediment. The weight of these layers might provide the needed pressure. Movements of the crust also twist, flatten, and stretch the mineral grains that make up the rock.

Heat is another agent that acts to change rocks. The interior of the earth is hot. Rock buried under several kilometers of earth can be affected by heat as well as by pressure. Magma may push into existing rock layers, causing great pressures as well as bringing heat with it. The magma, after giving up its heat, hardens into rock and is known as a **pluton** (**ploo**-tahn). Rock under high pressure can be heated to a high temperature without melting. When this happens, the rock can change. The minerals in that rock may grow into large crystals. New minerals may form as atoms rearrange. Sometimes gases trapped in the rock are released. New gases or liquids from surrounding rocks may mix with the rock material. Rocks changed in this fashion are often hard to trace to their original form.

Pluton Any large, intruded rock mass formed from magma.

Metamorphic rock is often found near mountains. If a mountain is the result of a *pluton,* hot magma comes in contact with rock material. During this kind of mountain building, rock changes result from both heat and pressure. Such changes formed most of the world's important deposits of ore minerals.

Folding and faulting often occur with mountain building. These processes twist and squeeze existing rock material. Metamorphic rock formed at the roots of mountains is exposed only after millions of years of erosion.

Quartzite (**kwort**-zite) is the metamorphic rock made from sandstone. Sand grains in quartzite are fused together more firmly than they are in sandstone. Quartzite is harder than sandstone. *Slate* is a metamorphic rock made from clay or shale. Slate can be found as black, red, green, or blue rock. *Marble* is another metamorphic rock that can vary in color. Marble can be made from limestone. The color of marble depends on the minerals mixed in with the limestone. *Gneiss* **(nice)** is a metamorphic rock with a streaked or banded appearance. See Fig. 9–21. Gneiss usually has the same minerals as granite. However, the changes that produce gneiss are so great that it is hard to determine the original rock material. Without the bands, most gneiss looks like granite. Many different rocks probably can form into gneiss. In the same way, fine-grained rocks can be formed into the metamorphic rock called *schist* **(shist).**

Fig. 9–21 (top left) *Quartzite;* (top right) *slate;* (bottom left) *marble;* (bottom right) *gneiss.*

THE ROCK CYCLE

Rock cycle The endless process by which rocks are formed, destroyed, and formed again in the earth's crust.

The formation of metamorphic rock is one step in an endless process that has been going on for billions of years. The process is called the **rock cycle.** This process consists of the mixing and reusing of the earth's rock material. For example, hot magma exists beneath the earth's surface. When magma is forced to the surface, it is called lava. Lava cools and hardens, forming extrusive igneous rock. Agents of erosion act on this rock. The magma in the crust may also harden below the surface, forming dikes, sills, or batholiths. These intrusive igneous bodies are brought to the surface by erosion or crustal movement. Agents of erosion act on these bodies as soon as they reach the surface.

Erosion results in sediments. Sedimentary rock is formed from deposits of sediments. This rock becomes buried by other layers. Sedimentary rock may remain buried, or may be brought to the surface, where it is eroded.

Any rock that is buried below the surface can be changed by heat and pressure into metamorphic rock. The new metamorphic rock eventually is brought to the surface. Erosion then acts on this form of rock.

Scientists believe that a large portion of the *rock cycle* is taking place at the edges of the crustal plates. Magma is coming to the surface along the mid-ocean ridge and near trench areas. In the trenches, old rock is being forced down into the mantle. In the mantle, it is melted and mixed with other rock material. This material may then make its way back to the crust as new lava or magma. Such recycling has been going on since the earth was formed. Fig. 9–22 represents the rock cycle.

Rocks can take shortcuts through the rock cycle. For example, igneous rock can be changed into metamorphic rock. It does not have to weather first. Exposed sedimentary rock may weather and become part of new sedimentary rock.

The rocks of the earth's crust have gone through the rock cycle many times. This is the reason that rocks near the earth's surface never have ages greater than 3 billion years, while the earth is probably 4.5 to 5 billion years old. All original rocks of the earth's crust have been changed by the rock cycle since they were first formed. The age of rocks in the crust shows only how long they were formed as part of the rock cycle.

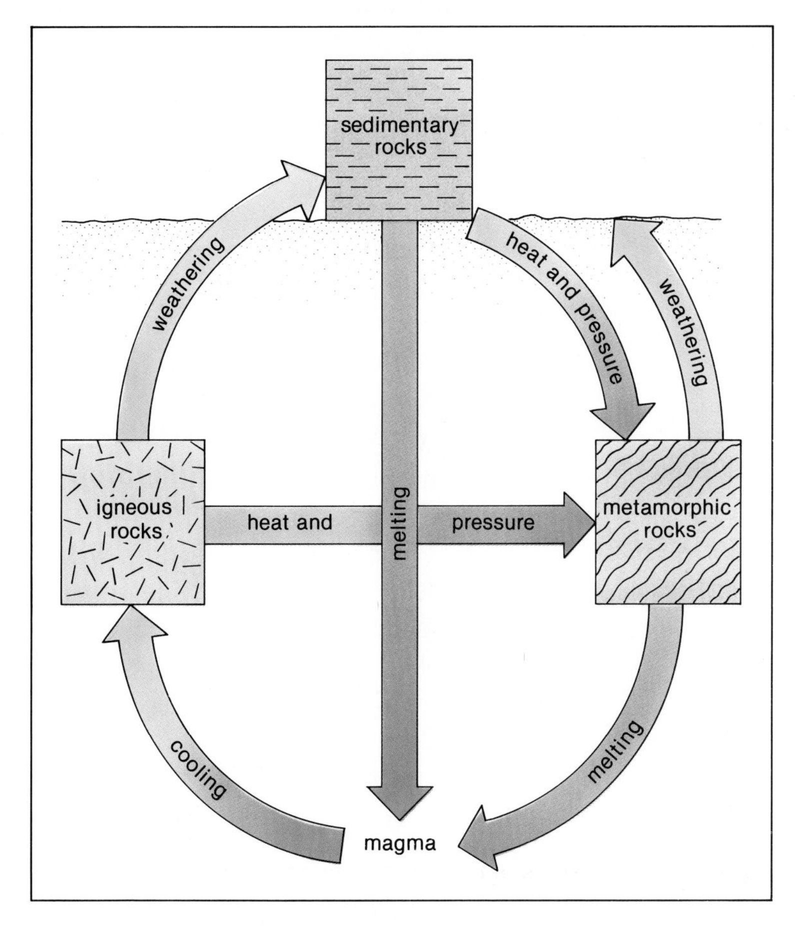

Fig. 9–22 The rock cycle.

SUMMARY

Metamorphic rock is formed by the action of heat and pressure on existing rock. Slate, gneiss, marble, and schist are some common metamorphic rocks. The formation of metamorphic rock is one step in the rock cycle. The rock cycle is the name given to the process that recycles the rock materials of the earth.

QUESTIONS

Use complete sentences to write your answers.

1. What does "metamorphic" mean?
2. Describe two ways that metamorphic rock may be formed.
3. Name two common metamorphic rocks and the original rock from which each was made.
4. Draw a diagram of the rock cycle.

INVESTIGATION

THE FORMATION OF METAMORPHIC ROCK

PURPOSE: To use clay to demonstrate one way that rocks are changed.

MATERIALS:

modeling clay, two 1-cm cubes (different colors)
bread, two 1-cm squares
waxed paper
eyedropper
water

PROCEDURE:

A. Form about ten small spheres from each piece of clay. These represent pieces of sediment. Each color represents a different mineral.

B. Put the spheres together loosely in a pile. Call this the original form of the sediment.

C. Press the pile of clay spheres with the palm of your hand, but not hard enough to remove all the space between the spheres.

1. Which would the pile represent, sedimentary or metamorphic rock?

D. Take apart a small part of the pile and inspect the spheres.

2. Describe the spheres now.
3. Compare the space between particles in this flattened model with the space in the original form.

E. Press the remaining pile of spheres until their shapes are completely destroyed.

4. Compare the appearance of the clay now with the form it was in before.

F. Look at Fig. 9–20, page 256.

5. What evidence do you see that this rock has been under great pressure?

G. Scientists often work with models and try to discover things in an indirect way. For example, to test the effect that great pressure may have on the amount of water a rock can hold, one might do the following experiment.

H. Place a 1-cm square of bread on waxed paper. Using an eyedropper, release water one drop at a time onto the bread. Count the number of drops the bread can soak up before water starts coming from the lower edges onto the waxed paper.

6. How many drops did you count?

I. Flatten a second 1-cm square of bread until you have pushed most of the air from it. This reduces the space between the particles of bread, just as pressure does in rock.

J. Using the procedure in step H, count the number of drops this piece of bread holds.

7. How many drops did it hold?

CONCLUSIONS:

1. What effect does pressure have on the grain size and shape of particles in rock?
2. What evidence would you look for to tell whether the rock you have is the sedimentary rock, sandstone, or its metamorphic form of quartzite?
3. Which would hold more water, sedimentary rock or its metamorphic form?

CAREERS IN SCIENCE

METALLURGICAL ENGINEER

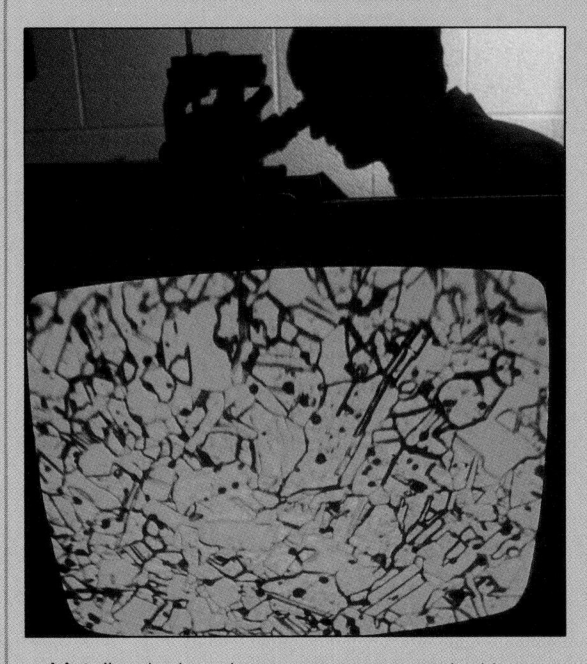

Metallurgical engineers are concerned with practical ways of changing raw metals into useful products. Some of these engineers study methods of removing metal from ores such as iron ores. Other engineers are interested in how these metals can be changed into finished products. These products include space exploration devices, medical equipment, and power plant machinery. Metallurgical engineers study the properties of metals and develop ways of casting metal and shaping it. These processes are used in the industrial production of steel and other metals, as well as in casting operations used by sculptors of metal.

Metallurgical engineers have at least a bachelor's degree, and many have advanced training. They are employed by mining companies and in industries that manufacture machinery, electrical parts, and other types of equipment. For more information, write: Engineers' Council for Professional Development, 345 East 47 Street, New York, NY 10017.

STONEMASON

Do you like the feel of a piece of marble or limestone in your hands? Do you enjoy building things? Stonemasons work with different types of stone, assisting in the construction of large buildings such as offices, and hotels, and churches. They also work on smaller jobs such as floors, fireplaces, and window sills. The artistry in the face of a building is often the work of the stonemason.To make sure the stones fit together, masons must measure them accurately. Then they must carefully position the stones, using mortar to hold them together. Masons must be concerned with the safety and stability, as well as the appearance, of their work.

Individuals who want to become masons usually start as apprentices. In the apprenticeship program, they learn skills such as sketching and how to read blueprints. Jobs for stonemasons are available with construction companies. For further information, write: Associated General Contractors of America, Inc., 1957 E Street N.W., Washington, DC 20006, or visit a construction site in your own area.

¡COMPUTE!

SCIENCE INPUT

Rocks, as they form and decay, go through constant changes. The specific changes depend on the environment in which the rocks are located. This ongoing effect of changing from one rock type to another is called the rock cycle. Understanding the nature of this cycle is useful in determining the geologic history of an area. Environmental conditions act on the features of the land to create their varying forms.

COMPUTER INPUT

Computer programmers draw an outline of their program using a system of symbols, lines, and terms called a flowchart. A flowchart provides a shorter, faster way of communicating than sentences. Most important, however, it lets both the programmer and other users of the software keep track of the steps of the program. A flowchart outlines a series of paths and processes that provide the instructions for the program.

The flowchart in this chapter represents the rock cycle. It outlines the various paths of rock originations and rock decay, providing a short review of some of the material you learned in this chapter. The program has been written on the basis of the flowchart. Examine the chart very carefully and try to answer the questions *before* entering and running the program. To make it easier to study, draw a larger version of the flowchart. The glossary will help you to keep track of what each step means.

FLOWCHART QUESTIONS

1. What type of rock is formed from molten rock? What types are formed from sediments? What form does partially melted rock take?
2. The future of a rock is greatly affected by whether or not it lies on the surface. How does that factor influence the rock's future?
3. The flowchart has no end or stop point. Why?

WHAT TO DO

Using your interpretation of the flowchart, try to answer the flowchart questions.

Enter Program Flowchart into your computer. When you run it, it will ask you questions about different types of rock. Choose one of the alternatives. Compare each screen step of the program with the flowchart. Compare the answers you gave to the flowchart questions with the answers the computer gives. Run the program several times to help you review the steps in the rock cycle.

GLOSSARY

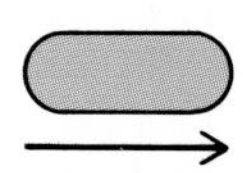

Start or stop.

Shows the direction the program will follow.

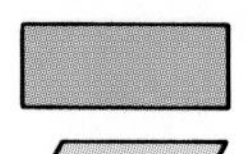

A single step in the program.

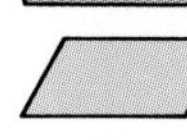

Input or Output.

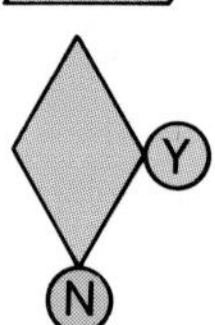

A decision point: A question must be answered, or two things are compared. "Y" means yes and "N" means no.

THE ROCK CYCLE: PROGRAM FLOWCHART

PROGRAM

```
100  REM FLOWCHART
110  REM REPLACE THIS STATEMENT
     WITH CLEAR SCREEN STATEMENT
120  PRINT "HOT MOLTEN ROCK"
130  PRINT "COOLS AND FORMS
     IGNEOUS ROCK"
140  PRINT "DIASTROPHISM OCCURS"
150  INPUT "IS THE ROCK AT THE
     SURFACE (Y/N)? ";A$
155  PRINT
160  IF A$ = "N" THEN GOTO 210
165  IF A$ <> "Y" THEN GOTO 150
170  PRINT "WEATHER AND EROSION
     OCCUR"
180  PRINT "SEDIMENTS FORM"
190  PRINT "SEDIMENTARY ROCK
     FORMS"
200  GOTO 140
210  PRINT "HEAT AND PRESSURE ARE
     APPLIED"
220  INPUT "DOES THE ROCK MELT (Y/
     N)? ";A$
225  PRINT
230  IF A$ = "Y" THEN GOTO 120
235  IF A$ <> "N" THEN GOTO 220
240  PRINT "THE ROCK IS CHANGED"
250  PRINT "METAMORPHIC ROCK IS
     FORMED"
260  GOTO 140
```

PROGRAM NOTES

A clear screen statement is an instruction to the computer to remove all data from the TV screen or monitor. Not all computers use the same commands. Consult your user's manual for the clear screen statement appropriate to your computer.

FLOWCHART

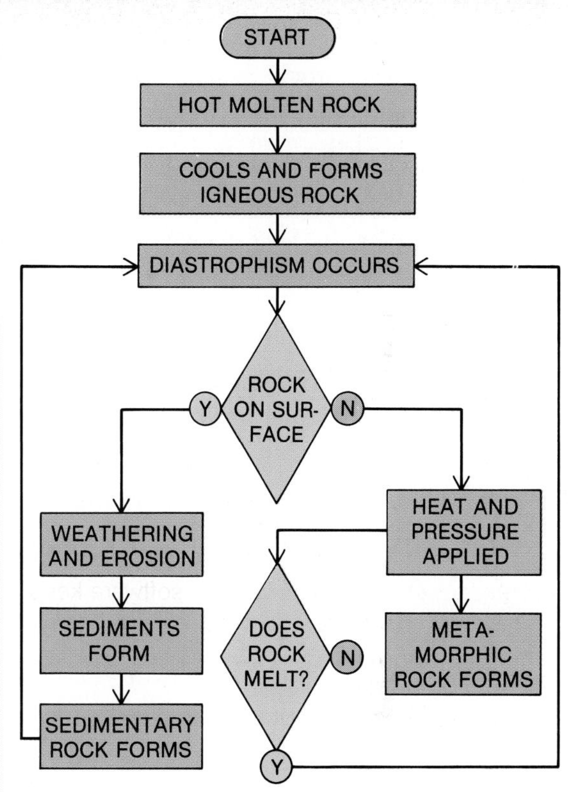

BITS OF INFORMATION

Can you imagine a computer small enough to sit on your lap, a computer that would let you write letters, or make notes, and even do programming? Such computers are available and can weigh as little as four kilograms. They are not yet a replacement for regular computers. They have small screens and can handle only a small range of software. But they certainly can be helpful for the many people who spend a great deal of time traveling and wish to work en route.

CHAPTER REVIEW

VOCABULARY

On a separate piece of paper, match each term with the number of the phrase that best explains it. Use each term only once.

rock	sedimentary rock	rock cycle	pluton
igneous rock	batholiths	sills	
sediment	metamorphic rock	dikes	

1. Underground bodies of igneous rock in cracks crossing layers of existing rock.
2. Any large intruded rock mass formed from magma.
3. A piece of the earth made up of two or more minerals.
4. May make up the cores of mountain ranges.
5. The endless process by which rocks are formed, destroyed, and formed again.
6. Rock made when a layer of sediment becomes solid.
7. Rock that has been changed by heat or pressure.
8. Material that settles at the bottom of water.
9. Rocks formed when molten material cools.
10. Bodies of igneous rock formed underground that run parallel to layers of existing rock.

QUESTIONS

Give brief but complete answers to each of the following questions. Unless otherwise indicated, use complete sentences to write your answers.

1. What is an igneous rock?
2. Name the two groups of igneous rock and tell how they differ in the way they form.
3. List three rocks formed from lava.
4. How can you tell whether an igneous rock has cooled rapidly or slowly?
5. Name and describe three kinds of igneous rock structures formed underground.
6. What is the most abundant of all sediments? Name and describe two other sediments from which rock is formed.
7. How does a sediment become sedimentary rock?
8. Name two common sedimentary rocks formed from rock fragments.

9. What is chemical sediment? Name one rock formed from chemical sediment.
10. How did coal form?
11. What is organic sediment?
12. How does heat change rock?
13. What two agents cause metamorphic rock?
14. How do igneous and metamorphic rocks differ in the way they form?
15. Does a rock cycle occur on the moon? Explain your answer.

APPLYING SCIENCE

1. Organize a rock-hunting party. This can be an actual outing to gather rocks or a hunt through magazines and other pictorial sources to locate pictures of rocks. Prepare a display of your results, grouping the minerals and rocks according to whether they are igneous, sedimentary, or metamorphic. Label as many as possible with their scientific names. Use a mineral identification key.
2. Make a list of the rocks and minerals that were used to construct your school building.
3. Make a poster-size drawing of the rock cycle. Attach to it, in the proper places, examples of igneous, sedimentary, and metamorphic rock.
4. Many public buildings are made of stone. Locate as many as you can in your area. Determine what stone each is made of, whether the stone is igneous, sedimentary, or metamorphic. Inside the buildings there may be other types of stone for floors, stairways, or walls. Be sure to list all types you find. Find out if any of the stones were taken from the local area.

BIBLIOGRAPHY

Bates, Robert L., and Julia A. Jackson. *Our Modern Stone Age*. Los Altos, CA: Kaufman, 1981.

Dietrich, R. V. *Stones: Their Collection, Identification, and Uses*. San Francisco: Freeman, 1980.

MacFall, Russell P. *Rock Hunter's Guide; How to Find and Identify Collectible Rocks*. New York: Crowell, 1980.

Radlauer, Ruth. *Grand Teton National Park*. Chicago: Children's Press, 1980.

CHAPTER

10

Everything we use, wear, eat, or drink comes from the earth. The earth has always been able to supply us with building materials, clothing, food, and water. Will this always be true?

EARTH RESOURCES

CHAPTER GOALS

1. Tell what is meant by a natural resource and explain why some are renewable and some are nonrenewable.
2. Discuss several important problems having to do with population growth.
3. Explain what a local water budget is and the importance of protecting the watersheds as well as preventing pollution of air and water resources.
4. Tell the difference between a rock and an ore.
5. Describe how mineral deposits form and why it is necessary to conserve our mineral resources.
6. Discuss our dependence on fossil fuels for energy.
7. Explain the role of nuclear energy and other alternative energy resources in our future needs for energy.

10–1. People and Resources

At the end of this section you will be able to:

- Define *natural resources*.
- Explain how the growth of the earth's population affects the food supply.
- Describe three kinds of pollution.

You are now living on a kind of space station. It is called the planet Earth. A space station like the one shown in Fig. 10–1, page 268, must supply all the needs of its passengers. Similarly, the earth is the source of the materials needed to support its population. See Fig. 10–2, page 269.

NATURAL RESOURCES

The earth has everything people need to live. It gives us food and the materials needed to make clothing and shelter. The fuels that furnish energy come from the earth's crust. Any material from the earth that can be used in some way by people is called a **natural resource.** *Natural resources* are made into various products. For example, this book was made from many natural resources. The paper came from forests. Chemicals

Natural resource Any of the substances from the earth that can be used in some way.

Fig. 10–1 In the future, the space shuttle will bring supplies to a space station like this. Planet Earth is also like a large orbiting space station.

from rock and air helped to change the wood into paper. Minerals supplied the metals used for printing the book. The energy used for printing came mostly from coal or oil. It is easy to forget that the raw, natural materials that are used for our basic needs come from the earth. The rows of food in a supermarket do not look like the fields in which the food grew. We do not think of cotton fields or herds of sheep when we shop for jeans and sweaters. Air in buildings is heated and cooled. Often, there is little thought of the fuels needed to make the artificial environments in which we spend much of our time. It is important to look at the natural environment to remember how natural resources are used in everyday life.

There are two kinds of natural resources. Some resources, such as forests, wildlife, and water, are *renewable*. A renewable resource can be replaced in a short time after being used. Other earth resources are *nonrenewable*. These resources are not replaced or are replaced very slowly. For example, it may take millions of years to replace the supplies of minerals and fossil fuels, such as oil, taken from the earth's crust. Thus nonrenewable resources must be used carefully.

THE EARTH'S POPULATION AND FOOD SUPPLY

Are there enough resources to feed the world's people? The world's population is growing rapidly. In 1700, there were about 600 million people on earth. One hundred years later, the population had increased to 1 billion. At that rate, the population of the earth doubled every 150 years. However, by 1980 the population was 4.5 billion. See Fig. 10–2. The earth's population is now growing at a rate that will cause it to double every

33 years. If the population continues to grow at the present rate, in 66 years there will be four additional people for each person alive today. This rapid growth has been brought about largely by better control of disease and a more dependable supply of food.

Thousands of years ago, our ancestors learned to use the land to grow their food. They used muscle power and simple tools. Those farming methods changed very little from ancient times until about 150 years ago. Then the machines that were developed in the Industrial Revolution were first used for agriculture. The machines made it possible for a single person to farm many acres of land. See Fig. 10–3. Farmers also began using artificial fertilizers. This greatly increased the amount of food they could grow. They also began using chemicals to control pests that damaged crops. In countries such as the United States, the average farmer today produces enough food for 30 people. But improved farming methods have not reached all parts of the world. For example, an average farmer in India produces only enough food for several people. One way of feeding the world's growing population is by using better farming methods in all parts of the world.

The amount of land that can be used for growing food is limited. About 60 square meters of farmland are needed to grow food for one person. This is an area about the size of a tennis court. It takes about 10 percent of the earth's land to feed the nearly five billion people now on the earth. Much of the land contains mountains, deserts, and land not useful for farming. A

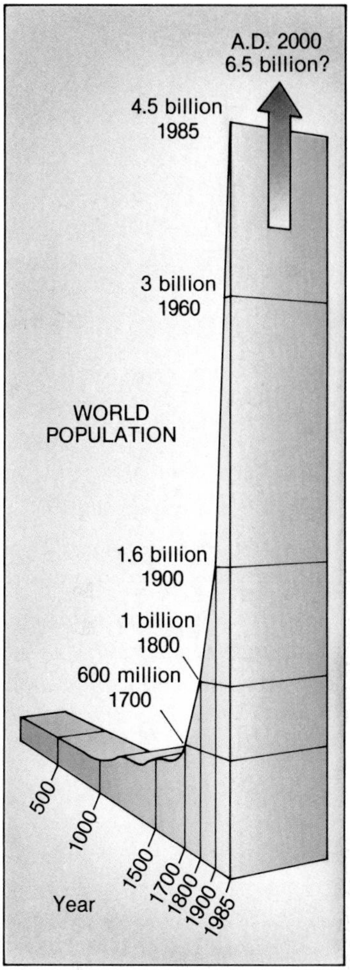

A.D. 2000
6.5 billion?
4.5 billion
1985
3 billion
1960
WORLD
POPULATION
1.6 billion
1900
1 billion
1800
600 million
1700
500
1000
1500
1700
1800
1900
1985
Year

Fig. 10–2 As this graph shows, the earth's population has more than doubled during the last century.

Fig. 10–3 Farm machines have increased our ability to grow food.

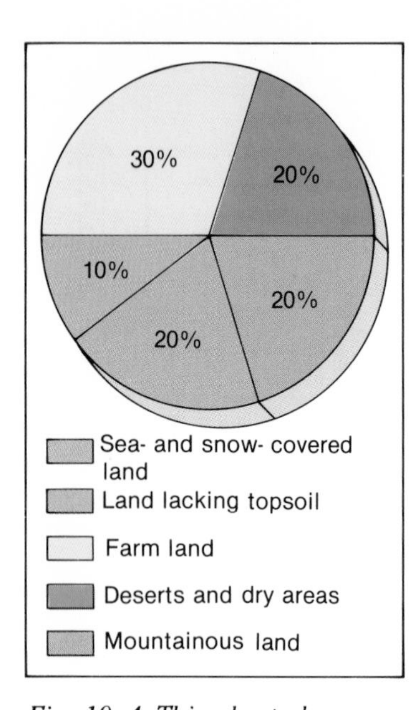

Fig. 10–4 This chart shows how much of the earth's surface is utilized for farming.

total of 30 percent of the earth's surface can be used for farming. See Fig. 10–4. However, many of these areas have climates or other conditions that require new farming methods to make the land productive. Scientists have been able to develop new kinds of crops that grow faster, resist disease, and have more food value. For example, rice is one of the most commonly grown foods in the world. But rice is difficult to grow, has diseases, and often produces a grain that lacks essential food values. By using improved crops, along with the best methods of farming of all available land, the earth could probably provide food for a population of 40 to 50 billion people.

What about resources from the sea? Oceans cover more than twice as much of the earth's surface as the land does. Plants grow as well in the sea as they do on the land. Therefore, the sea produces about twice as many plants as the land. But few people eat plants from the sea. Most of the seafood that people eat is fish. These fish make up only a small part of the food from the sea. But the plants from the sea that might be used for food are too small to be farmed. Increasing the food supply from the oceans would require a special kind of farming called "ocean farming." Ocean farming could be done by fencing off and fertilizing large areas of the sea.

The use of better farming methods on more of the land may also make it possible to feed the world's population for the next century. However, if the population continues to double every 33 years, there will be a time when enough food cannot be grown. Then population growth will stop, mainly because people will starve. As the world's population continues to grow, better farming methods will be needed to continue to feed the world's people.

PEOPLE AND THE ENVIRONMENT

Lack of food is not the only problem that is caused by increased population. One of the most serious problems is the pollution of our most valuable natural resources—air, water, and soil. The adding of harmful substances to the earth's environment is called **pollution.** There are three major kinds of *pollution:* air pollution, water pollution, and soil pollution.

Pollution The adding of harmful substances to the environment.

1. Air. Automobiles, power plants, factories, the burning of fuels for heating, and burning of trash are major causes of air pollution. Most coal and petroleum contain sulfur. When the

Fig. 10–5 The burning of fuels releases gases that cause air pollution.

coal or oil is burned, deadly sulfur dioxide gas is put into the air. See Fig. 10–5. Breathing air containing sulfur dioxide can cause health problems. Sulfur dioxide also causes *acid rain.* Rain and snow mix with sulfur dioxide to form an acid. The acid rain and snow cause damage to resources such as forests and lakes. Air pollution also comes from such things as bits of rubber that spin off automobile tires and gases given off in car exhaust.

2. Water. Water can be polluted in three ways. One of the most dangerous kinds of water pollution occurs when untreated sewage is dumped into streams, lakes, or the ocean. This waste may add disease-producing bacteria and viruses to the water. Sewage can also pollute water in another way. Chemicals in the sewage can help plants such as algae to grow. Algae float on the surface of ponds and streams. When the algae and other plants decay, they use up the oxygen from the water. Other plants and animals no longer can live in the water because of the lack of oxygen. Water can also be polluted by poisonous chemicals, such as those used on crops to control pests. These chemicals can be washed off the land and enter the water supply. These poisons affect the living things in the water. They also are harmful to the people who eat the food from the polluted waters.

3. Soil. A third form of pollution is caused by the buildup of harmful substances in the soil. Some soil pollution comes from the air. Most, however, is caused by the chemicals used to control pests and plant diseases. See Fig. 10–6. Small amounts of chemicals, such as *ethylene dibromide* (EDB) have been found

Fig. 10-6 Chemicals used to control pests and diseases on crops can remain in the soil.

in water wells in Florida. EDB has been found to cause cancer in laboratory animals. Many poisonous chemicals in containers are buried in landfills. Some of these chemicals have leaked into ground water. Ground water supplies drinking water for more than half of the population of the United States.

Chemicals are not the only pollutants. The trash you throw away adds to pollution. Trash is only one part of the pollution problem that you have some control over. Burning trash may pollute the atmosphere. Dumping it without careful planning may pollute the land around you. Unless pollution is controlled, our natural resources will be destroyed. It may take many years before they are renewed.

SUMMARY

Like a spacecraft, the earth must supply everything needed by the world's population. However, this planet may be in danger of becoming overcrowded. Even if the earth is able to feed a large population, adequate food will do little good if the environment is polluted to dangerous levels.

QUESTIONS

Use complete sentences to write your answers.

1. What is a natural resource?
2. What is a renewable resource? Give an example.
3. Why are minerals called nonrenewable resources?
4. What is the main problem with population growth?
5. List three major kinds of pollution and give an example of each kind.

INVESTIGATION

TRASH AND POLLUTION

PURPOSE: To determine the amount of trash produced in your town based on the measured amount of trash you make in a day.

MATERIALS:
pencil paper

PROCEDURE:

A. Make a list of all the things you and your family throw in the trash each day. Include lunch bags, food storage bags, candy wrappers, cans, bottles, notebook paper, paper towels, napkins, etc.

B. Find the weight, in kilograms, of each item on your list. If you must guess, use the table below for some rough weights.

Examples	Weight (kg)
bottle, soft drink	.500 kg
can, soft drink, soda	.023 kg
lunch bag	.008 kg
candy wrapper	.006 kg
notebook paper (ea.)	.005 kg
newspaper	.500 kg
milk carton	.020 kg
napkin	.002 kg
plastic fork or spoon	.004 kg

C. Find the total weight, in kilograms, of the things you throw out each day.
   1. What is the total weight?

D. Individually, your waste may not seem like much. But you live in a community of many people. To get an idea of the amount of waste your community produces each day, multiply the amount of your waste by the number of people in the community. Your teacher will tell you the population of your community.
   2. What is the total weight, in kilograms, of the trash thrown out daily by your community?
   3. What groups of people may throw out more trash per day than students? What groups would produce smaller amounts of trash?
   4. A metric ton is 1,000 kilograms. How many metric tons of trash are produced by the people in your community each year?

E. In some communities it costs $40 to get rid of each metric ton of trash. Multiply the number of metric tons of trash made by your community by $40.
   5. How much would it cost your community to rid itself of the trash at $40 per metric ton?

F. A trash truck can haul about five metric tons a day to the disposal area. Divide the total number of metric tons of trash by five to find the number of trash trucks needed each day.
   6. How many trash trucks are needed each day?

CONCLUSIONS:

1. How many metric tons of trash were produced in your entire community daily?
2. Do you think it is likely that your community produces more trash than this? Explain your answer.

Fig. 10-7 Aluminum cans can be recycled.

10-2. Mineral Resources

At the end of this section you will be able to:

- Compare ordinary rocks with *ores*.
- List three ways that mineral deposits may be formed.
- Using an example, show the need to conserve and *recycle* mineral resources.

Used aluminum cans are valuable. You can be paid for aluminum cans turned in to collection centers or aluminum recycling plants. See Fig. 10-7. However, no one will pay you for old "tin" cans. The "tin" cans are made up mostly of iron. Why are aluminum cans valuable but not cans made of iron? The answer to this question can be found by learning more about the way we find and use the earth's mineral resources.

WHAT IS A MINERAL DEPOSIT?

Ore A rock that contains useful minerals and can be mined at a profit.

Some minerals are useful because they contain certain chemical elements that are valuable. For example, a mineral that contains the element copper might be useful. The copper metal could be separated from the mineral and used to make electric wires. Other minerals are useful just as they are found. Limestone, for example, is used to make cement. Any kind of rock that is useful because it contains a valuable mineral is called an **ore.** To be called an *ore*, the rock must be able to be mined at a profit.

Some chemical elements, such as aluminum, are found in many different minerals. Aluminum is such a common part of minerals that it is likely to be found in almost any rock you may pick up. On the other hand, some elements, such as copper, are much less common in minerals. Thus ordinary rocks are likely to contain only very small amounts of this element.

FORMATION OF MINERAL DEPOSITS

Mineral deposit A body of ore.

Rocks in certain places in the earth's crust may contain much larger amounts of a particular element or mineral than are ordinarily found. A body of rock that contains a large amount of an element or a mineral is called a **mineral deposit.** A *mineral deposit* is formed when changes in the earth's crust cause minerals to separate from their parent rock. The minerals are first separated. Then they are deposited within another body of rock, forming an ore. Most important mineral deposits are formed in one of three ways.

1. Dissolving. First, minerals can become separated from rocks by dissolving in hot water. The hot water may come from ground water that is heated by bodies of magma below the surface. Some hot water may also come from the magma itself. As the hot water moves through the rocks, it dissolves minerals. This hot mineral mixture then enters cracks in cooler rocks. As the water cools, the dissolved minerals settle out as solid material. This may form a mineral vein, or streak, running through rocks. See Fig. 10–8. Ores of copper, gold, silver, lead, and tin are often found in veins.

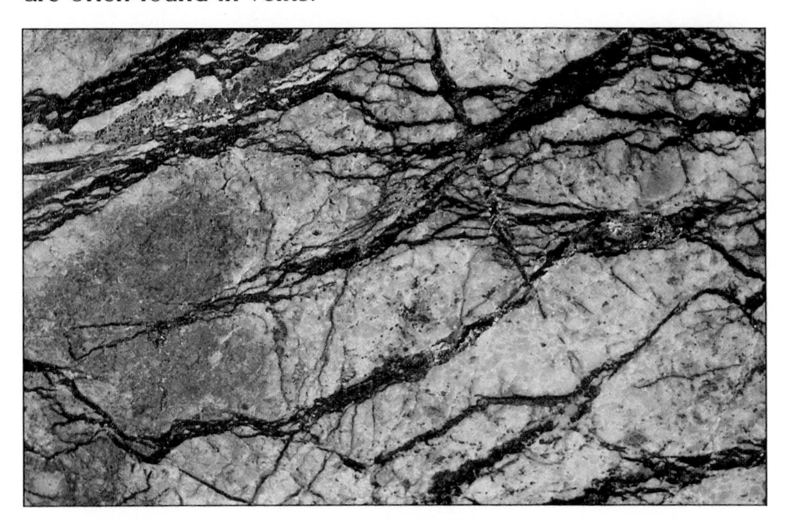

Fig. 10–8 This rock has numerous mineral veins of copper running through it.

In the past few years, scientists have found that the mid-ocean ridge is also a place where mineral deposits are forming from heated water. Research submarines exploring some parts of the mid-ocean ridge have found hot springs on the sea floor. Cold sea water enters through fractures. It is then heated by the magma rising beneath the ridge. The heated water dissolves minerals, then rises again to the sea floor. See Fig. 10–9, page 276. Hot water bursts through the sea floor at many vents near the rift in the center of the ridge. As the water cools, it deposits solid minerals in structures like chimneys. Many of the chimneys pour out a stream of black mineral-rich water. See Fig. 10–10, page 276. The mineral deposits built up along the mid-ocean ridge are carried away by the spreading sea floor. Scientists believe that many mineral deposits now found on land originally were formed along the mid-ocean ridge.

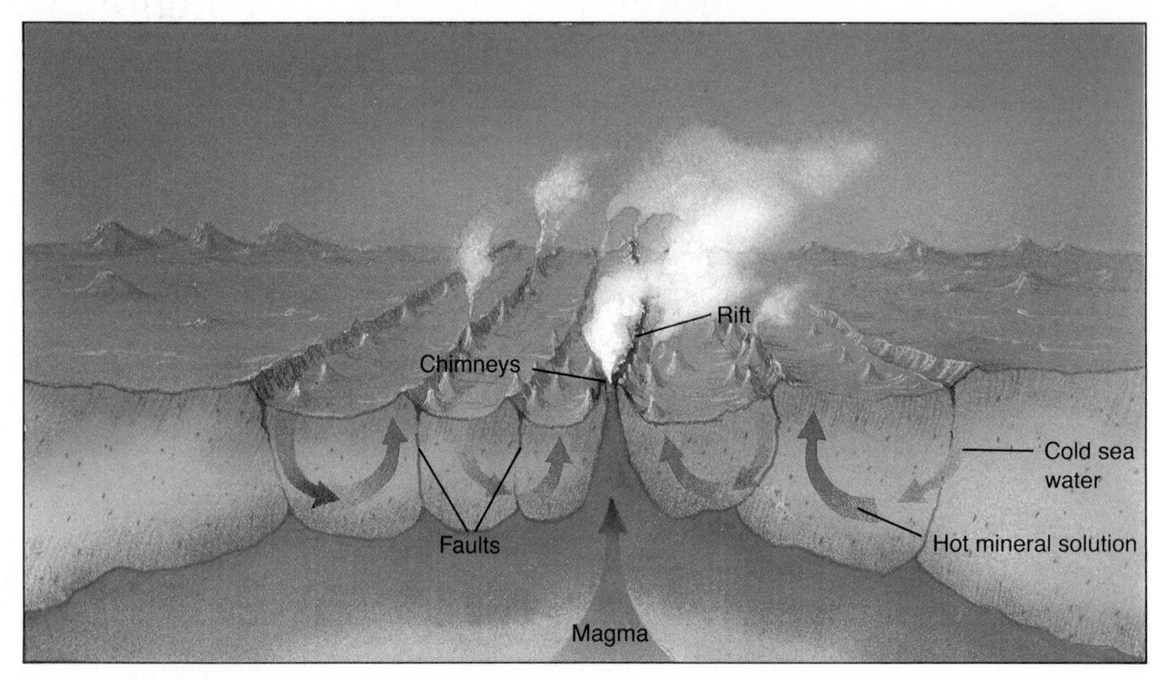

Fig. 10-9 Cold sea water is heated by magma along the mid-ocean ridge. The heated sea water carrying dissolved minerals rises back to the sea floor as many hot springs.

2. Weathering. A second way that ore deposits are formed is by weathering of rock. Minerals that make up the ore deposit are separated from the rock by the weathering process. For example, aluminum is found most often in minerals from which it cannot be separated profitably. Weathering of the aluminum-bearing minerals breaks them down into a form from which the aluminum can be taken. Deposits of aluminum ore are found in warm, humid regions. There the weathering processes have been most effective. Weathering may separate the denser minerals, such as gold, from their parent rocks. The denser mineral grains are moved by streams and often settle where the current is weak. Deposits built up by stream action in this way are called *placers*.

3. Cooling magma. Mineral deposits may also be formed in a third way. A body of magma may be pushed into the crust and cooled. Within the cooling magma, some minerals may separate and then sink toward the bottom. A layer of ore that is rich in some important metals can be formed in this way. The ores of chromium and nickel metals are examples of this. They are usually found within a mass of igneous rock formed from cooled magma. Hot magma may also change the rock it invades. Some mineral deposits may be caused by these changes around the edges of a body of magma.

Fig. 10-10 The hot, mineral-rich water is moving rapidly upward through this "chimney" in the mid-ocean ridge.

MINERALS AND THE FUTURE

Mineral deposits form very slowly. It has taken millions of years to form most of the ores that are now in the earth's crust. For this reason, minerals are a nonrenewable resource. As the world's population grows, the need for mineral resources also increases. In some countries, such as the United States, the need for minerals grows faster than the number of people. Each year, the amount of mineral resources needed by each person becomes greater. This has been caused by a demand for more new kinds of products and the use of minerals to make agricultural fertilizers. The need for mineral resources in other parts of the world is also growing each year. How can the world's need for minerals be met?

One way to increase the supply of minerals is by finding new deposits. There are still large areas of the earth's continents that have not been explored fully for mineral deposits. There are also large supplies of certain minerals on the sea floor. The prospect for finding new mineral deposits is good. However, many deposits will be expensive to mine. For example, the minerals on the sea floor are found mostly in deep water. It will be expensive to bring these minerals to the surface and then to the land.

The need for minerals can also be met by **conservation.** Controlling the way any natural resource is used is called *conservation*. One method of conservation of mineral resources is **recycling.** *Recycling* is the process by which a mineral is recovered and used over and over again. A large volume of metals, among them aluminum, is now recycled. One hundred used cans can make ninety new ones. Recycled aluminum also saves energy. It takes about one-fifth as much energy to melt used aluminum as it does to separate the metal from its ore. Iron is also recycled. About one-third of the iron used today comes from the recycling of larger items, such as wrecked cars. At this time, it does not save money to collect and recycle small amounts of iron, such as "tin" cans. At some time in the future, the need for conservation of all mineral resources may make it necessary to recycle everything made of iron.

Conservation Saving the natural resources by controlling how they are used.

Recycling The process of recovering a natural resource and using it again.

Metals are not the only valuable resources that can be recycled. For each metric ton of old newspaper that is recycled, about 17 trees are saved from being cut down to make new paper. Trash becomes a valuable resource if the used metals,

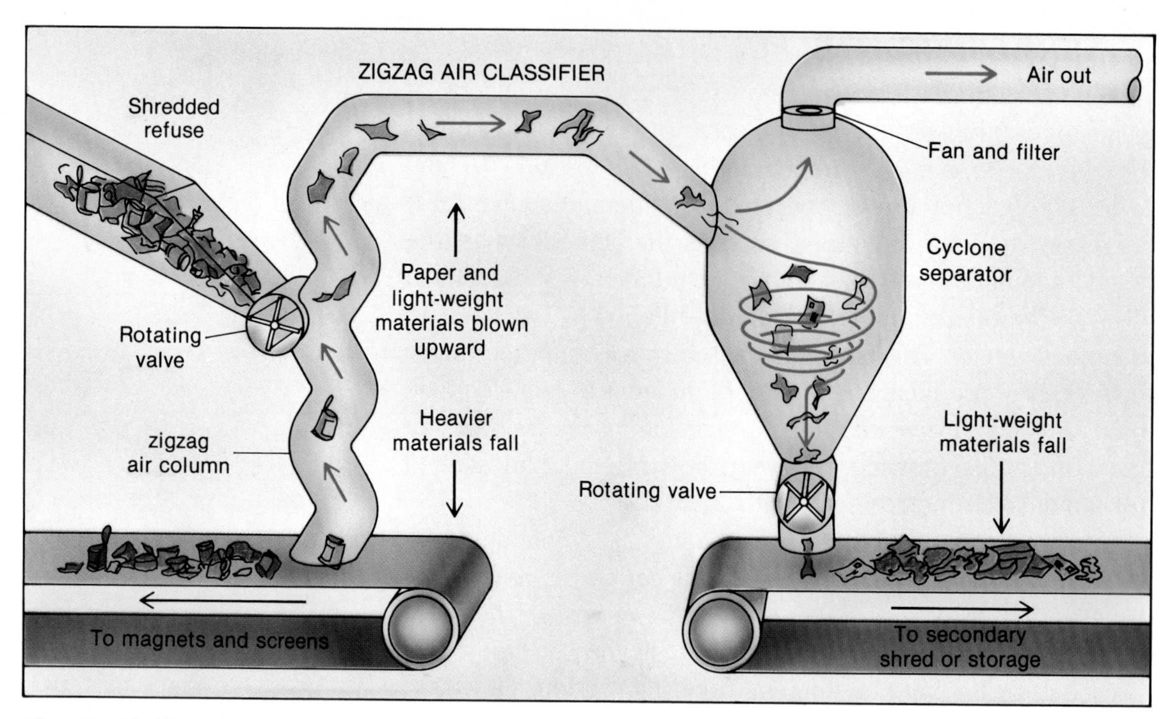

Fig. 10–11 This kind of system can be used to separate trash before recycling.

paper, and other materials can be separated and recycled. Fig. 10–11 shows a system that can be used to separate the useful materials in trash. Recycling plays an important part in the conservation of all natural resources.

SUMMARY

Some rocks contain enough useful minerals to be taken from the earth at a profit. Such rocks make up ores that are found as mineral deposits in the earth's crust. Mineral deposits form in the crust as a result of the action of hot water, magma, and weathering. Since mineral resources are nonrenewable, future needs can be met only by finding new deposits or by conservation.

QUESTIONS

Use complete sentences to write your answers.

1. What makes a rock an ore?
2. Describe three ways mineral deposits may form.
3. Give two reasons why we may run low on some minerals in the future.
4. How can the world's need for minerals be met?

INVESTIGATION

PANNING FOR GOLD

PURPOSE: To separate a mineral from other materials.

MATERIALS:
shallow dish or pan
container of water
large container for overflow water
sand, gravel, fool's gold mixture

PROCEDURE:

A. Place a handful of the mixture in the shallow pan. You are to separate the parts of the mixture by using the technique called panning. If your classroom lacks the necessary facilities, the panning may be done as a demonstration. Fool's gold, the mineral iron pyrites, resembles real gold so closely that, during the California gold rush, many mistook it for the real thing. Your main goal is to recover as much of this mineral from the mixture as possible.

B. Fill the shallow pan nearly full of water. Hold the pan with both hands and, using a circular motion, gently rock the contents back and forth. DO THIS OVER THE OVERFLOW WATER CONTAINER IN CASE OF SPILLS.

C. When the water becomes cloudy, gently and carefully tip the pan and allow the cloudy water to run off into the overflow container.

D. Set the pan down and pick out any large stones and set them aside.

E. Fill the pan with water and repeat the motion as in step B. This time, tilt the pan forward slightly so that the water and sand slowly flow over the side and fall into the overflow container. Continue this until most of the sand has been removed from the pan. Give the pan a quick backward motion to spread out the remaining material.

F. Pick out any of the gold-colored pieces you see. Continue until you think you have recovered all the mineral.

G. Look at Fig. 10–12, which represents a stream system. By panning, a prospector found placer gold at points A, C, E, and G. He found no gold at points B, D, and F.

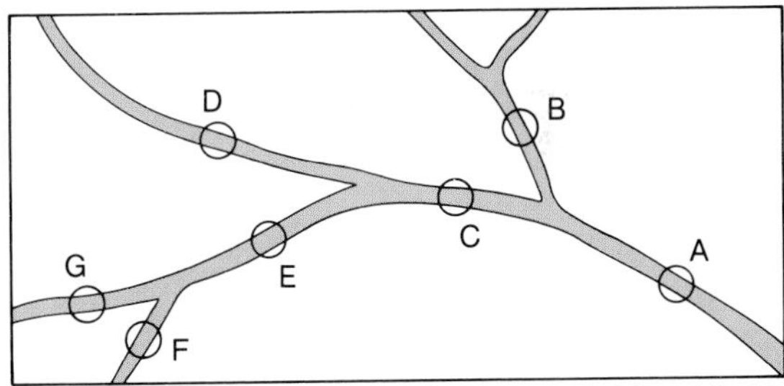

Fig. 10–12

CONCLUSIONS:

1. In Fig. 10–12, where would the source of the gold most likely be found?
2. Why did the fool's gold stay in the pan?
3. What could be done to increase the amount of mineral recovered each time you panned?

10-3. Water Resources

At the end of this section you will be able to:

- Identify three of the ways in which water can become polluted.
- Compare the amount of water used by households, industry, and agriculture.
- Explain the importance of local water budgets and *watersheds*.

Name something that you use every day and that costs about 15¢ a metric ton delivered to your house. You are correct if you answered that water fits this description. Turn on a faucet and out comes a stream of clean, running water. Will there always be enough? The answer is yes, if two problems can be solved. One problem comes from pollution of the water supply. The other problem is that nature does not always provide the water where it is needed.

PROTECTING THE WATER SUPPLY

Many millions of years ago the earth had more than nine million billion metric tons of water. Almost all of that water is still available today. Water is a renewable resource. Water is used, but it is not used up. Almost every drop is returned to the natural water cycle. In this cycle, water moves to the sea, evaporates, and falls as rain to the earth again. Water is made pure as it moves through the water cycle. In this way, nature gives us pure water.

Since 1900, the amount of water used in the United States has doubled every 25 years. People now use more water than ever before. See Fig. 10-13. If the population continues to grow at the present rate, three times as much water will be needed in the year 2000 as is needed now. There is plenty of water to meet those future needs. About 12 trillion liters of rain fall on the United States each day. We use only about one-tenth of it. However, none of that water can be used unless it is protected from pollution.

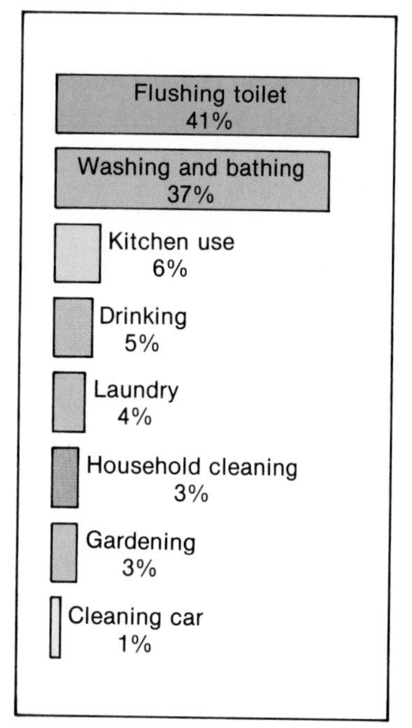

Fig. 10-13 This graph shows how water is used in the average household.

1. Sewage. One kind of pollution comes from untreated sewage wastes. See Fig. 10-14. Raw sewage in the water supply carries germs that cause diseases. However, the water flowing

Fig. 10-14 Raw sewage is being flushed into the water. This can cause serious pollution problems.

in streams can break up and destroy the dangerous wastes. Some of the wastes are dissolved and some settle to the bottom. The moving water also mixes with the air. The wastes in streams combine with oxygen. The dissolved oxygen in the water acts on the wastes, making the sewage harmless. But in an overloaded stream, waste materials use up all of the oxygen dissolved in the water. The stream cannot purify itself until more oxygen is taken in from the atmosphere. Increased amounts of waste material in the water prevent the natural action from catching up. Thus the stream becomes polluted. Pollution from sewage can be prevented in two ways. One way is to control the amount of wastes dumped into the water. A second way puts the raw sewage through a treatment plant. A sewage treatment plant destroys the germs. Many treatment plants also remove much of the material that would use the oxygen in the water.

2. Chemicals. A second kind of pollution is caused by dumping chemicals that cause the growth of too many plants. The oxygen supply of a stream or lake can be upset. One such chemical is called *phosphate* (**fahs**-fate). Phosphates are found in some detergents, cleaning materials, and fertilizers. They cause small forms of plant life, such as algae, in the water to grow rapidly. When these plants decay, they use the dissolved oxygen in the water. Many of the living things in that body of water are killed. This upsets the natural balance of life in the water. To maintain a pure water supply, materials that damage the natural balance of life cannot be allowed to enter streams and lakes.

Fig. 10–15 Wastes from industry may be one of the major sources of chemical pollution.

Chemicals that come from industrial wastes may be the most serious pollutants over a long period of time. This kind of pollution can be caused by chemicals directly entering the water supply. See Fig. 10–15. The chemicals used in agriculture can also be a source of pollution. Many of these chemicals wash down through the soil or run off in streams. See Fig. 10–16. Some chemicals enter the ground water supply and cause pollution in water taken from wells. Streams may carry the chemicals over a wide area. The polluted water then becomes part of the ground water or collects in lakes and reservoirs. Most of the water now used in homes contains very small amounts of chemicals from industry and agriculture. Over long periods of time, the effect of these chemicals on human health and other forms of life is not known.

3. Oil spills. A very serious kind of pollution comes from oil spills in the oceans. The oil can be released by accidents to tankers or offshore oil wells. In past years, much of the oil pollution came from tankers that cleaned their oil-storage tanks with sea water. Now there are international laws to prevent such pollution. However, accidental oil spills will continue to be a problem. Oil pollution in the sea can kill much of the natural life. The floating oil spreads out. Much of it finally sinks to the sea floor. Some may reach the shore, where beaches are ruined and the marine life often is killed. Spilled oil is very hard to remove from the sea surface or the land along the shore. The solution to the problem can be found only in measures that will prevent this dangerous form of pollution from happening.

Fig. 10–16 Chemicals used as fertilizers and for pest control in agriculture can enter the ground water supply.

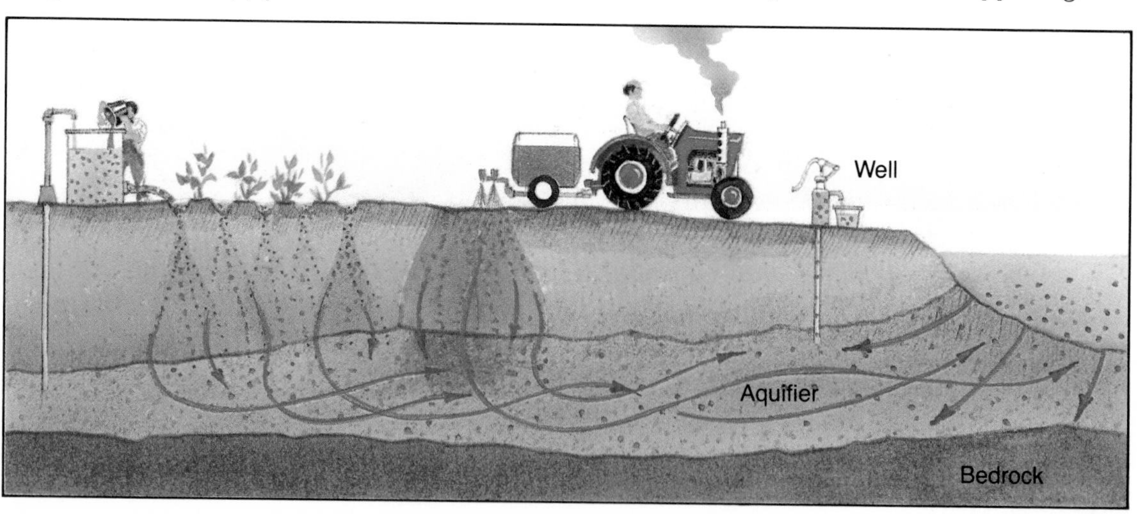

4. Thermal pollution. Another way in which water can become polluted is through its use as a cooling agent. Power plants, for example, produce a large amount of waste heat. Many power stations use water from rivers to remove this heat. Water used to cool power plants becomes heated. This heated water is sent back to nearby rivers, lakes, or oceans. The heated water may affect the natural balance of plant and animal life where it is released. The release of heat into streams, other bodies of water, or the atmosphere is called **thermal pollution.** *Thermal pollution* can kill the eggs or young of animals that live in water, such as fish. The increase in temperature also reduces the oxygen in the water. This can change the normal life cycle of plants and animals in the water.

Thermal pollution Release of heat into streams, lakes, oceans, or the atmosphere.

THE WATER SUPPLY

About 97 percent of the earth's water is found in the oceans. Over 2 percent of the remaining water is "locked up" in the polar icecaps and glaciers. The supply of water available for human use must come from the 1 percent that remains. This water moves between the sea and the land.

It may seem to you that the important uses of water are the household uses. Your life would certainly be different without a supply of water in your house. However, households use only a small part of the total water supply. Industry uses about three times more water each day than is used in all the households in the country. In order to make an automobile, for example, more than 100,000 liters of water are needed. In order to make one liter of gasoline, nearly 200 liters of water are needed.

Almost all of the water used in factories and homes can be cleaned and made pure. It can then be stored and used again before it is returned to the natural water cycle. Some water cannot be returned to the water cycle easily, such as water used for crops. In the United States, more water is used for farming than for all other uses combined. Water needed by crops is brought to the fields from deep wells or distant sources. The water used for farming may be taken up by plants. The rest runs over the surface, evaporates, or seeps into the ground. This water cannot be used by people until it makes its way through the water cycle. The water will continue to move through the cycle until it reaches a large body of water or becomes part of the ground water.

LOCAL WATER SUPPLIES

The water cycle describes how moisture moves between bodies of water, the air, and the land. It is also important to know how much water there is at a given location. A *local water budget* is the record of moisture income, storage, and outgo at a given location. The local water budget will tell you whether there is a lack of water or an oversupply of water at any particular time. Surplus water is stored in the soil, as part of the ground water, or becomes runoff. This runoff may be stored in lakes or reservoirs. Any water that cannot be stored flows away in streams to the ocean. Tables 10–1 and 10–2 give the amount of moisture income and outgo each month for a typical year in Savannah, Georgia, and Yuma, Arizona. Compare these two tables. Which place has the greatest yearly surplus of water?

WATER BUDGET FOR SAVANNAH, GEORGIA												
	Jan.	Feb.	Mar.	Apr.	May	June	July	Aug.	Sept.	Oct.	Nov.	Dec.
Income (mm) (rain)	70	80	88	73	80	135	175	180	150	75	50	70
Outgo (mm) (evaporation, use)	20	25	40	75	125	165	175	155	120	75	28	18
Monthly surplus (mm)	50	55	48	0	0	0	0	25	30	0	22	52
Monthly deficit (mm)	0	0	0	2	45	30	0	0	0	0	0	0

Table 10–1

WATER BUDGET FOR YUMA, ARIZONA												
	Jan.	Feb.	Mar.	Apr.	May	June	July	Aug.	Sept.	Oct.	Nov.	Dec.
Income (mm) (rain)	17	14	10	4	2	0	4	12	6	4	6	15
Outgo (mm) (evaporation, use)	18	19	20	90	140	190	210	200	165	100	30	15
Monthly surplus (mm)	0	0	0	0	0	0	0	0	0	0	0	0
Monthly deficit (mm)	1	5	10	86	138	190	206	188	159	96	24	0

Table 10–2

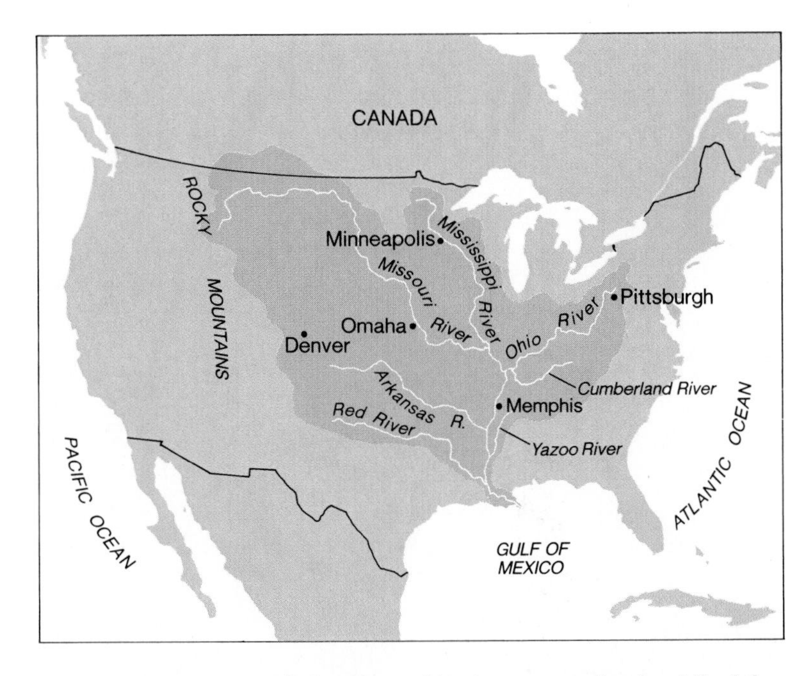

Fig. 10–17 The watershed of the Mississippi.

Watershed The land area that drains into a stream or river.

A local water supply is affected by its **watersheds.** All of the land area that is drained into a stream or river is a *watershed.* Every square meter of land is part of a watershed. Each stream has its own watershed. A watershed may be as small as a single field or as large as several states. The watershed of the Mississippi River, for example, includes almost half of the United States. See Fig. 10–17.

The flow of water in streams is determined mostly by its watershed. A watershed made up of land without plants will not soak up water. Muddy water can run off rapidly and cause floods. Since there are no plants to hold the soil together, soil is washed into the streams and reservoirs. This reduces the amount of water that can be stored. On the other hand, a watershed that is covered with plants will absorb most of the water falling on it. Floods and droughts are less common where watershed areas are covered with forests or other plants.

A watershed can be damaged if the protective plant covering is removed. This may happen because of poor farming methods, fires, or the cutting of trees for lumber. See Fig. 10–18. Today many watersheds are destroyed by the building of paved lots, such as shopping malls. If there is to be a steady flow of water into streams and rivers, all watersheds must be protected. Damaged watersheds must be repaired by replacing the plants.

Fig. 10-18 Forest fires are a common cause of damage to watersheds.

SUMMARY

The earth provides a water supply that is able to renew itself. However, the water supply must be protected from pollution in order for us to continue to have enough water to meet our needs. Most of the fresh water supply is used by agriculture and industry. Although there is plenty of water, it is not always found where needed. Many regions must plan carefully in order to have sufficient water to meet their needs. Destruction of watersheds can affect the water supply of streams and rivers.

QUESTIONS

Use complete sentences to write your answers.

1. Explain why water is a renewable resource.
2. Name three ways in which water can become polluted.
3. Describe the manner in which industrial and agricultural chemicals get into our home water supply.
4. Why is thermal pollution a problem?
5. Compare the amount of water used in industry and agriculture with that used in the home.
6. Why is it important to protect the watersheds of streams?

INVESTIGATION

LOCAL WATER BUDGETS

PURPOSE: To study the local water budgets for two different cities.

MATERIALS:
meter stick
two 10-cm tall containers
small pan
container of water (water source)

PROCEDURE:

A. Look at Table 10–1, page 284. It shows the water budget for Savannah, Georgia.
   1. What is the water income for January?
   2. What is the water outgo for January?

B. Set one of the 10-cm tall containers in the pan. This will represent Savannah's water supply. Fill the container to a depth of 70 mm (7 cm). This represents the 70-mm water income for the month of January.

C. Now remove 20 mm of water from the container. This amount of water, lost by use of plants, evaporation, and other ways, is the outgo for January.
   3. What is the depth of the water left in the container? This water represents the water stored in the soil, lakes, reservoirs, etc.
   4. How does this compare with the monthly surplus for January shown in Table 10–1?

D. Fill the second container to 80 mm (8 cm) for the income shown for February. Remove 25 mm of the water for outgo. The remaining 55 mm are the surplus for February. Pour this into the first container with the January surplus.
   5. Describe what happened.

E. Any water that overflows into the pan represents water that cannot be stored. It runs off to another place. The water left in the container represents stored water in the water budget for future use. Pour the water in the pan back into the water source.

F. Continue this for the remaining months. Notice that in April, May, and June, the outgo is greater than the income. You will need to pour some water out of the first container to meet this outgo.
   6. Was there any month during the year that the surplus ran out?
   7. In what months were the lakes and reservoirs full?

G. Look at Table 10–2. It shows the water budget for Yuma, Arizona.
   8. Describe three ways that Yuma's water budget differs from Savannah's.
   9. Why do you think millions of dollars have been spent to route water from the Colorado River to Yuma?

CONCLUSIONS:
1. In which city would you expect to find a desertlike environment? Why?
2. If the water budget shows a deficit for a given month, does that mean there is no water available for use? Explain your answer.

10-4. Nonrenewable Energy Resources

At the end of this section you will be able to:

- List the *fossil fuels* and describe how each is formed.
- Describe how the supply of fossil fuels will change in the future.
- Compare the processes of *fusion* and *fission* as sources of *nuclear energy.*

Think about a world in which most work had to be done by using energy supplied by human or animal muscles. See Fig. 10-19. How would your life be different? What happens when the electric power is shut off? What would happen if the gasoline supply were cut off? Without a steady supply of energy, there would be many changes in the way you live.

Fig. 10-19 Without fossil fuels for tractors, we would have to rely on horse power to plow fields. Think of other ways your life style would change without fossil fuels.

FUELS FROM THE EARTH

Two hundred years ago, muscle power and power from burning wood were the major sources of energy. Today, such sources supply only a tiny part of the world's energy needs. The Industrial Revolution brought machines that use large amounts of energy. Most of this energy has come from the burning of **fossil fuels.** These fuels have been formed from the decay of plant and animal fossils in the earth's crust. The *fossil fuels* are coal, petroleum, and natural gas.

Fossil fuel Mineral fuel found in the earth's crust.

1. Coal. The energy in coal came from the sun. Plants living in swamps long ago stored solar energy. The swamps became buried because of changes in the crust. Then the plant material was slowly changed to coal deep beneath the surface. Thus the huge deposits of coal that exist today were once living plants.

There are several forms of coal. Each form is a stage in a process by which plant materials are changed by increasing pressure and heat. The process changes plants into pure carbon. The first stage is *peat*. The plant material is only slightly changed in this stage. Peat looks like rotted wood and is found on or near the surface of the ground. It is used as a fuel in some parts of the world. The second stage in forming coal is *lignite* (**lig**-nite), often called "brown coal." Lignite is brownish black, catches fire easily, and burns with a smoky flame. Deposits of lignite are found beneath the surface. The third stage in the making of coal is *bituminous* (bih-**too**-mih-nus), or soft coal. It is dark brown or black in color and breaks easily into blocks. Bituminous coal burns with a smoky flame. It is the most abundant kind of coal. More heat and pressure on a coal deposit will change bituminous into a much harder kind of coal called *anthracite* (**an**-thruh-site). This is a hard, brittle, black form of coal that is almost pure carbon. It burns with a blue, nearly smokeless flame and produces great heat. Anthracite is a kind of metamorphic rock. The other forms of coal are sedimentary rock. The different forms of coal are shown in Fig. 10–20.

Fig. 10–20 Stages in coal formation.

Fig. 10–21 Strip-mining for coal is safer, but it destroys the landscape and the natural habitat for wildlife.

There are serious problems in using coal as a major source of energy. For example, much coal contains large amounts of sulfur. When the coal is burned, the sulfur produces sulfur dioxide gas. This gas causes a serious air pollution problem. Sulfur dioxide gas also is a cause of acid rain. Power plants that use coal must install special equipment to prevent the release of large amounts of sulfur dioxide from their smokestacks. Carbon dioxide gas, which is also released by burning coal, may build up in the atmosphere. This carbon dioxide may cause the earth's climate to change. Many scientists have predicted a warming of the earth.

Mining coal also creates serious problems. Coal is obtained from many deposits by strip mining. See Fig. 10–21. The damage caused to land by strip mining must be repaired to preserve the natural environment. Another problem with coal is the danger to the miners who work in underground coal mines. Lives are lost because of accidents. Diseases can be caused by breathing coal dust. However, these problems are not likely to prevent large amounts of coal from being used as a main source of energy. In the future, coal may be used as a liquid fuel. It then can help to replace **petroleum** (puh-**troe**-lee-um). Coal can also be used to make a gas that can be used in place of natural gas.

Petroleum A fossil fuel that is usually a liquid, that comes from the decay of ancient life in warm, shallow seas.

2. *Petroleum* is a liquid that comes from the decay of ancient life forms in warm, shallow seas. The remains of these plants and animals settled on the sea floor and were covered by sediments. After about 100 million years, this material was changed to substances made of hydrogen and carbon. These substances are called **hydrocarbons.** Petroleum is a mixture of many different *hydrocarbons*. Petroleum can be separated into different products, such as gasoline, kerosene, and motor oil. See Fig. 10–22 *(top)*. Some of the hydrocarbons in petroleum are used to make plastics, rubber, soaps, and some medicines.

Hydrocarbons The principal substances found in petroleum. They contain only hydrogen and carbon.

Petroleum was formed from the mud on the sea bottom. The mud was later changed into the sedimentary rock called shale. Pressure usually forced the petroleum out of the shale. The petroleum then moved into rock with more open spaces, such as sandstone. Petroleum may also have collected where layers of sandstone and limestone were joined. Water may also have been trapped in the sedimentary rock. The oil, being less dense

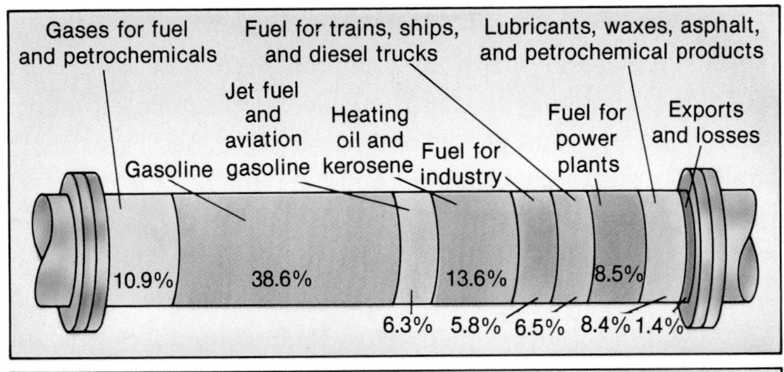

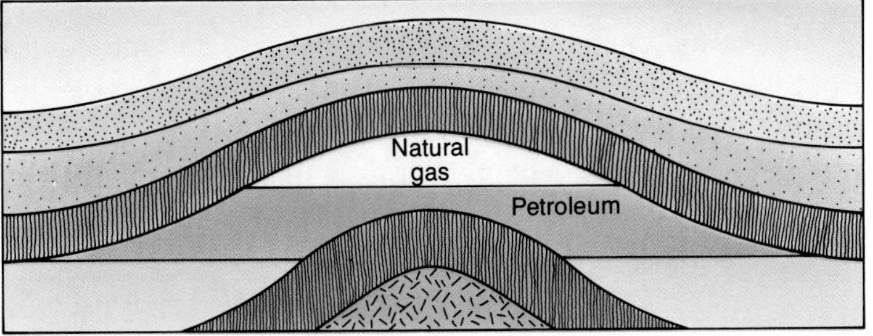

Fig. 10–22 (top) *Refining separates petroleum into these products.* (bottom) *The petroleum is trapped beneath a dome of non-permeable rock forming an oil pool.*

than water, floated upward. The oil stopped rising when it met a layer of rock that it could not flow through. The oil collected below this layer and formed an *oil pool.* Fig. 10–22 *(bottom)* shows one of the most common arrangements of rock layers that contain oil pools.

3. Natural gas is also a hydrocarbon. Petroleum and natural gas are usually found together. Natural gas is often trapped with the petroleum beneath the solid rock covering an oil pool. See Fig. 10–22.

FOSSILS FUELS IN THE FUTURE

The supplies of fossil fuels in the earth's crust are limited. It has taken millions of years to make them. Thus they are nonrenewable. When the supplies are used up, there will be no quick way to replace them. There are probably greater amounts of coal than of petroleum remaining within the earth. Coal comes from sediments formed in the shallow seas that once covered large parts of the continents. Because the continents are very old, enough time has passed for large deposits of coal to be formed. Petroleum, on the other hand, is found in sediments laid down along the edges of continents. As the crustal plates moved and collided, these sediments were destroyed. Petroleum and natural gas deposits are younger than coal and are less common. It is not known how much oil and gas are left.

Fig. 10–23 A piece of oil shale that has been cut and polished to show the brown streaks that yield oil.

Most of the petroleum and gas deposits still to be found in the United States probably are found in the shallow waters just off some parts of the coasts. There is a huge amount of petroleum found in a kind of rock called *oil shale*. See Fig. 10–23. Large deposits of oil shale are found in parts of Colorado and Utah. However, the petroleum in oil shale is not liquid. The rock must be mined and then heated to release the oil. This is a difficult and expensive process. Many problems must still be solved before oil shale can become a source for our petroleum needs.

Most of the energy that is used at the present time comes from the burning of fossil fuels, which are nonrenewable. Industry is the biggest energy user. Making steel, for example, uses large amounts of energy. Forms of transportation, such as automobiles, also use a large share of the energy supply. Industry and transportation together use about 60 percent of the energy produced in this country each year. The rest is used by homes and businesses.

The total amount of energy used in the United States has grown rapidly during recent years. If energy continues to be used at the present rate, the amount needed will double every 20 years. This increase in the need for energy is caused by the growing population. Also, the amount of energy used by each person is increasing every generation. You use twice as much energy as your parents did when they were your age. Table 10–3 shows how much energy is used by home appliances each year.

For the past hundred years, the growing need for energy has been met by using fossil fuels. However, fossil fuels cannot be renewed in a short period of time. This has created an energy crisis. At the present rate of use, the supply of petroleum will run out. It probably will be used up within the next 50 to 100 years. The supply of coal will last much longer. In fact, at the present rate of use, coal reserves will last about 200 years. However, burning coal can cause serious damage to the environment. In the future, renewable sources of energy may be able to replace the fossil fuels. These energy sources will not run out because they will constantly be replaced. Until this happens, however, the only answer to the present energy crisis may be careful conservation of the present energy supply.

ESTIMATED ANNUAL ENERGY CONSUMPTION OF ELECTRIC APPLIANCES UNDER NORMAL USE			
Appliance	Estimate of Energy Used Each Year* (kW-h)**	Appliance	Estimate of Energy Used Each Year* (kW-h)**
Blender	15	Washing machine (nonautomatic)	76
Broiler	100	Water heater	4,219
Carving knife	8	Air cleaner	216
Coffee maker	106	Air conditioner (room)	860*
Deep fryer	83	Electric blanket	147
Dishwasher	363	Dehumidifier	377
Egg cooker	14	Fan (attic)	291
Frying pan	186	Fan (circulating)	43
Hot plate	90	Fan (rollaway)	138
Mixer	13	Fan (window)	170
Microwave oven	190	Heater (portable)	176
Range		Heating pad	10
with oven	1,175	Humidifier	163
with selfcleaning oven	1,205	Hair dryer	14
Roaster	205	Heat lamp (infrared)	13
Sandwich grill	33	Shaver	1.8
Toaster	39	Sun lamp	16
Trash compactor	50	Toothbrush	0.5
Waffle iron	22	Radio	86
Waste disposer	30	Radio/Record Player	109
Freezer (15 cu ft)	1,195	Television	
Freezer (frostless 15 cu ft)	1,761	Black & white	
Refrigerator (12 cu ft)	728	tube type	350
Refrigerator (frostless 12 cu ft)	1,217	solid state	120
Refrigerator/Freezer		Color	
(14 cu ft)	1,137	tube type	660
(frostless 14 cu ft)	1,829	solid state	440
Clothes dryer	993	Clock	17
Iron (hand)	144	Floor polisher	15
Washing machine (automatic)	103	Sewing machine	11
		Vacuum cleaner	46

* Based on 1,000 hours of operation per year. This figure will vary widely depending on area and specific size of unit.
SOURCE: Edison Electric Institute
** kW-h = kilowatt-hour

Table 10–3

NUCLEAR ENERGY

Nuclear energy Energy that is produced by changes in the nuclei of atoms.

Nuclear energy has already begun replacing fossil fuels. *Nuclear energy* comes from changes in the *nuclei* (**noo**-clee-ie), or cores, of certain atoms.

Fission The splitting of the nuclei of certain kinds of atoms.

There are two ways to make energy from the nuclei of atoms. One method is called **fission** (**fish**-un). In *fission,* the nuclei of atoms are split into smaller nuclei by neutrons. See Fig. 10–24. When fission takes place, a large amount of heat is given off. The heat can be used to produce steam to run electric generators. See Fig. 10–25. Uranium is the fuel most often used in these power plants. A piece of uranium the size of a marble can produce the same amount of heat as 100 metric tons of coal. But there is not very much uranium in the earth. Uranium could be used up within the next 50 years.

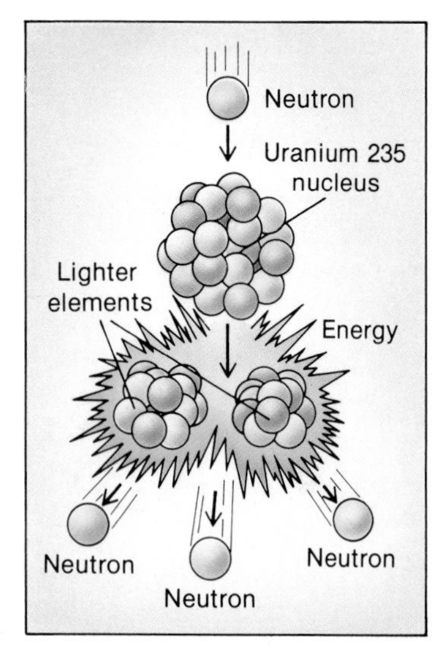

Fig. 10–24 A neutron can strike certain kinds of atomic nuclei causing fission with a release of energy.

Many people believe that nuclear plants are not safe. One problem is the radioactive wastes that come from the power plants. These wastes must be removed from time to time. They are stored until they lose their radioactivity. The wastes can remain dangerously radioactive for hundreds, even thousands, of years. A place must be found to put these dangerous substances where they cannot escape into the environment. Some people believe that nuclear power plants are unsafe because of the way they are cooled. Plants are cooled by water. If the cooling system were to stop working, the plant would overheat, and could even melt. Then the radioactivity would escape. However, the people who build and operate nuclear plants believe that the problems can be solved. All nuclear plants use safety systems, and research is reducing the risks.

Fig. 10–25 This power plant produces electricity by using the heat from nuclear fission.

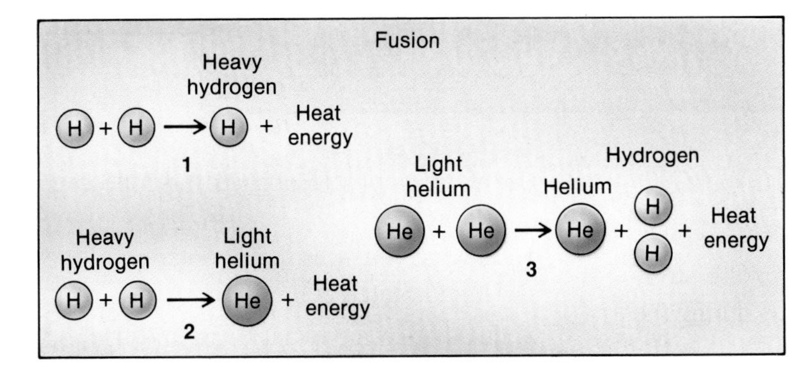

Fig. 10–26 Fusion is joining together atomic nuclei, causing energy to be released.

A second method of producing nuclear energy may solve some of the problems caused by nuclear plants. This method is called **fusion.** *Fusion* (**fyoo**-zhun) is the joining of atomic nuclei to make new and larger nuclei. See Fig. 10–26. When fusion of nuclei takes place, very large amounts of heat are released. This heat may be used to run electric generators. But nuclear fusion happens only at very high temperature and pressure. It is hard to make machines that can withstand these temperatures and pressures. Scientists are trying to find ways to use nuclear fusion as a source of energy. It would solve many of the world's energy problems. The fuel needed for nuclear fusion is a form of hydrogen that can be taken from sea water. At some time in the future, the sea may become an energy resource that will never run out.

Fusion The joining together of nuclei of small atoms to produce larger atoms with the release of energy.

SUMMARY

During the past century, people have been using more and more fossil fuels, such as oil and coal. These fuels take millions of years to form. The supply of fossil fuels will be running out unless we conserve energy. Nuclear energy may replace fossil fuels in the future.

QUESTIONS

Use complete sentences to write your answers.

1. Name and describe the four forms of coal.
2. Name two other fossil fuels and tell how they formed.
3. Explain why there will be a short supply of fossil fuels in the future.
4. Describe two ways to produce energy from the nuclei of atoms.

SKILL-BUILDING ACTIVITY

USING GRAPHS TO STUDY ENERGY USED IN THE UNITED STATES

PURPOSE: To prepare graphs for the study of energy used in the United States during the last half of the 20th century.

MATERIALS:
graph paper pencil

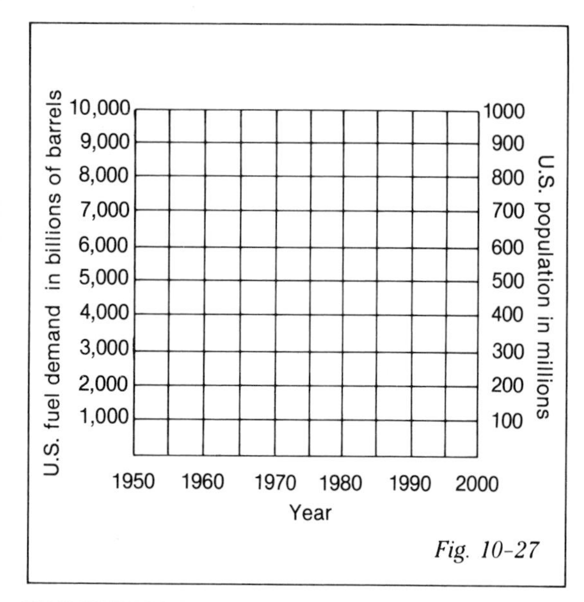

Fig. 10-27

PROCEDURE:

A. On graph paper, draw axes like those in Fig. 10-27 and plot the values shown below. Use a solid line for Fuel Demand/Year and a dotted line for Population/Year.

Year	Fuel Demand (billions barrels)	Population (millions)
1950	1,800	150
1960	2,800	180
1970	4,300	210
1980	5,600	220

B. Use your graphs to answer the following questions.
   1. In what year was the population of the United States 200 million?
   2. In that year, what was the United States' fuel demand?

C. One way to see national energy used is to note how much energy is required to support one person. To do this, divide your answer to question 2 by 200 million.
   3. What was the answer?

D. Using a dashed line, extend the solid line (fuel demand) and the dotted line (population) to the year 2000. Keep the shape of each graph the same. Then answer the following questions.
   4. What will the population of the United States be in 1990?
   5. What will the United States' fuel demand be in 1990?
   6. What will the fuel demand per person be for the year 1990?
   7. How does this compare to your answer to question 3?

E. The future population and demand for fuel involves many things not included in these graphs. The effects of alternate energy sources, conservation, and future weather conditions are not known.
   8. What will the United States' fuel demand per person be in the year 2000?

CONCLUSIONS:

1. Explain the relationship between population growth and fuel demand.
2. What must be done so that the future fuel demand can be met?

10-5. Renewable Energy Resources

At the end of this section you will be able to:

- List four ways that solar energy can be used.
- Explain how *geothermal* energy can be used.
- Describe how moving water, wind, and plants can provide energy.

Sunlight contains energy. However, this energy must be gathered and changed into a useful form such as electricity. The mirrors shown in Fig. 10-28 are helping to change sunlight into electricity.

Fig. 10-28 Solar mirrors can be used to turn solar energy into electricity.

SOLAR ENERGY

When sunlight falls on your skin, the warmth you feel is part of the *solar energy* that the earth receives from the sun. Each year the sun sends 500 times more energy to the United States than we use. The entire energy needs of the country could be met by using only a small part of that solar energy. However, solar energy must be gathered and changed into forms that can be used.

1. Passive. A house can be built in a way that allows it to gather solar energy. A solar-energy system that doesn't have any motors, pumps, or other moving parts, is called *passive.* "Passive" means "not active." A passive solar house has windows facing the sun. Sunlight enters the house through the windows and heats materials inside the house, such as stone or

Fig. 10–29 During the daytime, sunlight enters through the windows and heats the air and floor. The warm air rises while the cool air sinks. At night, the warm floor provides heat.

brick. See Fig. 10–29. The warmed materials then heat the living space. The materials can also store some of the heat for the night.

2. Collectors. Solar collectors are usually boxes with glass or plastic tops. They are placed on roofs facing the sun. Inside the boxes are tubes, which are usually filled with water and antifreeze. However, another liquid or air may also be used. The tubes and the inside of the box are painted black to absorb more heat. After the water is heated, it is moved to a storage tank where it is ready to be used. See Fig. 10–30. One problem with solar heating is the lack of sunlight on cloudy days.

Fig. 10–30 (left) *This drawing illustrates how this house uses solar energy directly to heat water. The water is heated as it passes through solar collectors on the roof.* (right) *Even older homes can have solar collectors installed.*

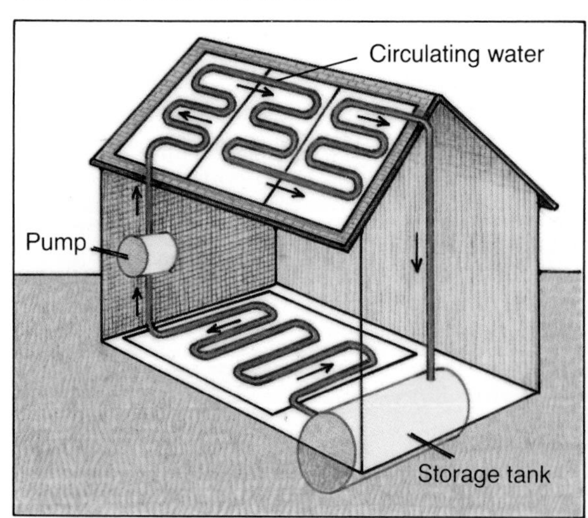

3. Solar cells. Sunlight can be changed into electricity by the use of **solar cells.** Each *solar cell* produces only a small amount of electricity. Therefore, a large number of cells are needed to create a useful amount of electricity. While solar energy itself is free, solar cells are presently very expensive to make. For this reason, it would be costly to use enough solar cells to provide the electricity needed in a house. Their use now is mostly in spacecraft and other places where no other source of electricity can be used.

Solar cell A device that is used to change sunlight directly into electricity.

4. Concentrators. Sunlight, if concentrated enough, can be used to boil water and produce steam. The steam can then be used to turn generators to produce electricity. Solar power plants have been built that use mirrors to reflect sunlight from a large area onto a smaller surface. See Fig. 10–31. This is similar to using a magnifying glass to concentrate sunlight.

Fig. 10–31 A central, solar power station uses mirrors to reflect sunlight onto steam generators.

GEOTHERMAL ENERGY

In some places it is possible to use **geothermal** (jee-oe-**thur**-mul) **power.** *Geothermal power* comes from heat within the earth's crust. The heat causes ground water to become hot enough to make steam. One source of geothermal power uses this steam when it is released through wells. Only a few of

Geothermal power Energy obtained from heat in the earth's crust.

the earth's locations can use geothermal energy. Some places have been found where bodies of magma are close to the surface. Another way to use the earth's heat is by drilling deep wells at these places to produce very hot water. The hot water can then be used as heat or to generate electricity.

WATER, WIND, AND PLANT POWER

We usually think of solar energy as just sunlight. However, solar energy also exists in other forms: It causes water to evaporate and fall as rain; by heating the atmosphere, it powers the wind; and sunlight is stored by plants as they grow. Water, wind, and plants are indirect and renewable sources of solar energy.

Hydroelectric power Electricity produced by water moving downhill.

1. Water power. Water power, or **hydroelectric** (hie-droe-eh-**lek**-trik) **power,** has been used for many years to produce electricity. See Fig. 10-32. *Hydroelectric power* comes from water that is flowing downward. As the water flows down, it turns a machine that makes electricity. The energy of falling water really comes from solar power. Heat from the sun causes water to evaporate, mostly from the sea. When the water falls as rain and runs back to the sea, stored solar energy is released. Hydroelectric power is constantly renewed as the water cycle goes on. The world still has large amounts of water-power resources that are not being used.

Fig. 10-32 Water flowing through this dam turns machines that produce electricity.

2. Wind. Wind also comes from solar power. Wind is caused by the heating of air. Like water power, wind is renewable. The energy of wind is never used up. There is a large amount of energy in wind. Some of this energy can be used to produce electricity by turning a windmill. See Fig. 10–33. Using even a small part of the total amount of energy in winds would meet the world's power needs. Windmills seem to work best on plains, the tops of mountains, and along some coastal areas. Small windmills have been in use in the United States since about 1920. These were mainly on farms. However, the wind cannot always be used as a source of power. There are not many places where the wind blows steadily enough to run a power station made up of windmills.

Fig. 10–33 Wind energy produces electricity by turning this modern windmill.

3. Biomass. Plants are like solar collectors. They absorb energy in sunlight. Leaves, wood, and other parts of plants are made when the plants use solar energy. Thus wood and other plant materials contain stored solar energy. This source of energy is called **biomass.** The plant materials that make up *biomass* can be changed into a liquid fuel. For example, plant material that contains sugar or starch can be made into alcohol. The alcohol can be burned as a motor fuel or mixed with gasoline to make *gasohol.* An acre of corn can produce about 1,045 liters of ethanol. But using farms to produce fuel also uses up land that could be used for growing food.

Biomass Plant materials that are used to produce energy.

SUMMARY

As the supplies of fossil fuels are used up, renewable energy sources must help to meet our energy needs. The most important renewable source of energy is the sun. Solar energy can be used directly. It also can be used in the form of moving water, wind, or plant materials. The earth's heat can be used as a source of energy in certain locations.

QUESTIONS

Use complete sentences to write your answers.

1. Describe four ways that solar energy can be used.
2. Give two ways in which geothermal energy can be used.
3. Explain how energy can be provided by hydroelectric power, wind, and biomass.

INVESTIGATION

MAKING A SOLAR COLLECTOR

PURPOSE: To make a solar collector and find out how it absorbs light to make heat.

MATERIALS:

box	2 thermometers
black paper	transparent wrap
tape	light source

PROCEDURE:

A. Make a copy of the following table:

Condition	Inside Thermom- eter	Outside Thermom- eter
After 5 min. with light		
After 5 min. without light		

B. Line the inside of the box with black paper.

C. Tape one thermometer inside the box. Be sure that the numbers on the thermometer are showing.

D. Cover the open side of the box with transparent wrap. Tape the edges to the sides of the box.

E. Place the box and the other thermometer under a light source as shown in Fig. 10–34. After five minutes, read both thermometers. Record the temperatures in your table.

1. Which thermometer had the higher temperature?

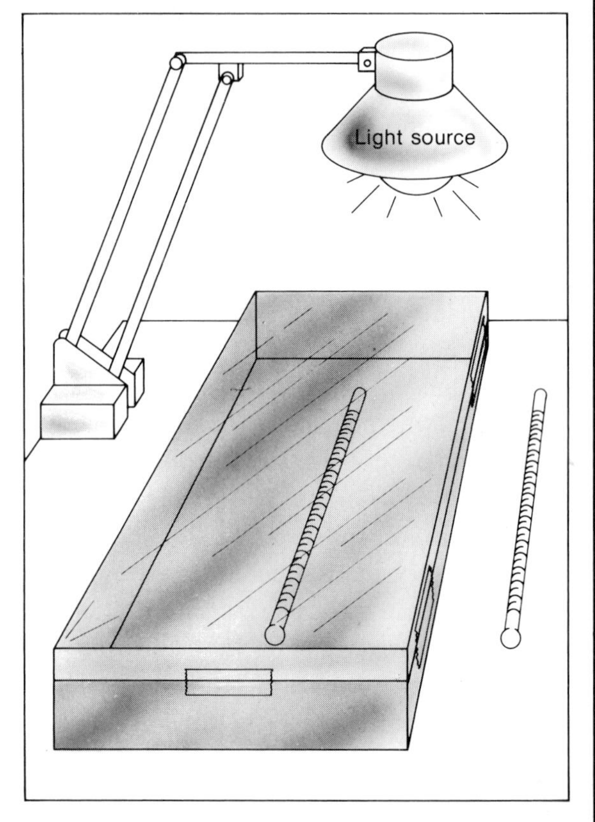

F. Remove the light source. After five minutes, read both thermometers. Record the temperatures in your table.

2. Which thermometer had the higher temperature?

CONCLUSIONS:

1. Which thermometer absorbed more heat when placed under the light source? Why?
2. Which thermometer lost more heat when the light was removed? Why?
3. From the results of this activity, explain why solar collectors have a transparent cover and are painted black inside.

TECHNOLOGY

CONVERTING GEOTHERMAL ENERGY TO ELECTRICITY

Heat energy from the earth may be free for the taking. What sounds like a dream may fast be becoming an economic reality. People have used the earth's heat, called geothermal energy, for cooking, recreational activities, and for heating homes since before the time of the Romans. Increasingly sophisticated technology has allowed us to tap heat sources deeper and deeper into the earth. In Reykjavík, the capital of Iceland, hot water from wells deep in the earth has been piped into homes, factories, and schools since 1930. Some of these wells are over 2 kilometers deep, and the temperature of the water from them is almost 225°C. Reykjavík is located over a gigantic reservoir of hot water, and can use geothermal energy directly. However, if geothermal energy is to be used more generally, it must be converted to a form of energy that can be transmitted from the source of heat to the place where its use is required. That form of energy is electricity.

The world's largest geothermal electricity project is located at Geysers, California, this facility produces enough electricity to meet the requirements of San Francisco. Geysers is unusual in that the water it uses is brought to the surface as steam. This steam can be used directly to turn turbines to generate electricity. More commonly, hot water comes to the surface under pressure, and energy must be added to change it to steam. This hot water usually contains dissolved minerals, which present one of the major problems to be solved if geothermal energy is to be used economically. Some water contains up to 35 percent dissolved minerals. When pressure is reduced and the water turns to steam, these minerals settle. This settling can reduce the diameter of a 7.6-centimeter pipe to as little as 1.3 centimeters in 100 hours and destroy parts made of normal metals.

A number of different approaches to converting geothermal energy to electricity are being tested today. At one facility in California, water at 210°C is pumped to the surface. As the pressure on the water is reduced, some of it "flash boils," or turns to steam. In this way, about 20 percent of the liquid is used to turn turbines. The rest of the water is pumped back into the ground. Another approach is the binary-cycle (two cycle) method. In this method, the hot water is used to heat another substance, which has a lower boiling point. This second liquid vaporizes and drives a turbine. With this method, scientists hope to eliminate the problem of dissolved minerals and to allow the use of lower temperature water.

With the potential radiation problems of nuclear plants, the pollution problems surrounding plants that burn coal, and the high cost of gas and oil, alternative energy sources are always being sought. Geothermal generation of electricity could be an important technological advance for the western United States and other parts of the world. (The photo shows a geothermal plant that is now in operation in Italy.) Geothermal energy, like wind energy, can provide a safe and clean source of fuel to bolster depleting reserves of fossil fuels.

CHAPTER REVIEW

VOCABULARY

On a separate piece of paper, match the number of the sentence with the term that best completes it. Use each term only once.

fossil fuel	watershed	conservation	fission
geothermal power	thermal pollution	mineral deposit	fusion
solar cell	pollution	natural resource	petroleum
nuclear energy	recycling	ore	hydrocarbons

1. A(n) ______ is any substance from the earth that can be used in some way.
2. The adding of harmful substances to the environment is called ______.
3. A kind of rock that contains useful minerals is called a(n) ______.
4. ______ is the saving of natural resources by controlling how they are used.
5. A(n) ______ is a body of ore.
6. The process of recovering a natural resource and using it again is ______.
7. When water is used as a cooling agent, ______ may occur.
8. All the land drained by a stream or river is called a(n) ______.
9. A mineral fuel found in the earth's crust is ______.
10. ______ is usually a liquid fossil fuel.
11. ______ are the principal substances found in petroleum.
12. The kind of energy produced from the nuclei of atoms is called ______.
13. ______ is the splitting of the nuclei of certain kinds of atoms.
14. ______ is the joining together of small atoms to make larger atoms.
15. A(n) ______ changes sunlight directly into electricity.
16. Energy obtained from heat in the earth's crust is called ______.

QUESTIONS

Give brief but complete answers to each of the following questions. Unless otherwise indicated, use complete sentences to write your answers.

1. Name a natural resource and tell why it is one.
2. What is the difference between a renewable resource and a nonrenewable resource? Give an example of each kind.
3. Describe two problems resulting from the earth's population growth.
4. Explain why it is important to prevent pollution.

5. Why are mineral deposits considered to be nonrenewable?
6. How can a damaged watershed cause floods, pollution, and a reduction of water supply?
7. List three problems of using coal as an energy source.
8. What is the difference between fusion and fission?
9. Why is solar energy a renewable energy source?
10. Compare using solar energy to using fossil fuels for energy.
11. How are petroleum and natural gas related?

APPLYING SCIENCE

1. You already know that aluminum cans and newspapers are recycled. Discuss recycling with a number of people to find out what other materials are being recycled. List at least five other materials being recycled and describe how they are being recycled.
2. Organize a trip to the facilities of your local water supplier. Find out where the water comes from and what is done to purify it for your use. Also ask, what plans have been made for any future population growth. Find out if the water supplier has a plan to prevent pollution of the water supply and how you and your classmates might help in the plan.
3. Write to the U.S. Committee for Energy Awareness, P.O. Box 37012, Washington, DC 20013, or to your local member of Congress, and ask to be sent the most recent update on "Energy Projections to the Year 2000," which is prepared from time to time by the U.S. Department of Energy. They will also send you the most recent information on alternate energy sources. Take the material to school and discuss the information with your teacher or class.

BIBLIOGRAPHY

Branley, Franklyn M. *Water for the World.* New York: Crowell, 1982.

Eldridge, Frank R. *Wind Machines.* New York: Van Nostrand Reinhold, 1980.

Findley, Rowe. "Our National Forests, Problems in Paradise." *National Geographic,* September 1982.

Hellman, Hal. "The Day New York Runs Out of Water." *Science 81,* May 1981.

Kraft, Betsy Haney. *Oil and Natural Gas.* New York: Watts, 1982.

Roberts, Leslie. "Reactors In Orbit." *Science 83,* December 1983.

UNIT 3

AIR AND WATER

CHAPTER

11

Our planet is surrounded by a layer of gases called the atmosphere. Without these gases, there would be no clouds or colorful sunsets. The earth's surface would be as barren as the moon.

AIR

CHAPTER GOALS

1. Name the main gases that make up the atmosphere.
2. Describe air pressure and how it is measured.
3. Explain how the air in your area may become polluted.
4. Compare the four layers of the atmosphere.

11–1. Composition of the Earth's Atmosphere

At the end of this section you will be able to:

- List the five most abundant gases that are found in the atmosphere.
- Describe two kinds of air pollutants.
- Explain how the most common man-made air pollutants are produced.

You could live about five weeks without food and about five days without water. However, you could live only five minutes without air. For this reason alone, the atmosphere might be called the most important of all earth resources.

GASES IN THE AIR

The air that you take into your lungs with each breath is not a single gas. It is a mixture of many gases. The amount of some of the gases changes from time to time. However, the five gases described below make up 99.99 percent of the total volume of the atmosphere. Air also usually contains small amounts of dust and other solid particles.

1. Nitrogen is the most plentiful of all the gases in air. It makes up about 78 percent of the volume of air. See Fig. 11–1. You do not use the nitrogen in the air. The nitrogen taken in with each breath is normally exhaled. Most plants, on the other hand, use nitrogen. However, plants cannot take the nitrogen directly from air. Some bacteria, called nitrogen-fixing bacteria, are able to take nitrogen from the air and change it into forms that plants can use. These bacteria live in the soil and in the roots of some plants, such as clover. Lightning makes nitrogen

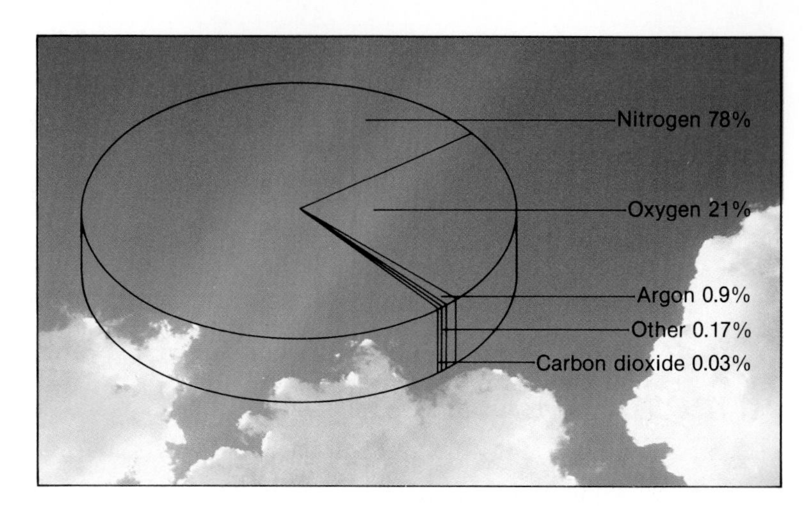

Fig. 11-1 Nitrogen and oxygen make up most of the volume of air.

compounds in the air. These compounds can be used by plants, too. The compounds are carried by rain into the soil and absorbed through the roots of the plants. Animals that eat plants get the nitrogen they need in this way. The waste products of animals and dead plants break down and release nitrogen into the soil. Other bacteria cause nitrogen to be returned to the atmosphere. The process by which nitrogen is taken from the air and then returned to the air is called the *nitrogen cycle*. See Fig. 11-2. The nitrogen cycle returns as much nitrogen to the air as is removed. This keeps the amount of nitrogen in the air at a constant level.

2. Oxygen is the second most plentiful gas in air. It makes up about 21 percent of the air by volume. Animals remove some of the oxygen from the air as they breathe. Oxygen is also removed by the decay of plant and animal matter and by the

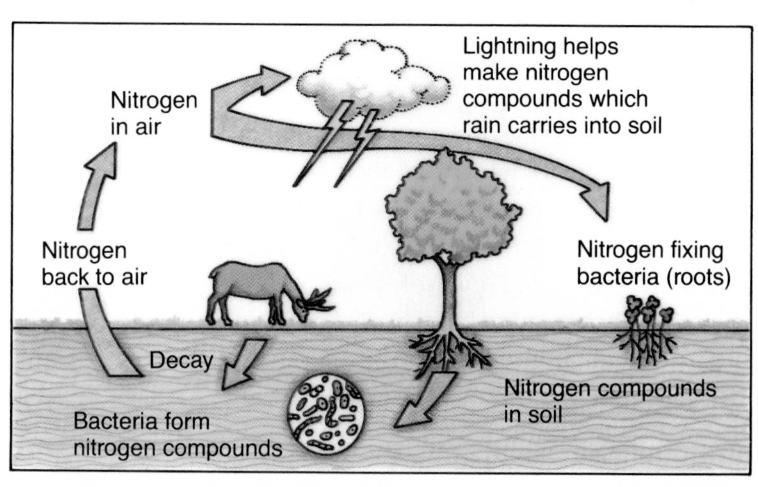

Fig. 11-2 The nitrogen cycle.

weathering of rocks. At the same time, green plants give off oxygen through a process called *photosynthesis* (foh-toe-**sin**-the-sis). Photosynthesis replaces the oxygen that is used. In this way, the amount of oxygen in the air stays the same.

3. Argon is a gas found in very small amounts in air. It makes up 0.9 percent of the air. Argon does not combine chemically with other substances. The atmosphere also contains tiny amounts of neon, helium, and other rare gases that are similar to argon.

4. Carbon dioxide makes up only 0.03 percent of the air, but it is an important gas. Carbon dioxide is used by plants during photosynthesis. The carbon dioxide is taken in through the plant leaves. Green plants use carbon dioxide to make their own food. Oxygen is released as a byproduct as plants carry on photosynthesis. For animals, carbon dioxide is a waste product. Animals give off carbon dioxide as they obtain energy from their food. The process by which carbon dioxide is taken from the air, then returned, is called the *carbon dioxide cycle*. See Fig. 11–3. The amount of carbon dioxide used by plants seems to be equal to the amount released by animals and other natural processes. However, this balance is upset by the burning of fuels, such as coal and oil. Burning these fuels dumps billions of metric tons of carbon dioxide into the atmosphere each year. As a result, the amount of carbon dioxide in the atmosphere is increasing slightly each year.

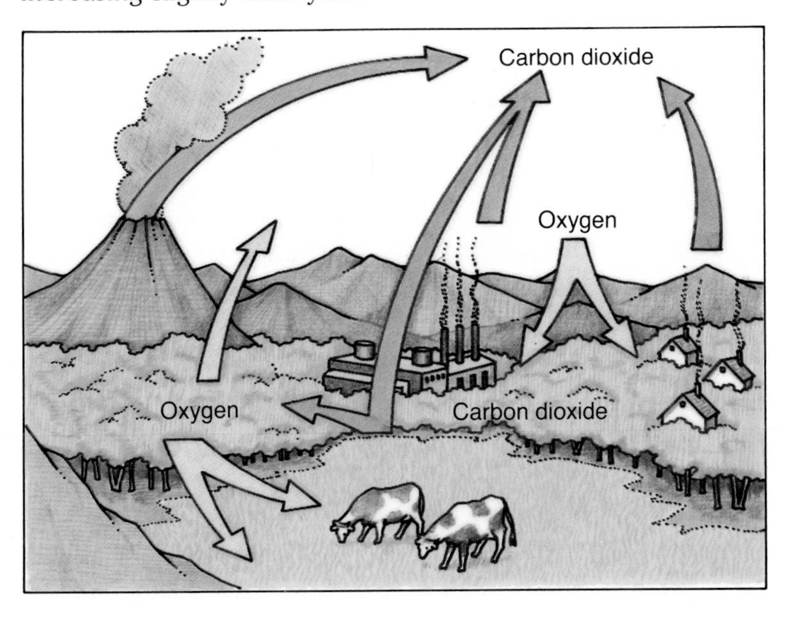

Fig. 11–3 The carbon dioxide cycle.

5. Water vapor. One of the most important gases in the atmosphere is *water vapor*. Unlike the first four gases described, the amount of water vapor in air varies greatly. The amount of water vapor in extremely cold air or in desert regions is very small. On the other hand, the warm air over warm oceans contains large amounts of water vapor. The amount of water vapor in air may be as much as 3 percent by volume. But most of the time it makes up less than 1 percent of the volume of air.

AIR POLLUTANTS

Every year millions of metric tons of substances not normally found in the air are dumped into the atmosphere. These substances in the atmosphere are *air pollutants*. There are two kinds of air pollutants: *solids* and *gases*. Some kinds of solid pollutants come from natural sources. These are dust and microscopic living things such as bacteria. Volcanoes are also a source of dust pollution in the atmosphere. Some volcanoes throw dust particles high into the atmosphere. It may take several years for the smallest of these volcanic dust particles to return to earth.

People, mainly in cities, also add large amounts of solid particles to the air. Smoke from the burning of fuels such as coal gives off particles of carbon and other solids, which enter the air. Burning wood in home heating stoves has caused a smoke pollution problem in some cities. Automobiles also give off small, solid particles of carbon. Automobiles that burn leaded gasoline give off tiny particles of lead. Industries may give off dust from operations that involve cutting, grinding, and crushing materials. One very harmful form of solid pollution is asbestos. See Fig. 11–4. Asbestos is a mineral used in making many things. It is found, for example, in some insulating materials and in some automobile parts. Breathing air that is polluted with small particles of asbestos is known to cause lung cancer.

Fig. 11–4 Asbestos was used as a fire retardant in buildings until it was deemed unhealthful.

CAUSES OF AIR POLLUTION

1. Automobiles. Gases cause more air pollution than solids do. The automobile is one of the major sources of pollutant gases. One of the most harmful gases given off is *carbon monoxide*. This very poisonous gas may build up in the air near heavy traffic or in any closed space where a motor is running. Automobiles also give off *nitrogen oxides* and *hydrocarbons*.

Fig. 11-5 (top) *Smog is often formed in cool air that has been trapped beneath a warmer layer.* (bottom) *A photo of smog over a city where cool air from the ocean is trapped by warm air from inland. The surrounding mountains also help to hold the smog over the city.*

These two kinds of gases together can produce smog. Smog can cause serious health problems and damage rubber and other materials. See Fig. 11-5.

2. Coal and oil. Not all pollutant gases come from automobiles. Burning coal and oil in power plants and industries can produce *sulfur dioxide*, along with nitrogen oxides. Sulfur dioxide is dangerous to breathe. It also combines with water to produce an acid. This acid damages metal, and even some kinds of stone. Sulfur dioxide and nitrogen oxides dissolve in raindrops to make acid rain. Acid rain may be causing serious harm to forests and to the life in lakes.

Most pollutants do not stay in the atmosphere for a long time. Many solid particles fall out of the air or are washed out by rain and snow within a few weeks. Most gases are broken down or absorbed by moisture within several months. However, pollutants that are carried to the upper parts of the atmosphere may remain there for up to three years. For example, a powerful volcanic eruption may throw dust to a great height. The volcanic dust will be trapped in the upper atmosphere. There are no weather changes at those high altitudes to help clean the air. The dust may affect the weather as it is carried all over the world.

Serious air pollution problems are usually caused by people in cities or other areas with large populations. Laws have been passed to help control the sources of air pollution in such areas. For example, automobile makers are required by law to provide special converters in all automobile exhaust systems. The converter keeps carbon monoxide and hydrocarbon gases from being given off. Automobile makers also must make engines that give off smaller amounts of pollutants than earlier engines. Other laws control power plants and industries that give off pollutants. These laws have already helped greatly in reducing air pollution almost everywhere. More laws and public information programs that show the benefits of clean air are still needed. Polluted air cannot be purified and reused as water can. But pollutants put into the air can be reduced.

SUMMARY

The atmosphere is made up of a layer of gases that surrounds the solid earth. This layer is made up of a mixture of gases. The most abundant gases are nitrogen and oxygen, along with small amounts of carbon dioxide, water vapor, and some rare gases. Air pollutants can be solids or gases. Human-made gases cause the most serious kinds of air pollution.

QUESTIONS

Use complete sentences to write your answers.

1. Name the five most abundant gases in the atmosphere. List them in order and describe the amount of each that is found in the air.
2. List the sources of solid air pollutants.
3. Name four gases that are common air pollutants. Explain how each is produced.

INVESTIGATION

ACIDS FROM THE AIR

PURPOSE: To test for gases in the air that form acids.

MATERIALS:

graduated cylinder	matches
limewater	flask
beaker	stirring rod
straw	vinegar
candle	blue litmus paper
pie tin	

PROCEDURE:

A. Pour 10 mL of the limewater into the beaker. CAUTION: LIMEWATER WILL DAMAGE SKIN IF LEFT ON IT. RINSE OFF WITH WATER. Use the straw to blow your breath through the limewater solution until a change occurs in the limewater. Do NOT get limewater in your mouth.
   1. What happens to the limewater?

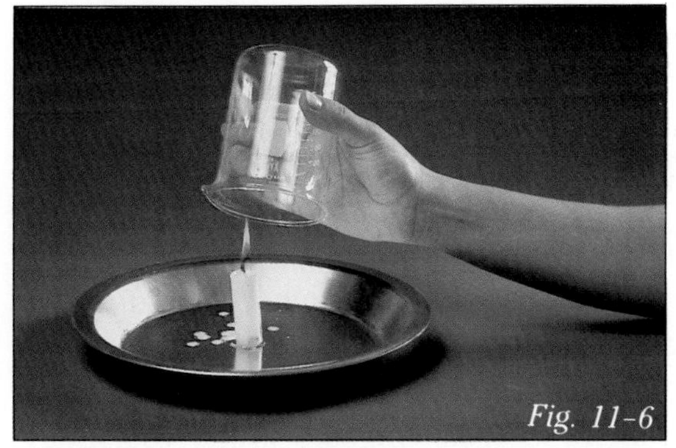

Fig. 11-6

B. The result in procedure A is the test for carbon dioxide. Carbon dioxide enters the air in large amounts by the combustion of fuels. Attach the candle to the pie tin by using a few drops of molten wax from the burning candle. Place the flask upside down over the burning candle. See Fig. 11-6. Allow the candle to go out. Then place the flask upright and add 10 mL of limewater. Swirl it around to mix with the gases in the flask.
   2. Describe what happens to the limewater.
   3. What gas must have been in the flask?

C. Carbon dioxide dissolves in water to produce carbonic acid. Another gas found in air as a result of the combustion of fuel oil, gasoline, and coal is sulfur dioxide. It, too, forms an acid when dissolved in water. Sulfur dioxide is also made by burning match heads. It has a noticeable odor. Light a match and notice the odor that comes from the burning match head.
   4. Describe the odor.

D. Using a stirring rod, place a drop of vinegar on a piece of blue litmus paper.
   5. What happens to the litmus color?

E. The result in procedure D is the test for an acid. Both carbon dioxide and sulfur dioxide are given off when a match burns. These gases must dissolve in water to become an acid. Moisten a piece of blue litmus paper with distilled water. Light a match and hold the moistened blue litmus above the flame of the match until you see a color change.
   6. What happens to the litmus color?

CONCLUSIONS:

1. Why did the moist litmus change color when held over a burning match?
2. Explain how drops of water falling through the air may produce acid rain.

11-2. Atmospheric Pressure and Temperature

At the end of this section you will be able to:

- Explain how air pressure is measured.
- Predict how pressure and temperature will change as you move upward from the earth's surface.
- Name four layers of the atmosphere beginning at the earth's surface.

Fig. 11-7 (right) *Hang gliders depend on the upward motion of warm air.*
Fig. 11-8 (below) *The column of air pushing down on you weighs more than 1,000 kg!*

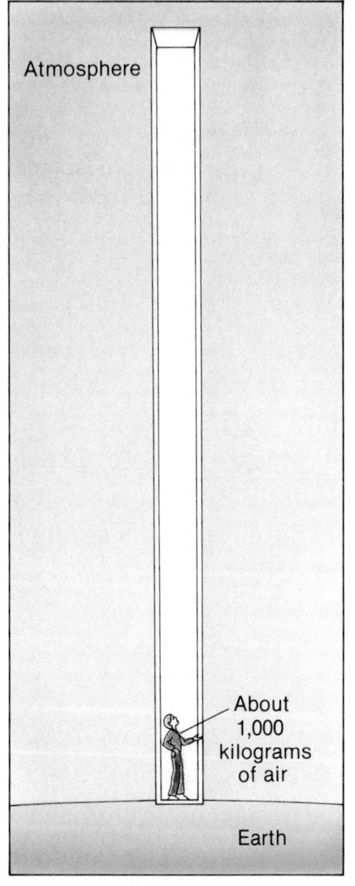

Air cannot be seen. Only its effects can be seen. The hang glider shown in Fig. 11-7 has just soared off a cliff. A column of warm air lifts the glider and keeps it in the air. How can you tell that you live at the bottom of an ocean of air?

MEASURING AIR PRESSURE

As you walk around, you are carrying almost 1,000 kilograms of air on your head and shoulders. See Fig. 11-8. The weight of this air pushing down causes air pressure. You do not feel this pressure because there is equal pressure inside your body. Sometimes you can see the effect of air pressure. For example, air pressure makes it possible for you to drink through a straw. The liquid moves up the straw because you remove the air from inside the straw by breathing in. This causes the air pressure inside the straw to drop. The air pressure on the liquid then pushes it up the straw. See Fig. 11-9.

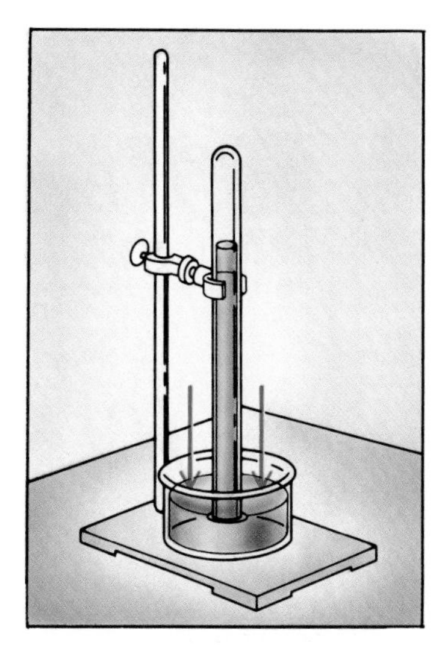

Fig. 11-9 When you drink through a straw, you are causing a pressure change.
Fig. 11-10 A simple mercury barometer.

Air pressure can be measured by using the weight of air. The air pushes a liquid up into a tube from which the air has been removed. Atmospheric pressure is measured by an instrument called a **barometer** (buh-**rom**-uh-tur). One kind is a mercury *barometer*, shown in Fig. 11–10. A glass tube about one meter long is closed at one end. It is then filled with mercury. Mercury is used because it is a very dense liquid. The tube is then placed, open end down, into a dish of mercury. The mercury runs out of the tube until its weight is equal to the weight of air on the mercury in the dish. In other words, the mercury is pushed up into the tube by the air pressure.

Barometer An instrument that is used to measure atmospheric pressure.

At sea level, the mercury runs out of the tube until the height of the mercury left in the tube is equal to 760 millimeters. For this reason, the air pressure at sea level is said to equal 760 millimeters of mercury. If air pressure increases, the height of mercury in the tube will rise. If air pressure decreases, the height of the mercury in the tube will fall. A measurement of the height of mercury in a barometer is used to express air pressure.

Mercury barometers are not easy to move from one place to another. A more useful kind of barometer is an *aneroid* (**an**-uh-roid) barometer. It does not use a liquid. Aneroid means "without liquid." An aneroid barometer uses a round, sealed metal can. One side of the can squeezes in or springs out as the air pressure changes. This movement causes the pointer on a dial to move. The air pressure can then be read. For example, on an average day at sea level, the dial might give a reading of 760 millimeters or 29.92 inches.

Fig. 11-11 An aneroid barometer shows air pressure on a dial.

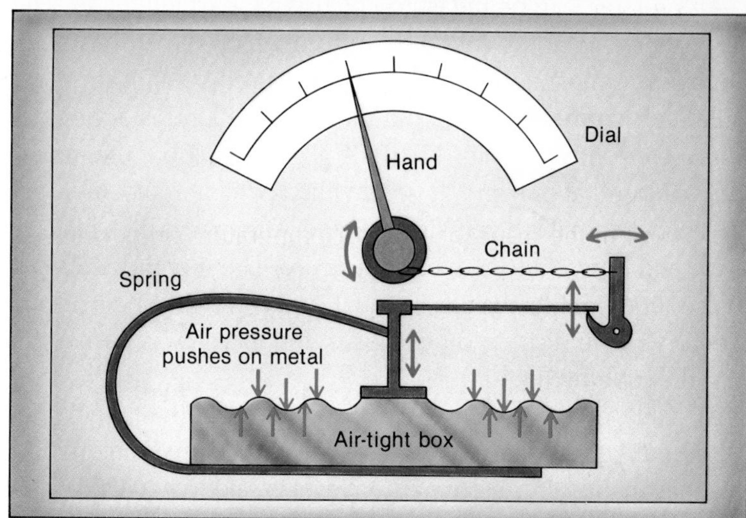

Fig. 11-12 This diagram shows how an aneroid barometer works.

Some aneroid barometers, called *barographs*, keep a continuous record of air pressure. The recordings made by this kind of barometer appear as a line on paper held on a revolving drum. The line traced on the paper shows how the air pressure changes over a period of time.

AIR PRESSURE AND TEMPERATURE

When you take a trip to the mountains, you may notice that your ears pop. This happens because of a slight drop in the air pressure. If you were at the top of a mountain five kilometers high, about half of all the atmosphere would lie below and half above. Thus on the mountain there is less air above you than there is at sea level. This causes the air inside your ear to have a higher pressure than the air outside. When pressure inside and outside of your ears becomes equal, a popping sound is heard.

In addition to lower air pressure than at sea level, the temperature on the mountain top is colder. At an altitude of five kilometers, the temperature averages 30 °C colder than at sea level. Pressure and temperature continue to drop at higher altitudes. At a height of about 11 kilometers, the pressure drops to 25 percent of that at sea level. At the same altitude, the temperature is about −55 °C. This altitude marks the boundary between two layers of the atmosphere.

LAYERS OF THE ATMOSPHERE

1. Troposphere. Temperature changes at certain altitudes divide the atmosphere into layers. Closest to the earth is the **troposphere** (**troe**-puh-sfear), the densest layer of the atmosphere. This is the layer in which you live. Almost all weather takes place in the *troposphere*. Gravity causes the gases of the atmosphere to be pulled toward the earth's surface. This causes the troposphere to have about half of the total weight of the atmosphere. In the troposphere, the temperature drops an average of 6.5 °C for each kilometer of increasing altitude. At a height of about ten kilometers the temperature stops dropping. See Fig. 11–13, page 320. This marks the upper boundary of the troposphere. The upper boundary of the troposphere is higher over the equator than at the poles.

Troposphere The most dense layer of the atmosphere. It lies closest to the earth's surface.

2. Stratosphere. Above the troposphere, at an altitude of about 20 kilometers, is the clear, cold layer called the **stratosphere** (**strat**-uh-sfear). It extends upward to an altitude of about 50 kilometers. Temperatures in the lower *stratosphere* are very cold. However, above the altitude of 20 kilometers, the temperature begins to rise again. At the height of about 50 kilometers, the temperature rises to about 0 °C. A small amount of a special form of oxygen, called *ozone,* is the reason for the temperature rise. This small amount of ozone is very important. Ozone is able to filter out the sun's most dangerous ultraviolet rays. If the full strength of those rays were to reach the earth's surface, they would harm living things.

Stratosphere The atmospheric layer above the troposphere, between about 20 km and 50 km, showing a temperature increase with increasing altitude.

3. Mesosphere. The next layer above the stratosphere is called the **mesosphere** (**mes**-uh-sfear). See Fig. 11–13. It is the coldest layer of the atmosphere. The *mesosphere* extends

Mesosphere The atmospheric layer above the stratosphere, extending from about 50 km to 85 km, where temperature decreases with increasing altitude.

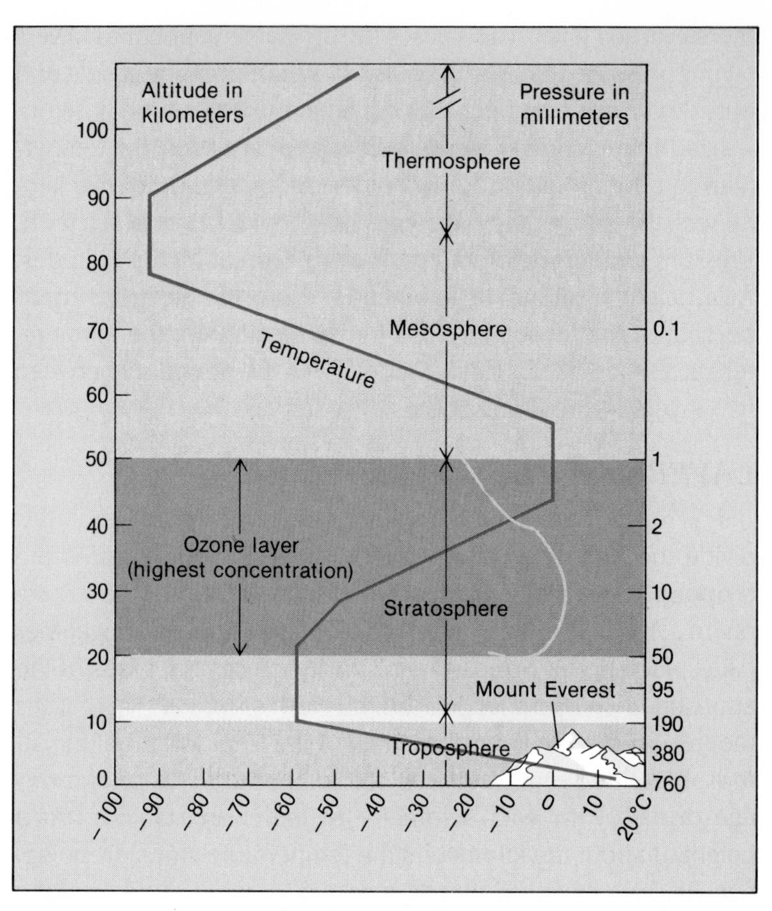

Fig. 11-13 Changes in temperature mark the boundaries between layers of the atmosphere.

upward from about 50 kilometers to about 85 kilometers, where the temperature decreases to about −90°C.

Thermosphere The layer of the atmosphere above the mesosphere, extending from about 85 km to about 600 km, where temperature increases with increasing altitude.

4. Thermosphere. Above the mesosphere is the **thermosphere** (**thur**-muh-sfear). The *thermosphere* reaches to an altitude of about 600 kilometers, with the temperature going up to 2,000°C. Air in the thermosphere is very thin. It is only one ten-millionth (0.0000001) as dense as air at sea level. Beyond the thermosphere, the earth's atmosphere blends gradually into the very thin gases found in deep space.

Ionosphere A region of the upper atmosphere that contains many electrically charged particles.

Within the thermosphere is a region called the **ionosphere** (ie-**on**-uh-sfear). It begins at a height of about 80 kilometers and goes up to about 400 kilometers. The *ionosphere* also makes up the upper part of the mesosphere. This region contains electrically charged particles called *ions.* Ions are atoms or molecules that are charged when electrons are gained or lost. This happens in the ionosphere when solar energy is absorbed by the gases. The ionosphere plays an important part in

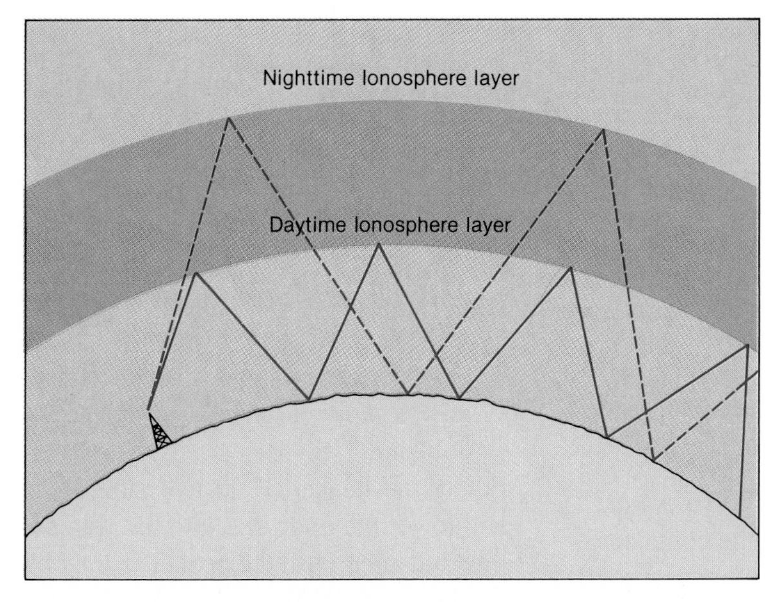

Fig. 11-14 Radio waves are reflected from the ionosphere.

the way radio waves can be sent between different places on the earth. The ionosphere can reflect many kinds of radio waves back toward the earth, just as a mirror reflects light. This allows many kinds of radio waves to travel long distances by bouncing back and forth between the earth and the ionosphere. See Fig. 11-14. Radio signals often cover greater distances at night than during the day. This happens when the ionosphere is at a higher altitude at night.

SUMMARY

Air pressure is caused by the weight of the atmospheric gases. Air pressure is measured by barometers. Moving upward through the atmosphere, pressure decreases as the temperature changes. The atmosphere is divided into layers because of these temperature changes.

QUESTIONS

Use complete sentences to write your answers.

1. Explain how two kinds of barometers work for measuring air pressure.
2. Describe how pressure and temperature change as you move upward in the lower atmosphere.
3. List the four layers of the atmosphere starting at the earth's surface and going upward.

INVESTIGATION

ATMOSPHERIC PRESSURE

PURPOSE: To show the effect of air pressure.

MATERIALS:

large container	water
small beaker	drinking straw

PROCEDURE:

A. Fill the large container nearly full of water. Lay the beaker on its side in the container and let it fill completely with water. Turn the beaker upside down underwater. Carefully lift the beaker until its open end is just below the surface of the water.

1. What happens to the water in the beaker?
2. How does the water level in the beaker compare to the water level in the container?

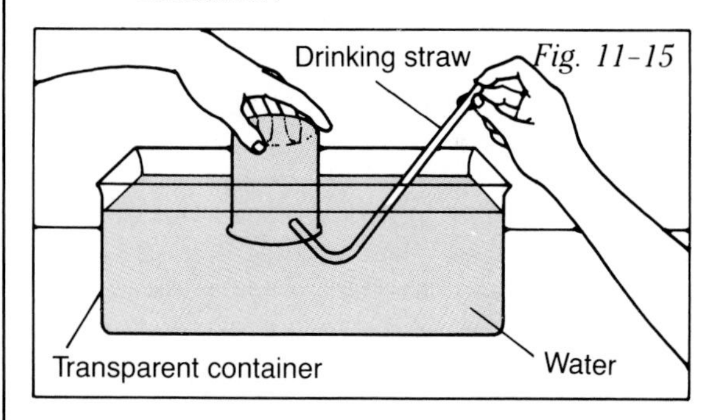

Fig. 11-15

B. Hold the beaker as you did above with its open end just underwater. Put your finger over one end of the straw. Keep your finger over the outside end of the straw as you place the other end into the water in the beaker. See Fig. 11-15. Release your finger and observe what happens. (If nothing happens to the water in the beaker, it is because some water got into the straw. Remove the straw, shake it dry, and try again.)

3. What happens to the water inside the beaker?
4. How does the water level in the beaker now compare to the water level in the container?

C. Empty the beaker. Hold it upside down and lower the open end into the water in the container. Push the beaker to the bottom of the container.

5. What happens to the air inside the beaker?
6. How does the water level in the beaker compare to that in the container?

D. Put your finger over the outside end of the straw. Hold the beaker as you did for procedure C and insert the other end of the straw into the beaker. Remove your finger from the outside end of the straw.

7. What happens to the air inside the beaker?
8. How does the water level in the beaker now compare to the water level in the container?

CONCLUSIONS:

1. How does air pressure explain your observations in procedures A and B?
2. How does air pressure explain your observations in procedures C and D?
3. A stoppered bottle half-full of water is set upside down in the water. What happens when the stopper is removed with the mouth of the bottle underwater?

CAREERS IN SCIENCE

AIR ANALYST

Air analysts monitor the pollution levels in the atmosphere. They take air samples and analyze them for pollutants. For example, analysis may show that pollution levels in a city are too high. Analysts may also be asked to take samples of the air in industrial worksites and other locations such as school buildings, where the health of large numbers of persons might be affected by pollutants in the air. Devices are used that collect dust so it can be examined for particles of lead, rock, coal, or asbestos. After the samples are analyzed, reports of the findings are issued. If the pollution level is unsafe, analysts suggest methods that can be used for reducing pollution.

For a position as an air analyst, you would need a college degree, with courses in chemistry, meteorology, and mathematics. For more information, write: Air Pollution Control Association, P.O. Box 2861, Pittsburgh, PA 15230. Your local branch of the Environmental Protection Agency is also a good source of information.

METEOROLOGICAL TECHNICIAN

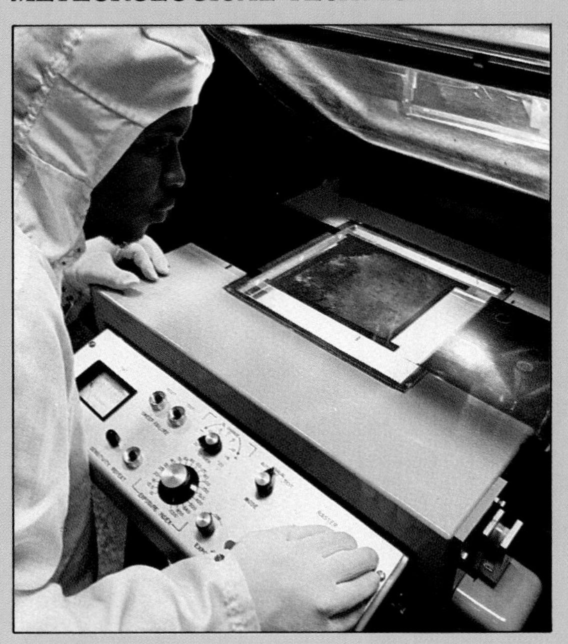

There's an old saying that everyone talks about the weather but no one does anything about it. As a meteorological technician, you would be able to "do something" by working with meteorologists in the field of weather forecasting. Technicians work at weather stations where they make observations and record data important to forecasting, such as precipitation (levels of rainfall), wind speeds, cloud cover, and barometric pressure (the pressure of the atmosphere). They may also be involved in receiving information transmitted from weather satellites about wind patterns and ocean currents. This information is sent to the National Oceanic and Atmospheric Administration's computer center where national weather maps are prepared. Technicians are also involved in weather research projects and in developing instruments.

Specialized training is offered at community colleges and technical institutes. For more information, write: American Meteorological Society, 45 Beacon Street, Boston, MA 02108.

¡COMPUTE!

SCIENCE INPUT

Some weather factors are valuable predictors of weather and some are not. Two commonly considered weather variables are temperature, measured with a thermometer, and air pressure, measured with a barometer. Before the use of weather satellites, which give more up-to-the-minute information for forecasters, the thermometer and barometer, along with past experience, were the chief tools for weather prediction. In this exercise you will use these "old-fashioned" tools to gather data. The modern computer will then be used to analyze the data.

COMPUTER INPUT

The computer's ability to manipulate and rearrange information is especially important. When data are collected, great quantities of information can be provided. If the information is not organized according to some system, none of it will make sense.

For example, imagine someone asking you to collect data about the students in your class. First you would have to know what kinds of questions you intended to answer with the data. Otherwise, you wouldn't know what kinds of information to collect. Once you collected the data, you would then have to arrange it or it would seem like a hodgepodge of information. If you sort the data and provide order, however, patterns become clear and it becomes easier to draw conclusions.

A computer can manipulate data easily. It can let the facts tell a story from many points of view. In this exercise, you will sort weather data to see if any patterns emerge that can help predict weather in the future.

WHAT TO DO

First, you must collect the data for the computer to sort. For 14 days, at approximately the same time each day, record the temperature (degrees Celsius), the air pressure (millibars), and the present state of the weather (fair, rain, possible rain). Make a chart of that data. A sample data chart is provided for your reference.

Enter Program Forecaster into your computer. Enter your data in statements 401 to 415. Use the form given in the example shown in statement 401. When you run the program, it will ask you how you would like to sort the data, by temperature or air pressure. Try both methods. After you've analyzed the data, decide which instrument was more useful for predicting the weather: the thermometer or the barometer.

GLOSSARY

HIGH LEVEL LANGUAGE — A programming language that uses simple terms from everyday vocabulary for its commands.

BASIC, LOGO, PASCAL, FORTRAN — Examples of high level languages using the alphanumeric system.

LOW LEVEL LANGUAGE — Machine language. It uses electronic data, not words, to execute the program.

BINARY NUMBER SYSTEM — Example of low level machine language.

WEATHER MEASUREMENT: PROGRAM FORECASTER

PROGRAM

```
100  REM FORECASTER
110  DIM T$(15), AP$(15), SW$(15)
120  FOR X = 1 TO 15
130  READ T$(X),AP$(X),SW$(X)
132  H = X
135  IF T$(X) = "0" THEN H = X - 1:
     GOTO 150
140  NEXT
150  PRINT: PRINT "1) SORT
     TEMPERATURE & STATE OF
     WEATHER"
160  PRINT "2) SORT AIR PRESSURE &
     STATE OF WEATHER"
170  PRINT: INPUT "    WHICH CHOICE
     (1 OR 2)? ";C$
180  C = VAL(C$)
190  IF C < 1 OR C > 2 THEN GOTO 150
200  K = 0
210  FOR X = 1 TO H - 1
220  IF C = 2 THEN GOTO 240
230  IF VAL(T$(X)) < = VAL(T$(X + 1))
     THEN GOTO 290
235  GOTO 250
240  IF VAL (AP$(X)) < = VAL (AP$(X +
     1)) THEN GOTO 290
250  T$ = T$(X):T$(X) = T$(X + 1): T$(X
     + 1) = T$
260  AP$ = AP$(X):AP$(X) = AP$(X + 1):
     AP$(X + 1) = AP$
270  SW$ = SW$(X):SW$(X) = SW$(X +
     1): SW$(X + 1) = SW$
280  K = K + 1
290  NEXT
300  IF K > 0 THEN GOTO 200
310  PRINT "DAY"; TAB( 8)"TEMP."; TAB
     (15)"PRES."; TAB( 25)"STATE OF
     WEATHER"
320  FOR X = 1 TO H
330  PRINT X;
332  IF C = 1 THEN PRINT TAB( 8)T$(X);
335  IF C = 2 THEN PRINT TAB
     (15)AP$(X);
337  PRINT TAB(25)SW$(X)
340  NEXT
350  GOTO 150
400  REM PLACE DATA STATEMENTS
     HERE
401  DATA 55, 1023, FAIR
402  DATA 45, 1010, RAIN
500  DATA 0,0,0
999  END
```

SAMPLE DATA CHART

Day	Temperature (Degrees C)	Air Pressure (Millibars)	State of Weather
#1	45°	1010	rain
#2	55°	1023	fair

BITS OF INFORMATION

Would you like to set up computer communication between your school and one in another neighborhood? It's possible if both locations have computers and modems. A modem (short for modulator-demodulator) changes the electronic impulses inside your computer into vibrations that a telephone can understand and back again into electronic impulses to the other computer.

CHAPTER REVIEW

VOCABULARY

On a separate piece of paper, match each term with the number of the statement that best explains it. Use each term only once.

barometer	mesosphere	thermosphere	photosynthesis
ionosphere	stratosphere	troposphere	pollutant

1. The outermost layer of the atmosphere.
2. A region of the upper part of the atmosphere containing many electrically charged particles.
3. An instrument used to measure atmospheric pressure.
4. The atmospheric layer just above the troposphere.
5. The layer of the atmosphere closest to the earth's surface.
6. The upper atmospheric layer that filters the sun's ultraviolet rays.
7. A substance not normally found in air.
8. The process by which plants release oxygen.

QUESTIONS

Give brief but complete answers to each of the following questions. Unless otherwise indicated, use complete sentences to write your answers.

1. What two gases make up 99% of the air by volume?
2. Why is it difficult to list the percentage of water vapor found in air?
3. How is air pollution causing serious damage to forests and to the life in lakes?
4. What is meant by the *carbon dioxide cycle?*
5. Which pollutants from auto exhausts produce smog?
6. Explain why you should never run an auto engine in a closed garage.
7. Pollutants that get into the upper atmosphere may become a much greater hazard than those in the lower atmosphere. Why is this?
8. Which would you rather take with you on a trip to the mountains, a mercury barometer or an aneroid barometer? Explain your choice.
9. Which layer of the atmosphere contains about half the total weight of air?
10. In which atmospheric layer does the highest temperature exist?
11. Why do your ears pop when you go from a lower to a higher level?

12. Describe what happens to the temperature and pressure as you go higher and higher into the troposphere.
13. Predict what the temperature would be at an altitude of two kilometers above a location on the ground where the temperature is 20 degrees Celsius.
14. In which atmospheric layers is ozone found? Why is it important?
15. Just after dark you notice that the radio station you are listening to, which is some distance from your town, begins to fade away, and you start receiving a station even farther away. Why is this happening?

APPLYING SCIENCE

1. Write down the location of a distant radio station that you can receive at night. Make a diagram showing the location of the station and your location. Show on the diagram how the radio waves were able to reach your location at night and not during the day.
2. Locate an aneroid barometer. On a weekend day, make readings on it every two hours from the time you get up in the morning until you go to bed that night. Note the kind of weather at the time of each reading. Make a graph of the readings and try to account for any pattern you find in the data.
3. If you were to climb Mount Everest, an 8,845-meter-high mountain, what changes in the atmosphere would you expect? What materials would you bring with you so that you could better survive these changes?
4. The automobile as a producer of air pollutants has been a target for state and federal governments. Find out what kinds of gases are being regulated as auto exhaust. A good source of information is a local automobile dealer. Most would be willing to allow you to find out how their autos rate against the state and federal regulations, if you explain that you are doing a class project.

BIBLIOGRAPHY

Allen, Oliver. *Atmosphere* (Planet Earth Series). Chicago: Time-Life Books, 1983.

Franklin, Deborah. "From Bust to Dust." *Science News*, June 1984.

Gedzelman, Stanley D. *The Science and Wonder of the Atmosphere.* New York: Wiley, 1980.

Golden, Frederick. "Lady in a Cage." *Discover,* July 1984.

CHAPTER

12

Every day energy from the sun affects our lives. During the day, we can see the sun's light. On sunny days we can feel its warmth. In the summer many of us make use of its tanning rays.

ENERGY AND WATER IN AIR

CHAPTER GOALS

1. Describe how the sun's energy reaches the earth and what its effect is on the earth's atmosphere.
2. Explain the role of convection in producing winds.
3. Describe the system by which air moves between the equator and the poles.
4. Explain how water enters and leaves the atmosphere.

12–1. Atmospheric Energy

At the end of this section you will be able to:

- Name three kinds of radiant energy.
- Explain how the shape of the earth and the presence of water affect the heating of the earth.
- Explain what is meant by *convection cells* and give several examples.
- Name and describe the wind belts of the earth.

Have you ever seen the sun at midnight? The left-hand page shows how the sun looks at midnight near the North Pole at certain times during the year. The colored rings might represent all the forms of energy given off by the sun. Some of that energy is trapped by the atmosphere.

ENERGY FROM THE SUN

The sun releases a great amount of energy. About two-billionths (0.000000002) of that energy is received by the earth. Even this tiny part of the sun's energy is a very large amount. It is 13,000 times greater than all the energy used by everyone on earth.

If the earth had no atmosphere, its surface would be like the surface of the moon. The days would be very hot and the nights very cold. The atmosphere prevents this from happening. During the day, the atmosphere reflects some energy back into space. At night, it holds some of the sun's energy from the day. Thus the atmosphere is able to prevent the earth from getting too hot during the day and too cold at night.

Fig. 12–1 Moving air, which gets its energy from the sun, keeps this sailboard moving.

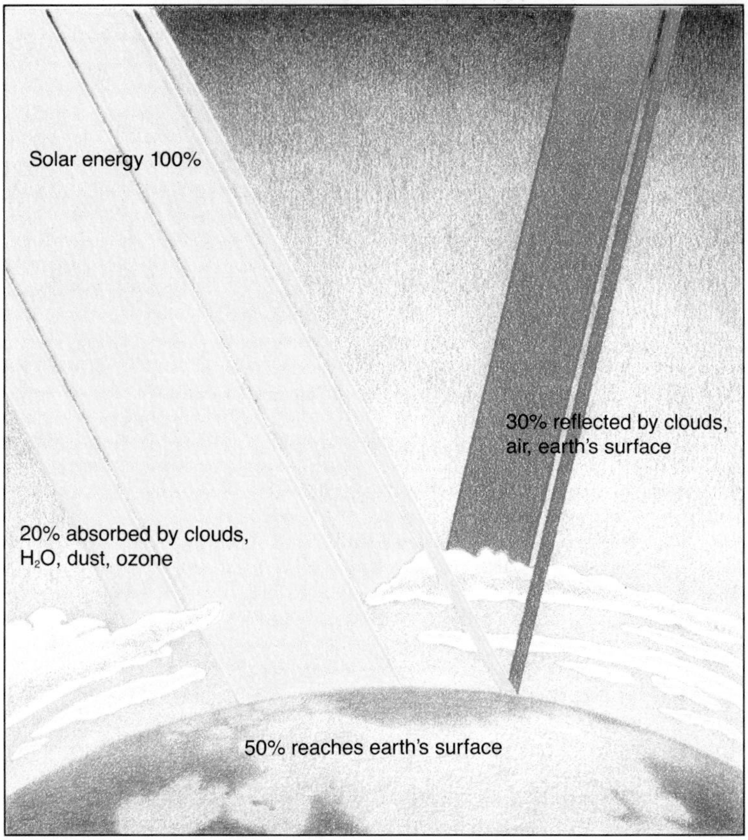

Solar energy 100%

30% reflected by clouds, air, earth's surface

20% absorbed by clouds, H_2O, dust, ozone

50% reaches earth's surface

Fig. 12–2 The atmosphere allows less than half of the sun's energy to reach the earth's surface.

About 30 percent of the solar energy that reaches the earth is reflected back into space. Clouds are the main reflectors of solar energy. About 20 percent of the solar energy falling on the earth is absorbed directly by the air, and only about 50 percent reaches the surface. See Fig. 12–2.

ATMOSPHERIC EFFECTS

Radiant energy A form of energy that can travel through space.

The sun's energy reaches the earth as **radiant** (**rade**-ee-unt) **energy.** *Radiant energy* is energy that can travel through space. *Light* is one of several kinds of radiant energy. It is the only kind that can be seen. When sunlight falls on your skin, you feel warmth. This is caused by a kind of radiant energy called *infrared* (in-fruh-**red**) *energy.* Sunburn is caused by another kind of radiant energy, *ultraviolet* (ul-truh-**vie**-uh-let) *energy.* Unlike visible light, infrared and ultraviolet energy cannot be seen. Fig. 12–3 shows the kinds of radiant energy given off by the sun.

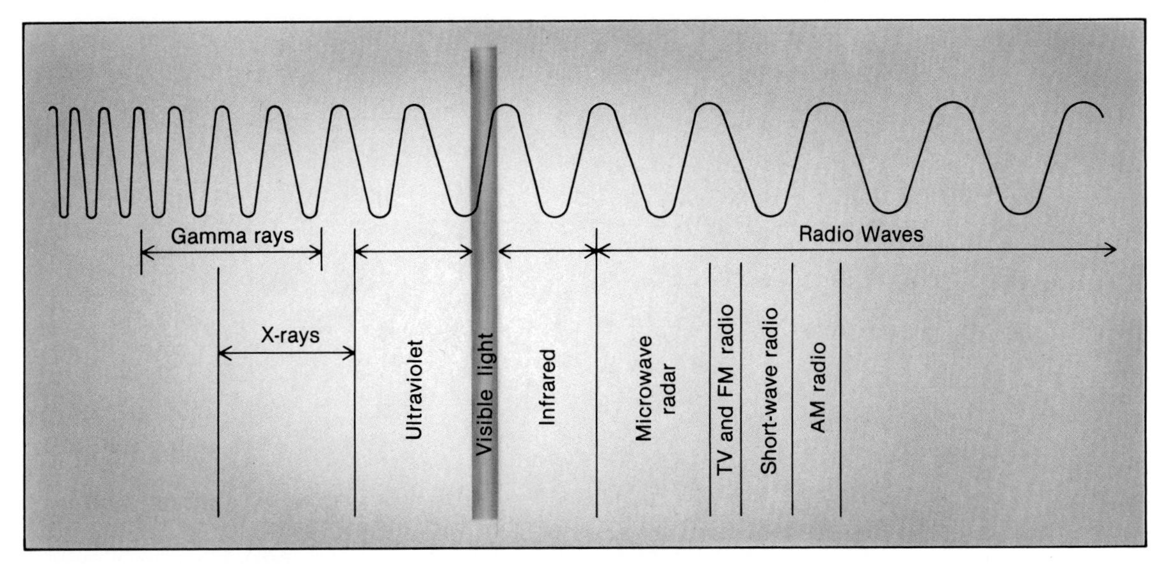

Fig. 12-3 The different forms of radiant energy are shown on this chart. We can see only the narrow band of visible light.

Clouds block infrared rays, along with most of the light of the sun. Thus the air feels cooler when a cloud covers the sun. Most of the sun's ultraviolet energy is absorbed by a layer of ozone gas in the upper atmosphere. The small amount of ultraviolet rays that get through this layer can cause sunburn. However, ultraviolet rays pass through clouds. Therefore, you can get sunburned on a cloudy day.

Gases in the air catch rays of blue light from the sun. While other colors pass through, most of the rays of blue light are spread out in all directions. These blue rays are what make the sky look blue. Above the atmosphere, space appears black since there are no gases to catch the light. When the atmosphere contains large particles of dust or water vapor, the other colors of visible light are spread out. This causes the sky to look less blue and more milky, or white. The blueness of the sky shows how free it is of dust and other particles.

Have you ever noticed how hot a car with its windows closed becomes when it is left in the sun? The atmosphere is heated in much the same way as the air in the car. The inside of the car gets hot when radiant energy passes through the car windows. The energy is absorbed by the interior of the car. The energy is then given off again, but it has been changed into longer waves. The glass in the car windows will not allow the longer-wave radiant energy to escape. Energy becomes trapped inside the car, causing the materials and the air inside the car to get hotter and hotter.

Fig. 12-4 (top) *The greenhouse effect occurs in a greenhouse because the glass windows trap the air.* (bottom) *The same thing happens to a city, but the warmer air is trapped by the atmosphere.*

The land and water of the earth are also warmed by absorbing radiant energy. This shorter-wave energy is changed by the land and water, and sent back into the air as longer waves. The longer waves cannot pass out of the atmosphere. This energy is blocked and absorbed by the carbon dioxide and water vapor in the air. The trapped energy heats the air. This heating is called the **greenhouse effect.** This term is used because the atmosphere acts like the glass in a greenhouse, or the windows of a car. The atmosphere allows radiant energy to enter, but prevents it from leaving. The *greenhouse effect* heats the atmosphere. See Fig. 12-4. By adding more carbon dioxide to the atmosphere, more heat will be trapped. The earth already may be getting warmer. The burning of fossil fuels has already put large amounts of carbon dioxide into the air.

Greenhouse effect The name given to the method by which the atmosphere traps energy from the sun.

HEATING THE EARTH'S SURFACE

1. Water and land. Large amounts of heat energy can be absorbed by water without changing the temperature of the water very much. You could confirm this by heating water in a metal pot. The metal of the pot and its handle will soon become too hot to touch, while the water inside is only warm. However, once the water is heated and removed from the heat source, the pan will cool quickly, while the water remains warm for a long period of time. Because water heats slowly, air near oceans and lakes stays cool during the day. The air in these areas also remains warm at night because water cools slowly. But the land is like the metal pot. When the sun is out, land heats quickly. At night, land quickly loses the heat it absorbed during the day. Thus the air around land areas is warm during the day and cool at night.

2. The shape of the earth. The heating of the earth's surface also is affected by the round shape of the earth. If the earth were flat, all areas would receive the same amount of energy from the sun. But because the earth's surface is curved, polar regions receive less energy than the regions near the equator. See Fig. 12-5. Regions around the equator receive direct rays from the sun, while polar regions receive slanting rays. Slanting rays spread the energy over a larger area than direct rays do. The direct rays falling near the equator have more energy than slanting rays. Near the equator, the air gets hotter than it does at the poles. Without some way to transfer heat from the equator to the poles, the equator would become hotter and the poles would become colder.

Fig. 12-5 (left) *Areas near the equator, which receive direct sunlight, have little temperature change year-round.* (below) *Areas near the poles receive slanting sun rays and have large temperature changes throughout the year.*

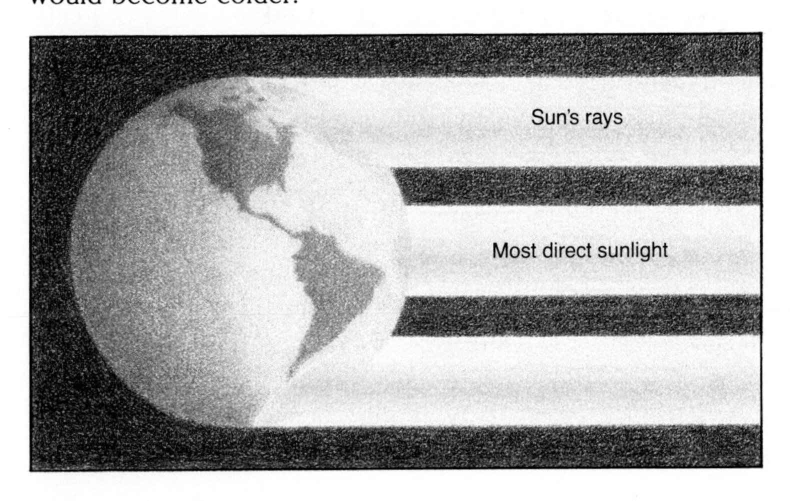

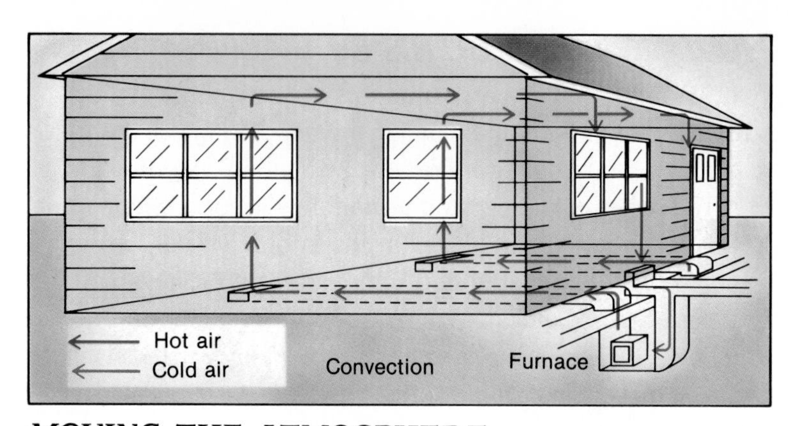

Fig. 12-6 Many homes are heated by convection currents.

MOVING THE ATMOSPHERE

Air moves when it is not heated evenly. When air is heated, it becomes less dense and lighter. The surrounding cooler air, which is more dense and heavier, will force the warmer air upward. Cool air always moves toward warm air because of this difference in density. The movement of air because of the difference in temperature and density is called **convection.**

Convection Movement of air caused by differences in its temperature.

Many heating systems in houses use *convection.* In this system, heated air usually comes into a room near the floor. This warm air is pushed up and moves toward the ceiling. Cool air moves downward toward the floor, where it is mixed with warm air. See Fig. 12-6. The air keeps moving until the air in the room is the same temperature everywhere.

Convection also causes air to move between the earth's poles and the equator. Air near the poles is cooled, becoming dense and heavy. This cold, heavy air settles downward. Air near the equator is warmed, causing it to become less dense and lighter. The warmer air rises and streams from the equator toward the poles. Cold air from the polar regions moves toward the equator. See Fig. 12-7. Heat is moved from areas of high temperature to areas of low temperature. Thus the sun's energy is spread more evenly over the earth. This has made it possible for life to exist on much of the land surface of the earth.

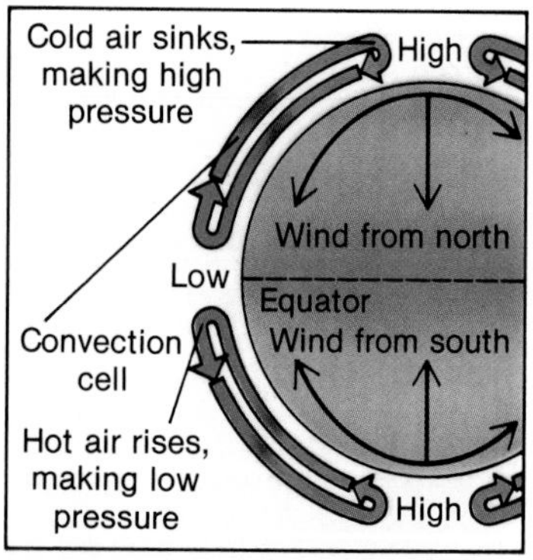

Fig. 12-7 Cold air moves toward the equator and warm air moves toward the poles.

MOVING AIR

When air moves horizontally, it is called *wind.* Wind almost always is caused by differences in temperature in the atmosphere. Temperature affects air pressure. Warm air has a lower air pressure than cool air. As cool air moves into an area of low pressure, the warm air is lifted. As the warm air rises, it is

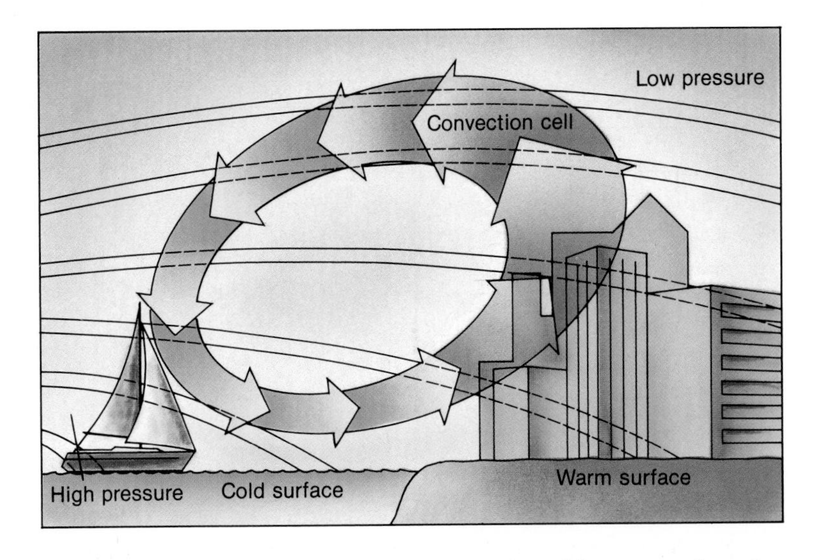

Fig. 12-8 Wind moves air toward a low-pressure area, forming a convection cell.

cooled. It sinks back toward the ground and is warmed again. The complete circle of moving air is called a **convection cell.** Movement of air in *convection cells* causes winds along the ground to flow from areas of high pressure to areas of low pressure. See Fig. 12-8.

Convection cell A complete circle of moving air caused by temperature differences.

1. Mountain and valley breezes. An example of a convection cell is found on mountains. During the night, the air around mountain tops cools more rapidly than the air in the valleys. The cooled air has a higher air pressure than the warmer air. The cooled air moves down the mountain slopes, forming what is called a *mountain breeze.* During the day, air near the mountain tops heats more rapidly than air in the valleys. Because of this uneven heating, the direction of the airflow is reversed. A *valley breeze* is formed. See Fig. 12-9.

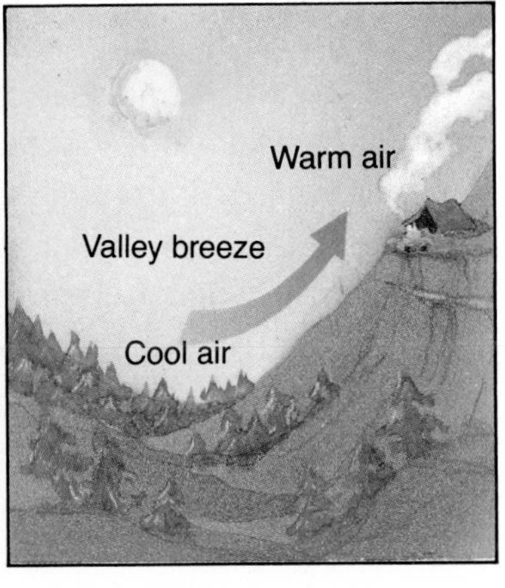

Fig. 12-9 (left) *How do mountain breezes form?* (right) *How do valley breezes form?*

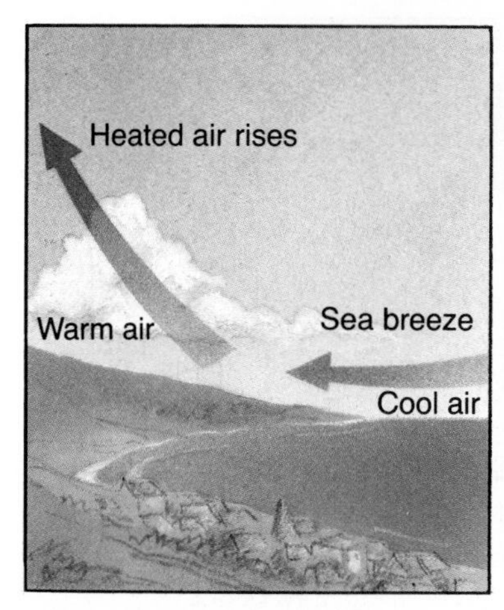

Fig. 12-10 (left) *How do sea breezes form?* (right) *How do land breezes form?*

2. Land and sea breezes. Convection cells are also formed during the day and night in coastal areas. During the day, the land heats faster than the water. The air above the land becomes warmer than the air above the water. An area of low pressure is created above the land. Cool air moves from over the water toward the land. This creates a *sea breeze* in the late afternoon. At night, the land and the air above it cool more rapidly than water. This causes the air movement to be reversed. Air moves from land to water, forming a *land breeze*. See Fig. 12-10. Sea and land breezes occur almost every day along coastlines in warm climates.

3. Monsoons. A *monsoon* is a giant land and sea breeze. A monsoon affects large parts of entire continents. Instead of following the temperature changes of day and night, the monsoon is caused by the differences in summer and winter temperatures. High summer temperatures cause air from the cooler sea to move to the warmer land. In the winter, the air moves from over the cooler land to the warmer sea. Monsoon winds are found all over the world. One of the best known is the monsoon wind that affects India. During the summer months, warm, moist air sweeps in from the Indian Ocean. Much of the food needed for the huge Indian population depends on the rains brought by the summer monsoon.

GLOBAL WINDS

While many winds are caused by local conditions, there is a pattern of winds that covers the entire earth. It is caused by the uneven heating of the earth. The belt of hot air around the equator is a low-pressure area. Caps of cold air at the poles are high-pressure areas. If the earth did not turn, one giant convection cell would carry cold air from the poles to the equator. Another giant convection cell would carry warm air from the equator to the poles. But the earth is turning. It turns once every 24 hours. Because of this movement, places on the equator move at a speed of over 1,600 kilometers an hour. A place very near one of the poles, however, is hardly moving!

Fig. 12–11 shows what happens because of this difference in speed. Anything moving over the earth follows a curving path caused by the earth's rotation. This is called the **Coriolis** (kore-ee-**oe**-lus) **effect,** after the person who first described it. The *Coriolis effect* makes a moving object in the Northern Hemisphere drift to the right when facing in the direction it is moving. In the Southern Hemisphere, a moving object drifts to the left when facing in the direction it is moving. The Coriolis effect is seen most when something moves a great distance. Air moves in a curved path between the equator and the poles.

Coriolis effect Bending of the paths of moving objects as a result of the earth's rotation.

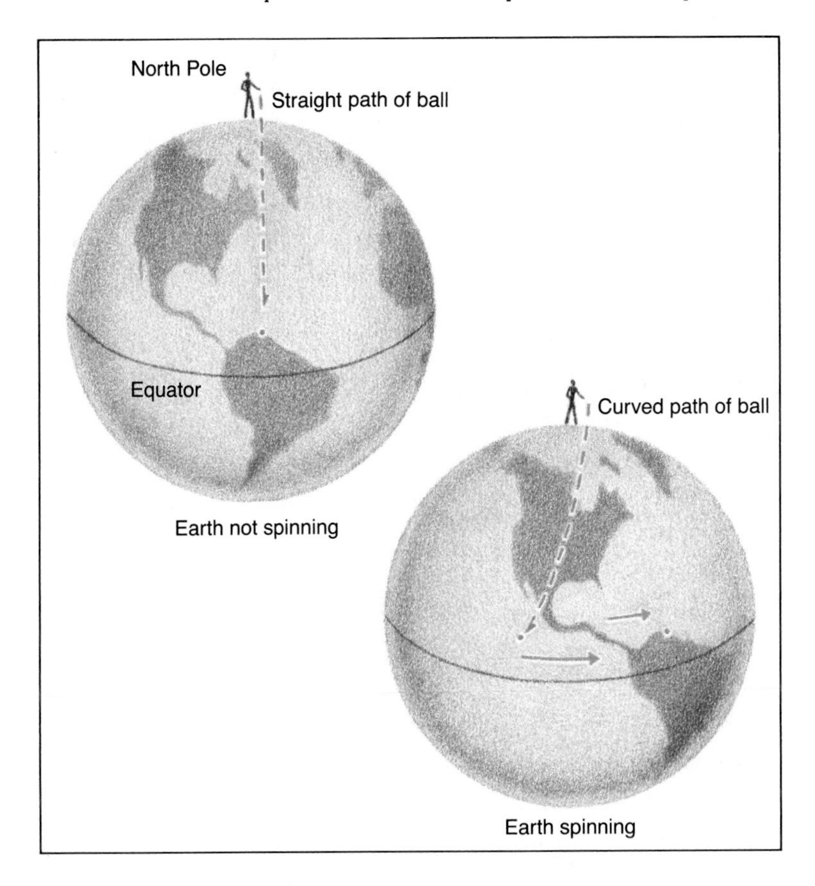

Fig. 12–11 The Coriolis effect is noticeable over great distances.

Near the equator is a belt of air that is strongly heated by the sun's nearly vertical rays. This heated air rises, thus creating a belt of low pressure called the *doldrums* (**dole**-drumz). The air in this belt moves mostly upward. There are only weak winds near the earth's surface. This region got its name, doldrums, from the days when sailing ships were stalled there for lack of wind. The rising air currents from the doldrums turn north and south toward the poles as they become upper-level winds. This movement, or circulation, of air between the equator and the poles, combined with the effects of the earth's rotation, produces a series of *wind belts* in each hemisphere.

1. Trade winds. In the upper parts of the troposphere, the air is cooled as it moves toward the poles. About one-fourth of the way to the poles, some of the air sinks back to the surface. At the surface, the sinking air divides into two parts. In the Northern Hemisphere, one part moves toward the North Pole. The other part moves back toward the equator and creates the warm, steady belt of *trade winds*. The Coriolis effect bends these winds so that they appear to come from a northeasterly direction. The trade winds blow steadily and gently, completing a giant convection cell that began in the doldrums.

2. Westerlies. The story is different, however, for the air that moves from the area where the air sinks toward the poles. These are bent, by the Coriolis effect, to the right in the Northern Hemisphere and to the left in the Southern Hemisphere. Since this makes the winds appear to come from the west, they are called the *westerlies*. Notice that winds are named according to the direction from which they come. The westerlies make up the second of the earth's wind belts. Unlike the trade winds, the westerlies are often strong winds that were once feared by crews of sailing ships.

3. Polar easterlies. The westerlies blow toward the poles until they meet very cold air moving toward the equator from the poles. This band of cold air moving away from the poles is called the *polar easterlies*. The polar easterlies make up the last of the earth's wind belts. The complete pattern of wind belts for the entire earth is shown in Fig. 12–12.

The wind belts describe an overall picture of the movement of air in the earth's atmosphere. At any particular place, the wind patterns are formed from many things. There are local effects, such as land or sea breezes, along with the general air-

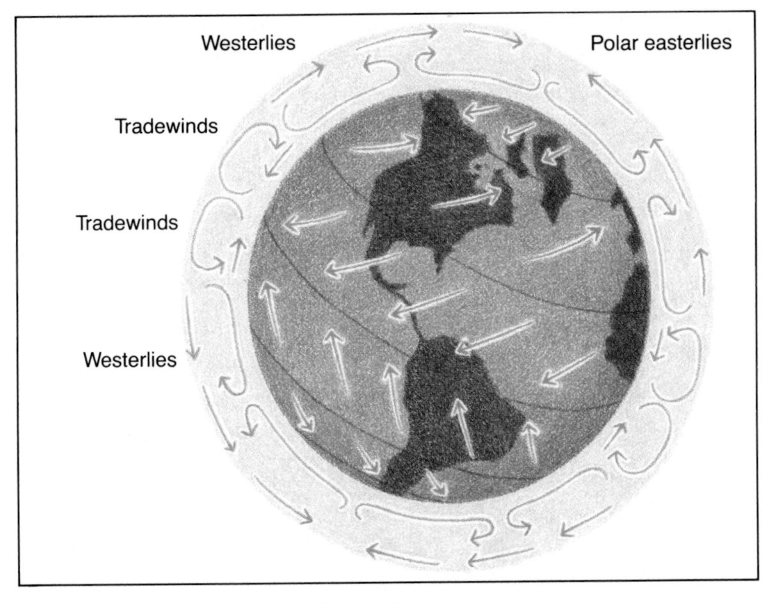

Fig. 12-12 The earth's wind belts.

flow of the wind belts. Winds do not always flow at the same speeds or along the same paths.

SUMMARY

Sunlight delivers a very large amount of energy to the earth. Without the atmosphere, this energy would quickly escape back into space. But some gases in the atmosphere trap the solar energy that has reached the earth's surface. The unequal heating of the atmosphere causes winds to blow from high-pressure areas to low-pressure areas. Rotation of the earth causes the wind path to bend. Each hemisphere has a group of major wind belts that combine to move air between the warm equator and the cold poles.

QUESTIONS

Use complete sentences to write your answers.

1. List three kinds of radiant energy that come from the sun. Identify which one can be seen, which one is felt as heat, and which one causes sunburn.
2. What does the shape of the earth have to do with the air near the equator becoming hotter than the air at the poles?
3. Give three examples of winds occurring in places where the earth's surface is not heated equally.
4. Name and describe the wind belts of the earth.

SKILL-BUILDING ACTIVITY

USING A THERMOMETER

PURPOSE: To develop skill in the use and care of thermometers.

MATERIALS:
laboratory thermometer
clinical thermometer

PROCEDURE:

A. There are many types of glass thermometers. You will look at two that are in common use—the laboratory (Celsius) thermometer and the clinical thermometer. The clinical thermometer is used for taking body temperature, is small in size, has a narrow range, and usually has a Fahrenheit (F) scale.
  1. Which of your thermometers is the clinical thermometer?
  2. What are the lowest and highest temperatures shown on it?
  3. What part of a degree is each smallest division on the scale?

B. Look at the laboratory thermometer.
  4. What are the lowest and highest temperatures shown on it?
  5. What part of a degree is each smallest division on the scale?

C. Regardless of the kind of glass thermometer you work with, all must be handled with great care because they break easily. Clinical thermometers are usually triangular in shape and do not tend to roll. Look at the laboratory thermometer.
  6. Describe what you may do to keep it from rolling off the table.

D. In order for heat to affect the liquid in the thermometer, the bulb holding the liquid must be as thin as possible. This makes it the weakest part of the thermometer. Look at the bulbs of both thermometers.
  7. Why should you never use a thermometer as a stirring rod?

E. Make a gentle fist around the bulb of the laboratory thermometer. Hold it this way for at least three minutes. BEWARE, THERMOMETERS BREAK EASILY.
  8. Look at the smallest division on the scale. What is the temperature shown? The units are as important as the number! A reading is incomplete without the unit. (Use °C with the laboratory thermometer and °F with the clinical thermometer.)

F. Another difference in the clinical thermometer is that it contains a wiggly constriction in the liquid column near the bulb. This keeps the liquid column from returning automatically to lower readings. Before making a reading, it is necessary to shake the liquid down by a few quick movements of the wrist. Your teacher will show you this motion. Repeat the activity described in procedure E, using the clinical thermometer.
  9. What is the reading on the scale?

CONCLUSIONS:
  1. Describe the laboratory thermometer and give several reasons why care is needed when it is used.
  2. How does the laboratory thermometer differ from the clinical thermometer?

INVESTIGATION

ABSORBING RADIANT ENERGY

PURPOSE: To study the rate at which soil, water, and air absorb radiant energy.

MATERIALS:

sheet of paper
2 small containers
soil
water
3 thermometers
electric lamp

PROCEDURE:

A. Place the sheet of paper on a table and arrange the two small containers side by side near the middle of the paper. Fill one container nearly full with soil and the other nearly full with water. Put a thermometer in the soil so that the bulb is just barely below the surface of the soil. Place another thermometer in the water so the bulb is just barely below the water surface. Lay the third thermometer on the paper near the edges of the two containers.

B. Copy the following table. You will record all your measurements in your table.

	Beginning Temp. °C	Final Temp. °C	Temp. Difference °C
Soil			
Water			
Air			

C. Read each of the thermometers and record the results in your table in the column labeled "Beginning Temp. °C."

D. Arrange the lamp so that it is about 15 centimeters above the table and will shine equally on the thermometers in the soil, water, and air.

E. Turn on the lamp and let it shine on the three thermometers for two minutes.

F. Turn off the lamp and immediately read the three thermometers. Read the one in air first, the one in soil second, and the one in water last. Record your results in the column labeled "Final Temperature °C." Leave the thermometers undisturbed until later.

G. Find the difference between the beginning and final temperatures for each of the three thermometers. Record this information in your table in the column labeled "Temperature Difference °C."

   1. Which substance—air, water, or soil—absorbed the radiant energy from the lamp with the greatest temperature change?
   2. Which absorbed the least?

H. After a few minutes, look again at the three thermometers.

   3. Which has cooled the fastest?
   4. Which has cooled the slowest?

CONCLUSIONS:

1. List the three substances studied in this investigation in the order in which they changed temperature when absorbing radiant energy. List the one that showed the the greatest change first.
2. List the three substances studied in the order in which they cooled. List the one that cooled fastest first, the slowest last.
3. Explain your answers to conclusions 1 and 2.

Fig. 12-13 Water that is exposed to air evaporates.

12-2. Water Enters the Air

At the end of this section you will be able to:

- Explain why you feel cool when water evaporates from your skin.
- Define *humidity, absolute humidity,* and *relative humidity,* and explain how they are related.
- Describe how dew, fog, and frost form.

It is not difficult to see that water evaporates. Wet clothes dry when hung outside; the streets dry after a rain; any trapped water exposed to the air slowly disappears. See Fig. 12-13. If you have ever felt cold when the water evaporates from a wet bathing suit, you have made an important discovery about the evaporation of water. This discovery will help to explain why evaporated water, or water vapor, is a small but very important part of the atmosphere.

HOW WATER ENTERS THE AIR

The chill you feel from a wet bathing suit or when you step out of a shower demonstrates an important fact about the evaporation of water. Heat energy is absorbed when water evaporates. The chill you feel when water evaporates from your skin is caused by the loss of heat from your body. The heat energy your body loses goes into the air with the water vapor.

Water vapor that enters the atmosphere from the surface of the sea uses energy from the sun. Radiant energy causes some water molecules at the sea's surface to move faster than the rest of the water molecules. These fast-moving water molecules leave the liquid. They become water vapor in the air. See Fig. 12-14. About one-third of all the solar energy received at the earth's surface is spent evaporating sea water.

Around the world, millions of tons of water are evaporated every minute. Most of the water vapor in the atmosphere comes from the sea. But lakes, rivers, moist soil, plants, and animals also release water vapor into the air. Industrial plants, cooling towers, and a number of other human activities are important sources of water vapor in certain areas.

Air moves the water vapor once evaporation has taken place. The winds of the atmosphere act like a giant transport system that moves huge amounts of the water vapor. The winds carry more water than the largest streams found on the surface of the earth.

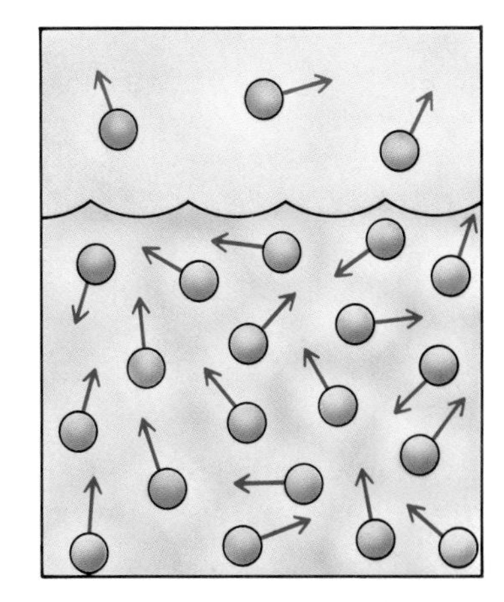

Fig. 12–14 Fast-moving molecules move out of a liquid.

MEASURING MOISTURE IN THE AIR

The moisture in the air is called the **humidity** (hew-**mid**-ih-tee). Air does not always have the same *humidity*. The amount of moisture in the air may be anywhere from 0 to about 4 percent of the volume of the air. The amount of water contained in a volume of air can be measured. This measurement is called the *absolute humidity*. For example, if one cubic meter (1 m^3) of the air contained ten grams of water, the absolute humidity of the air would be expressed as ten grams per cubic meter (10 g/m^3).

Humidity The amount of water vapor in the air.

Air at a given temperature never holds more than a certain amount of water vapor. For example, when the air in your classroom is 25°C, it can hold no more than 22 grams of water in every cubic meter. Air that contains all the water vapor that it can hold at that temperature is said to be *saturated*. Thus, if each cubic meter in your classroom contained 22 grams of water at a room temperature of 25°C, the air would be said to be saturated. The warm air near the earth's equator can hold about ten times as much water vapor as the cold polar air.

Air is not usually saturated. Most of the time, the air contains less water vapor than it can hold. One measurement that often is used to describe the amount of water vapor in the air compares how much water vapor the air contains to how much the air can hold at its temperature. The measurement is called the **relative humidity.** *Relative humidity* is always given as a percentage. For example, one cubic meter of air at 25°C can hold 22 grams of water. If that volume of air is holding only 11 grams of water, its relative humidity is .50 or 50 percent (11 g/m^3 divided by 22 g/m^3). A relative humidity of 50 percent means that the air contains half of the water vapor it can hold at that temperature. The relative humidity of the air has a strong effect on how comfortable you feel. When the relative humidity is high, moisture does not evaporate quickly from the skin. This makes you feel sticky and warmer.

Relative humidity The amount of water vapor in the air compared to the amount of water vapor the air can hold at that temperature.

As air is cooled, a temperature is reached at which the amount of water vapor in the air is equal to the amount of moisture it can hold. Suppose, for example, that the cubic

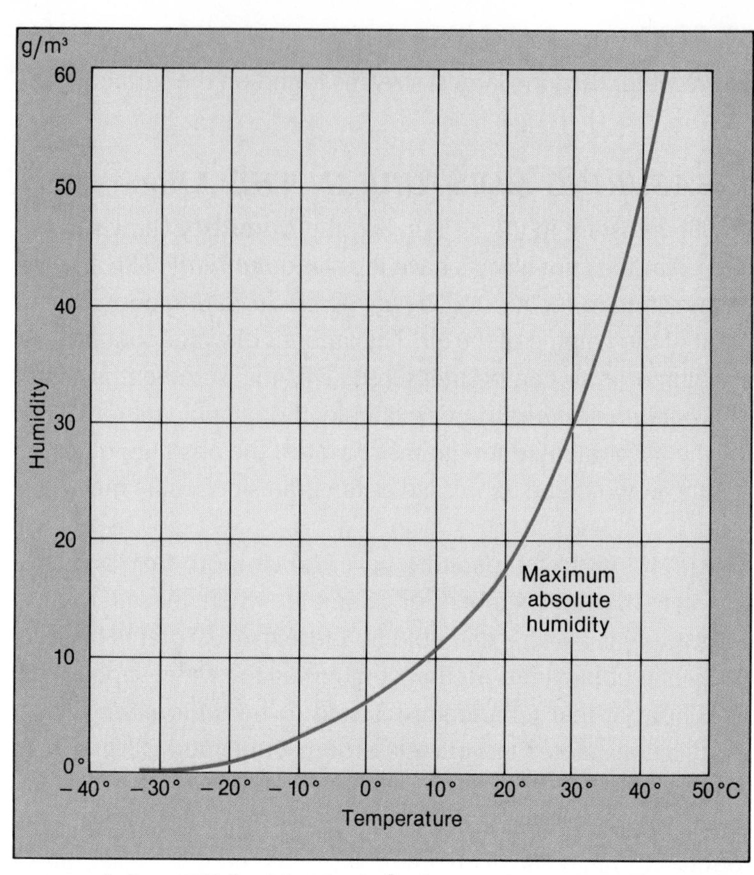

Fig. 12–15 According to this graph, how much water is needed to saturate the air at 30° C? at 20° C?

meter of air at 25°C with 11 g/m³ of water is cooled. The graph in Fig. 12–15 shows that when air reaches a temperature of about 12°C, it will be able to hold only 11 g/m³. The air will become saturated. Its relative humidity will become 100 percent.

CONDENSATION

The cooling of air will often cause it to become saturated with water vapor. The temperature at which the air becomes saturated is called its **dew point.** When air reaches its *dew point,* its relative humidity is 100 percent. Cooling to a temperature below the dew point causes the water vapor to begin to change back into a liquid. The process by which a gas is changed into a liquid is called **condensation.**

Dew point The temperature at which air becomes saturated with water vapor.

Condensation The process by which a gas is changed into a liquid.

The water that collects on the outside of a glass of ice water is an example of *condensation.* The air touching the cold glass is chilled below its dew point. Water that has condensed on the cold surface is *dew.* On cool nights with no wind, dew may form on cool surfaces of automobiles, rocks, and plants. When

Fig. 12-16 (left) *How does frost form?* (right) *How does dew form?*

the dew point is below 0°C, water vapor forms a solid instead of a liquid, called *frost*. Frost forms when the temperature is below freezing. See Fig. 12-16.

In some cases, the cooling of air below its dew point causes small droplets of water or ice crystals to condense on dust in the air. When this occurs near the earth's surface, it is called a *fog*. If it occurs higher in the air, it is called a *cloud*. Fog also forms when air is blown across a cold surface. This kind of fog often occurs when warm, moist air is blown from bodies of water across cold land surfaces.

SUMMARY

Energy is absorbed when liquid water evaporates. The water vapor carries this energy with it into the air. Most of the moisture in the air comes from the sea. The amount of moisture that can be held by air is greater as the temperature rises. When air is holding all the water vapor it can, the air is saturated. Lowering the temperature of air will reduce its ability to hold water vapor. Water may condense from cooled air.

QUESTIONS

Use complete sentences to write your answers.

1. How could you demonstrate that evaporation of water requires energy?
2. Explain the relationship between humidity, absolute humidity, and relative humidity.
3. Explain how dew, fog, and frost are similar.

INVESTIGATION

TEMPERATURE CHANGES DUE TO EVAPORATION

PURPOSE: To find out how the temperature changes when liquids evaporate.

MATERIALS:

2 thermometers
cotton cloth (3 pieces)
string
container with water
container with glycerine
container with alcohol
index card
paper towels

PROCEDURE:

A. Allow both thermometers to lie on the table for five minutes.

B. Copy the following table.

Thermometer	Temp. Before Fanning	Temp. After Fanning
Dry bulb Bulb with water		
Dry bulb Bulb with glycerine		
Dry bulb Bulb with alcohol		

C. Read one of the thermometers. This is the temperature of the air.
   1. What is the temperature of the air?

D. Wrap a small piece of cotton cloth around the bulb of one of the thermometers. Tie with string. Dip the cloth in water until completely wet.

E. Hold both thermometers up and fan them gently with an index card for exactly three minutes.

F. Read the thermometers, wet bulb first, immediately after fanning them. Record the temperatures in your table.

G. Remove the cloth from the thermometer and wipe the thermometer dry. Repeat steps D through F, using alcohol in place of water.

H. Remove the cloth from the thermometer and wipe dry. Repeat steps D through F. This time use glycerine instead of water.

I. Remove the cloth from the thermometer. Place the pieces of cloth and string in the containers provided.
   2. What is the reason for using a dry bulb in each case above?
   3. In each case, did evaporation cause a cooling effect or did it cause a heating effect?
   4. Which liquid gave the greatest difference in temperature when it evaporated from the surface of the thermometer?
   5. Which liquid gave the least difference in temperature when it evaporated from the surface of the thermometer?

CONCLUSIONS:

1. What effect on temperature does evaporation of a liquid have?
2. Explain why you feel cold when you get out of a swimming pool into the windy air.
3. Why is using alcohol to cool a fever better than using water?

CAREERS IN SCIENCE

CLIMATOLOGIST

Climatologists study weather conditions of the past and try to make forecasts for the future. To make such forecasts, they analyze records showing past humidity (percent of water in the air), temperatures, precipitation (levels of rainfall), and other elements.

Today, some climatologists are involved in studying the "greenhouse effect"—the gradual warming of the earth's atmosphere—as well as the effects of warm ocean currents and volcanic eruptions on changing weather patterns.

Climatologists have college degrees, with courses in meteorology, physics, and advanced mathematics. These scientists are employed by the National Oceanic and Atmospheric Administration (NOAA), by organizations such as the hurricane center in Florida, and by private industry. They may also act as consultants to radio and television weather broadcasting stations. For further information, write: American Meteorological Society, 45 Beacon Street, Boston, MA 02108.

FORESTRY TECHNICIAN

You may have seen news broadcasts showing forestry technicians at work fighting forest fires. Those firefighters are only one type of forest management technician assisting foresters in the job of maintaining the nation's forests. Some technicians help to oversee the proper cutting of trees, as well as the replanting of timber in areas where trees have been removed. Others help to plan what kinds of trees will be planted. Forestry technicians are also involved in building roads and cutting trails through forests for hikers and campers. In addition, the increasing use of our national parks and forests for recreation has created the need for persons with skills in forestry, conservation, and public relations.

The field is very competitive, and openings are limited. Formal training, which is available at community colleges or technical schools, will help you to obtain a job. For further information, write: The U.S. Department of Agriculture, Forest Service/Human Resource Program, P.O. Box 2417, Washington, DC 20013.

CHAPTER REVIEW

VOCABULARY

On a separate piece of paper, match the number of each blank with the term that best completes each statement. Use each term only once.

condensation	Coriolis effect	humidity
convection	dew point	radiant energy
convection cell	greenhouse effect	relative humidity

Energy that the earth receives from the sun is called __1__. The energy then becomes trapped by the earth's atmosphere in a process called the __2__. When the earth's surface is heated unequally, air moves by the process of __3__. Sometimes air will move in a complete circle called a(n) __4__. The __5__ will cause the moving air to change direction.

Warm air will absorb water vapor called __6__. If the temperature is lowered to its __7__, the air will become saturated. At that temperature, the __8__ will become 100 percent. Some of the water vapor may turn back into liquid water in a process called __9__.

QUESTIONS

Give brief but complete answers to each of the following questions. Unless otherwise indicated, use complete sentences to write your answers.

1. What happens to the amount of solar energy reaching earth as it enters and passes through the atmosphere?
2. Explain why you can get a sunburn on a cloudy day.
3. How does the atmosphere trap radiant energy?
4. How does the atmosphere spread solar energy over the earth?
5. Where are the doldrums located and what causes them?
6. What causes land and sea breezes?
7. Describe a monsoon as a convection cell.
8. Compare the Coriolis effect in the Northern Hemisphere with the Coriolis effect in the Southern Hemisphere.
9. Describe one of the three major wind belts as a very large convection cell.
10. Why do you feel cool in a wet bathing suit?

11. Explain how water evaporates from the seas.
12. What is the difference between humidity and relative humidity?
13. How are dew and frost related?
14. Describe how fog forms.
15. Compare fog with a cloud.
16. Why does the air near oceans and lakes stay cool during the day and warm during the night?
17. A sample of room air has a temperature of 20°C and holds 9 g/m^3 of water. (a) Use the table in Fig. 12–16 to find out how much moisture the air can hold at this temperature. (b) Calculate the relative humidity of the sample.

APPLYING SCIENCE

1. Compare the placement of air conditioners near the ceiling with the placement of heaters, which are usually near the floor.
2. Why does fog form after clear nights more often than after cloudy nights?
3. Explain how the Coriolis effect would affect a flying airplane.
4. It is common practice, in winter, for the weather service to give the "chill factor" for cold temperatures. The chill factor is the temperature we feel due to the effect of wind. In the summer, the weather service often will give a temperature that includes the effect of humidity when the temperature goes above 32°C (90°F). Why would high humidity make us feel hotter at higher temperatures? Contact your local Weather Service Office to find out how this kind of temperature is determined.

BIBLIOGRAPHY

Allen, Oliver. *Atmosphere* (Planet Earth Series). Chicago: Time-Life Books, 1983.

Moses, L., and John Tomikel. *Basic Meteorology*. California, PA: Allegheny Press, 1981.

Reck, Ruth. "The Albedo Effect." *Science 81*, July/August, 1981.

Schaefer, Vincent J., and John A. Day. *A Field Guide to the Atmosphere*. Boston: Houghton Mifflin, 1983.

CHAPTER

13

Lightning lights up the sky during a thunderstorm. Lightning can strike the earth with 100 million volts of electricity and with temperatures five times hotter than the sun's surface.

WEATHER

CHAPTER GOALS

1. Explain how clouds form and precipitation occurs.
2. Relate the movement of air masses to weather and climate.
3. Explain how fronts and severe storms are produced.
4. Explain how precipitation forms.

13–1. Water Leaves the Air

At the end of this section you will be able to:

- Name the conditions necessary for clouds to form.
- Describe three types of clouds.
- List four kinds of precipitation.

Some mornings when you go outside, you walk into a cloud. A fog has settled in during the night. The fog may seem to come from nowhere. But it is a visible part of the cycle in which water enters and leaves the air.

HOW CLOUDS FORM

Fog is a kind of cloud that forms at ground level. Fog, like all other clouds, is made of drops of water so small that they can float in the air. For example, water vapor enters the air when you take a hot shower. When the warm, moist air strikes a cool surface, the water vapor leaves the air and changes back to a liquid. This process is called *condensation*. A bathroom mirror becomes coated with a film of condensed water. Two conditions are necessary for water vapor in air to condense. First, the air must be cooled. At lower temperatures, air holds less moisture. Fog often forms at night because the temperature of air near the ground drops.

A second condition necessary for water vapor in the air to condense is the presence of some kind of solid surface. For example, the solid surface of a mirror provides a place on which water vapor can condense. Water condenses on a surface such as a mirror because the mirror is cooler than the warm, moist air in the bathroom. Water molecules that make up water vapor tend to stick to solid surfaces. More water is then attracted until

Fig. 13-1 During a shower, water vapor condenses on the cooler surface of the mirror.

Fig. 13-2 At the dew point, warmer, moisture-filled air will also condense.

drops form. See Fig. 13–1. Without a surface on which the molecules can land, water vapor usually will not condense.

In order for water vapor to condense from the air, the same conditions must exist in nature as in a bathroom. The first step is the lowering of the temperature of the air to its *dew point.* The dew point is the temperature at which the water vapor in air begins to condense. For example, air can be cooled to its dew point when it touches a cool surface. For this reason, dew forms at night on surfaces such as grass. Air is also cooled when it mixes with colder air. See Fig. 13–2. This happens when you open the door of a freezer. Air is also cooled when it moves upward in the atmosphere. As the air rises, it expands because the atmospheric pressure decreases. The air cools as its molecules move farther apart as a result of the drop in pressure.

The second step needed for water vapor to condense is to have a solid surface available. Just as dew condenses on grass, water vapor high in the atmosphere can condense on solid particles in the air. The air is filled with tiny solid particles. There are salt particles from the sea and dust from soil and rocks. These solid particles are carried all over the earth by winds. Smoke also adds particles to the air. All of these kinds of solid particles provide places for liquid water to collect when water vapor from the air condenses.

When air is lifted and cooled, its water vapor may condense and form clouds. This can happen in several ways. For example, when air moves over mountains, it is lifted. Clouds are often found around mountaintops. See Fig. 13–3 *(left).* Air also is lifted when dense, cold air pushes beneath warmer air. As the

lighter, warm air rises, clouds are formed. Also, when air is heated, the warm air rises. The rising air then cools and forms clouds. See Fig. 13–3 *(right)*. Thus, in order for clouds to form, the air temperature must first drop to its dew point. Then tiny solid particles must be present to provide a surface on which the water vapor can condense.

Fig. 13-3 (left) *A cloud may form when air is cooled as it is lifted over a mountain.* (right)*Clouds may also form when warm air rises and then is cooled at higher altitudes.*

KINDS OF CLOUDS

Most clouds are formed high above the earth's surface. As the rising air is cooled, tiny droplets of water gather around small solid particles in the cooling air. These cloud droplets are so small that they float in the air. The slightest rising air currents will keep these tiny droplets suspended. As more water vapor condenses around the droplets, they become larger and larger. Each cloud droplet reflects some of the sunlight falling on it. It is this reflected light that gives clouds their familiar white color.

Clouds come in many shapes and sizes. Observing clouds can provide clues to changes in the atmosphere and the weather to come. A system of names, based on the height and appearance of clouds, is used in *meteorology* (meet-ee-uh-**rol**-uh-jee). Meteorology is the study of the atmosphere.

Often the sky is a mixture of different clouds that are carried in several directions by winds at various altitudes. Careful observation of cloud shapes shows that there are three separate kinds of clouds. Layers or flat patches are called *stratus* (**strat**-us) *clouds* (spread out). See Fig. 13–4 *(left)*. Rising mounds that develop individually, often with tops shaped like

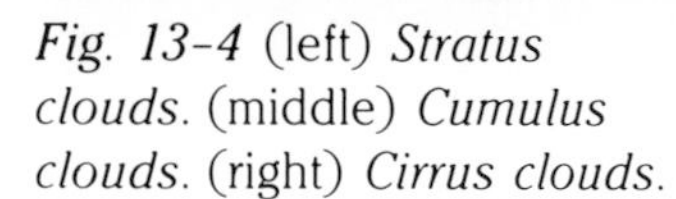

Fig. 13-4 (left) *Stratus clouds.* (middle) *Cumulus clouds.* (right) *Cirrus clouds.*

cauliflower, are called *cumulus* (**kyoo**-myoo-lus) *clouds* (piled up). See Fig. 13-4 *(middle).* Feathery clouds made of ice crystals formed at high altitudes are called *cirrus* (**sear**-us) *clouds* (curled). See Fig. 13–4 *(right).* Sometimes a cloud has two of these basic shapes at the same time. Then the cloud is identified as a combination of the two. For example, a layer of clouds with a piled-up surface is called *stratocumulus.*

The system of naming clouds also takes their height into account. Low clouds are found at an altitude below two kilometers. Middle clouds are found between two and seven kilometers. These clouds are always named by adding *alto* before the basic name. Thus a stratus cloud at an altitude of three kilometers is called altostratus. Above seven kilometers are high clouds that have *cirro* added before the basic name. For example, a stratus cloud at that altitude is called cirrostratus. Cumulus clouds have vertical development, since they may grow upward to high altitudes.

Rain or snow may come from stratus or cumulus clouds. In either case, the term *nimbus* is used with the basic name to form the names nimbostratus or cumulonimbus. Fig. 13–5 shows the way clouds are named according to their shapes and altitudes.

PRECIPITATION

Any liquid or solid water that falls to the earth's surface is called precipitation. Precipitation begins with the cooling and condensing of water vapor. A cloud produces precipitation when its droplets or ice crystals become large enough to fall as

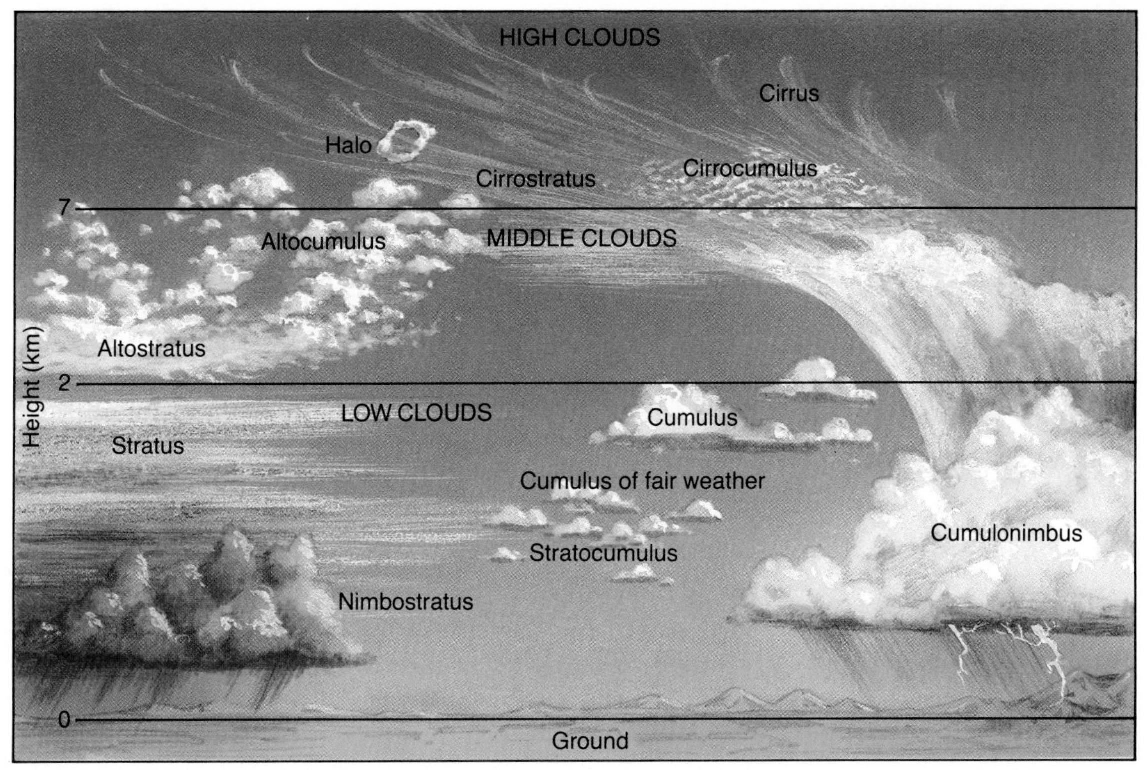

Fig. 13-5 Classification of clouds according to height and form.

rain or snow. There are four kinds of precipitation: rain, snow, sleet, and hail. The form in which precipitation reaches the earth's surface depends on the amount of water vapor in the air, the air temperature, and air currents.

1. Rain. Droplets of water remain suspended in the air as long as they are small. In some clouds, droplets become too large to remain suspended. These larger droplets begin to fall. As they move downward, they collide with other droplets. As a result of the collisions, they combine with other drops and grow even larger. These drops can grow to be a hundred or a thousand times larger than the original cloud droplets. Many of these drops become heavy enough to fall as raindrops. See Fig. 13-16. Some scientists think that electrical charges may attract cloud droplets to one another in the formation of raindrops. Rain probably forms this way in warm areas.

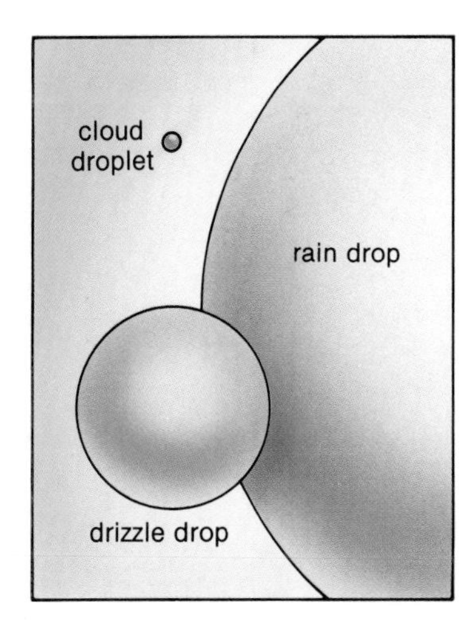

Fig. 13-6 Raindrops are large drops. The sizes here are exaggerated.

2. Snow. In cooler parts of the earth, precipitation can be formed in a different way. Most clouds in these cooler regions have ice crystals in their upper parts. Water from the air condenses on the cool surfaces of the ice crystals and freezes. The process continues until the ice crystals become snowflakes

Fig. 13-7 Hail is precipitation in the form of lumps of ice.

large enough to fall. Usually, the air that is below the clouds is warmer than in the clouds. This causes the snowflakes to melt and fall as rain to the earth. However, if the air is cold all the way to the ground, snow falls.

3. Sleet. Sometimes snowflakes strike a layer of warm air and melt into raindrops. They may pass through a cold layer, refreeze and become frozen rain. Precipitation that reaches the ground in the form of frozen rain is called *sleet*.

4. Hail. Hail usually begins as a frozen raindrop. As the frozen raindrop falls, it may be blown up by strong, rising air currents. There more water droplets collect on it and freeze. The larger frozen droplet falls again. Strong currents of air may carry the frozen water droplet up again, where more water freezes on it. Each time the droplet is lifted, a new layer of ice is added, increasing its size. Finally, it falls to the ground as a hailstone. Hail usually falls as small ice particles, but may be as large as walnuts, or even baseballs. See Fig. 13–7.

Knowledge of the role played by ice crystals in the forming of rain has led to the first scientific attempts to make rain. The method is called *cloud seeding*. In cloud seeding, substances that form ice are put into clouds. Two substances commonly used are dry ice and *silver iodide* (**eye**-uh-died). Dry ice is solid carbon dioxide. Its low temperature causes ice crystals to form from the water vapor in the air. These ice crystals then produce snow or rain. Crystals of silver iodide also act as seeds on which ice crystals can grow.

SUMMARY

Clouds form when certain conditions are found in the atmosphere. Precipitation occurs when cloud droplets or ice crystals are too large to remain suspended in the air. Clouds are seeded by adding substances that cause ice crystals to form inside clouds.

QUESTIONS

Use complete sentences to write your answers.

1. Give an example of how clouds might form.
2. Name three kinds of clouds, based on shape, and describe the appearance of each kind.
3. Describe four kinds of precipitation.

INVESTIGATION

DEW POINT

PURPOSE: To find the dew point of the classroom air and to predict the height at which clouds form.

MATERIALS:

shiny can, water, stirring rod, thermometer, ice

PROCEDURE:

A. Measure the air temperature. Hold the thermometer in the air for two minutes (do not touch the bulb of the thermometer). Read the temperature.
   1. What is the temperature of the air?

B. Fill a clean, dry can about half full with water. Do not breathe on it. If moisture forms on the can, refill it with slightly warmer water and wipe the outside dry.

C. Add a piece of ice to the water. Put in the thermometer. Stir with a stirring rod. DO NOT USE YOUR THERMOMETER AS A STIRRING ROD. Be careful not to breathe on the can. Look carefully at the can to see if dew or moisture is collecting on the surface. It may help to run your finger over the surface to feel the dew. See Fig. 13–8. When dew forms, remove the ice and measure the water temperature.
   2. At what temperature did dew first appear on the can?
   3. Compare the air temperature to the dew point.

D. Clouds usually do not form near the ground because the temperature is above the dew point. As the height increases, the air temperature decreases. When the temperature equals the dew point, clouds can form. To find this temperature, subtract the dew point from the air temperature.
   4. What is the temperature difference?

Fig. 13–8

E. The air temperature decreases 1°C for every 100 m above the ground. Find the height where clouds will form by multiplying the temperature difference by 100 m.
   5. At what altitude is the temperature equal to or just below the dew point?

CONCLUSIONS:

1. Describe what would happen if the temperature in the classroom were lowered to the temperature you found as the dew point.
2. To predict the height at which clouds would form, it is assumed that the dew point remains the same, but it actually decreases slightly with increasing height. How does this affect your prediction for the cloud height?

13–2. Air Masses

At the end of this section you will be able to:

- Define *air mass* and *front.*
- Describe the four air masses that affect North America.
- Describe three major climate zones.

Have you ever gone into a cave or a cellar on a hot day? In those places, the air is cool and damp. It feels much different from the air outside. Air trapped inside a closed space, such as a cave or cellar, takes on the temperature and humidity characteristics of that place. What happens when parts of the earth's atmosphere remain stationary, like air held in a cave?

Fig. 13–9 Air trapped inside a closed space, such as a cave, takes on the temperature and humidity of that place. That is why caves are usually cool and damp.

HOW AIR MASSES ARE MADE

When air remains still for a time over one part of the earth's surface, it takes on the properties of that region. For example, air that sits over warm ocean waters becomes warm and moist. On the other hand, air that stays over the cold regions near the poles becomes very cold. Because low temperatures prevent air from holding much moisture, the air there is also dry. A large body of air that takes on the temperature and humidity of a region is called an **air mass.** An *air mass* usually covers millions of square kilometers. It reaches from the surface of the earth to a height of three to six kilometers. At any particular altitude within this huge body of air, the temperature and humidity are the same or nearly the same. Once an air mass begins to move, it moves as a body.

Air mass A large body of air that has taken on the temperature and humidity of a part of the earth's surface.

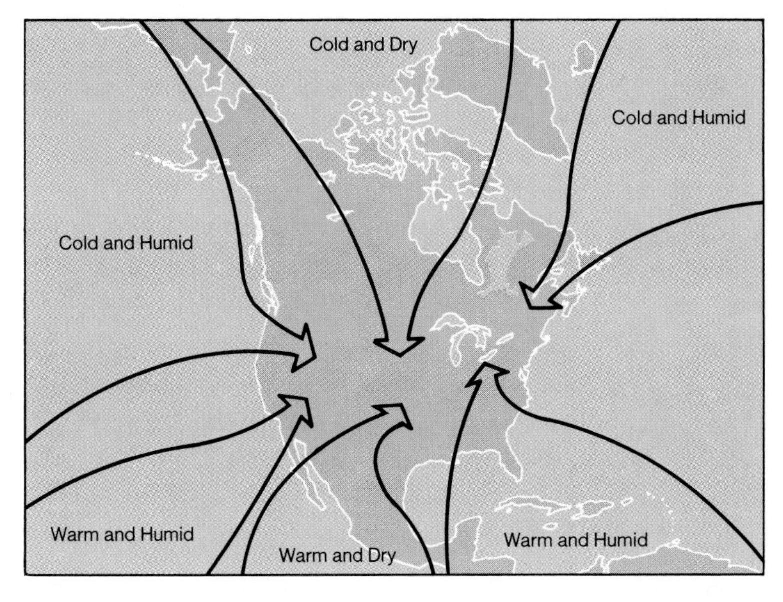

Fig. 13-10 The source areas shown here affect weather in North America.

Air masses can only form at places where the earth's surface is almost the same over large areas and the winds are not strong. In this way, the air's temperature and moisture levels become similar to those of the area below it. The oceans near the equator and the regions near the poles are places where air masses can form. Areas on the earth's surface where air masses are formed are called *source regions*. There are two general source regions for the air masses that affect North America. In these source regions, four kinds of air masses are formed. See Fig. 13-10.

1. Cold and dry polar air. Northern Canada and Alaska are the source regions for cold, dry polar air. During the winter, these polar air masses bring very cold weather as they are carried by the prevailing westerlies across the eastern part of the country.

2. Cool and moist air. Cool, moist air masses are formed in the northern Pacific Ocean. These masses bring rain and snow to the Pacific Northwest. As the air masses move across the country and pass over the mountains, they usually lose most of their moisture. These now-drier air masses bring mild, fair weather to the eastern United States. Moist polar air masses also form in the northern Atlantic Ocean. They usually drift toward Europe. However, these cold, moist air masses sometimes invade the New England states, bringing rain in the summer and heavy snow in the winter.

3. Warm and moist air. Air masses coming from the south are mostly warm and moist. The ones that have the greatest effect on the weather in North America are formed over the Gulf of Mexico, the Caribbean Sea, and parts of the Atlantic Ocean. These air masses move north, bringing warm and often stormy weather to the central and eastern United States. Warm, moist air from the south may also move into the desert regions of Arizona and southern California during the summer, causing heavy thunderstorms.

4. Hot and dry air. Hot, dry air masses also form during the summer in northern Mexico and the southwestern part of the United States. Northward movement of these air masses is responsible for much of the hot, dry summer weather of the American Southwest.

FRONTS

As air masses move, they meet each other. See Fig. 13–11. The air in these masses does not mix easily. This is mainly because of the differences in density between the two bodies of air. Cold air has a greater density than warm air. Mixing air with different densities is like trying to mix oil with water. Because they don't mix easily, the boundary between the two air masses tends to be sharp and distinct. Boundaries separating air masses are called **fronts.** The temperature, pressure, and wind direction of the air are generally different on opposite sides of the *front*.

Front The boundary separating two air masses.

Most fronts form when one air mass moves into a region already occupied by another air mass. If a cold air mass moves

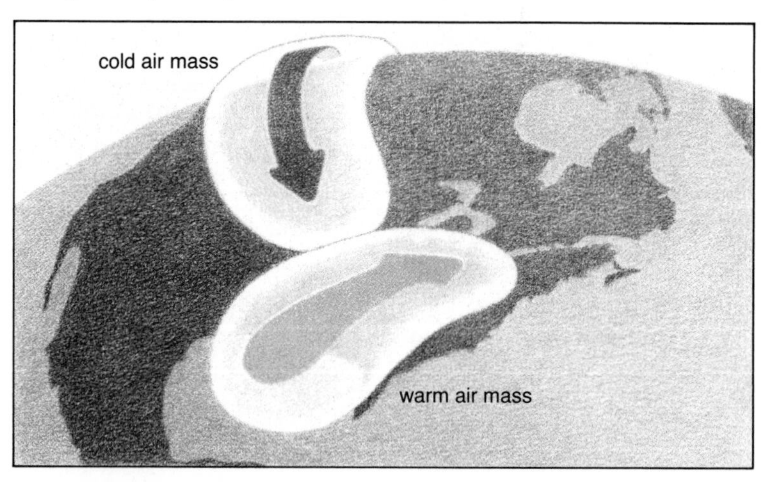

Fig. 13–11 When air masses meet, they do not mix.

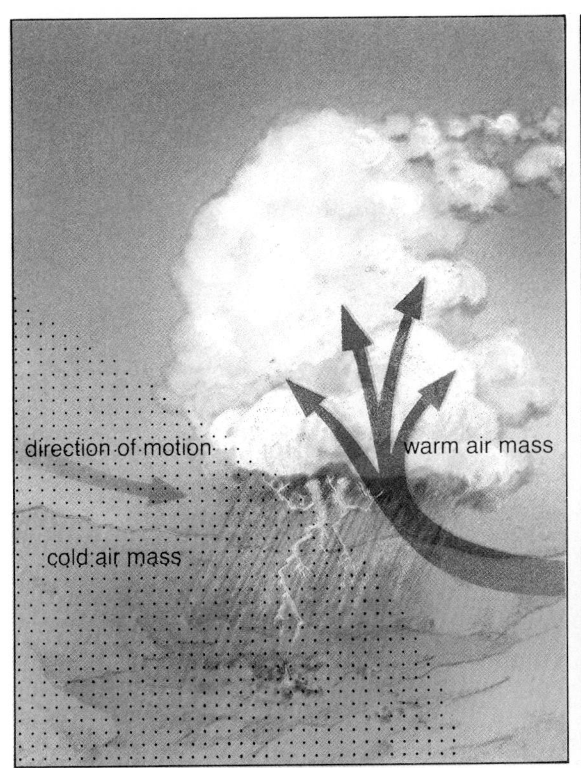

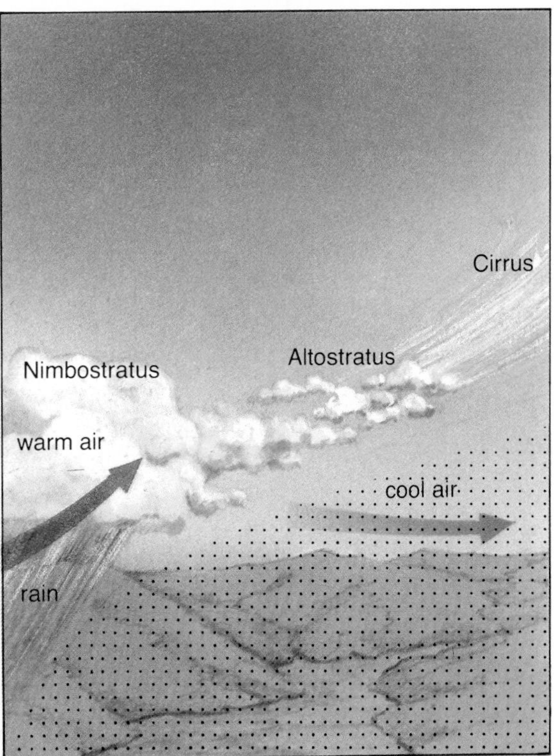

Fig. 13-12 (left) *A cold air mass moving in on a warm air mass, forming a cold front.* (right) *A warm air mass moving in on a cold air mass, forming a warm front.*

into an area occupied by warmer air, *a cold front* is created. The cold air does not mix with the warmer air. Instead, the colder, denser air pushes under the warmer air like a wedge. See Fig. 13-12 *(left)*. The lifting of the warm air usually causes cumulus clouds to form along the cold front. The clouds along a cold front usually bring heavy rain. Cold fronts move quickly because the dense, cold air easily pushes the warmer air. The stormy weather that comes with a cold front usually passes quickly. The air mass behind a cold front brings cool, dry weather.

When a warm air mass pushes into an area occupied by cold air, a *warm front* is formed. Because it is less dense, the warm air rises up over the colder air. See Fig. 13-12 *(right)*. As the warm air is lifted, cirrus clouds are formed high in the sky. As time passes, the cloud cover becomes lower and thicker, until a solid sheet of stratus clouds covers the sky. A long period of gentle rain may occur, followed by slow clearing and rising temperatures. Warm fronts do not move as quickly as cold fronts. Thus the weather changes that follow a warm front are not as noticeable as the changes following a cold front.

Sometimes air masses move parallel to the front that separates them. Then the front does not move and is called a *stationary front.* A stationary front may be either a warm or a cold front.

Air masses can be located by using the temperature readings of many areas. When the readings are recorded on a map, lines can be drawn through all the points of equal temperatures. Such lines are called **isotherms** (**ie**-suh-thurms).

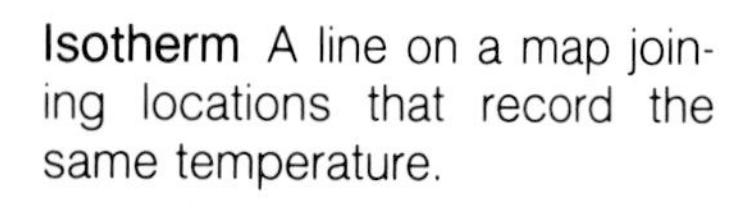

Isotherm A line on a map joining locations that record the same temperature.

On the map shown in Fig. 13–13, the *isotherm* through all the points of equal temperature (15°C) represents an air mass.

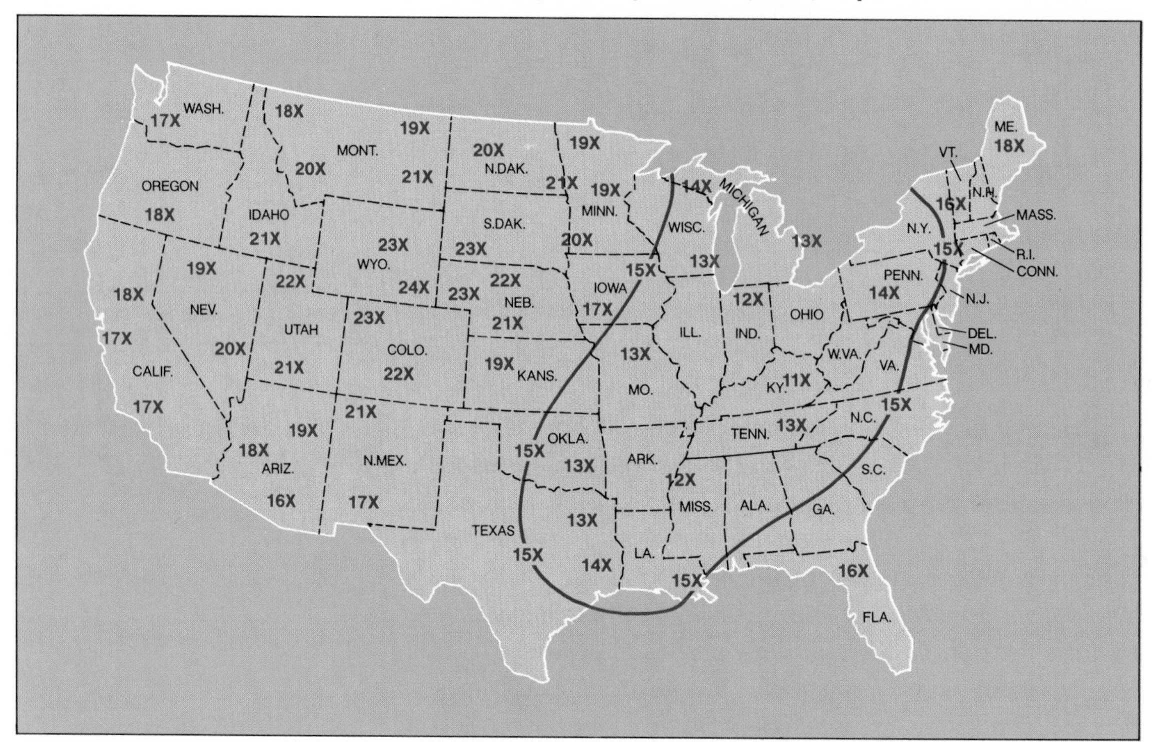

Fig. 13–13 A line called an isotherm is drawn on this map to connect all locations having a temperature of 15° C.

Climate The average weather at a particular place.

CLIMATES

Moving air masses and fronts are the main causes of day to day weather changes. **Climate,** however, is the average weather at a particular place over a long period of time. The *climate* of an area is determined by the type of air masses that are found over its location. There are three major climate zones. See Fig. 13–14.

1. Tropical. The climate of regions near the equator is caused by warm air masses. The average temperature during the year is 18°C and above. This climate zone is called *tropical.*

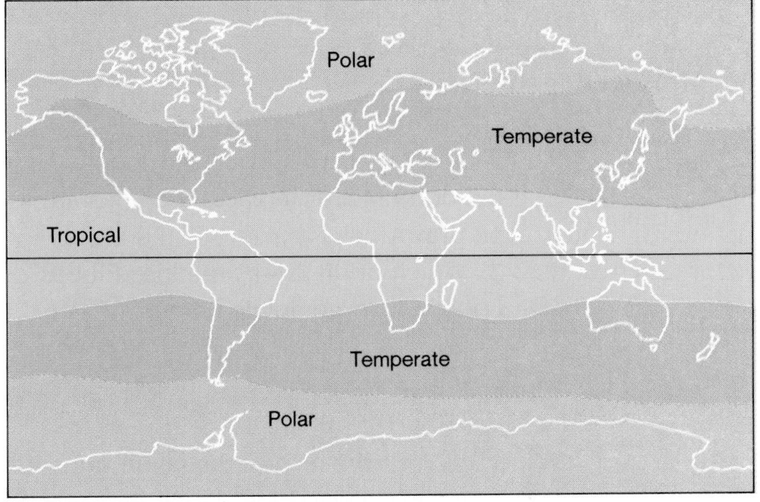

Fig. 13-14 The three main climate regions are caused by movements of warm and cool air masses.

2. Polar. *Polar* climates are caused by air masses formed near the poles. Average yearly temperature for regions with a polar climate is 10°C or less. The summers are short, and the winters are long and harsh.

3. Temperate. Between the equator and the poles are regions with *temperate* climates. This kind of climate is affected by both warm and cold air masses. The weather in the temperate climatic regions can change from day to day and from one part of the year to another. Most of the United States is found in this kind of climatic region.

SUMMARY

Air masses are produced when air remains over one area for a long period of time. Boundaries between different air masses can be identified because different air masses do not mix easily. These boundaries, or fronts, bring certain kinds of weather as they pass over a place. The kinds of air masses that usually move over a region determine its climate.

QUESTIONS

Use complete sentences to write your answers.

1. What is an air mass? How is an air mass usually made?
2. Describe the four air masses that affect North America. Indicate where each forms.
3. What is a front? How is a front usually located?
4. Describe the three major climate zones.

INVESTIGATION

LOCATING AIR MASSES

PURPOSE: To use temperatures for locating cold and warm air masses.

MATERIALS:
tracing paper pencil

PROCEDURE:

A. Trace the map in Fig. 13–13, page 362, and write the temperatures and X's as shown.

B. Study the solid line that has been drawn through all stations (X's) that have a temperature of 15°C. Notice that the line sometimes goes between stations of 14°C and 16°C. This is because there is a 15°C temperature about halfway between those readings even though there is no reporting station at that point.

1. If temperatures were available from several stations in Canada near the Great Lakes, do you think the 15°C line would form a closed loop?

C. On your map, find a station where the temperature is 13°C. Starting at that station, draw a light line to show where 13°C locations would be. Remember, if there is no 13°C station nearby, you will need to draw halfway between the 12°C and 14°C stations. Continue the line until it has gone through all the stations and all locations that are at 13°C. Your line should form a closed loop. Smooth it out and darken it similarly to the 15°C line. This is the 13°C line. Check with your teacher to be sure it is drawn accurately.

D. Draw lines for 17°C, 19°C, 21°C, and 23°C. Not all will be closed loops. Only the odd number temperature lines are to be drawn.

2. Which of the lines are closed loops?

E. Look at your completed map.

3. What is the lowest temperature found in the smallest closed loop in the eastern half of the map?
4. What is the highest temperature found in the smallest closed loop in the western half of the map?

F. The closed loops show where the air masses are located. Cold air masses have isotherms that decrease in temperature as you move toward the center. Look at your completed map.

5. Where is the center of the cold air mass? (Name the region of a state near the center.)

G. Warm air masses have isotherms that increase in temperature as you move toward the center.

6. Where is the center of the warm air mass? (Name the region of a state near the center.)
7. Where would you expect a front to occur for the conditions given on this map?
8. For a warm front, what weather might you expect to find along the front? Would the weather change severely after the front passed?

CONCLUSIONS:

1. Explain how air masses on a map can be located.
2. How can you tell where the center of a cold or warm air mass is found?

13–3. Weather Changes

At the end of this section you will be able to:

- Trace the steps in the development of a *cyclone.*
- Describe an *anticyclone.*
- Explain how a *hurricane* forms.
- Describe the conditions that cause thunderstorms and tornadoes.

Fig. 13–15 Tornado winds caused this damage.

Weather changes are caused by moving air masses. Whenever you see the weather change from sunshine to rain, or from hot to cool, it is because a new air mass has taken the place of an old air mass. Sometimes the weather change can be very sudden and violent, as in the case of a tornado. Tornado winds can cause damage like that shown in Fig. 13–15. They cause more deaths annually in the United States than any weather disturbance except lightning. Like all weather changes, tornadoes begin with the formation of an air mass.

WEATHER FRONTS

During World War II, high-flying bombers were met by a river of fast-moving winds of about 120 to 240 kilometers per hour. These winds form the *jet stream.* See Fig. 13–16. The jet stream occurs at an altitude about 10 to 15 kilometers above the surface.

The jet stream is formed along the boundary of a cold and a warm air mass called the **polar front**. The huge mass of cold air

Polar front The boundary where cold polar air meets warm air.

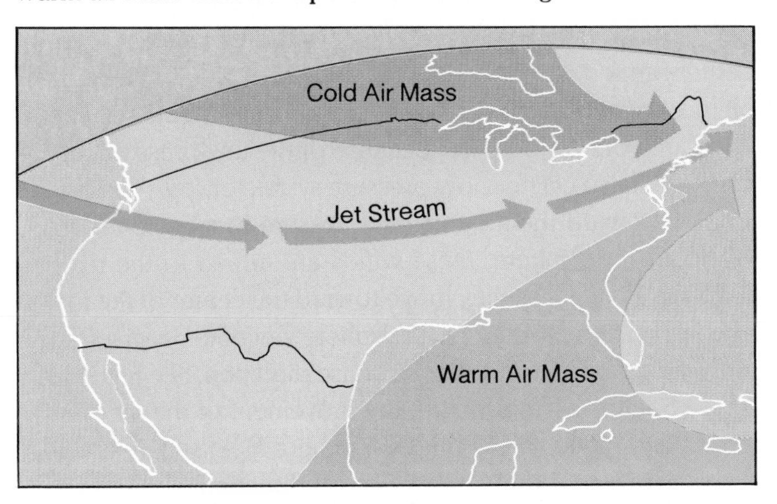

Fig. 13–16 Jet streams are strong upper-level winds formed when cold, polar air meets warmer air.

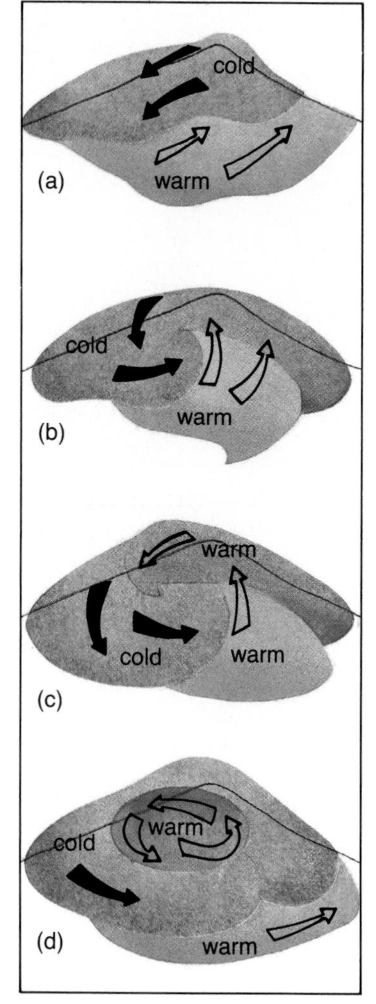

Fig. 13–17 The formation of a cyclone.

Cyclone A kind of storm that is formed along the polar front.

forms around the North Pole. This cap of dense air eventually begins to push slowly southward. The earth's rotation then causes the cold air mass to move west, forming the wind called the *polar easterlies*. At the same time, a warm air mass moves toward the poles and to the east, forming the winds called the *westerlies*. The polar easterlies and the westerlies meet and move past each other, like traffic on opposite sides of a highway, forming the *polar front*. At high altitudes near the polar front, a band of very strong winds that moves from west to east forms the jet stream. Pilots in aircraft traveling eastward often use the jet stream to save fuel. They can thus increase their speed by as much as 160 kilometers per hour. Traveling west, they follow a path that avoids the jet stream or they fly below it.

Often the smooth flow of the winds past each other at the polar front is disturbed. A disturbance, or interference, in the flow can be caused by many things. For example, interference can be caused by mountains or by the differences in temperature between land and sea surfaces. Any interference can produce a small wave, or bulge, in the polar front. See Fig. 13–17. At the bulge, the warm air pushes into the cold air, making a warm front. The cold air moves around and behind the warm air, forming a cold front. The faster-moving cold front soon catches up with the warm front. The warm, moist air is squeezed upward, leaving only cold air near the ground.

CYCLONES

Lifting of the warm, moist air almost always causes a storm center with dark clouds and heavy precipitation. A storm center formed in this way is called a **cyclone** (**sie**-klone). The air in a *cyclone* tends to have an upward movement. This results in an area of low pressure. Because of the low-pressure area, a cyclone is also called a low-pressure system or *low*. Winds tend to blow toward the area of low pressure in a cyclone. In the Northern Hemisphere, these winds are turned to the right by the Coriolis effect as they move toward the center of the low. A cyclone in the Northern Hemisphere becomes a giant swirl, slowly twisting in a counterclockwise direction. See Fig. 13–18. In the Southern Hemisphere, the movement of the air is in the opposite direction. Lows in the Southern Hemisphere are giant swirls, slowly twisting in a clockwise direction.

A mass of cool, dry air may also develop a spinning motion. The greater density and higher pressure of cool air causes an outward flow of air. Winds created by this air movement are given a clockwise twist by the spinning motion of the earth or the Coriolis effect. The result is a large, slowly spinning body of air called an **anticyclone** (an-tee-**sie**-klone). In the Northern Hemisphere, *anticyclones* turn in a clockwise direction. Their motion is counterclockwise in the Southern Hemisphere. Because they are formed by cool air that is usually not very moist, anticyclones often bring fair, dry weather. Anticyclones are also called high pressure systems or *highs*.

Fig. 13-18 A cyclone seen from an orbiting spaceship.

Anticyclone A large mass of air spinning out of a high-pressure area.

HURRICANES

Hurricanes (**hur**-ih-kanes) are cyclonic storms formed over warm ocean water near the equator. *Hurricanes* are usually smaller, but more intense, than cyclones formed near the polar front. Hurricanes that affect North America are produced in the late summer and early fall. During that time of the year, conditions in the atmosphere cause rapid upward motion of very warm, moist air over the warm part of the ocean. As the warm air rises, its moisture condenses, releasing huge amounts of energy. As more moist air is drawn into the developing storm, a great column of rapidly rising air is created. The rotation of the earth twists the rising column. See Fig. 13-19.

Hurricane A small but intense cyclonic storm formed over warm parts of the sea.

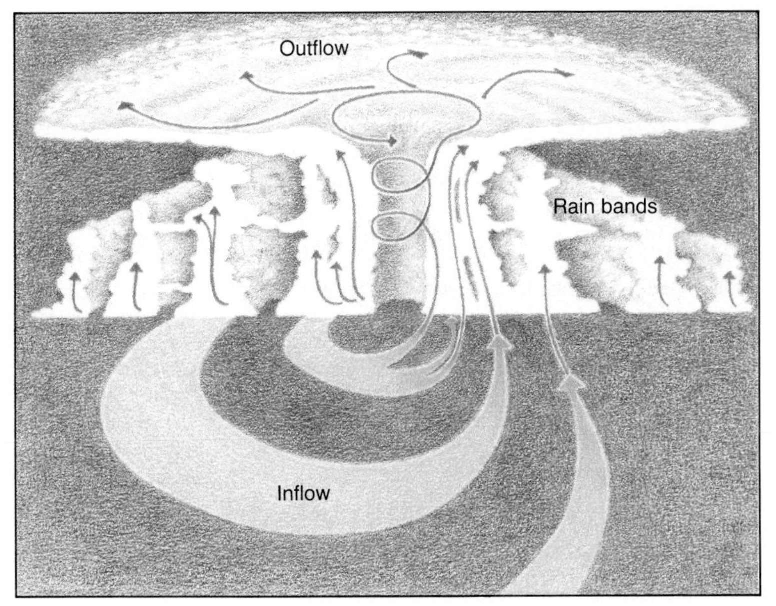

Fig. 13-19 Formation of a hurricane.

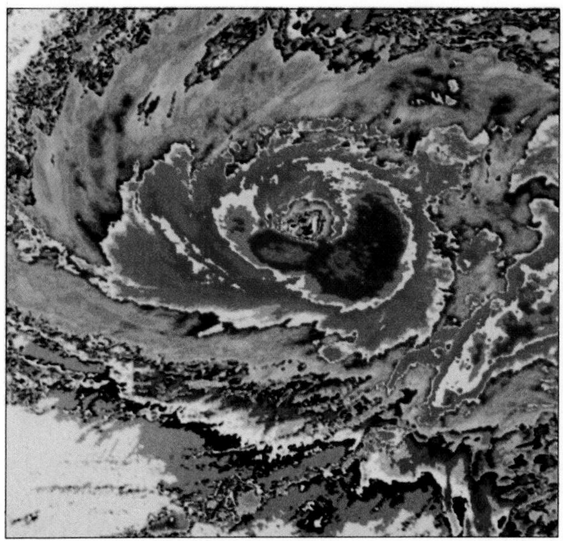

Fig. 13-20 (left) *Different colors show the different temperatures of this hurricane.* (right) *Satellite photograph of hurricane Alicia. Clouds can be seen to circle the "eye."*

When they are fully developed, hurricanes in the Northern Hemisphere have bands of clouds spinning in a counterclockwise direction around a center of calm air. See Fig. 13-20. The center of a hurricane is called the *eye.* Winds circling the eye may reach speeds greater than 100 kilometers an hour. As hurricanes come near land, winds fling destructive waves onto the shore. Much of the damage caused by hurricanes occurs when they strike the shore with their huge waves. As the storm moves inland, heavy rain causes flooding. Hurricanes usually die out as they move over land or cooler water as their supply of warm, moist air is cut off.

THUNDERSTORMS

Fig. 13-21 A thunderhead.

Most people are familiar with thunderstorms. These violent storms are produced when warm, moist air is lifted very rapidly. Towering clouds develop quickly. These clouds belong to the type called cumulonimbus. They are often called *thunderheads.* See Fig. 13-21. Thunderstorms often accompany advancing cold fronts. The cold air wedges under the warm air and lifts it. Cold fronts are sometimes marked by a continuous line of thunderstorms and strong winds. Such a line is called a *squall* (**skwawl**) *line.* See Fig. 13-22 *(left).* Local conditions can give rise to thunderstorms. For example, warm, moist air blown up the side of a mountain, or heated by an area of warm ground, may form small thunderstorms.

Fig. 13–22 (left) *A squall line is made of many thunderstorms.* (right) *Lightning.*

Lightning occurs in a thunderstorm because the ice crystals and water droplets in the cloud become electrically charged. Electricity builds up in the thundercloud until it discharges as a lightning flash. Most often, the lightning jumps within the clouds, but it may pass between the clouds and the earth. See Fig. 13–22 *(right).* The passage of lightning through the air causes heating. The heating expands the air and creates a tremendous sound wave that we hear as thunder.

Very heavy rain can fall during a thunderstorm. Thunderstorms also produce precipitation in the form of *hail.*

TORNADOES

The most destructive of all weather disturbances are *tornadoes* (tor-**nade**-oez). Tornadoes are small, but extremely violent storms that form along cold fronts. A tornado usually drops from a thunderstorm as a thin, white, funnel-shaped cloud. Scientists are not really sure how a tornado forms. But they do know the kinds of weather conditions that are likely to produce tornadoes. Most tornadoes occur in the spring and in the central and southeastern U.S. Everything needed for a tornado-producing thunderstorm is present. The conditions are most likely to occur along the polar front where cool, dry air and warm, moist air meet. The cool, dry air forms over Canada and moves south. The Rocky Mountains cause the air to move in a southeasterly path. At the same time, warm, moist air from the Gulf of Mexico moves northward, where it meets the cool, dry air. The two air masses collide, mostly in the central U.S. The difference in temperature on each side of the front is usually very large, especially in the spring. This condition creates the strong thunderstorms that produce tornadoes.

Fig. 13-23 The funnel cloud of a tornado.

Up to 1,000 tornadoes hit the United States each year. Most of these take place in a belt from Texas to Michigan, called "Tornado Alley." Tornadoes vary in size. Some are small, with diameters of tens of meters. Others can be hundreds of meters in diameter. Tornadoes usually last only a few minutes, but are very destructive. Their winds, sometimes moving as fast as 800 kilometers an hour, can destroy almost everything in their paths. See Fig. 13-23. Today, scientists are doing experiments in order to understand better the cause of tornadoes.

SUMMARY

A very large, spinning weather system can result when cold air from the poles meets warmer air. Moist air that is made to rise rapidly can create thunderstorms. Hurricanes affecting North America usually come from the warm parts of the air over the oceans near the equator. The collision of warm and cool air sometimes provides the conditions that are necessary to make tornadoes.

QUESTIONS

Use complete sentences to write your answers.

1. List the steps in the development of a cyclone.
2. Compare an anticyclone to a cyclone.
3. How does a hurricane differ from a cyclone?
4. Trace the development of a thunderstorm.
5. How are tornadoes related to thunderstorms?

SKILL-BUILDING ACTIVITY

PREDICTING THE PATHS OF HURRICANES

PURPOSE: To use the National Weather Service method of tracking hurricanes to predict their paths and possible danger zones.

MATERIALS:
tracing paper   pencil

PROCEDURE:

A. Trace the map in Fig. 13–24 on the next page.

B. The table below gives the position of two hurricanes for each day of a hurricane watch. On your map plot the position of hurricane Vera for each day. For example, on Day 1 this hurricane was where the line labeled J meets the line labeled 23. Put a small dot at this point.

HURRICANE VERA		HURRICANE WANDA	
Day	Position	Day	Position
1	R–16	1	R–14
2	P–15	2	O–14
3	N–14	3	L–15
4	J–11	4	I–14
5	J–7	5	G–12
6	I–11	6	G–9
7	K–9		

C. Label each point with the number of the day. Draw a dotted line that joins the locations. Label it "Vera."

1. What is hurricane Vera's path?
2. Where and on what day does Vera reverse its direction?
3. Which cities shown on the map would have been damaged most by Vera?
4. Where do you think hurricane Vera will be near on Day 8?
5. On what days and to what cities shown would you have issued hurricane warnings for hurricane Vera?

D. Plot hurricane Wanda. Use a dot for each point, number each one, and connect them with a solid line. Label it "Wanda."

6. Describe the path of Wanda.
7. Which hurricane traveled faster?
8. What other place would be in danger from hurricane Wanda?
9. On what day and to what cities shown would you issue hurricane warnings?
10. Where do you think hurricane Wanda will be on Day 7?
11. What will happen to the hurricane as it continues inland?
12. What is the direction of the winds around a hurricane?
13. The sun provides some energy for hurricanes, but what is another source of their energy?

CONCLUSIONS:

1. What general path do hurricanes take as they form off the southeastern coast of the United States?
2. What valuable service can be given by those who plot the paths of hurricanes?

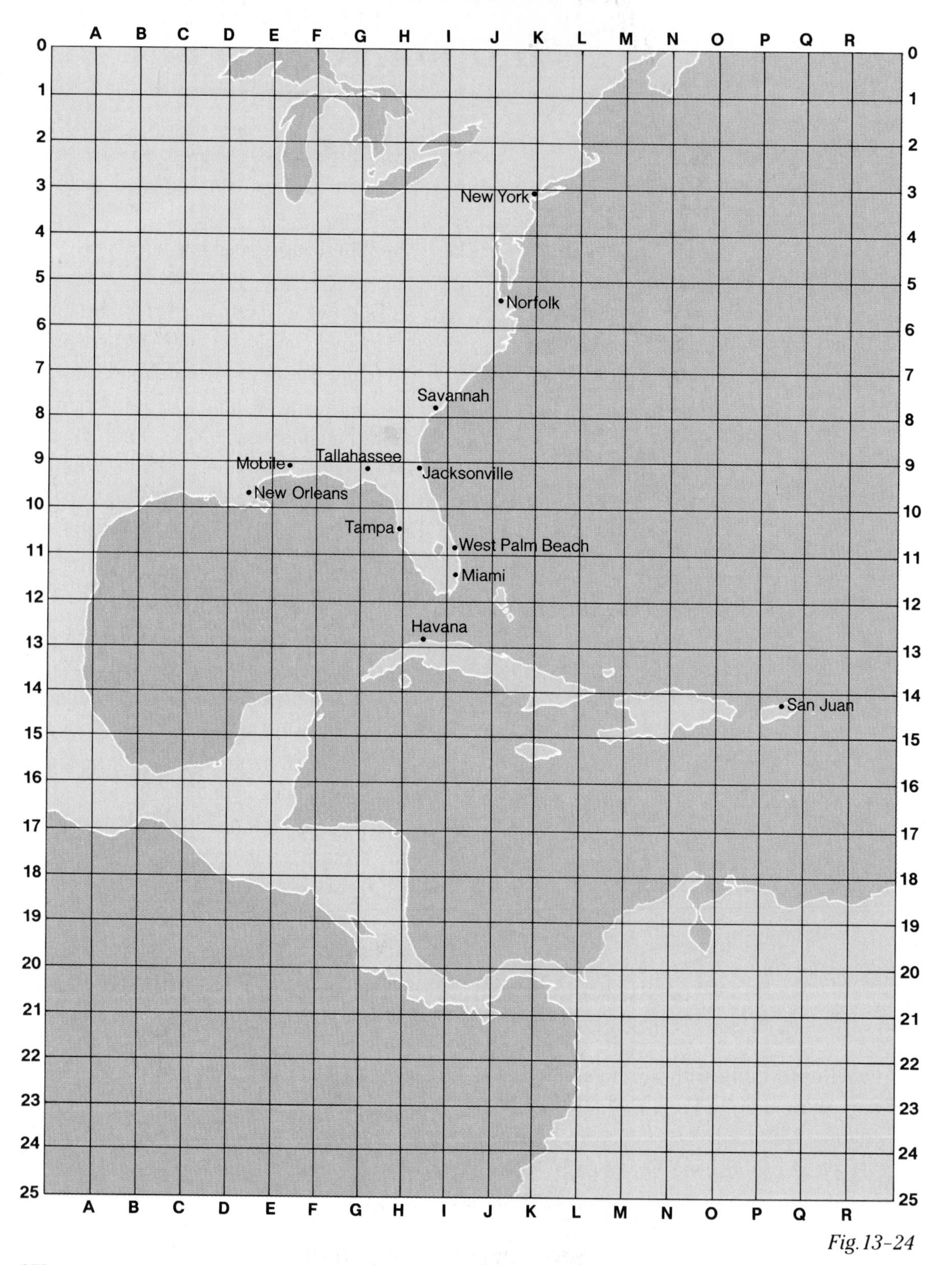
A B C D E F G H I J K L M N O P Q R
0 1 2 3 4 5 6 7 8 9 10 11 12 13 14 15 16 17 18 19 20 21 22 23 24 25
New York
Norfolk
Savannah
Mobile
Tallahassee
Jacksonville
New Orleans
Tampa
West Palm Beach
Miami
Havana
San Juan

Fig. 13-24

13-4. Predicting Weather

At the end of this section you will be able to:

- Name three weather instruments.
- Explain how a weather forecast is made.

On a cold winter morning, a television weather announcer received a phone call from an angry viewer: "Please come out and shovel your 'partly cloudy' off my driveway!" The evening before, the reporter had predicted no snow. That weather prediction was wrong. Twenty-five centimeters of snow had fallen. Why is weather prediction so difficult? In this section you will learn how a weather forecast is made.

WEATHER INSTRUMENTS

Weather prediction begins with measurements of the atmosphere. In Chapters 11 and 12 you learned about measuring temperature, relative humidity, and air pressure. Other measurements must also be made. For example, the direction and speed of the wind is an important measurement. Wind direction is measured by a *wind vane*. See Fig. 13-25. A wind vane is turned by the wind so that it points into the wind. Winds are named for the direction from which they come. A wind blowing from the south to north, for example, is called a south wind. The speed of the wind is measured by an **anemometer** (an-uh-**mom**-uh-tur). See Fig. 13-25. An *anemometer*

Anemometer An instrument used to measure wind speed.

Fig. 13-25 Identify the wind vane and the anemometer shown in this photograph.

Description of Wind	Wind Speed (km/h)	Description
Calm	0.0–0.7	Still; smoke rises vertically
Light air	0.8–5.4	Smoke drifts, but wind vane remains still
Light breeze	5.5–12.0	Wind felt on face; leaves rustle; wind vane moves
Gentle breeze	12.1–19.5	Leaves and small twigs move constantly, flags extended
Moderate breeze	20.0–28.5	Raises dust; moves twigs and thin branches
Fresh breeze	29.0–38.5	Small trees in leaf begin to sway
Strong wind	39.0–50.0	Large branches move; wires whistle; umbrella difficult to control
Stiff wind	50.5–61.5	Whole trees sway; somewhat difficult to walk
Fresh gale	62.0–74.5	Breaks twigs from trees; walking difficult
Strong gale	75.0–83.0	Blows off roof shingles
Whole gale	83.5–102.5	Trees uprooted; much structural damage
Storm	103.0–117.5	Widespread damage
Hurricane	118.0 and above	Extreme destruction

Table 13–1

usually has a small cup at the end of each of three long arms. The cups catch the wind, spinning the arms of the instrument. The speed at which the arms turn gives a wind-speed measurement. The wind speed is usually given in meters per second or miles per hour. Wind speed can also be described by the terms shown in Table 13–1. A wind vane and an anemometer are often used together, as shown in Fig. 13–25, page 373.

Precipitation is measured by an instrument called a *rain gauge*. A rain gauge can be any container with straight sides, such as a jar. The depth of rain that falls into the container is measured with a ruler. Official weather stations use a rain gauge that allows 0.01 inch of rain to collect in a small bucket. In the United States rain is measured in inches. In other countries the measurements are in millimeters. When the bucket is filled, it tips over to spill the collected water and causes 0.01 inch of rain to be recorded. At the same time, another bucket tips into place and begins to collect another 0.01 inch of rain. This method allows a measurement of how fast the rain is falling as well as the amount. Solid precipitation, such as snow, is melted in order to record its amount. The depth of snow is also measured. Radar can also show approximately how much precipitation is falling over a very large area.

Not all measurements of the conditions in the atmosphere are made at the earth's surface. Balloons are sent up carrying instruments that send back their readings by radio. See Fig. 13-26. Temperature, humidity, and pressure are measured as the balloon rises. Those measurements are transmitted back to a ground station. The position of the balloon is also tracked to give measurements of upper level winds.

Fig. 13-26 A weather balloon.

Weather satellites give information about weather that cannot be obtained in any other way. See Fig. 13-27. For example, satellites are able to provide pictures of clouds covering large parts of the earth's surface. Daytime pictures of clouds are made with television cameras using ordinary light. At night, the clouds are photographed by cameras that record the heat given off by the clouds. Positions of clouds can show low-pressure

Fig. 13-27 A weather satellite.

systems and fronts. Some satellites also carry instruments that can measure air temperature and humidity at different levels above the earth's surface. Thus satellites can provide information about atmospheric conditions in regions where there are few weather stations.

MAKING WEATHER FORECASTS

Each day the National Weather Service, a government agency, receives a great deal of weather information. Daily, it gets thousands of reports from weather stations, ships and aircraft, radar stations, satellites, and balloon measurements of the upper atmosphere. This information is used to prepare charts and maps that show present weather conditions. The kind of weather map that you have seen might look like the one in Fig. 13–28. Symbols show areas of low and high pressure, fronts, and precipitation. Weather maps may also show wind direction and speed, atmospheric pressure, and clouds.

The first step in preparing a weather forecast is to study the most recent weather maps. From those maps, the meteorologist making the forecast can get a general picture of the present

Fig. 13–28 Weather maps use symbols to show the weather.

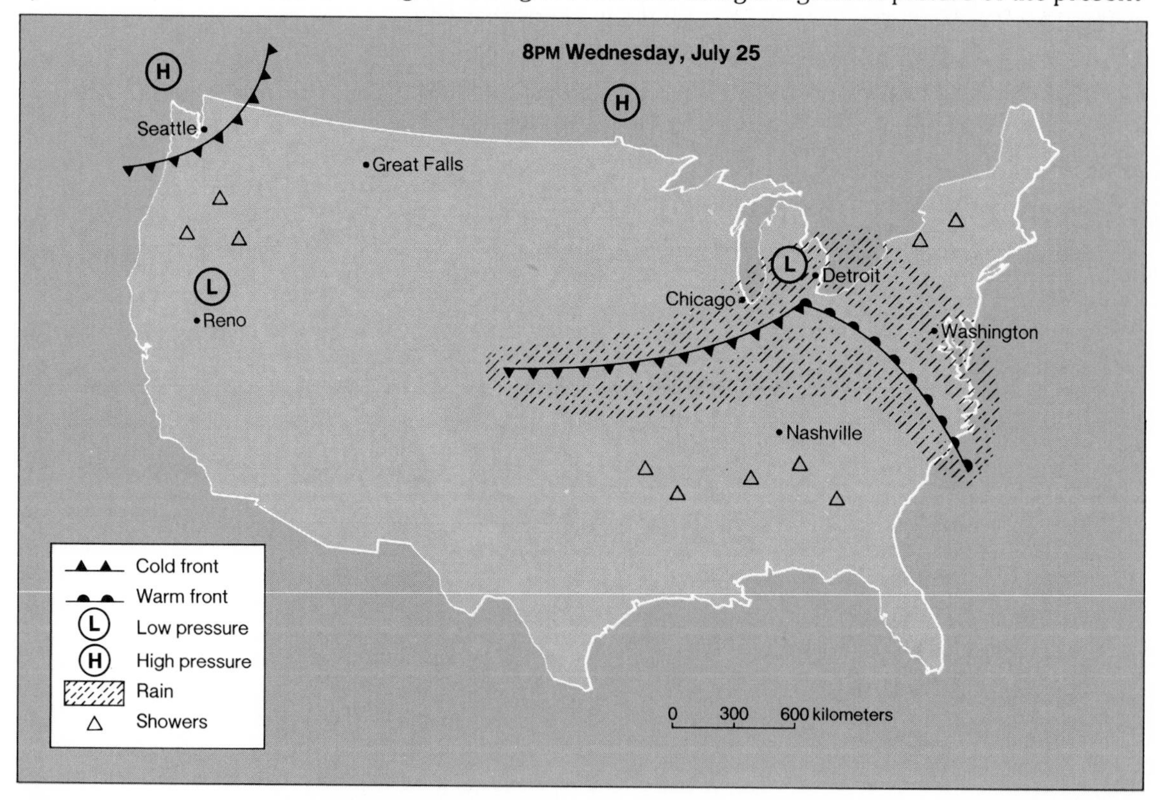

weather conditions. The second step is to compare the present map with others made 24 hours before. Maps showing weather over the entire Northern Hemisphere may be used. This gives a general picture of the existing weather systems, along with their direction of movement and speed. Since satellite photographs show the actual weather, they can be used to check the maps. Reports from radar stations also give information about the location of precipitation. The third step is to use computers to prepare maps that show how weather conditions can be expected to change within the next 24 hours. By putting all of this information together, the meteorologist makes a forecast. The forecaster must decide where the fronts, lows, and highs will move. Usually the forecast is made by a local weather office, using information supplied by the National Weather Service. Special forecasts are also prepared to give warnings of floods, tornadoes, large thunderstorms, and hurricanes. Some weather offices also provide frost warnings for farmers and air pollution forecasts.

A forecast is usually most accurate for the next 6 to 24 hours following the forecast. A forecast made for a period of 2 or 3 days is usually fairly accurate. However, the accuracy becomes less with each passing day. Beyond 5 days, the forecast is usually not dependable.

SUMMARY

Weather prediction depends on observing the conditions in as much of the atmosphere as possible. This information is used to prepare weather maps that show the main weather characteristics at a particular time. Forecasts are made by predicting how the weather map may look as time goes on.

QUESTIONS

Use complete sentences to write your answers.

1. List three measurements made in the atmosphere that are used to predict weather changes.
2. What measurements are made by weather balloons?
3. What information about the weather can be obtained by weather satellites?
4. Describe the three steps involved in making an accurate weather forecast.

SKILL-BUILDING ACTIVITY

COMPARING AND CONTRASTING

PURPOSE: To use information on a weather map to predict weather.

PROCEDURE:

A. Look at the weather map in Fig. 13-28, page 376. Use the legend to answer the following questions.
   1. Which cities shown have overcast skies and rain?
   2. What kinds of fronts and pressure systems are shown near those cities?

B. Look at the weather map in Fig. 13-29 below. Compare and contrast this weather map with the one in Fig. 13-28 to answer the following questions.
   3. How many hours elapsed between the time each of the weather maps was made?
   4. Using the scale on the map, how far did the cold front move along a line from Seattle to Great Falls?
   5. In what direction did the low pressure near Detroit move?
   6. In what direction did the cold front near Chicago move?
   7. Which cities on the map had the weather change from cloudy and rainy to clear?
   8. Which cities on the map had the weather change from clear to cloudy with showers or rain?
   9. Which cities on the map should have the greatest change in weather in the next 24 hours?
   10. Describe this change in weather.

CONCLUSIONS:

1. What information is needed in order to forecast the weather?

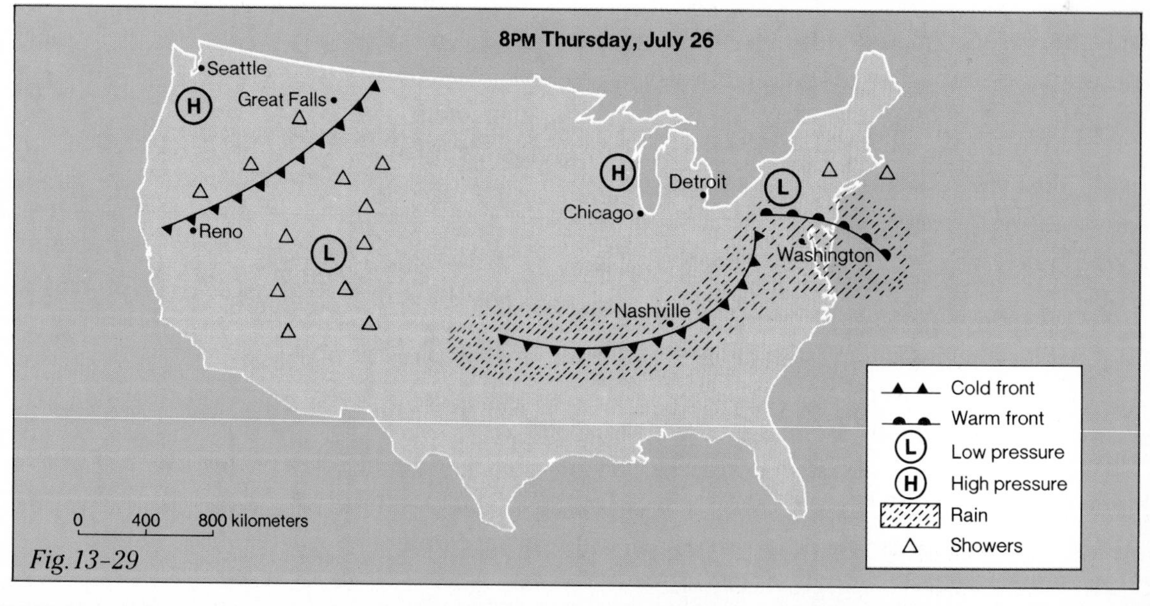

Fig. 13-29

TECHNOLOGY

WINDSAT JOINS THE WEATHER SATELLITES

At the present time, data collected by several different kinds of weather satellites help meteorologists make their forecasts. NOAA-7 and NOAA-8 are in orbit over the poles. Each spacecraft, called a Television and Infrared Observation Satellite (TIROS), covers the whole earth twice a day. Using television devices and infrared (heat) sensors, they send back information to help predict the global movement of storms. Two other satellites, called Geostationary Operational Environmental Satellites (GOES), are in circular orbits about 35,800 kilometers above the equator. GOES are different from TIROS in that they maintain their orbits from approximately the same points near the equator. This makes it easier to track them from earth and to collect data. Receivers can be directed to one point, whereas TIROS must be tracked as they move in their polar orbits. Both types of satellites carry instruments to monitor the sun's activity, to collect and relay information radioed from remote parts of the world, and to send satellite images and National Weather Service maps to both amateur and professional users.

These satellites have already made important contributions. They have, for example, increased the accuracy of weather prediction, especially through observations of weather patterns. Winds over the continental United States blow mostly from the west; therefore, observations of weather patterns over the Pacific Ocean are important to forecasters. GOES satellites, with their heat sensors, can make observations on the progress of freezing temperatures. Warnings are then issued to fruit growers in Florida and southeast Texas so that measures can be taken to prevent or limit damages to crops. Weather satellites are now helping to save lives as well. The World Weather watch was established in 1968. The United States was the first to commit spacecraft to this program of global weather observations, which, in recent years, has launched a new service—sea search and rescues. NOAA-8—equipped with instruments from France and Canada—and two Russian polar orbiters can detect distress signals from anywhere on the globe and relay them to rescue forces via local ground stations provided by many other countries. More than 185 lives have been saved since the program began in June 1982.

Even greater benefits may be gained from *Windsat,* a proposed new weather satellite. An artist's conception of *Windsat* can be seen in the illustration provided by NASA. *Windsat* was designed by RCA Commercial Communications Laboratories. Its function will be to produce maps of the earth's winds using infrared lasers to detect wind patterns. Information about winds is essential if long-range weather forecasts are to become more accurate.

Windsat will generate two laser pulses per second into the atmosphere. These beams will bounce off pollen and dust that are carried by winds and then return to the satellite. Sensors on board will then compute the direction and speed of the wind.

Windsat will, no doubt, be a more-than-welcome addition to the satellite family.

¡COMPUTE!

SCIENCE INPUT

In this chapter you learned how clouds were formed. Because of the need for rain, it is critically important to understand the conditions necessary for clouds to exist. Perhaps with more research about cloud development, techniques other than cloud seeding will be found that will enable us to use the water vapor in the air for agricultural purposes whenever the need exists. In this exercise, you will use your knowledge of the relationship between the altitude of rising air, air temperature, and dew point. (You may want to review Chapter 11, which discusses atmospheric pressure, and Chapter 12, which explains dew point.)

COMPUTER INPUT

As you have already seen in previous chapters, computer programs can be written to produce quizlike questions. Sometimes the questions, or data for the questions, are entered by the user, as was done in Program Plate Game and Program Erosion. Other times, different questions are created by using the computer's capacity to generate random numbers. When something is random it has no particular order or reason. A random number is one that has just as much of a chance to be chosen as any other. In a lottery, great efforts are made to have the numbers picked at random so that every ticket has an equal chance of being chosen. In a computer game, random numbers are sometimes used to create constantly changing situations that cannot be controlled by the player. In Program Cloud Formation, the computer's ability to generate random numbers is used to set up different atmospheric conditions. You must then decide if clouds will be formed or air will rise under each set of conditions.

In this exercise, the computer is used in three ways: (1) to produce a number of different situations; (2) to create a series of questions that test your knowledge of how clouds form; and (3) to give you feedback.

WHAT TO DO

After reviewing the necessary chapters, enter Program Cloud Formation into your computer. Save it on disk or tape and run it. The program may be used to practice independently or to test your knowledge in group competition. The program can best be used to produce data that you can then study to gain a better understanding of the relationship between all the variables the program contains. On a separate piece of paper, record the data produced.

PROGRAM NOTES

There are two kinds of variables: dependent and independent. An independent variable is one that remains constant in the particular problem you are considering. A dependent variable changes according to the character of the independent variable. In this program, cloud formation and rising air

CLOUD FORMATION CHART

altitude of rising air	*air temperature*	*dew point*	*Does cloud form?*	*Does air rise?*

are the dependent variables. Altitude, air temperature, and dew point are independent variables. You are given information about the three independent variables and asked to predict changes in the two dependent variables.

PROGRAM

```
100  REM CLOUD FORMATION
110  FOR X = 1 TO 10: PRINT : NEXT
120  FOR X = 1 TO 10: PRINT : NEXT
130  T = INT ( RND (1) * 30 + 1)
135  D = INT ( RND (1) * 30 + 1): IF D >
     T THEN D = T
140  T1 = INT ( RND (1) * 30 + 1)
142  E = INT ( RND (1) * 10 + 1)
145  PRINT "ALTITUDE OF RISING AIR"
150  PRINT "KM        AIR
     TEMP     TEMP    DP"
170  PRINT E; TAB( 10)T1; TAB( 20)T;
     TAB( 27)D
173  PRINT
175  INPUT "DOES CLOUD FORM (Y/N)?
     ";A$
176  IF A$ = "Y" AND D = T THEN
     PRINT "CORRECT!!"
177  IF A$ = "N" AND D = T THEN
     PRINT "INCORRECT"
178  IF A$ = "Y" AND D < > T THEN
     PRINT "INCORRECT"
179  IF A$ = "N" AND D < > T THEN
     PRINT "CORRECT!!"
180  IF D = T THEN PRINT "TEMP = DP
     - CLOUD FORMS"
185  IF D < > T THEN PRINT "TEMP <>
     DP - CLOUD DOES NOT FORM"
190  INPUT "DOES AIR RISE (Y/N) ? ";A$
191  IF A$ = "Y" AND T1 < T THEN
     PRINT "CORRECT!!"
192  IF A$ = "Y" AND T1 > = T THEN
     PRINT "INCORRECT"
193  IF A$ = "N" AND T1 < T THEN
     PRINT "INCORRECT"
194  IF A$ = "N" AND T1 > = T THEN
     PRINT "CORRECT!!"
200  IF T1 < T THEN PRINT "TEMP1 <
     TEMP2 - AIR RISES"
210  IF T1 > = T THEN PRINT "TEMP1
     >= TEMP 2 - AIR DOES NOT RISE"
220  PRINT
240  GOTO 120
```

GLOSSARY

RND A command in BASIC that instructs the computer to generate a random number.

VARIABLE Any factor or thing or number that can change. In this program, altitude, air temperature, dew point, rising air, and cloud formation are variables.

BITS OF INFORMATION

A program that creates models of different kinds of situations is called a simulation. Most often, these programs come with graphic displays; that is, images (pictures) that look like the real situations appear on the screen. An important area where computer images are used is the field of flight simulation for pilot training. The trainee is placed in an exact copy of the cockpit of an aircraft. Every dial, switch and knob present in a real cockpit are there to give the pilot-in-training the feeling of being in an actual plane. Images made by a computer portray what the pilot would see if he or she were flying. Thus, without endangering any lives, the person in training can gain a great deal of flight experience.

CHAPTER REVIEW

VOCABULARY

On a separate piece of paper, write TRUE next to the number of each statement that is true. Next to the number of each false statement, write FALSE, then make the statement true by writing the correct term in place of the underlined incorrect term.

1. Meteorology is the study of the atmosphere.
2. An anticyclone is a large body of air that has taken on the temperature and humidity of a region of the earth.
3. A small, intense cyclonic storm formed over warm parts of the oceans is called a hurricane.
4. A cyclone is a large mass of air spinning out of a high-pressure area.
5. The boundary separating two air masses is called a polar front.
6. An isotherm is a line on a map joining all the locations that record the same temperature.
7. The climate is the average weather at a particular place.
8. A tornado is a kind of storm formed along the polar front.
9. The boundary where cold polar air meets warm air is called a warm front.

QUESTIONS

Give brief but complete answers to each of the following questions. Unless otherwise indicated, use complete sentences to write your answers.

1. Describe several ways in which air is lifted and cooled to form clouds.
2. Explain why just cooling moist air may not cause clouds to form.
3. What is the main difference between stratus, altostratus, and cirrostratus clouds?
4. In what way are nimbostratus clouds related to cumulonimbus clouds?
5. Compare the way precipitation forms in warm areas with the way it forms in cold areas.
6. Describe the air masses that form in the northern Pacific Ocean and northern Atlantic Ocean and how they affect the weather in the United States.
7. Compare the two air masses that come from the south and the effect they have on the weather of North America.
8. What is the difference between weather and climate?
9. Explain how fronts and severe storms are produced.

10. Compare the kind of air and its movement in a cyclone to the kind of air and its movement in an anticyclone.
11. Compare and contrast a hurricane and a tornado.
12. How are tornadoes related to thunderstorms?
13. What do a wind vane and an anemometer tell about wind?
14. Describe how a weather station rain gauge works.
15. What kinds of information are given by the National Weather Service in a weather forecast?

APPLYING SCIENCE

1. Use the information about air masses that are common to your area and the general wind patterns to explain why your region has the kind of weather that it does during the summer and during the winter. Write a report describing the reasons for your explanation.
2. Make daily observations of the atmosphere for a week. Note, by name, the kinds of clouds present, whether there is rain, smog, or fog associated with the clouds, and any other weather events that might have occurred during the week. Report your results to the class.

BIBLIOGRAPHY

Canby, Thomas Y. "El Nino's Ill Wind." *National Geographic*, February 1984.

Gedzelman, Stanley D. *The Science and Wonder of the Atmosphere*. New York: Wiley, 1980.

Linn, Alan. "The Earth Spins, So We Have the Coriolis Effect." *Smithsonian*, February 1983.

Moses, L., and John Tomikel. *Basic Meteorology*. California, PA: Allegheny Press, 1981.

Moyer, Robin. "The Great Wind." *Science 80*, November 1980.

Olesky, Walter. *Nature Gone Wild*. New York: Messner, 1982.

Olson, Steve. "Computing Climate." *Science 82*, May 1982.

Stommel, Henry and Elizabeth. *Volcano Weather: The Story of 1816, the Year Without a Summer*. Newport, RI: Seven Seas Press, 1983.

UNIT 4

THE OCEANS

CHAPTER

14

Instruments have become vital to studying the sea. However, nothing can replace the first-hand knowledge that comes from direct observation of conditions beneath the sea.

THE OCEANS

CHAPTER GOALS

1. Name and describe the main features of the sea floor.
2. Compare the deep sea floor with the sea floor near the edges of the continents.
3. Identify the kinds of sediments found on the sea floor and describe how they are formed.
4. Describe how a record of the earth's history can be found in sea floor sediments.

14–1. The Earth Hidden Beneath the Sea

At the end of this section you will be able to:

- ☐ Name four ocean basins and describe how they were formed.
- ☐ Explain how the sea floor is explored by scientists.
- ☐ List three groups of features found on the ocean bottom.
- ☐ Describe the sea floor near the edges of the continents.

A world ocean covers seven-tenths of the earth's surface. While people speak of separate oceans and seas, it is actually one great ocean. The water in the ocean is not a very large part of the earth. If the planet were the size of an orange, the ocean would be only about as thick as a sheet of paper. However, without its thin covering of water the earth's surface would be very different. The weather would be different: The poles would be much colder and the tropics much hotter, and most scientists believe that there would be no life.

THE OCEAN BASINS

Suppose that the surface of the earth had no continents. Then the water in the ocean would cover the entire earth to a depth of thousands of meters. But the continents rise above sea level, leaving the water to fill the lower parts between them. The continents do not divide the ocean into parts. Instead, the ocean separates the land masses into huge islands. See

Fig. 14-1 From space the earth's surface can be seen to be mostly covered by water.

Fig. 14-1. The water-filled regions surrounded by continents and other raised parts of the crust make up the *ocean basins.*

There are three large and two smaller ocean basins. Each of them is called a separate ocean. The largest are the Pacific, Atlantic, and Indian oceans. The Arctic Ocean, at the north end of the globe, is smaller. The southern parts of the Atlantic, Pacific, and Indian oceans that surround the continent of Antarctica are often called the Antarctic, or Southern Ocean. However, the Antarctic Ocean does not occupy a separate ocean basin like the four other oceans.

The Pacific Ocean is the largest and deepest of all oceans. It covers more than one-third of the earth's surface. It is so large that all of the earth's lands could fit into the Pacific Ocean. Smaller water-covered parts of the earth are called *seas.* A sea may be a part of an ocean, such as the part of the Atlantic called the Caribbean. Seas may also be separate from the ocean, as the Mediterranean is. Since all oceans and seas are connected with each other, there is actually only one global ocean.

Ocean basins are formed when the moving crustal plates cause continents to break apart. The oceans and seas of today fill basins that began to be formed when the ancient supercontinent broke apart about 200 million years ago. Before the breakup of the supercontinent, there was only a single superocean. Since the continents are still moving and breaking apart, there will be different ocean basins millions of years from now.

EXPLORING THE SEA

People have lived by the ocean and sailed on it for thousands of years. However, what lay beneath the water was a mystery for a long time. Only in the past hundred years has the part of the earth's surface covered by water become known to scientists. Most of the knowledge about the ocean basins has been discovered within the past 30 to 40 years.

During early years of sea exploration, there was no accurate way of determining what the ocean bottom looked like. Explorers tried to find out by lowering a weight on a line or wire until it hit bottom. See Fig. 14-2 (*left*). But this was not an accurate method. Ships often drifted, causing the weight to drop down at an angle. Sometimes the line or wire was not long enough to reach bottom. Also, these measurements were made in only a few places. From these incomplete and often inaccurate measurements, maps of the ocean bottom were drawn. These early maps showed the sea floor as an almost level plain.

During World War I, a way to locate submerged submarines was found. This was done by using sound waves. The method is called *echo sounding*. In this method, sound waves sent from a ship through the water bounced off solid objects and returned to the ship. See Fig. 14-2 (*right*). The bouncing effect of sound waves is similar to the way a ball bounces. The farther away the object that the ball strikes, the longer it takes the ball to return. This principle also applies to sound waves. The time it takes for the sound waves to return to the ship shows how far away from the ship a submerged submarine can be found.

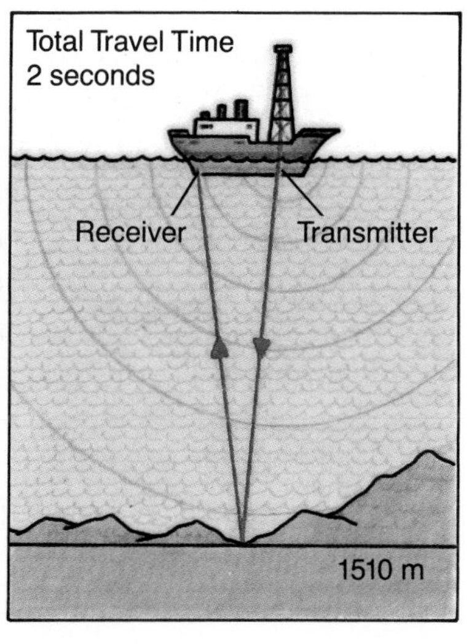

Fig. 14-2 (left) *Measuring the depth of the ocean by sounding was often inaccurate.* (right) *Echo sounding uses sound signals that are bounced off the sea floor.*

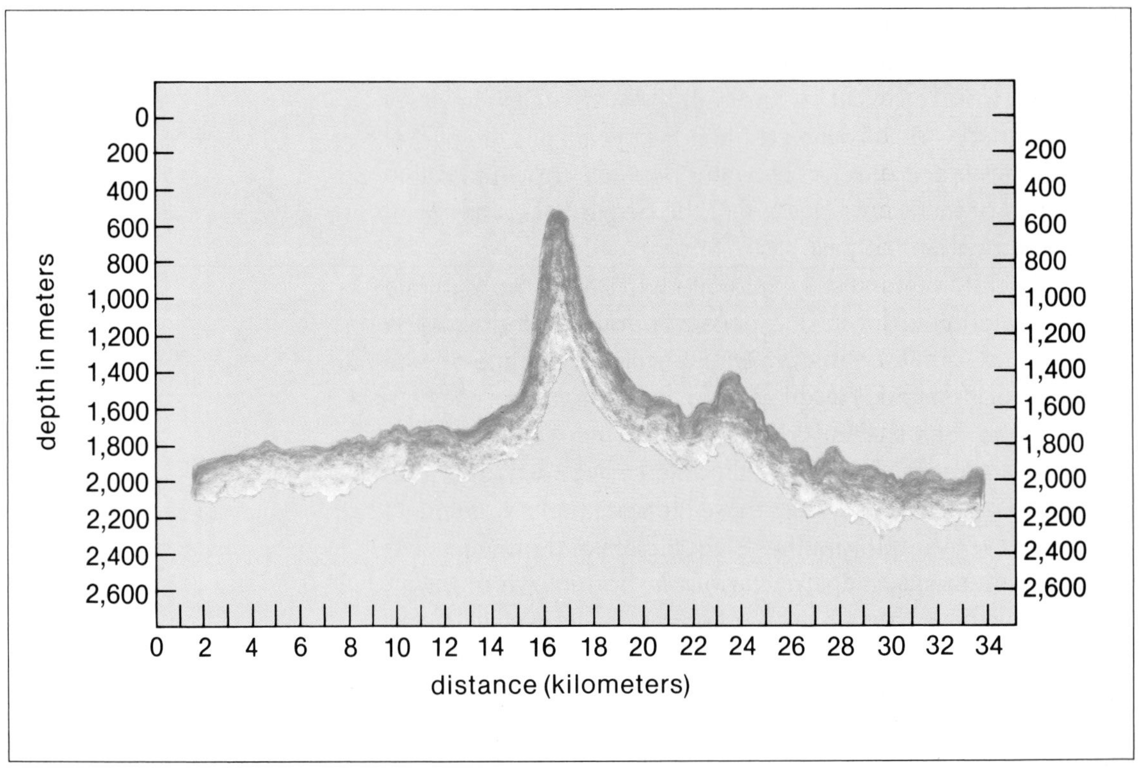

Fig. 14-3 The echo sounder produces a picture of the shape of the sea floor. The tracing shows hills and valleys much higher or lower than they actually are.

Oceanographers (oe-shuh-**nog**-ruh-furs), scientists who study the sea, soon realized that sound waves could be used to explore the ocean floor. The *echo sounder* was developed. The echo sounder sends out sound signals that bounce off the ocean bottom. For example, the time it takes for the sound signal to travel from the ship to the bottom and back again is measured to be two seconds. The amount of time it takes for the sound signal to reach the bottom can be found by dividing this measurement by two, which gives one second. By multiplying the speed of sound waves in water, 1,510 meters per second, by the time measurement of one second, the depth of the ocean at a given point is found to be 1,510 meters. See Fig. 14-2 (*right*), page 389. The differences in depth from point to point, recorded by the echo sounder, create a picture of what the sea floor looks like. See Fig. 14-3. Much of the sea floor has now been explored with echo sounders.

FEATURES OF THE SEA FLOOR

The pictures show that earth's surface beneath the water has hills, mountains, steep cliffs, and smooth plains. Oceanographers have divided these forms into three groups: the continental margins, the mid-ocean ridge, and the ocean basin floor. Many of these features are now seen to be related to the forces that shape all parts of the crust.

1. Continental margins. The areas between the sea floor and the continents are marked by the *continental margins.* Most continents are part of the same crustal plate as the nearby sea floor. Thus the continental margin usually is not the edge of a crustal plate. Instead, it is the part of the ocean where the continent begins to rise above the level of the sea. A continental margin is made up of three parts: the continental shelf, the continental slope, and the continental rise. The first of these parts is a region of shallow water called the **continental shelf.** See Fig. 14-4. The *continental shelf* is a part of the continent that has been flooded. Most parts of a continental shelf are nearly flat and slope gently downward toward the ocean basin floor. The depth of the ocean over a continental shelf is usually about 120 meters. The width of a continental shelf can vary from narrow to wide. It averages about 75 kilometers. During the last ice age, large amounts of water were locked up in glaciers. This caused the sea level to drop by 90 to 120 meters and exposed large areas of the continental shelves. Evidence of the land animals and plants that lived on these once exposed parts can now be found on the sea floor of the continental shelves.

Continental shelf The flat submerged edge of a continent.

Fig. 14-4 The shape of the sea floor in the North Atlantic. The differences in elevation are not as large as shown in this diagram.

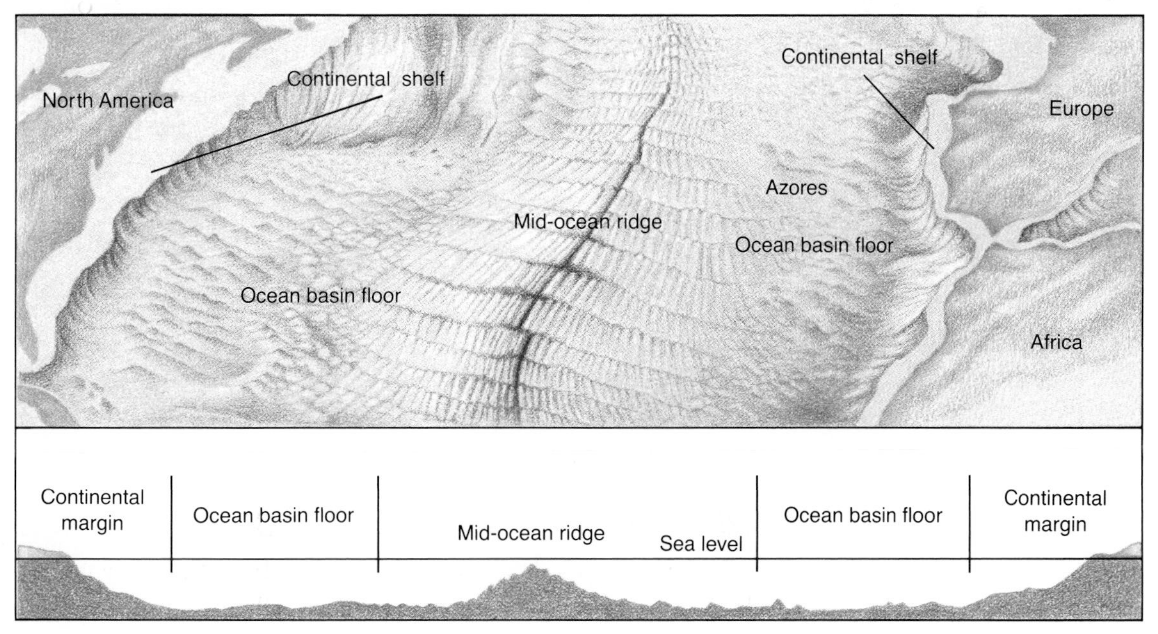

Submarine canyon Any deep valley cut into a continental shelf.

Deep canyons called **submarine canyons** have been cut in many places along the continental shelves. See Fig. 14-5. Some of these canyons were cut by rivers when the sea level was lower during the glacial periods. Other *submarine canyons* seem to have been formed by underwater landslides. These landslides occur when sediments that have built up on the continental shelves suddenly break loose and slide down toward the ocean basin floor. The moving sediments erode the continental shelves in the same way that rivers erode the land.

While the continental shelves make up less than 10 percent of the sea floor, they have been studied in detail by oceanographers. This is not only because they are close to land, but also because valuable resources are found there. The continental shelves are rich in a large variety of life. Most of the world's supply of fish is caught there. Large amounts of petroleum and natural gas are found in the rock layers that make up the continental shelves. About 25 percent of all oil and gas now produced comes from wells drilled from ocean platforms on the continental shelves. This percentage will increase in the future as other supplies are used up.

Continental slope The sloping part of the sea floor that marks the boundary between the sea floor and continental shelf.

The water depth increases rapidly where a continental shelf ends. This marks the beginning of the **continental slope.** See Fig. 14-4. The *continental slope* is the boundary between the

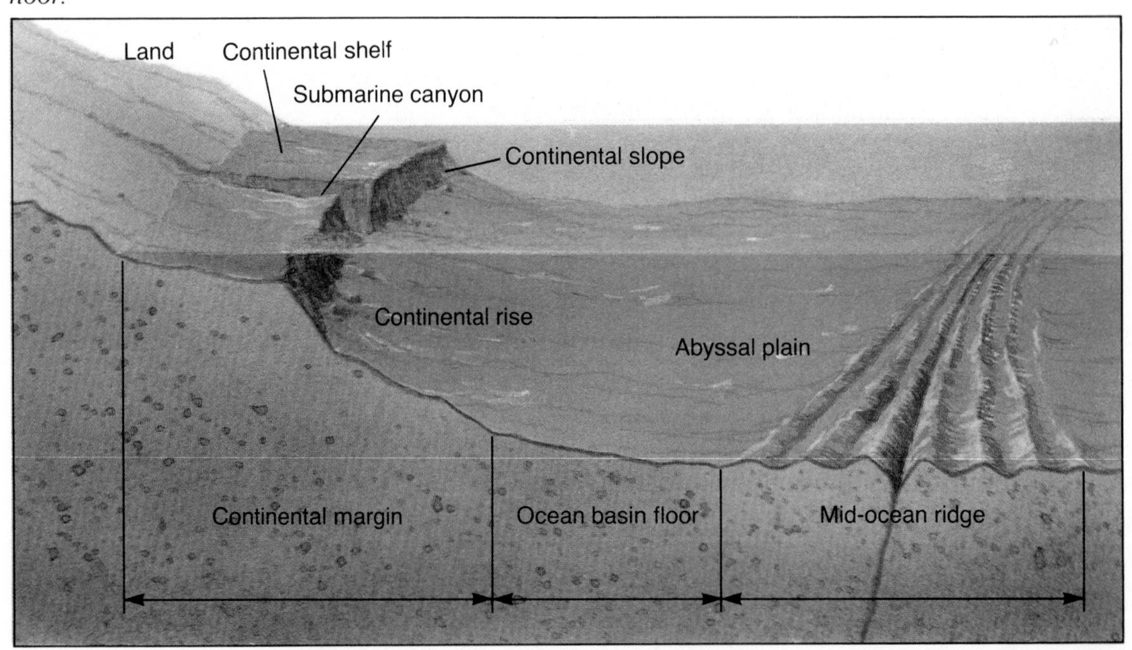

Fig. 14-5 The continental slope and the continental rise separate the continental shelves from the ocean basin floor.

continental crust and the oceanic crust that makes up the ocean basins. If the bottom of a pie pan were the ocean basin floor, the sides of the pan would be the continental slopes.

Beyond the continental slope is the third part of the continental margin. This area, where the slope becomes less steep, is the **continental rise.** See Fig. 14–5. The *continental rise* may reach out for hundreds of kilometers until it finally blends with the ocean basin floor.

Continental rise A gently sloping area at the base of the continental slope.

2. Mid-ocean ridge. An important feature found on the ocean floor is the *mid-ocean ridge* system. See Fig. 14–5. The mid-ocean ridge system is a continuous chain of wide mountains rising two to four kilometers above the ocean floor. The ridge is crossed by many faults. According to the theory of plate tectonics, the mid-ocean ridge system is where new crust is formed. Hot liquid magma from below the crust pushes up and out near the center of the ridge. As the magma hardens, new sea floor is created. The newly made crust cools and shrinks and settles to a lower elevation as it moves away from both sides of the ridge. This causes the sea floor to slope away from the sides of the mid-ocean ridge.

Parts of the mid-ocean ridge have been seen and photographed. Scientists are able to descend for short periods of time in specially built submarines called deep submersibles. However, this exploration is limited to small areas. These limited explorations have revealed places along the mid-ocean ridge where hot water pours out. These *hydrothermal vents,* or warm-water sea floor springs, provide energy, minerals, and food to support smaller sea organisms. In the total darkness of the sea floor these fountains of hot water are like oases in the desert. See Fig. 10–10, p. 276.

3. Ocean basin floor. On either side of the mid-ocean ridge is the ocean basin floor. See Fig. 14–5. On the deep ocean floor are very flat areas called **abyssal** (uh-**bis**-ul) **plains.** See Fig. 14–5. *Abyssal* means deep. The *abyssal plains* are formed by deposits of sediments. These sediments cover most of the hills and valleys that were made as new crust was produced along the mid-ocean ridge. Abyssal plains are the most level areas on the earth. They are flatter than any plains on the continents. The abyssal plains make up more than 60 percent of the ocean basin floor. Close to the continents, the abyssal plains contain sediments that were carried off the continental shelves and

Abyssal plain A flat, wide deposit of sediments that covers the deeper parts of the ocean basin floor.

down the submarine canyons. Farther out to sea, the sediments come from different sources.

Seamount A submerged volcano rising from the ocean floor.

Not all of the ocean basin floor is flat. Scattered over the entire sea floor are thousands of submerged volcanoes called **seamounts.** See Fig. 14–6. Some are in groups, while others stand alone. A large *seamount* can rise three to four kilometers above the sea floor. Many volcanic islands, such as Hawaii and Tahiti, are the exposed tops of huge seamounts that have grown from the sea floor.

Guyot A submerged volcano whose top has been flattened by wave erosion before it sank below the surface.

Some submerged seamounts have flat tops. These are called **guyots** (gee-**yos**). See Fig. 14–6. A *guyot* appears to have been an island at one time. As the sea floor spread away from the mid-ocean ridge and sank, causing the guyot to submerge, wave erosion flattened its top.

Trench A deep valley on the ocean floor found along the edges of some ocean plates.

The deepest parts of the sea floor occur where one crustal plate collides with another. As one plate moves beneath the other, an ocean **trench** is created. *Trenches* are found along the edges of ocean basins, particularly in the Pacific. See Fig. 14–6. The deepest of these trenches reach almost 11 kilometers below the sea surface.

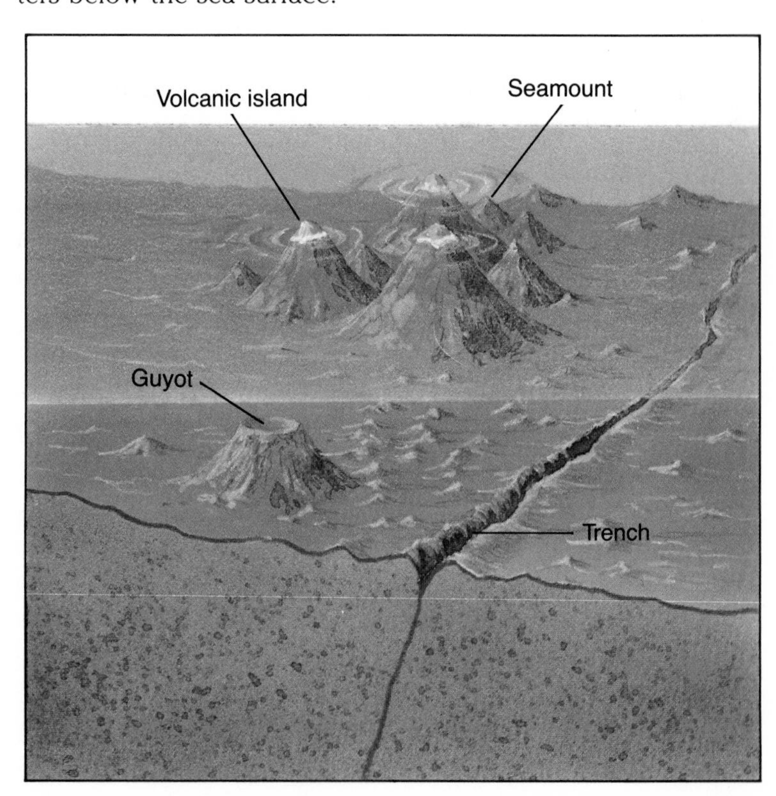

Fig. 14–6 The ocean basin floor has many unusual characteristics.

CORAL REEFS AND ATOLLS

Large numbers of small animals called **coral** (**kor**-uhl) may live in warm, shallow sea water. Each of the *coral* animals has a chalky skeleton, which is usually connected to its neighbors. The coral skeletons make a framework that also contains the remains of other small animals and plants. In time, a large structure called a *coral reef* is formed. See Fig. 14-7. Some coral reefs grow out from an island and remain attached to it. The islands that make up the Florida Keys have this kind of coral reef. A *barrier reef* is separated from land and usually lies some distance from the shore. Part of the east coast of Australia is protected by the Great Barrier Reef. Another kind of coral reef, called an *atoll* (**ay**-tawl), is circular and often not connected to any visible land. An atoll is formed in stages. First it grows as a submerged reef around a volcanic island. The island slowly sinks as the sea floor moves away from the mid-ocean ridge. As the island disappears, the reef grows upward until it finally remains as a doughnut-shaped atoll.

Coral A small animal that lives in great numbers in the warm, shallow parts of the sea.

Fig. 14-7 Coral is an animal that, in partnership with other small plants and animals, builds large reefs in shallow parts of the ocean.

SUMMARY

The earth's surface is divided into elevated continents and submerged ocean basins. Scientific exploration of the part of the earth's surface that is covered by water shows it to be made up of three kinds of features. The continental margins make up the shallow parts of the sea that border the land. The mid-ocean ridge occupies the central regions of the sea floor. The ocean basin floor makes up the remaining parts of the sea floor. Small coral animals build the foundations for reefs found in warm, shallow parts of the ocean.

QUESTIONS

Use complete sentences to write your answers.

1. Name the present ocean basins and explain how they were formed.
2. Explain how an echo sounder is used to measure ocean depths.
3. List three kinds of features found on the ocean floor.
4. Why are continental shelves so important?

INVESTIGATION

MAPPING THE OCEAN FLOOR

PURPOSE: To make a map of a section of the ocean floor.

MATERIALS:
graph paper
pencil

PROCEDURE:

A. The depth of the ocean at any place can be found by using a special instrument called an echo sounder. A sound is made by the echo sounder and the sound then travels through the water. When the sound strikes the ocean bottom, it is reflected back to a receiver on the ship. The time taken for the round trip is measured. In part B of this lab, you will see how the depth of the ocean bottom is calculated.

B. In Table 14–1, the average speed of sound in sea water and the time taken for a round trip is given. The distance the sound travels can be found by multiplying the speed of sound and the travel time (column 1 × column 2). Do this for each of the travel times given. The distance the sound travels is the distance from the ship to the ocean bottom and back. To find the depth of the ocean bottom, divide the distance traveled (column 3) by two.

   1. What does a longer travel time for the sound tell you about the bottom?

C. Draw a graph with the distance from shore on the horizontal axis and the depth of water on the vertical axis. Number the horizontal scale from 0 to 2,000. Beginning with 0 at the top, number the vertical scale from 0 to 4,000. Plot the points for the pairs of numbers in Table 14–2. Then connect the points.

Speed of Sound(m/s)	Time (s)	Distance (m)	Depth (m)
1,510	2		
1,510	3		
1,510	4		

Table 14–1

Distance (km)	Depth (m)	Distance (km)	Depth (m)
0	0	1,000	2,000
200	249	1,200	1,903
250	1,004	1,300	249
500	2,001	1,400	2,250
550	3,903	1,700	2,000
600	2,001	1,800	1,004
800	1,736	1,850	1,500
850	491	1900	1,019
950	491	2,000	2,197

Table 14–2

   2. Describe the shape of the ocean floor.

CONCLUSIONS:

1. Explain how the depth of the ocean floor is measured by an echo sounder.
2. Find and label the following features on your graph: guyot, mid-ocean ridge, continental shelf, continental slope, continental rise, seamount, and ocean trench.

14-2. Sea Floor Sediments

At the end of this section you will be able to:

- Identify two sources of sediments in the sea.
- Describe the kinds of sediments found in different parts of the sea floor.
- Explain how part of the earth's history is recorded in sea floor sediments.

A strange kind of rain occurs in the sea. It is a gentle rain of solid particles that slowly fall through the water and settle on the bottom This slow, but steady, solid rain causes almost all the sea floor to become covered with a layer of sediments.

Fig. 14-8 The visible load of this river can be seen when it empties into the sea.

SEDIMENTS ON THE CONTINENTAL SHELVES

Most rivers empty into the sea. Carried along with the water of the rivers are the products of rock weathering. Part of the material carried by the rivers is made up of small pieces of rock that can be seen. These particles are called the *visible* load. See Fig. 14-8. They consist of very fine particles of silt and clay as well as larger fragments of rock that are rolled or bounced along the stream bed. Some of the material carried by the stream is dissolved in the water. Dissolved material makes up the *invisible load.* Dissolved materials brought to the sea by rivers help to make the sea "salty." Both the visible load and the

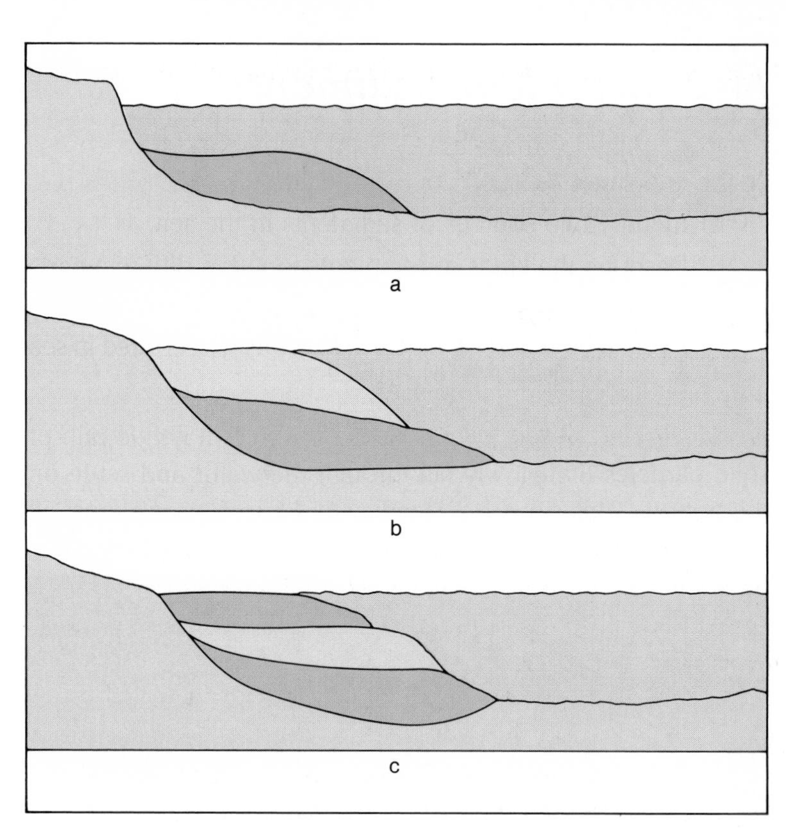

Fig. 14-9 Thick layers of sediments can build up on the continental shelf.

invisible load end up as sediments on the sea floor. These sediments from rivers are a major source of sediments in the sea.

Much of the visible load of a river is deposited close to the shore when the river empties into the sea. Large particles are dropped near the mouths of rivers. Waves and currents caused by waves may carry much of this material along the shore. The material is later deposited, forming sandy beaches. Sometimes, huge underwater, fan-shaped deposits are also formed where rivers empty into deep water. These deposits are like deltas hidden beneath the water. Finer particles are carried farther out and deposited on the continental shelves. Almost all of the sediments found on the continental shelves come from the land. Some continental shelves have very thick layers of sediments that have been deposited there.

As the edge of a continent moves away from the mid-ocean ridge, it tends to sink. Thus the continental shelf may also slowly sink at the same time as sediments are being deposited. This allows many layers of sediments to be built up, as shown in Fig. 14-9. Sediment layers many kilometers in thickness can be built up. Some layers of sediments formed this way are now found as part of the rock within today's continents. Apparently, these sediments were laid down in ancient times along the edges of continents that existed at that time. The movement of crustal plates pushed these thick layers of sediments into the bodies of the continents. They are found there today, bent and broken, as parts of mountains.

TYPES OF DEEP SEA SEDIMENTS

The sea floor is almost completely covered by sediment. Part of the sediment is carried down from the continental margins. The rest has slowly settled down to the bottom from the water above. Sea floor sediments are grouped according to where they came from.

1. Remains of living things. There is a type of sediment formed from the remains of living things. Near the equator and along the west coasts of continents, the conditions in the sea favor the abundant growth of living things. The most common forms of life in these waters are floating plants and animals that are so small that they can be seen only with a microscope. See Fig. 14–10. When these living things die, their remains sink to the bottom. These remains make up most of the sediment found on the sea floor in the deep parts of the ocean far from the continents. The sediment formed from remains of living things is called **ooze.** Beneath the ocean areas where sea life is abundant, a thick layer of *ooze* covers the entire sea floor.

Ooze A kind of sea floor sediment formed from remains of living things.

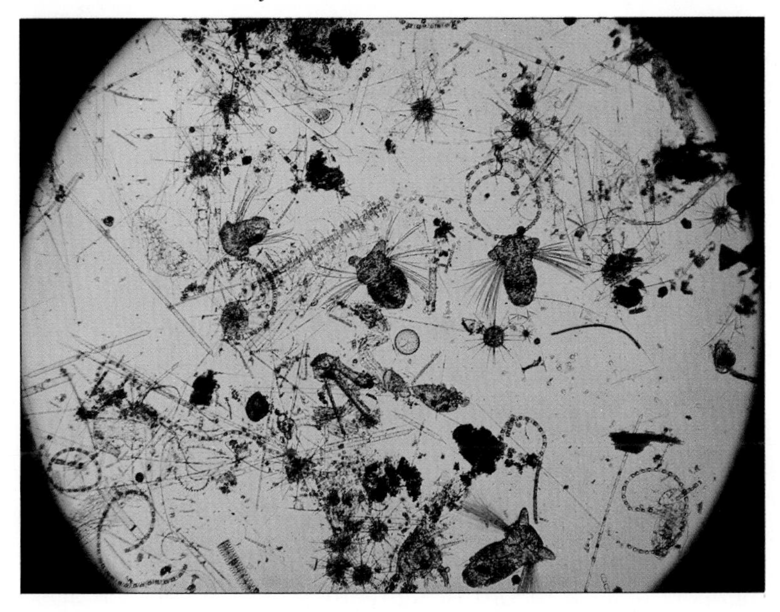

Fig. 14–10 Many parts of the ocean contain large amounts of microscopic plants and animals.

Red clay A kind of sediment found on the floor of most of the open sea.

2. Red clay. Most of the sea does not support enough life to produce a heavy layer of ooze on the bottom. Most of the open sea floor is covered with a sediment called **red clay.** However, this sediment is usually not red and does not contain much of the kinds of minerals that make up ordinary clay found on land. Instead, it is made up of the smallest particles carried by rivers into the sea. *Red clay* also contains windblown dust, volcanic dust, and even tiny particles from meteorites. Red clay is deposited very slowly. It may take 1,000 years for one millimeter of red clay to settle. Oozes, on the other hand, can be deposited at the faster rate of one centimeter every 1,000 years. On the continental shelves, sediments build up at the rate of one meter every 1,000 years.

Until the spreading of the sea floor from the mid-ocean ridge was discovered, scientists were puzzled about the lack of very thick sediments on the deep sea floor. During the hundreds of millions of years of earth's history, the deep sea floor should have built up a very thick layer of sediments. However, the exploration of the ocean showed that much of the sediment layer was missing. Now scientists understand that the layer was removed by moving crustal plates on the sea floor. The sea floor and the sediments carried on it were destroyed where the crustal plates collided. Thus no part of the ocean bottom is old enough to have built up a very thick layer of sediment.

A RECORD OF ANCIENT CLIMATES

Scientists can obtain samples of sea floor sediments for study. One of the most common methods uses a sharp-edged pipe that is dropped to the sea floor from a ship. When it hits the bottom, the pipe is driven into the sediments. Then it is brought back to the surface and a long cylinder of sediments, called a *core*, is taken from it. See Fig. 14–11. The layers of sediments can then be seen in the core.

Because the sediments build up over millions of years, they contain clues to the conditions in the ocean in the past. For example, the remains of certain animals and plants that were known to be alive at a particular time tell the age of a sediment layer. Because some of the animals and plants lived only in warm water or cold water, the temperature changes at the sea surface in the past can be found. In this way, the sea floor sediments contain a record of the earth's most recent ice ages.

Fig. 14-11 A core sample of the sea floor reveals the layers of sediments.

Even changes that take place on land are recorded in the ocean sediments. For example, dust from volcanoes sometimes reaches the ocean bottom. Sediments carried to the oceans from streams and rivers tell us what the soil on the earth's surface was like in the past.

SUMMARY

Almost all the weathered rock material that is carried into the sea by rivers remains there for a long time. Most of this material becomes part of the sediments that cover the sea floor. In some parts of the ocean, most sediments are the remains of living things. Heavy deposits of sediments are laid down on the sea floor near the edges of the continents. A fairly accurate record of the past temperature conditions on the earth can be found in sea floor sediments.

QUESTIONS

Use complete sentences to write your answers.

1. What two kinds of materials are brought to the sea by rivers?
2. Describe the various kinds of materials found on the sea floor at the following locations: (a) continental shelves; (b) deep seas where life is abundant; (c) deep seas where life is not abundant.
3. How do scientists obtain samples of sea floor sediments?
4. What kind of information is obtained from sea floor sediment samples?

SKILL-BUILDING ACTIVITY

SEQUENCING AND A MODEL OF A CORE SAMPLE

PURPOSE: To make a model of a core sample and use the idea of sequence in the study of it.

MATERIALS:

straw	marking pen
sea floor model	razor blade

PROCEDURE:

A. Push the straw straight down into the clay model of sediments of the ocean floor that your teacher has prepared. Be sure to push it all the way to the bottom.

B. Remove the straw. Using the marking pen, mark the straw to indicate the top of the soil sample.

C. Using the razor blade, carefully slice the straw and clay core lengthwise to expose the core sample. Leave the core sample in the straw to protect it. The layers in the sample represent layers of sediments.

D. Make a drawing of the core sample on your paper. Label the top and bottom of the sample. Shade in the different layers and label each according to its color.

E. Use your drawing and the model core sample to answer the following questions.

1. How many sedimentary layers are there in the core sample?
2. According to the sequence of the layers, which layer in the core sample was deposited last?
3. In the sequence of layers, which layer in the sample was deposited first?
4. Which layer is the thickest? If the rate at which materials enter the ocean has been nearly the same, what does the thickness of this layer tell about the relative time it took for it to form?
5. Starting with the oldest, describe each layer in the order in which the layers were deposited.
6. If your core sample was taken in the deep part of the ocean, how would you expect the particle sizes to compare with a sample made near shore?

D. Look at Fig. 14-11 on page 401.

7. What differences do you see between the two core samples being studied?
8. Could these core samples be from the same area of the sea floor? Explain.
9. Name four materials you might find in a core sample.

CONCLUSIONS:

1. Explain why a sea floor core sample is made up of a sequence of layers.
2. Describe how to tell the relative age of sediment layers in a sequence of them found in a core sample.
3. In a sequence of sediment layers found in a core sample, an unusually thick layer is found. What explanation can you give for it being so thick?
4. A core sample is made up of several layers of sand, with the top layer having some slightly larger rock pieces. From what part of the sea floor could this sample have come?

TECHNOLOGY

FORECASTING EL NIÑO

Today, satellites with computerized devices that send back images of ocean currents and weather patterns are helping us to understand how closely affected we are by events that seem far away.

Every few years, air pressure drops over the southeastern Pacific. The trade winds that usually blow across the Pacific toward Asia slow down or disappear. The warmer water that these trade winds normally hold back begins to move toward South America. (Since these warm waters occur off Peru around Christmas, Peruvian fishermen named this effect El Niño, meaning "the child.") In 1982–1983, this warming extended to one-third of the Pacific Ocean. In some places the ocean temperature rose 11 degrees. The warmer water heated the atmosphere above it. The speed and position of the jet stream was affected. It was pushed farther south. Unusual weather resulted. There were tornadoes in Los Angeles and 20 centimeters of snow in Atlanta in one day in April. Record-breaking rains and heavy flooding took place in Ecuador, northern Peru, California, Louisiana, and Cuba. The worst droughts in history were recorded in 1983. Fish moved to colder waters, causing the fishing industry to lose tremendous amounts of money. Because of the decrease in fish and other marine life, birds and other animals did not reproduce.

What was the cause of these natural disasters? The answer to this question lies in a series of complicated patterns of high- and low-pressure systems, ocean currents and countercurrents, and jet streams called El Niño and the Southern Oscillation (ENSO).

Many meteorologists and oceanographers believe El Niño, the warming of the water, is either caused by or is closely related to the Southern Oscillation, a giant rocking of tropical air masses over the Pacific. This rocking causes pressure systems to seesaw back and forth. When there is low pressure over the southeastern Pacific, there is high pressure over Australia, which usually brings dry weather. This occurred in 1983, when Australia suffered a great drought, the results of which can be seen in the photo.

Scientists are carefully analyzing the data collected during the 1982–1983 El Niño. Because of the worldwide effects that both El Niño and the Southern Oscillation have on weather climate, these scientists hope to be able to predict the return of El Niño. Mathematical models are being prepared that will allow the use of computers to study the effects of weather changes and food production. Already in use are computerized images transmitted from satellites. These photos are made with scanners sensitive to concentrations of chlorophyll. High levels of chlorophyll indicate the presence of plant life, which is an indication that the water temperature has risen. The goal of this research is to help us prepare for the changes in weather, climate, and ecological balance that the currents of El Niño and the winds of the Southern Oscillation bring. An improved understanding of these currents and winds may help to reduce the damage they can cause.

CHAPTER REVIEW

VOCABULARY

On a separate piece of paper, match each term with the number of the statement that best explains it. Use each term only once.

continental shelf	continental rise	guyot	ooze
submarine canyon	abyssal plain	trench	red clay
continental slope	seamount	coral	

1. A deep valley in the ocean floor.
2. The sloping part of the sea floor that separates the sea floor and continents.
3. The kind of sediment found on the floor of most of the open sea.
4. A kind of sea floor sediment made from living things.
5. A submerged volcano rising from the sea floor.
6. The flat, submerged edge of a continent.
7. A submerged volcano whose top has been flattened.
8. A deep valley cut into the continental shelf.
9. The flat, wide deposit of sediment that covers the deep ocean floor.
10. A gently sloping area at the base of the continental slope.
11. Small animals that live in great numbers in the warm, shallow seas.

QUESTIONS

Give brief but complete answers to the following questions. Unless otherwise indicated, use complete sentences to write your answers.

1. Name five common ocean basins.
2. Explain why there is really only one global ocean.
3. It takes 3.2 seconds for a sound to go to the sea floor and return. Sound travels 1,510 meters per second in the sea. How deep is the ocean at this location?
4. Starting with the continent coastline, list the following ocean floor features in the order they would be found as you moved toward the deep ocean: (a) mid-ocean ridge; (b) abyssal plain; (c) continental shelf; (d) continental rise; (e) continental slope.
5. Describe how each of the following is the result of a volcano in the ocean: (a) guyot; (b) seamount; (c) island.
6. What are the two main sources of sediment in the sea?

7. What is "ooze," and where is it found?
8. What are "hydrothermal vents," and why are they like oases in the desert?
9. What part of the ocean is mined for oil? What other resource is found there?
10. Describe how an atoll forms.
11. What finally happens to a layer of sea floor sediment?
12. How are submarine canyons made?
13. Compare the sediments found on continental shelves with that found on the abyssal plains.
14. Where is the sediment called "red clay" found, and of what is it composed?
15. What do sea floor sediments tell about the earth's history?

APPLIED SCIENCE

1. Using the information from your study of sea floor sediments, explain why a large puddle of muddy water leaves a layer of very fine sediments near the middle of it when it evaporates.
2. Write a paper discussing why there are so many coral reefs in the middle of the Pacific Ocean near the equator. You may want to look up additional information in the library.
3. Prepare a poster to show to the class the kinds of studies an oceanographer might do. Use the information presented in this chapter and any other information you may find in the library.
4. Make a poster-size drawing, to show the kinds of features and materials found on the ocean floor. Label all parts. Be sure to include the following: continental shelf, continental slope, submarine canyon, trench, guyot, seamount, abyssal plain, mid-ocean ridge, coral, ooze, red clay. Draw in some of the resources found in the sea and on the sea floor.

BIBLIOGRAPHY

Britton, Peter. "Cobalt Crusts: Deep Sea Deposits." *Science News*, October 30, 1982.

Britton, Peter. "Deep Sea Mining." *Popular Science*, July 1981.

Mathews, Samuel W. "New World of the Oceans." *National Geographic*, December 1981.

Time-Life Editors. *Restless Oceans* (Planet Earth Series). Chicago: Time-Life Books, 1984.

CHAPTER

15

Great demands are placed on the ocean's supply of fish. Entire schools of fish are being taken every hour by factory ships. Will the sea continue to provide food in the future?

WATER IN THE OCEAN

CHAPTER GOALS

1. Describe the properties of sea water.
2. Explain how the sun's heat affects the sea.
3. Identify and describe the resources that can be obtained from the sea.

15–1. Sea Water

At the end of this section you will be able to:

- Describe the composition of sea water.
- Explain how the sea became salty.
- Compare the temperature in the upper and lower parts of the ocean.

There is an old folk legend that explains why the sea is salty. According to this tale, somewhere on the bottom of the sea there is a magical salt mill. This mill is supposed to be grinding away all the time and pouring salt into the water of the ocean. In this section you will find out why this old story is partly true.

SEA WATER

A glass of sea water taken from almost any place in the open sea looks like a glass of ordinary drinking water. But the salty taste of sea water would quickly show that they are not the same. If you boiled the sea water until it all evaporated, white crystals would be left. A sample of sea water would leave behind an amount of solid material equal to about 3.5 percent of weight of the sea water. In other words, 1,000 grams of sea water contains about 965 grams of water and 35 grams of dissolved salts. See Fig. 15–1 (a), page 408. The number of grams of dissolved salts in 1,000 grams of water is called the **salinity** of the water. The *salinity* of sea water is 35, which is about 300 times the salinity of most ordinary drinking water.

Salinity The number of grams of dissolved salts in 1,000 grams of sea water.

The two most abundant dissolved substances in sea water are sodium ions and chloride ions. Ions are atoms or groups of atoms that have lost or gained electrons. Many substances are made of ions. For example, sodium ions and chloride ions

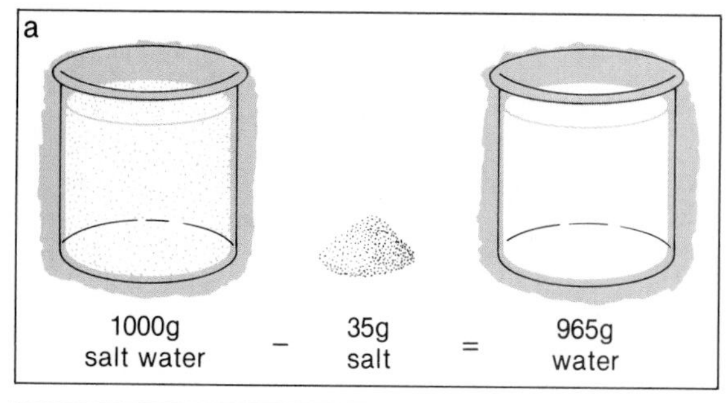

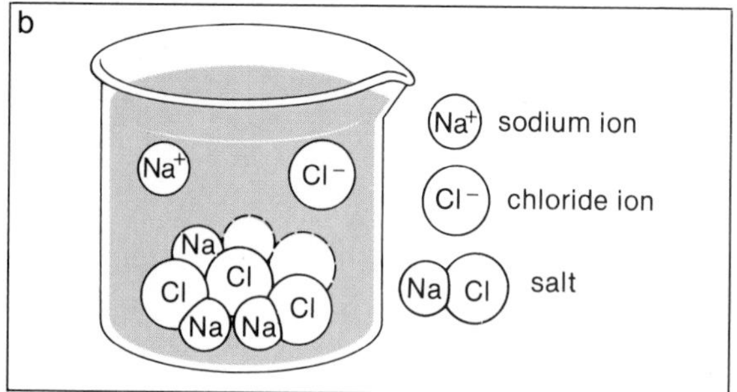

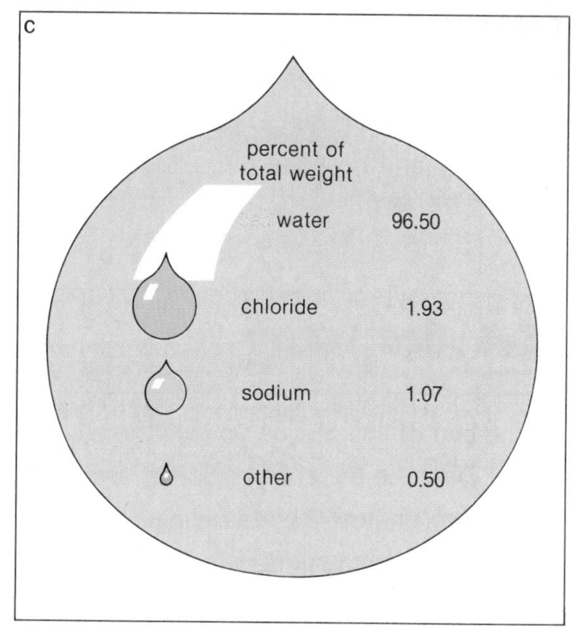

Fig. 15-1 (a) Sea water contains 3.5 percent dissolved salts by weight. (b) Salts such as sodium chloride separate into ions in sea water. (c) How does the weight of chloride ions compare to the weight of sodium ions?

make up ordinary table salt. Ordinary salt is the chemical compound called sodium chloride. When salt dissolves in water, it breaks up into sodium ions and chloride ions. See Fig. 15-1 (b). Thus when we refer to dissolved salt in the ocean, we are really referring to the dissolved ions coming from the salt.

Sea water is not the same as table salt dissolved in tap water. See Fig. 15-1 (c). This shows the percent by weight of sodium and chloride ions in sea water. You can see that there are also other ions present. Fig. 15-1 (c) shows that sea water contains more than sodium and chloride ions. Thus, sea water is more than just table salt dissolved in water.

There are other kinds of ions found in sea water. There are sulfate ions, magnesium ions, calcium ions, and potassium ions. See Fig. 15-2. Other substances, such as gold, are also present in very small amounts. Oceanographers believe that all of the chemical elements found in the earth's crust are present in sea water, mostly in the form of ions.

The salinity of water taken from different places in the ocean is not always the same. Water near the equator has a higher salinity than water closer to the poles. One cause of this difference is the warmer temperatures. This is caused by a higher rate of evaporation near the equator. Water that evaporates leaves the salts behind. Thus the water that remains after evaporation has occurred becomes more salty.

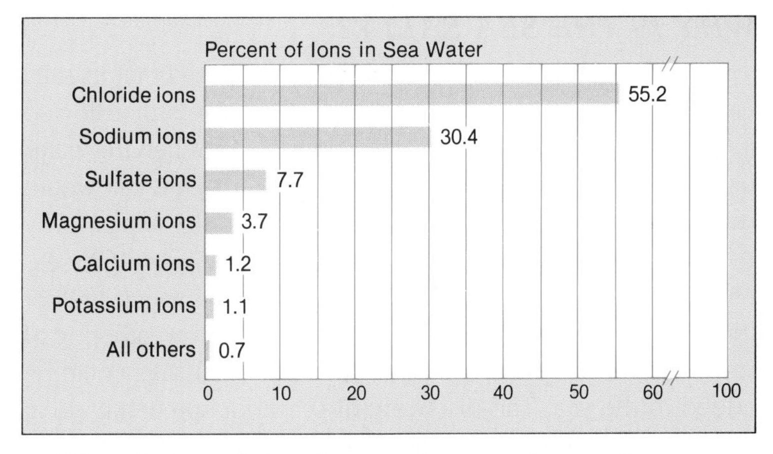

Fig. 15-2 This bar graph shows the most abundant kinds of ions in sea water.

Although the salinity of sea water may change, the proportion of the salts in sea water does not change. The salts in sea water are always present in the same proportion. The reason for this is shown in Fig. 15-3. This figure shows two glasses of water. When a spoonful of salt is dissolved in one, the water in that glass then has a fixed number of sodium and chloride ions. Adding water does not change the number of each ion. In the same way, adding or removing water from the ocean does not change the proportion of dissolved salts in sea water.

Sea water also contains large amounts of dissolved gases. These gases come from the atmosphere. The temperature of the water affects the amount of gases that will dissolve. Warm water is less able to hold dissolved gases than cold water.

The amount of oxygen dissolved in sea water is also affected by living things. Usually the amount of dissolved oxygen in the ocean is highest in the upper water layers. This is because plants living in these sunny upper layers make oxygen. In the darker lower depths, where there are fewer plants, more oxygen is used up by living and decaying things than is produced.

Fig. 15-3 The mixing of A and B does not change the proportion of dissolved materials.

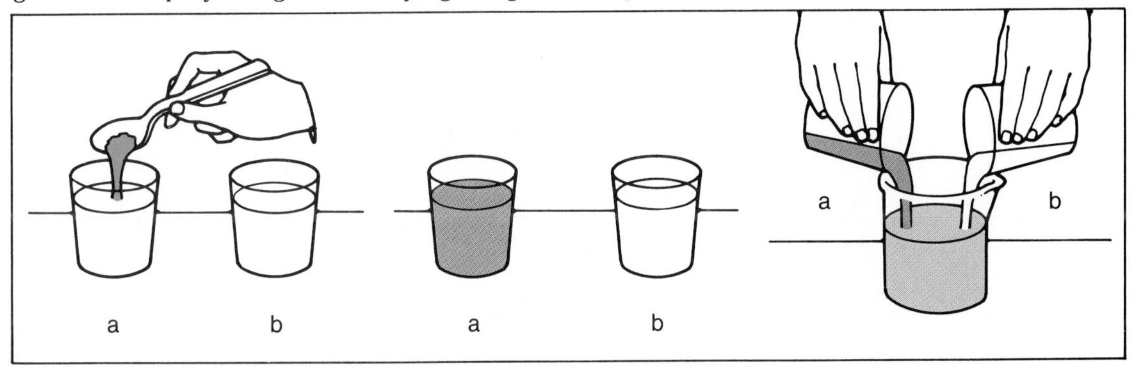

WHY IS THE SEA SALTY?

The scientists who first studied the sea believed that its salts came from the land. They thought that the rivers that flowed into the sea carried salts that came from rocks. When the water evaporated from the ocean, these dissolved materials would remain in the sea. However, this theory could not explain why some ions were found in sea water in large amounts. One example was the amount of chloride ions. It was much greater than the amount that could have come from the weathering of rocks. Scientists had to look for other ways that salts could be added to the sea. The discovery of sea floor spreading along the mid-ocean ridge helped to solve the mystery. When magma flows up into the ridge from the inside of the earth, chloride ions are given off. Thus scientists now believe that the volcanic activity is an important source of the materials dissolved in the sea. The magical "salt mill" of the old folk tale is actually the volcanic activity that creates new sea floor. Volcanic eruptions on land also produce materials that are washed into the sea by streams. Since the water in the ocean originally came from volcanic activity that took place during the earth's early history, sea water has always been salty. Since then, more salts have been added. Some salts have been added by rock weathering on the land. Also, many of the salts have come from rocks and sediments on the sea floor. Presently the makeup of sea water is not changing. Ions are slowly being removed by living things and the formation of sediments. But these are being replaced at about the same rate. Thus the makeup of sea water is not changing very much.

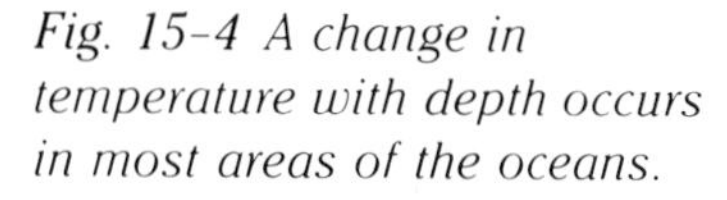

Fig. 15-4 A change in temperature with depth occurs in most areas of the oceans.

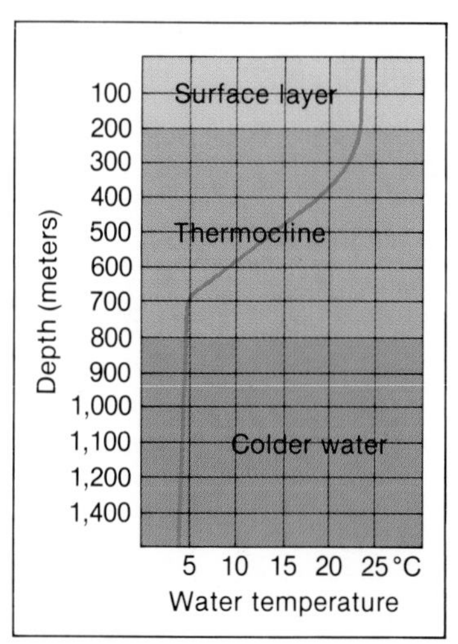

TEMPERATURE

The temperature of ocean water changes with depth. The ocean can be divided into three layers according to their temperatures: surface layer, thermocline, and a cold, deep layer.

1. Surface layer. The sun's rays falling on the sea surface heat the water. However, the rays do not reach more than a few meters into the water. Only the water very close to the surface can be heated. Waves cause the warm water to be mixed with cooler water from below to form a *surface layer.* Within the surface layer the temperature of the water remains the same. In most parts of the sea, the surface layer has the same temperature down to about 200 meters.

2. Thermocline. Below the surface layer, the temperature begins to drop rapidly. This layer of decreasing temperature is called the **thermocline.** The *thermocline* is the boundary between the warm surface water and a layer of cold deep water.

Thermocline A zone of rapid temperature change beneath the sea surface.

3. Cold, deep layer. In this layer, temperatures fall only a few more degrees. At depths greater than about 1,500 meters, the water temperature is usually less than 4°C. See Fig. 15-4. Since the world's oceans average about 4,000 meters in depth, most sea water is less than 4°C. Thus, most sea water is very close to freezing.

Near the poles, the sea receives little heat from the sun. Thus the sea surface in polar waters does not usually have a warm surface layer. Parts of the ocean near the poles are covered by a floating ice layer called *pack ice.* See Fig. 15-5 and Fig. 15-6. Because of its dissolved salts, sea water will not freeze at 0°C, as fresh water does. It must reach a temperature of −2°C in order to freeze. As sea water begins to freeze, the dissolved salts are not included in the ice that is formed. Thus pack ice is not salty. At the same time, the unfrozen water near the pack ice becomes more salty than normal. Usually, the layer of pack ice is no more than five meters thick because the ice protects the water below from freezing.

The ocean acts like a storage tank for heat on the earth's surface. There is more heat stored in the upper three meters of the

Fig. 15-5 Pack ice floating on the sea surface near the North Pole.

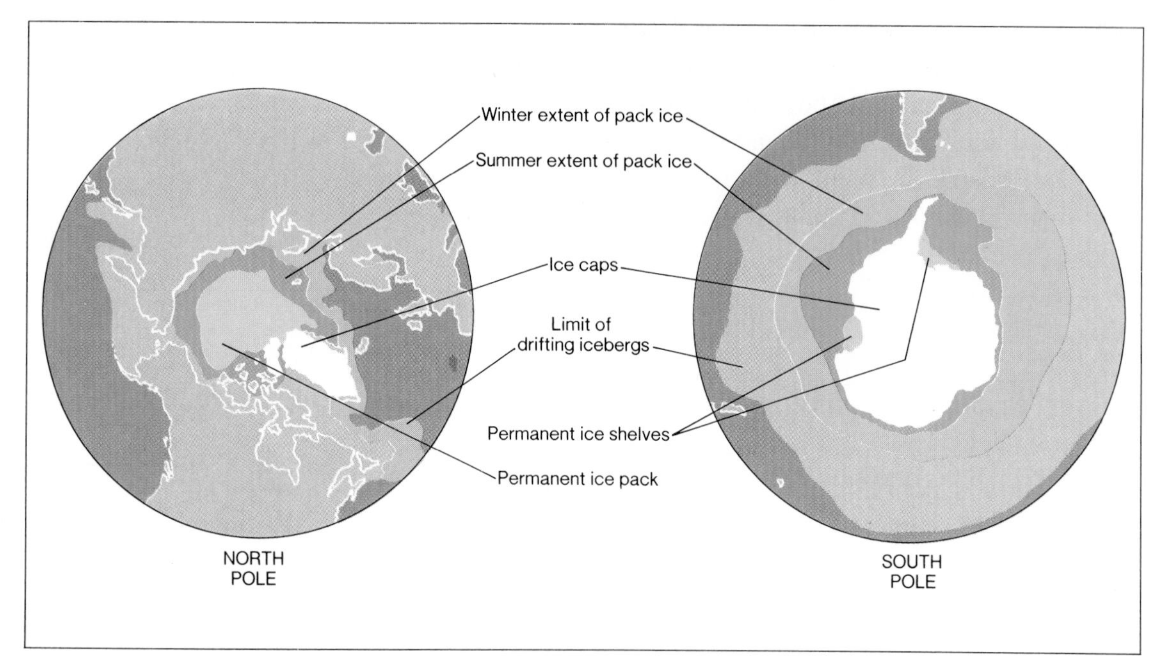

Fig. 15-6 Pack ice on the sea surface in polar regions is shown on these maps.

sea than in the entire atmosphere. Water also absorbs and releases heat much slower than air does. For this reason, the sea affects climates all over the world. In the winter, the water does not cool as quickly as the air. This causes the water to be warmer than the air. Thus, water has a warming effect on the atmosphere in the winter. In the summer, the water does not heat as quickly as the air. This causes the water to be cooler than the air. Because of this, the ocean tends to cool the atmosphere in the summer.

The oceans absorb nearly 75 percent of the sun's energy that reaches the earth. Scientists have been looking for ways to use this energy. One system uses a liquid such as ammonia, which turns into vapor at a low temperature. The warm ocean water would be used to vaporize the ammonia. The ammonia vapor would then drive a turbine to make electricity. A cable might be used to carry the electricity to the shore.

The warmest surface waters are found near the equator. There the sun's rays strike the earth most directly. In those tropical parts of the sea, surface temperatures of 21°C to 26°C are common. Those high temperatures cause evaporation of sea water into the atmosphere. The evaporation from tropical seas has two important effects on the atmosphere. First, the air becomes very moist. The warm, moist air masses formed over the oceans near the equator have a strong influence on the world's weather and climates. The second effect of the evaporation of water from the sea is the addition of heat to the atmosphere. The water vapor leaving the sea surface removes heat energy from the sea and adds it to the atmosphere. It is this energy that is released by hurricanes and other storms that form over the warm parts of the ocean.

PRESSURE

When you dive deep in a swimming pool, you will often feel an increase in pressure on your ears. The weight of the water above you causes this increase in pressure. In the same way, the weight of water in the ocean causes the pressure to become greater with increasing depth. If you were diving in the ocean, at a depth of 10 meters you would feel a pressure that was about double that at the surface. Because of the increase in pressure, divers can go down only about 100 meters with scuba equipment and about 300 meters using diving suits. In the deepest parts of the sea floor, the pressure is more than 1,000 times greater than the pressure at the surface.

SUMMARY

Sea water contains many dissolved substances. These materials have become dissolved in the water as a result of volcanic activity and the weathering of rock on the land. Sea water also contains dissolved gases from the atmosphere and living things. Most sea water is very cold. Only the upper layers of the ocean are warmed by the sun. The weight of water causes water pressure to increase with depth.

QUESTIONS

Use complete sentences to write your answers.

1. What is meant by the salinity of water?
2. Why is sodium chloride dissolved in water not the same as sea water?
3. List, in order of abundance, the six most abundant substances found dissolved in sea water.
4. Describe the temperature of each of the three layers of sea water as depth increases.

INVESTIGATION

THE EFFECT OF SALINITY ON FLOATING OBJECTS

PURPOSE: To find out what effect salts dissolved in water have on floating objects.

MATERIALS:

modeling clay
plastic straw
masking tape
ruler
tall 1-L container
water
sodium chloride
pencil

PROCEDURE:

A. Make a ball of modeling clay about two centimeters in diameter (about the size of a marble). Push the straw into the ball of clay far enough so that it sticks firmly, making a watertight seal.

B. Cut a piece of masking tape 8 centimeters long and 1.5 centimeters wide. On the nonsticky side, draw a pencil line down the center the full length of the tape. Make a scale on the line by placing marks across it at one-millimeter distances. Make the first mark and each fifth mark longer than the others. Number the longer marks 0, 5, 10, 15, and so on.

C. Attach the scale to the straw so the end marked "0" is very near the open end of the straw. See Fig. 15–7.

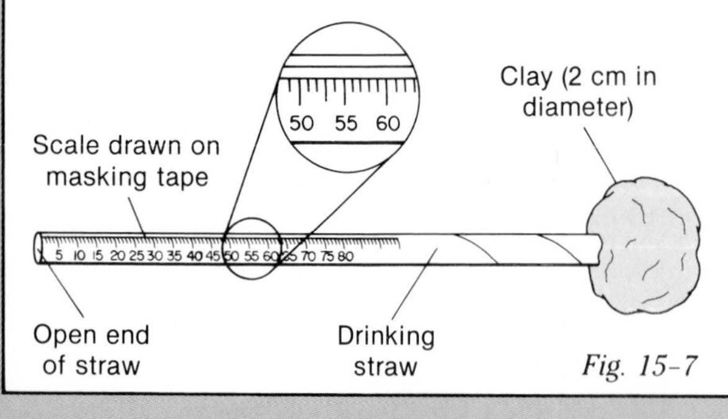

Fig. 15–7

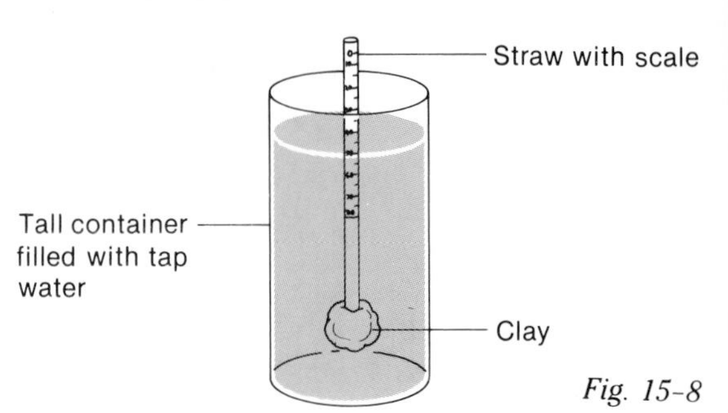

Fig. 15–8

D. Fill the tall container nearly full with tap water. Carefully place the straw in the container, clay end down. See Fig. 15–8. The straw should float in an upright position, with part of the scale sticking out of the water. If necessary, add or remove clay to make it float.
   1. What mark does the water come to on the scale? (It does not have to be zero.)

E. Remove the straw from the water. Make the water similar to sea water by dissolving 35 grams (5 level teaspoons) of sodium chloride (salt) in the water.

F. Place the straw in the salt water. Take a reading of where the water comes to on the scale.
   2. What is the reading?
   3. Would an ocean-going vessel float higher in a freshwater river or in the ocean?

CONCLUSIONS:

1. What effect do salts dissolved in water have on floating objects?
2. The Great Salt Lake in Utah has about 25 percent dissolved salts in it due to many years of evaporation. Would an object float higher in the Great Salt Lake or in the ocean?

SKILL-BUILDING ACTIVITY

THE HEAT ABSORPTION OF WATER AND ROCK

PURPOSE: To use the ideas of comparing and contrasting in a study of heat absorption.

MATERIALS:

beaker	graduated cylinder
water	2 styrofoam cups
wire screen	test tube
ring stand, ring	string
Bunsen burner	thermometer
balance	test tube holder
rock sample	

PROCEDURE:

A. Copy the following table.

Mass of rock	
Mass of water in test tube	
Beginning temp. of water, cup A	
Final temp. of water, cup A	
Temp. change due to rock	
Beginning temp. of water, cup B	
Final temp. of water, cup B	
Temp. change due to water	

B. Fill the beaker half-full of water and place it on the wire screen on the ring stand over a Bunsen burner. Heat to boiling. Go on to steps C and D while water is heating.

C. Use the balance to determine the mass of the rock. Record in your data table.

D. Using a graduated cylinder, measure an amount of water equal to the mass of the rock (volume of 1 mL of water equals mass of 1 g of water). Transfer this water to a test tube. Place the test tube of water in the water being heated.

E. After the beaker of water has come to a boil, adjust the burner flame so that it keeps the water barely boiling. Use the string to hang the rock in the boiling water. Make sure that it does not rest on the bottom of the beaker. After about five minutes, the rock and the test tube of water will be at the same temperature.

F. Using the graduated cylinder, pour the same amount of cold water into each of two styrofoam cups. The cups should be about one-third full. Label one cup A, the other B.

G. Measure the temperature of the water in cup A. Record in your table. Hold the thermometer in cup A. Quickly transfer the rock from the boiling water to the cup. Wait until the temperature stops rising. Record this final temperature of cup A.

H. Measure the temperature of the water in cup B. Record this in your table. Hold the thermometer in cup B. Using the test tube holder, transfer the test tube of water from the boiling water to cup B by pouring the test tube of water directly into cup B. Wait until the temperature stops rising. Record this final temperature of cup B.

CONCLUSIONS:

1. Compare the temperature change of the water in cup A with that in cup B.
2. Explain why land, which is mostly rock, cools faster at night and heats faster in the daytime than a body of water.

15–2. Resources from the Sea

At the end of this section you will be able to:

- Explain why certain parts of the sea produce more fish than others.
- Describe two resources other than food that can be obtained from the sea.
- Identify the major problems that prevent the orderly development of ocean resources.

Fig. 15–9 Much of the food we eat comes from the sea.

Look at some of the food that can be found in a supermarket. See Fig. 15–9. How much of the food that you eat comes from the sea? Today, many fishing boats sweep across the sea, taking every living thing. Will the sea continue to provide food in the future?

FOOD FROM THE SEA

On the earth's surface, the sea covers more than twice the area covered by land. The amount of plant growth on each square meter of the surfaces of the land and sea is about the same. Therefore, the sea is as able as the land is to produce food. Then why does less than 10 percent of the world's food supply come from the sea?

Part of the answer is found in the way we collect food from the sea. Almost all plant life in the sea is found in **plankton.** *Plankton* are plants and animals that float near the sea surface. Most are so small that they can be seen only with a microscope. The microscopic plants in plankton are called *phytoplankton.* They are found mostly in the upper few meters of the sea, where there is enough light to supply the energy needed for growth. Tiny animals, called *zooplankton,* then feed on the phytoplankton. Both kinds of plankton are eaten by larger animals. These, in turn, become food for larger fish and other marine animals. Thus phytoplankton are the beginning of a *food chain* that supports the animal life in the sea. If plankton could be harvested easily, the sea could provide a much larger amount of food. However, no practical way to harvest plankton has been developed. It is the other end of the food chain that supplies the harvest from the sea.

Plankton Small plants and animals that float near the surface of the sea.

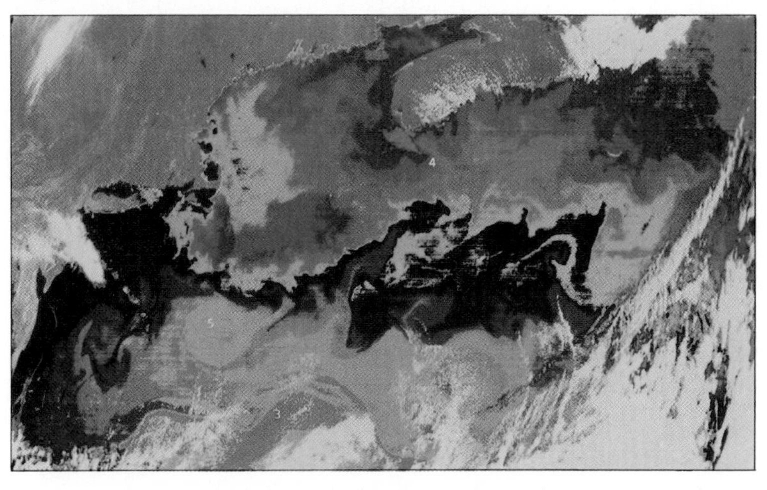

Fig. 15-10 This photograph taken from a satellite shows the sea surface off the northeast coast of the United States. Cape Cod is visible in the upper middle. Red and orange colors show a large amount of phytoplankton, whereas blue indicates water that is low in life.

Today the food taken from the sea is mostly fish. Fish are easily caught in nets and are found in large numbers in some parts of the ocean. Most of the fish are caught in the parts of the ocean over the continental shelves. The water over the continental shelves is rich in the phytoplankton that are the beginning of the food chain that supports the fish. See Fig. 15-10. Some areas near the coasts are natural fish farms, partly because of **upwelling** of sea water that occurs there. *Upwelling* is the rising to the surface of cold water from below. This carries minerals needed for plant growth to the surface. Upwelling also occurs in some parts of the open ocean. See Fig. 15-11.

Upwelling The rise of water to the sea surface.

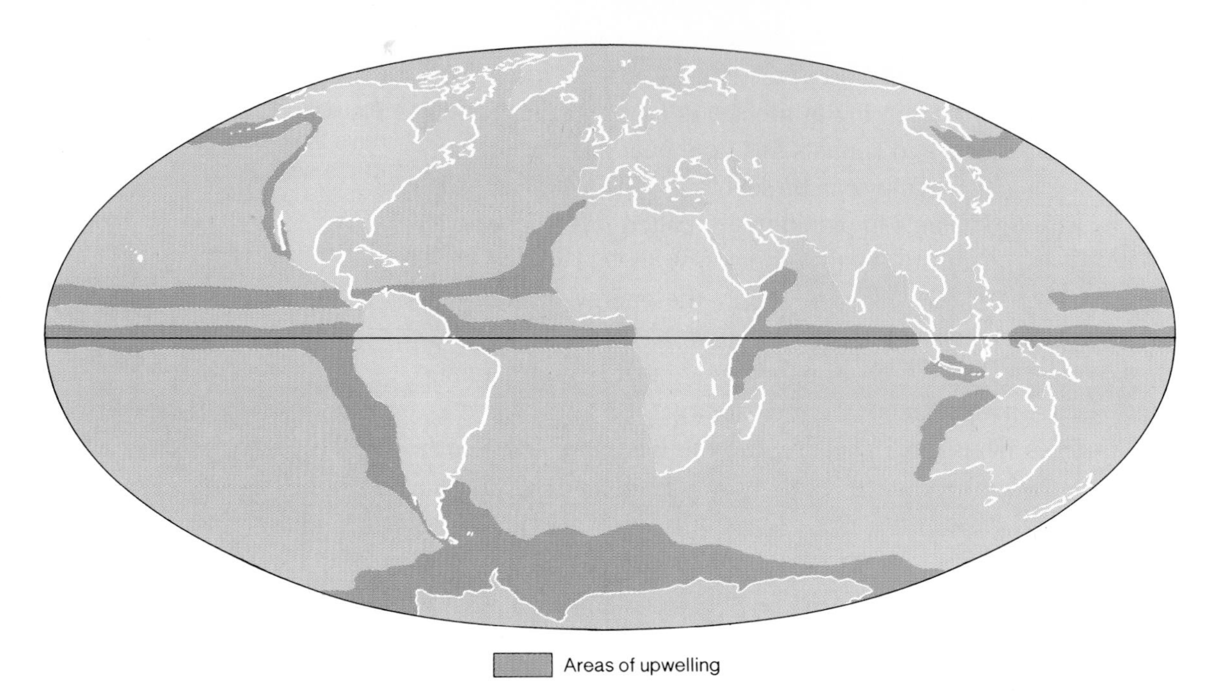

Fig. 15–11 Parts of the ocean where upwelling occurs are shown on this map. Fish are usually most abundant in these regions.

This situation is commonly found off the west coast of South America, the California coast, and the areas of the equator.

Today we are catching nearly one-third of all the fish large enough to be taken in nets. There is a limit to the number of fish that can be taken without destroying the source of supply. Increasing the catch might cause the fish population to drop. If the number of many kinds of fish in the sea is reduced to a low level, we may destroy a large part of the entire fish population. Some kinds of fish that were once plentiful, such as cod, are now much less abundant. Some fish populations that have nearly disappeared from the oceans are now being protected from fishing crews by laws. Scientists are trying to find all of the conditions in the sea that affect the fish populations. Then they will know how much fish can be harvested without affecting the number produced. Using that information would help to insure that there would always be a supply of fish. Animals that are lower in the food chain could also be added to the catch. For example, a possible new food source is *krill*. This is a small, shrimplike animal that is found in large numbers in the cold waters of the Southern Hemisphere. See Fig. 15–12. Krill can be used directly for food or as part of other foods. Other ocean animals, such as krill, that are not now used in large amounts may become an important new food source.

Fig. 15-12 Krill is a small animal found in huge numbers in some parts of the ocean.

There may also be another way to increase the amount of food we get from the sea. Farming methods can be used in the sea, just as they have been used on land. Making sea farms, however, is not a simple task. Large parts of the oceans would have to be fenced. The water would need to be fertilized. Huge floating tanks might have to be built to hold mineral-rich water brought up from the bottom. Until such ocean farms are in operation, most food from the sea will come from fishing. See Fig. 15-13.

Fig. 15-13 Fishing is the main way that food is obtained from the sea.

Fig. 15-14 Salt is the most abundant mineral taken from the sea.

MINERALS FROM THE SEA

The sea contains huge amounts of dissolved minerals. Although there are large amounts, most of these minerals are dissolved in a vast amount of water. Thus, it is very expensive to take these minerals from sea water. For example, it is possible to take a valuable mineral, such as gold, from sea water. But the process costs more than the gold is worth. At the present time, salt, bromine, and magnesium are the only minerals taken from the sea. See Fig. 15–14.

Nodules Lumps of sediment found on the sea floor that contain valuable minerals.

In the future, another way to obtain minerals from the sea may be used. In some places, large areas of the deep ocean floor are covered with lumps of minerals called **nodules** (**nod**-yoolz). The *nodules* are about the size of potatoes or eggs. See Fig. 15–15. Scientists do not know all the details about how nodules are formed. However, it is known that they grow very slowly. It may take a million years to form one of the larger nodules. They often contain large amounts of manga-

Fig. 15-15 These nodules were taken from the deep sea floor.

nese or iron. They also contain smaller amounts of copper, zinc, silver, gold, and lead. Thus the nodules are a vast reserve of these metal ores. Taking the nodules from the sea bed will require special mining methods. The nodules might be swept up by a machine that works like a giant vacuum cleaner. They also might be gathered by automated submarine vehicles that would travel between the sea floor and a ship. Nodules may be an important source of some minerals in the future.

Sea water is also a resource. In some parts of the world, fresh water supplies are so small that water is taken from the sea. The dissolved salts can be removed by either of two methods. The water can be heated and evaporated, leaving the salts behind. The vapor is then cooled and changed back into fresh water. A second method uses a plastic film that allows pure water, but not dissolved salts, to pass through it. Fresh water supplied by these methods is expensive. The cost is too high for the water

Fig. 15-16 Desalination plants take the salts out of the sea water.

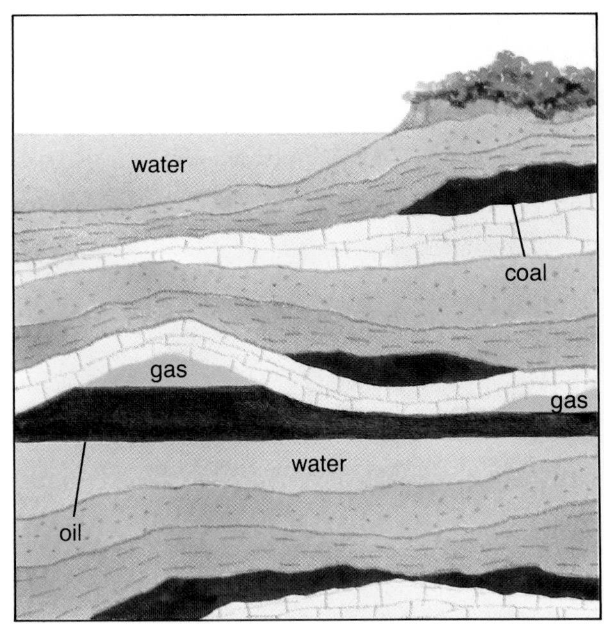

Fig. 15-17 (left) *Layering beneath the ocean showing the oil pools.* (right) *Offshore oil rigs drill into the ocean floor.*

to be used for irrigation. Now there are about 1,500 plants in the dry parts of the world that produce fresh water from the sea for human and industrial use. See Fig. 15-16, page 421.

ENERGY FROM THE SEA

The most important source of energy from the sea is petroleum. A large part of the world's reserves of petroleum and natural gas lie beneath the sea. Most of these offshore deposits are found along the continental shelves. It is difficult to reach these offshore deposits. It is also more expensive to drill for oil or gas on the sea floor than it is on land. Offshore drilling has become common because supplies of gas and oil are becoming harder to find on land. The oil and gas from the sea will have an even greater importance in the future. See Fig. 15-17.

The ocean is also filled with another source of energy. Most of this energy comes from the sun. Heat supplied by the sun is stored in the water and also causes the winds that produce waves and currents. Some of the sun's energy is also captured by the plants that grow in the sea. Motion of the water caused by the tides also can be an energy source. However, all of this energy is spread through the entire ocean. It is very expensive to build the equipment that can take large amounts of this energy from the water. The waves striking about 100 kilometers of shoreline have enough energy to supply electric power to a million homes. But it would be very difficult and expensive to build the equipment necessary to do this. It is likely that for many years the most important source of energy from the sea will continue to be the fuels taken from the sea floor.

PROTECTING THE OCEAN RESOURCES

The food, minerals, and the fuels in the sea are more difficult to obtain than those found on land. However, the difficulty in obtaining the resources is not the biggest problem. The real problem is deciding who owns the sea's resources. Countries cannot agree on questions about using the resources from the sea. How far out to sea do the borders of a country extend? Do nations have the right to take fish without limit in the open sea? Who has the right to collect minerals from the deep sea floor? Do countries without a coastline have any rights to ocean resources? Until there is agreement among nations there cannot be full use of the sea's resources.

Pollution of the sea is also a problem that must be solved among nations. Chemicals and other wastes that are dumped into the sea cause pollution that can affect the entire ocean. Only if all nations work together will the resources of the sea be protected.

SUMMARY

Food taken from the sea is mostly in the form of fish. The ability of the sea to provide food can be increased by using other forms of life and by creating sea farms. Most of the ocean's mineral wealth lies on or beneath the sea floor. The ocean is an energy resource. The development, use, and preservation of ocean resources depends on cooperation among nations.

QUESTIONS

Use complete sentences to write your answers.

1. Why are most of the ocean's fish found over the continental shelves?
2. Name three minerals presently being taken from the sea.
3. What is the most important energy resource that is taken from the sea?
4. List four questions that must be answered before much of the ocean's resources can be used.

INVESTIGATION

CHANGING SEA WATER TO FRESHWATER

PURPOSE: To use the process of distillation for changing sea water to freshwater.

MATERIALS:

graduated cylinder	wire screen
colored salt water (artificial sea water)	Bunsen burner
	clamp
flask	beaker
ring stand	glass tubing
ring	test tube
tap water	goggles

PROCEDURE:

A. Using a graduated cylinder, measure 25 mL of the colored salt water and place it in the flask. Place a drop of the salt water on your tongue. The salt content should be apparent. CAUTION: NEVER TASTE ANY SUBSTANCE IN THE LABORATORY WITHOUT PERMISSION.

B. Set up the equipment as in Fig. 15-18.

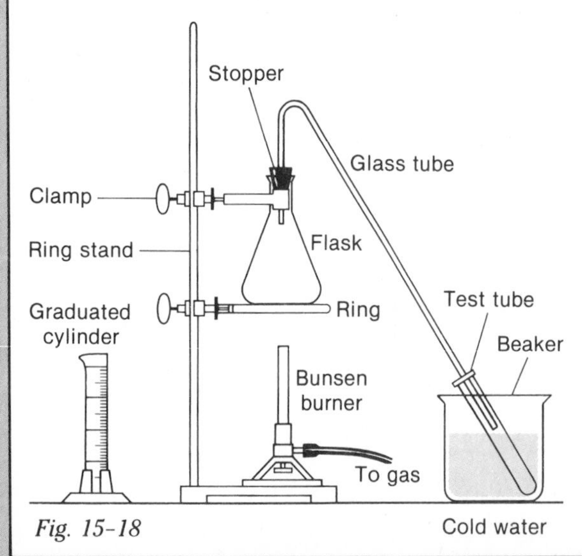

Fig. 15-18

C. CAUTION: WEAR GOGGLES FOR THIS PROCEDURE. *Gently* boil the salt water until there is almost none left in the flask. If steam escapes from the test tube, slow the heating. This process is called distillation. When the first signs of dryness or solid material appear, stop heating the liquid. NEVER BOIL A LIQUID TO COMPLETE DRYNESS IN A FLASK. Turn off the burner. Be careful of the hot apparatus. Let it cool.

1. How does the liquid in the test tube differ from the original liquid?
2. What do you think the liquid is?

D. When the apparatus has cooled, remove the flask. NEVER PLACE WATER IN HOT GLASSWARE.

3. Describe the material left in the flask.
4. What do you think is left in the flask?

E. Measure the amount of liquid collected in the test tube.

5. How does the amount in the test tube compare with what was in the flask?
6. Account for any difference found in your answer to question 5.

F. Add some tap water to the cooled flask. Shake the water around in the flask.

7. Does the material in the flask dissolve when tap water is added?
8. Taste the solution that results and describe the taste.

CONCLUSIONS:

1. Describe how salt can be removed from sea water by distillation.
2. Compare the properties of the sea water used in this investigation to the water obtained in the test tube.

CAREERS IN SCIENCE

COAST GUARD ENLISTED PERSONNEL

Each year, hundreds of lives are saved by the efforts of the U.S. Coast Guard. The Coast Guard, a branch of the armed services, conducts search and rescue missions along the coastline and on inland waterways, using both ships and helicopters. It is also the Coast Guard's duty to patrol the nation's coastlines and protect them from attack.

To join the Coast Guard, you must be at least 18 (17 with parents' consent), but no older than 26. You must also be in good physical condition and have a high school diploma. The term of enlistment includes four years of active service and two years in the reserves. You may move frequently or stay at one duty station.

Jobs in the Coast Guard vary from a seaman or seaman's apprentice, who might do labor on board ship or in the shipyard, to trained personnel such as electricians, mechanics, communication specialists, and radar technicians. Training is provided by the Coast Guard. For further information, write: Commandant, United States Coast Guard, Washington, DC 20590.

TECHNICAL WRITER

Technical writers translate difficult scientific concepts into language that most people can easily understand. A technical writer may write a magazine article on the newest underwater diving gear. She or he may prepare manuals on how to operate computers, or write advertising copy to describe new medicines. Technical writers may also prepare audio-visual aids on everything from the latest satellite to the most recent discovery in cancer research. There is a wide range of available work.

To become a technical writer, a college degree and strong writing skills are generally required. Competition is high, and potential employers usually look for people with a good educational background. Jobs are available in publishing companies and in high-tech companies, in advertising firms and with scientific organizations that produce publications. Many writers also work independently on a free-lance basis. For more information, write: Society for Technical Communication, Inc., 1010 Vermont Avenue N.W., Suite 421, Washington, DC 20005.

CHAPTER REVIEW

VOCABULARY

On a separate piece of paper, match each term with the number of the statement that best explains it. Use each term only once.

salinity	pack ice	phytoplankton	upwelling	nodules
thermocline	plankton	zooplankton	krill	

1. The rise of bottom water to the sea surface.
2. A small shrimplike animal that is often found in the cold waters of the Southern Hemisphere.
3. The floating layer of ice on the ocean near the poles.
4. The number of grams of dissolved salts in 1,000 grams of sea water.
5. A region of rapid temperature change beneath the sea surface.
6. Very small plants and animals that float near the sea surface.
7. Tiny animals that feed on phytoplankton.
8. Lumps of sediment found on the sea floor that contain valuable minerals.
9. Very small plants found in the upper few meters of the sea.

QUESTIONS

Give brief but complete answers to each of the following questions. Unless otherwise indicated, use complete sentences to write your answers.

1. How does the salinity of sea water compare to that of drinking water?
2. What is the salinity of a 1,000-gram sample of water if only 944 grams of it is pure water?
3. Give the names and the amounts, in percentages, of the two most abundant substances found in sea water.
4. Compare the salinity of surface sea water at the equator with that near the poles.
5. Why is the sea salty?
6. Compare the conditions found in the ocean surface layer with those found in the deep sea.
7. What two important effects on the atmosphere result from the evaporation of water from the tropical seas?
8. Why can the oceans produce twice the food that land provides?

9. What is the beginning of the food chain for life in the ocean?
10. Where are the best places for fishing located?
11. Why is upwelling necessary for abundant life in the ocean?
12. The ocean contains a large amount of gold and other minerals, but gold is not taken from sea water. Why is this?
13. What minerals do nodules contain that make them so valuable?
14. Describe the ways energy may be taken from the sea.
15. What is the biggest problem with taking resources from the ocean?

APPLYING SCIENCE

1. Determine the salinity of a sample of drinking water by doing the following: (a) Weigh a clean, dry 500-mL beaker. (b) Measure into the beaker 250 grams of water. (c) Boil until nearly dry. CAUTION: NEVER BOIL UNTIL COMPLETELY DRY! (d) Allow beaker to sit, away from heat, until completely dry. (e) Weigh beaker. The difference in original weight and final weight of beaker is the amount of dissolved material in the water for .25 kilogram of water. (f) Calculate the salinity.
2. Make up a sample of artificial sea water by dissolving 35 grams of rock salt in 965 grams of water. Try to get drinkable water from the salt water by freezing it.
3. In 1984 much of the krill usually present in the Antarctic waters disappeared. Find out why this happened and its effect on the life in the ocean.
4. Use a biology book to find out what plankton look like. Write a list of the ocean food chain in which plankton are involved.
5. Find out the details of recent developments or uses of sea resources such as: sea farming; sea mining; energy from the sea; seaweed products; etc. Report to the class on your findings.

BIBLIOGRAPHY

Britton, Peter. "Cobalt Crusts: Deep Sea Deposits." *Science News*, October 30, 1982.

Britton, Peter. "Deep Sea Mining." *Popular Science*, July 1981.

Matthews, Samuel W. "New World of the Oceans." *National Geographic*, December 1981.

Time-Life Editors. *Restless Oceans* (Planet Earth Series). Chicago: Time-Life Books, 1984.

THE MOTIONS OF THE SEA

CHAPTER GOALS

1. Explain the different ways that underwater currents can form in the oceans.
2. Describe the movement of the surface currents in the oceans and the factors that cause them.
3. Explain how surface currents can affect climates.
4. Describe the parts of a wave, and explain how a wave forms on the sea surface.
5. Explain what causes tides and describe the normal pattern of these giant waves.

16–1. Underwater Currents

At the end of this section you will be able to:

- Explain how differences in the temperature and salinity of sea water produce ocean currents.
- Describe how *turbidity currents* can be formed.
- Describe the path followed by deep currents of the Atlantic and Pacific oceans.

Much of the world's population depends on food from the sea. Fishing boat captains find that they can usually depend on great catches at certain places. Studies of the ocean currents show that deep water rises in these areas. This supplies the food needed for sea life. Knowing where the ocean currents rise, the oceanographer can now guide sea captains to places that are full of sea life.

TEMPERATURE

When the water in the sea moves in a certain direction, a **current** is formed. One type of *current* is formed when water is heated unequally by the sun's energy. For example, water near the equator is heated. This lowers the density of the water.

Current Water moving in a particular direction.

Water near the poles is cooled. This makes the water more dense. The cold water sinks and moves toward the warmer water near the equator. For example, the cold water in the Arctic and Antarctic sinks to the ocean floor. Then, the cold

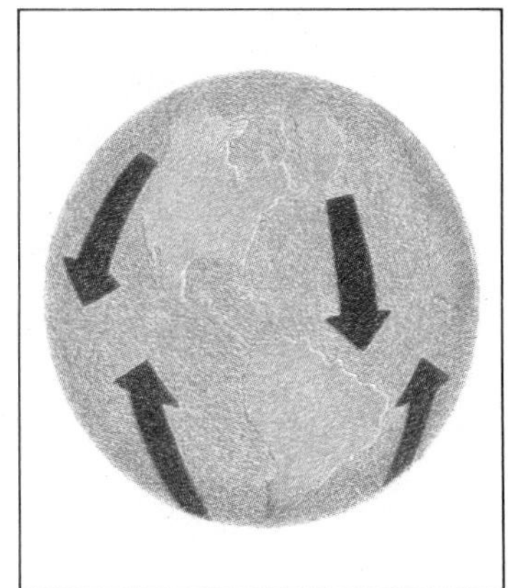

Fig. 16-1 *Deep currents move generally from the poles toward the equator.*

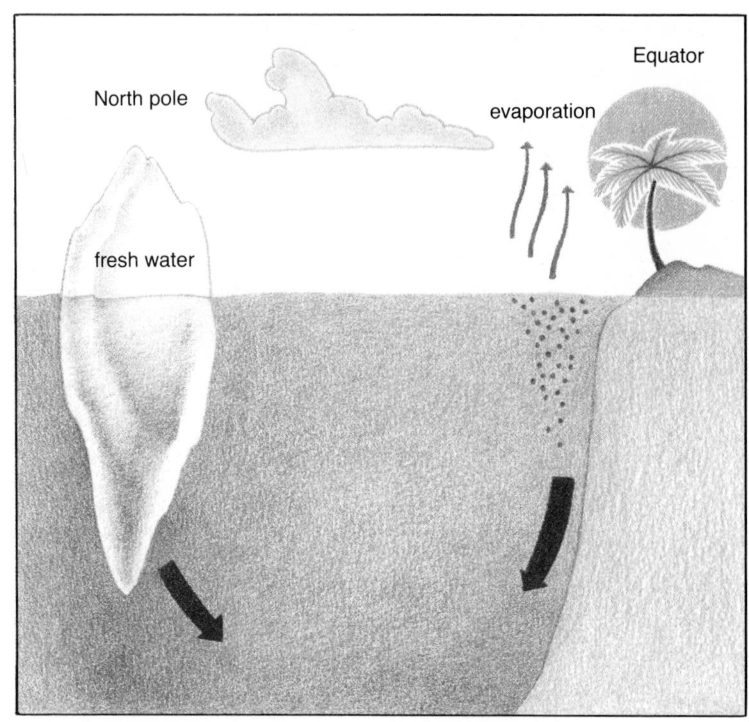

Fig. 16-2 *The salinity of the sea water can be increased by freezing or evaporation.*

water moves very slowly along the bottom of the ocean toward the equator. Some of the cold water mixes with the water above and moves back toward its starting place. The rest of this very cold water slowly moves through the deep oceans away from the polar regions where it was formed. See Fig. 16-1. Most of this icy water flows deep in the ocean for many years before finally making its way back to the surface.

SALINITY

Currents are also formed by changes in the salinity, or amount of dissolved salts in sea water. When salt dissolves in water, the density of the water increases and the water sinks. One way salinity is increased is when water freezes. When ice is formed in the sea, only the freshwater freezes. The salt in the sea water is forced out and goes into the water that does not freeze. Therefore, water surrounding newly formed ice has more salt, or a higher salinity. Another way salinity can increase is when water evaporates at the surface. See Fig. 16-2. In both cases, the surface water sinks because it has a higher density than the water under it.

Fig. 16-3 shows a current caused by differences in salinity. It is found at the entrance to the Mediterranean Sea at the Strait of

Gibraltar (ji-**brawl**-tar). The dry, warm climate of the Mediterranean causes evaporation from the surface of the water. This causes the salinity of the water to be much greater in the Mediterranean than it is in the Atlantic Ocean. The dense water from the Mediterranean sinks below the surface and flows westward past Gibraltar into the Atlantic. At the same time, the lighter Atlantic water flows eastward, past Gibraltar, on top of the heavy outgoing flow. During World War II, some submarines slipped past the defense at Gibraltar by shutting off their engines and riding this density current.

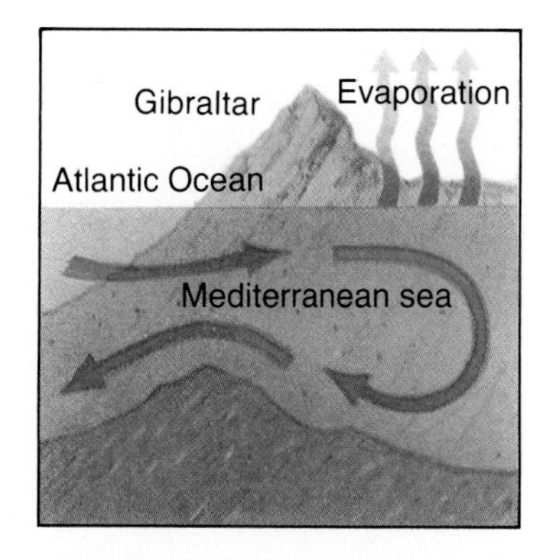

Fig. 16-3 A density current is found at the entrance of the Mediterranean Sea because of differences in salinity.

TURBIDITY CURRENTS

Another type of current caused by changes in water density is called a **turbidity** (tur-**bid**-ih-tee) **current.** *Turbidity currents* are strong, downslope movements of muddy water. They form from the sand and mud that build up on the continental slope. Sometimes they suddenly fall under their own weight and mix with water. Since the muddy water is more dense than the surrounding water, it rushes down the slope, causing a deep current. Turbidity currents may also be started by severe storms, earthquakes, or floods. Submarine canyons in the continental slope are probably carved out by turbidity currents where rivers carrying sediments flow into the ocean. These currents also move large amounts of sediment from the shallow continental margins to the deeper sea floor.

Turbidity current A current that is caused by the increased density of water as a result of mixing with sediments.

DEEP SEA CIRCULATION

Currents that are very deep in the sea are hard to study. Thus they are not understood as well as the currents near the surface. However, the main currents in the deepest parts of the Atlantic, Pacific, and Indian oceans have been charted.

Along the floors of these oceans, there is a northward flow of cold, dense water that comes from the Antarctic region. This current is formed when the cold water freezes. The freezing action leaves extra salt in the unfrozen water. This causes the water around the continent of Antarctica to be very dense. This water sinks and then moves northward along the bottoms of the three oceans that surround Antarctica. The speed of this icy cold bottom current is quite slow. It may take at least several

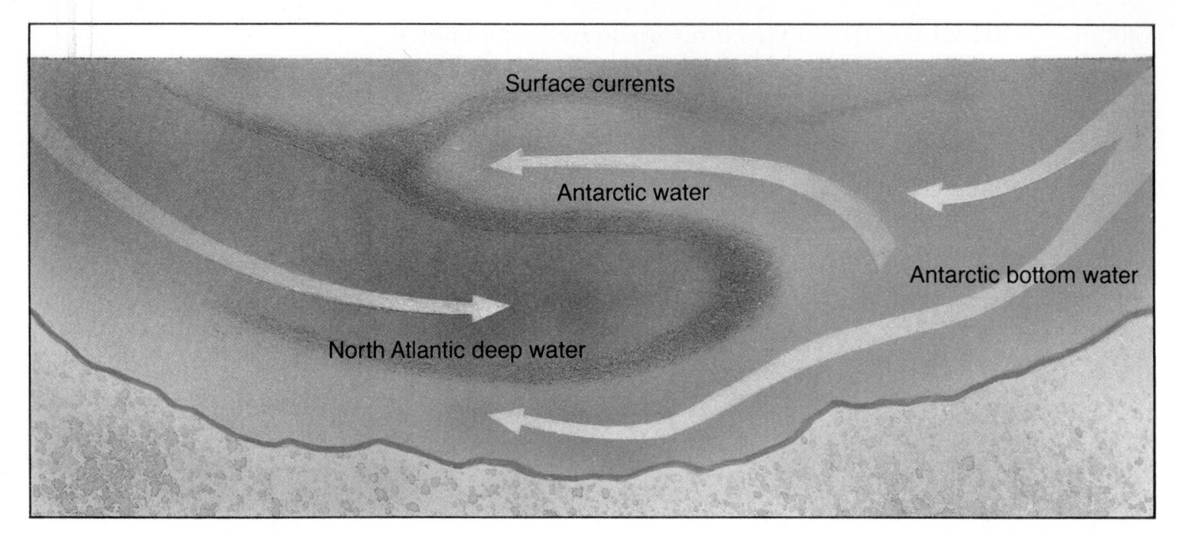

Fig. 16-4 Deep currents in the Atlantic result from water sinking near Antartica in the south and Greenland in the north.

hundred years for this water to travel from Antarctica to the northern Atlantic and northern Pacific oceans.

In the Atlantic, there is a deep current that moves south. This current is caused by the cold water that sinks near Greenland. This water near Greenland is less dense than the water that comes from the Antarctic. These two currents flow in opposite directions, meeting a little north of the equator. The less dense, North Atlantic water rides up over the heavier, Antarctic water. See Fig. 16-4. However, neither of these currents is affected by the surface movement of ocean water.

SUMMARY

Deep currents are caused by differences in the temperature or salinity of the water. Turbidity currents are caused by downslope movements of sand and mud. All of the world's oceans have slow currents of near-freezing water moving close to the bottom, away from the poles and toward the equator.

QUESTIONS

Use complete sentences to write your answers.

1. Give an example of how differences in temperature cause ocean currents.
2. Explain how changes in salinity cause ocean currents.
3. Explain how turbidity currents form.
4. Why do the deep currents in the Atlantic and Pacific oceans begin at the poles?

INVESTIGATION

DENSITY CURRENTS

PURPOSE: To see what happens when waters of different densities are put together.

MATERIALS:

liter container
water
sodium chloride (salt)
2 250-mL beakers
food coloring
balance

PROCEDURE:

A. Prepare a salt water solution by dissolving 35 grams (5 level teaspoons) of sodium chloride in 1 liter of water.

B. Fill one of the beakers half-full with tap water. Allow this to sit undisturbed for several minutes. Go on to step C.

C. Fill the other beaker half-full with the salt water. Add four or five drops of food coloring to the salt water and mix well.

D. Slowly pour one-third of the colored water down the side of the beaker of tap water. Make sure the two solutions do not mix when poured together. Do not move the beaker of tap water. See Fig. 16–5. If the colored water mixes with the tap water, pour out the mixture and repeat steps B, C, and D.

Fig. 16–5

E. Let the mixture sit a minute. Just as with air masses, water masses of different densities do not tend to mix.
   1. What do you observe?

F. Without lifting it from the table, slowly tilt the beaker of colored water and tap the beaker on one side. Hold it this way until the colored water settles in a new position. Then *gently* place the beaker back in its original position. Watch the colored water for a few minutes.
   2. What do you observe?

G. Repeat step F. This time watch carefully for waves at the border between the salt water and the tap water. These are called *internal waves.*
   3. Do the waves bounce off the side of the beaker?

H. Again repeat step F. Watch the waves on top of the tap water as well as the internal waves. It may be more difficult at this time since the colored water may be mixing into the tap water. If that happens, you may need to prepare another sample.
   4. Compare the size, shape, and speed of the internal waves with those of the surface waves.

CONCLUSIONS:

1. Which is denser, tap water or salt water?
2. Explain how internal waves are formed.

16-2. Surface Currents

At the end of this section you will be able to:

- Describe how the continents and the Coriolis effect change the motion of surface currents in the oceans.
- Describe the general pattern of surface currents in the oceans.
- Name and describe the paths of the main surface currents of the North and South Atlantic.

The beaches in southern California are much like the beaches in southern Florida. See Fig. 16-6. The air temperature is warm all year and the sun shines much of the time. However, if you entered the water, you would immediately feel one big difference. The water at the beaches in California is cold, with temperatures rarely reaching higher than 21 °C. In contrast, the water at the beaches in Florida is warm, with temperatures rarely lower than 21 °C. How can the water off the coasts of two sunny, warm regions be so different?

Fig. 16-6 Southern California beaches have water temperatures colder than southern Florida.

MOVING WATER

Water near the surface of the sea is moved mainly by winds. We know this because of the winds over the northern Indian Ocean. There the winds move in opposite directions in summer and winter. When the winds reverse, so do the surface currents. However, there are other factors that influence water motion near the ocean's surface. For example, the continents act as barriers, blocking the free movement of the ocean water. Also, the earth's rotation influences the motion of the sea water. The Coriolis effect causes the currents in the Northern Hemisphere to turn to the right from the direction of water flow. Thus surface currents do not necessarily move in the same direction as winds. In the Southern Hemisphere, the Coriolis effect causes a twist to the left from the direction of the water flow. In both hemispheres, this causes the water at and near the surface to form huge circles of moving water. In the Northern Hemisphere, the direction of this movement is clockwise. In the Southern Hemisphere, the motion is counterclockwise. See Fig. 16-7. There are also many smaller circles in each ocean.

Winds also may cause *upwelling,* or the rising of water from the deep layers of the ocean to the surface. This happens when wind moves warmer surface water away from the shore. As the warm surface water is moved away, it is replaced by cold water from below. This slow upward flow of cold water creates the area of cold water off the shore of California.

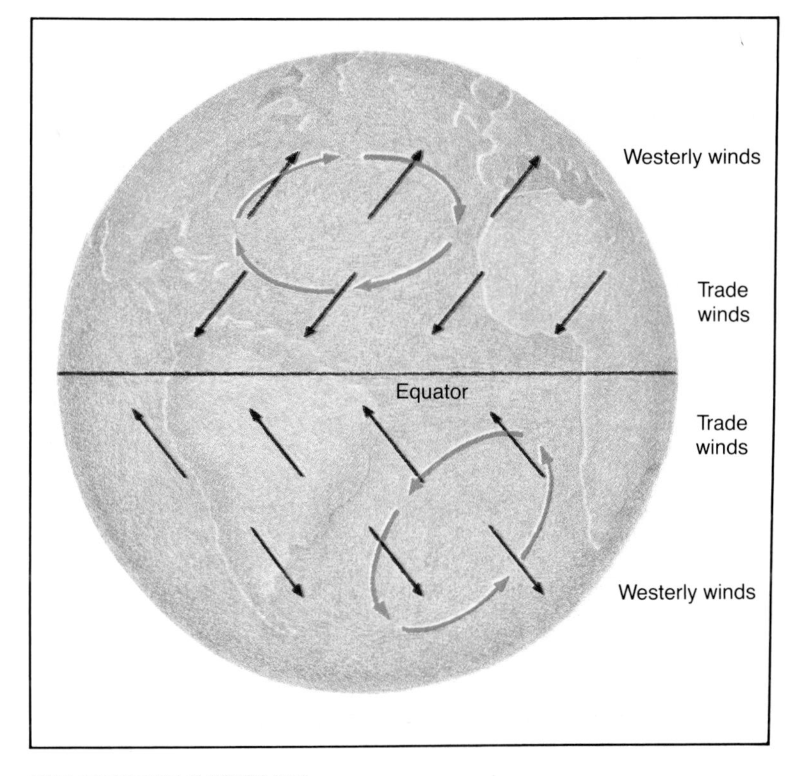

Fig. 16-7 The Coriolis effect and the action of wind produce huge circles of moving water.

CURRENTS OF THE NORTH ATLANTIC

1. The Gulf Stream. Around 1770, Benjamin Franklin learned that some ships had crossed the Atlantic from England to America in about two weeks less than the expected time. The captains of these ships described a swift, eastward current flowing in the North Atlantic. By sailing south to avoid this current, they made a faster voyage. Franklin measured the water temperature and the speed of the current during several trips between America and England. In 1779, he published the first scientific chart of an ocean current. This current is known as the Gulf Stream.

The Gulf Stream begins with waters moved by the trade winds. These winds blow from the northeast and southeast near the equator. Because of the Coriolis effect, the surface water moves westward in the Atlantic just north of the equator. This flow is called the North Equatorial (ee-kwuh-**tor**-ee-ul) Current. The North Equatorial Current pushes water against the coast of Central America into the Gulf of Mexico. This causes water to spill out of the Gulf of Mexico between Florida and Cuba, forming the Gulf Stream. See Fig. 16–8. In the Atlantic Ocean, the Gulf Stream flows around the Florida coast, bringing warm water to its beaches. The result is that the water temperature rarely drops below 21°C throughout the year. The Gulf

Fig. 16–8 This satellite photograph of the southeastern United States coast clearly shows the Gulf Stream.

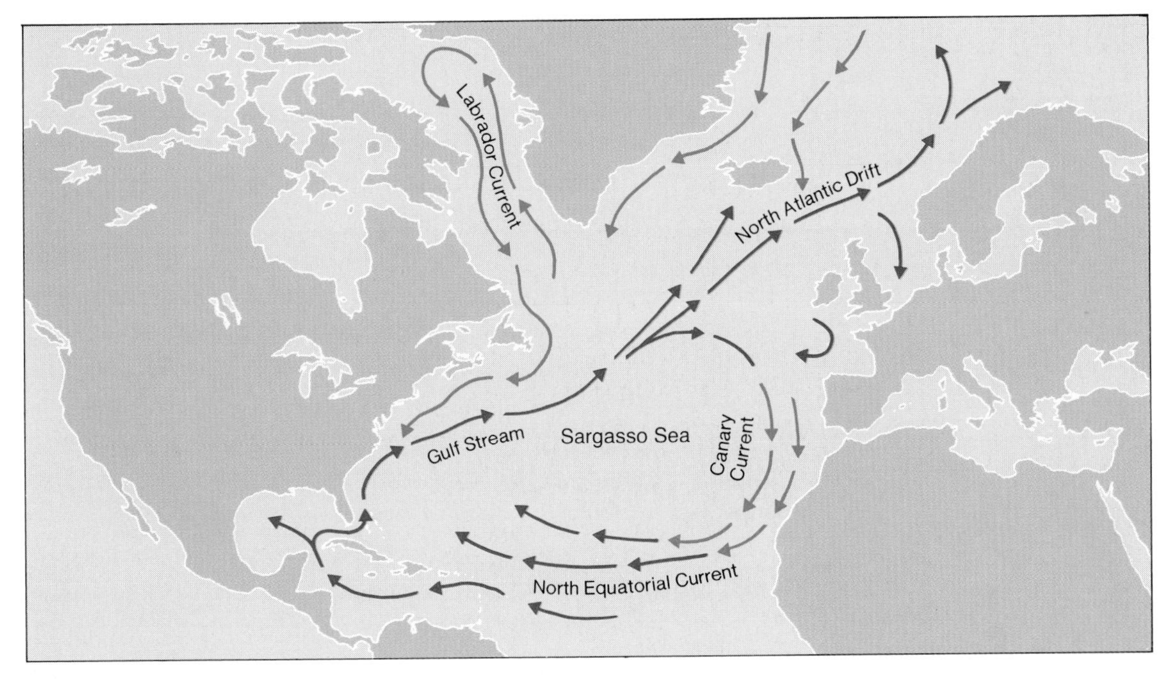

Fig. 16-9 Currents in the North Atlantic. Warm currents are shown in red.

Stream moves northward along the eastern coast of the United States and then turns east toward Europe. The Gulf Stream moves huge amounts of water. It sweeps along more than four billion tons of water each second. This is more than a thousand times the amount of water moved by the Mississippi River.

2. Labrador Current. Near Greenland, the Gulf Stream meets the cold, south-flowing Labrador (**lab**-ruh-dor) Current. The collision of the warm Gulf Stream and the cold Labrador Current causes the ocean in that area almost always to be covered by a thick fog.

3. North Atlantic Drift. Beyond Greenland, the path of the Gulf Stream is not as easily identified. Part of the current moves on toward northern Europe, forming the North Atlantic Drift. A slow-moving current is called a "drift." Another branch of the Gulf Stream turns south, forming the Canary Current. The Canary Current moves toward the equator and begins to complete the huge circle of currents that move around the North Atlantic.

4. North Equatorial Current. Off the coast of Africa, the Canary Current joins the North Equatorial Current to complete the circle formed by the North Atlantic currents. See Fig. 16-9. The center of the North Atlantic has no major currents. In an area called the Sargasso Sea, large amounts of seaweed are found because no major current sweeps it away.

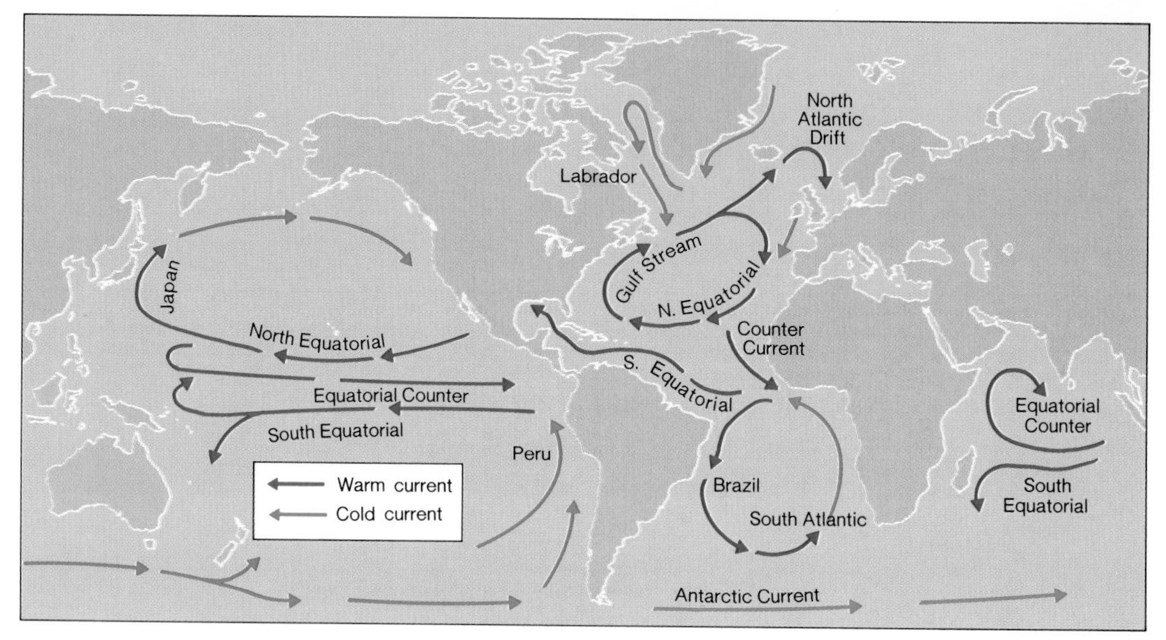

Fig. 16-10 The principal ocean currents of the world.

CURRENTS OF THE SOUTHERN OCEANS

The currents in the North Pacific move in a huge clockwise circle, very much the same as currents do in the North Atlantic. However, in the southern parts of the Atlantic and Pacific oceans, the currents flow in a counterclockwise direction. The Antarctic Current follows an unusual path. It completely circles Antarctica; no landmass blocks its movement.

Currents in the Indian Ocean are controlled by two factors. One factor is the South Equatorial Current, which divides as it moves toward Africa. One part turns to the north and the other part moves to the south. The southern branch turns back toward Australia. Near Australia the current breaks up, forming many small currents that move among the South Pacific islands. The other factor affects the northern area of the Indian Ocean. There the currents follow the strong monsoon winds. During different seasons of the year, the currents change their directions. The principal ocean currents of the world are shown in Fig. 16-10.

OCEAN CURRENTS AND CLIMATE

Ocean currents, such as the Gulf Stream and the North Atlantic Drift, have a strong influence on temperature and climate. Wind blowing across these currents and onto land can raise the land's average temperature. For example, northwestern Europe has a higher average temperature than other areas at the same latitude. This occurs because of the steady westerly winds. These winds blow over the North Atlantic Drift toward northwestern Europe.

Ocean currents also help transfer heat from areas near the equator to the poles. Otherwise the regions near the equator would become warmer, while the poles would become colder.

Scientists believe that changes in ocean currents also may cause weather patterns to change. For example, a warm current of equatorial water appears about once every ten years along the west coast of South America. It is called "El Niño," which means "the child," because it usually appears around Christmas. Fish that normally inhabit the waters along the coast of Peru disappear because they cannot live in the warm water. In 1982, a very strong El Niño current developed, causing the surface water in the eastern Pacific to become very warm. Unusual winter storms and a very hot summer in North America resulted when the atmosphere was heated by the warm water in the Pacific.

SUMMARY

Although winds are the cause of surface currents in the sea, other influences such as land barriers and the Coriolis effect change the direction the water moves. Major surface currents in the northern and southern parts of the Atlantic and Pacific oceans form huge circles. Other oceans have their own pattern of currents. Surface currents in the sea are an important influence on the atmosphere, and therefore affect the climates of the continents.

QUESTIONS

Use complete sentences to write your answers.

1. Explain how ocean currents are affected by land and the Coriolis effect.
2. Describe the general pattern of surface currents in the North and South Atlantic.
3. Name the five main currents of the North Atlantic and briefly describe each.
4. Name two main currents of the South Atlantic and briefly describe each.

INVESTIGATION

THE CORIOLIS EFFECT

PURPOSE: To demonstrate how the Coriolis effect occurs.

MATERIALS:

sheet of paper	ball bearing (1 cm or larger)
cardboard	pencil
carbon paper	
pin	

PROCEDURE:

A. On a level table, place the sheet of paper on the cardboard and put the carbon paper, carbon side down, on the paper. Then stick a pin in the center of the carbon paper to hold the paper, carbon paper, and cardboard in place.

B. Starting in one corner of the paper, roll the ball bearing across the carbon paper to the other corner. Lift the edge of the carbon paper and label the track No. 1. Draw an arrow on the track showing the direction of the roll.

   1. Describe the track made by the ball as it rolled over the carbon paper.

C. The ball bearing represents moving water. The paper and carbon represent the earth. When viewed from the North Pole, the earth turns in a counterclockwise direction. Hold a corner of the paper and carbon with one hand and the cardboard with the other hand. Keep the cardboard still while you turn the paper and carbon counterclockwise slowly around the pin. While you are doing this, have your partner carefully roll the ball bearing across the carbon paper from one corner of the paper to the other.

D. Lift a corner and label the track No. 2. Place an arrow on the track to show the direction the ball bearing moved.

   2. Describe the track made by the ball as it rolled over the carbon paper.

E. When viewed from the South Pole, the earth turns in a clockwise direction. While you carefully turn the papers in a clockwise direction, have your partner slowly roll the ball bearing across the carbon paper again. Label this track No. 3. Indicate the direction the ball bearing moved with an arrow.

F. Remove the pin and carbon. Inspect the tracks.

   3. Describe the track the ball made when the papers were turned in a counterclockwise direction (track No. 2).
   4. Describe the track the ball made when the papers were turned in a clockwise direction (track No. 3).

CONCLUSIONS:

1. What caused the difference between tracks 1 and 2?
2. What caused the difference between the tracks marked 2 and 3?
3. Look at Fig. 16–10 on page 438. Use the information from this investigation to explain why the large circles of currents in the North Atlantic and South Atlantic curve in the directions shown.

16–3. Waves

At the end of this section you will be able to:

- Describe a wave in terms of its height, wavelength, and period.
- Explain how waves are formed.
- Describe the motion of water in waves.

If you have ever seen surfing, or been in the ocean, you have seen or felt the power of the waves. See Fig. 16–11. Many people can feel the up-and-down motion of waves while in a boat. Anyone who has been caught by surprise and thrown over by a breaking wave knows that waves carry energy. Where do waves get their energy?

Fig. 16–11 Large waves are every surfer's wish. Where do these waves come from?

WHAT IS A WAVE?

Wind produces the familiar waves that are almost always present on the sea. Even a gentle breeze blowing across the surface of still water causes waves. Air never moves in smooth streams, but has a certain amount of whirling motion. The whirling motion of moving air makes small dents in the water's surface. This whirling action also raises a part of the water. The wind then pushes on the raised part of the water's surface and a wave is created. As long as the wind continues to blow, the wave will move and grow larger.

If you toss a stone into a quiet pool of water, waves are created. The falling stone carries energy into the water. Anything

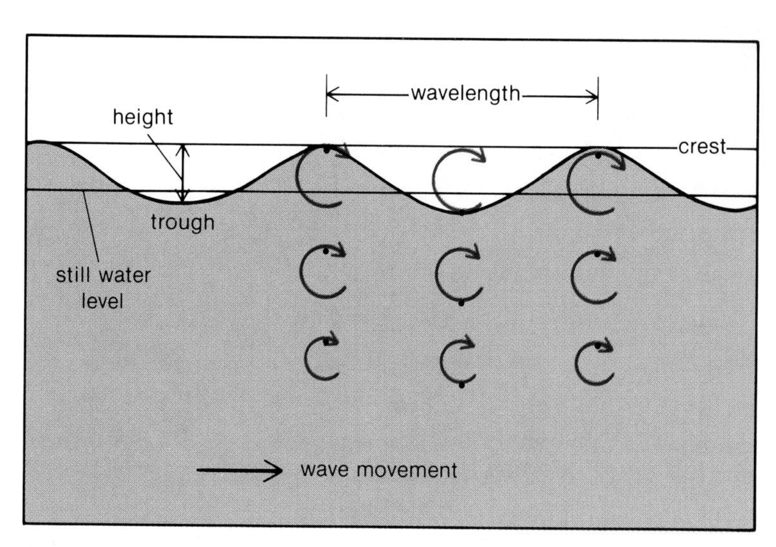

Fig. 16–12 The parts of a wave.

that adds energy to water can make waves. All waves have two major parts. The raised parts are called *crests*. Between the crests are low areas called *troughs* (**trawfs**). Waves can be described by measurements of their crests and troughs. For example, the *wave height* is the distance between the lowest part of a trough and the highest part of the next crest. The distance between two neighboring crests, or two neighboring troughs, is called the *wavelength*. See Fig. 16–12. The time it takes for two crests to pass a given point is the *wave period*.

When the wind begins to blow over calm water, patches of ripples appear. These small waves are called "cat's paws." If the wind continues at the same speed, the little waves will grow. The shape and size of the waves that are produced depend on three things: the speed of the wind, how long the wind blows, and the **fetch.** *Fetch* is the distance over which the wind blows steadily. Strong winds blowing for many hours over a long fetch produce waves with the greatest heights, wavelengths, periods, and speeds. However, waves can only grow to heights equal to about one-seventh of their wavelengths. Waves that build up higher than that break and tumble over. Thus waves that have short wavelengths cannot get very big. Really large waves can only be formed on the open sea, where strong winds blow across fetches of thousands of kilometers.

Fetch The distance over which the wind blows steadily, producing a group of waves.

The wind usually makes waves of many sizes. Wind-caused waves tend to move in the direction of the wind. The larger waves grow even larger since they are better able to take energy from the wind. These large waves move away from the

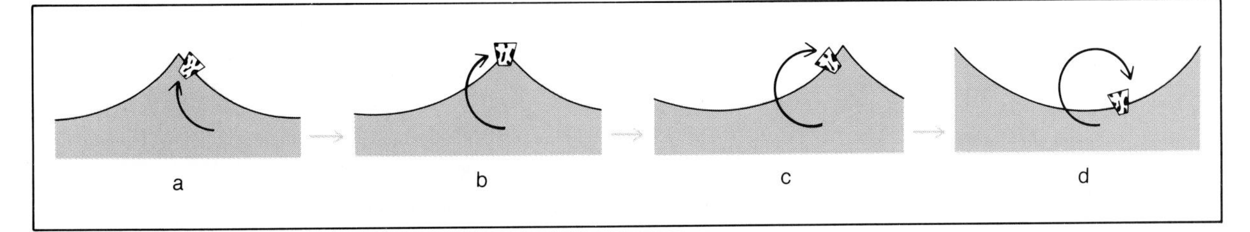

Fig. 16-13 Water particles move in circles as the wave passes.

place where they were formed and travel great distances across the ocean. Long, rolling waves moving across the open sea are called **swells.** If you watch the waves that reach the shore, it is possible to make good guesses about their origins. The long *swells* that roll in slowly were probably formed by storms some distance away. Small, choppy waves are usually young. They were probably made by local winds.

Swells Long, regular waves that move great distances across the oceans.

MOTION OF WATER IN WAVES

A swell formed in the middle of an ocean does not move water from the middle of the ocean along with it. Water does not move over great distances with the waves. It is the energy in the waves that moves. You can see this for yourself. Watch the action of a floating cork as a wave passes. The cork is lifted up, carried forward, but then back again. The cork moves in a circle. See Fig. 16–13. The cork moves ahead only slightly as the wave passes. Water particles also move through circles as a wave passes. The movement of the wave across the water is the result of these circular motions. Below the surface, water also moves in circles as the waves go by. However, the size of these circles gets smaller with greater depth. See Fig. 16–14. Submarines traveling at depths greater than 100 meters escape most of the circular movement of wave action.

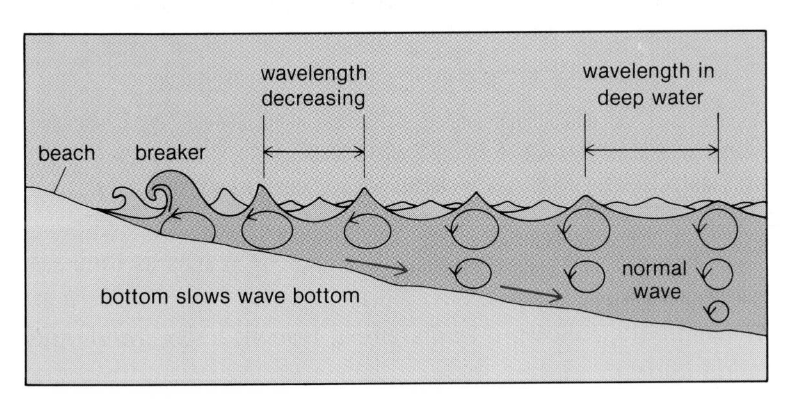

Fig. 16-14 Breakers form when the lower parts of a wave hit the sand on the shore.

Fig. 16–15 A large breaker coming to shore.

Breaker A wave whose top has curled forward and crashed down against the shore.

When waves come close to the shore, they enter shallow water. Near the shore, the circling water particles in the waves begin to rub against the sea floor. The bottoms of the waves are slowed. As the bottoms of the waves slow, the tops may lean too far forward and tumble over in foamy crashes. See Fig. 16–15. Such waves are called **breakers.** Usually water thrown up on the shore by *breakers* runs back as a weak current below the surface. Sometimes the water piles up on the shore and then suddenly rushes back toward the sea in strong, swift *rip currents*. Rip currents are dangerous to swimmers. A person caught in such currents can be carried out to deep water.

SUMMARY

Sea waves are usually caused by the wind blowing across the water. The size of a wave is determined by the strength and duration of the wind and the fetch. Waves can be described by measurements such as height, wavelength, and period. Water particles move in circular paths as a wave passes. When waves reach shallow water near the shore they may form breakers.

QUESTIONS

Use complete sentences to write your answers.

1. What causes most surface waves?
2. Describe the meaning of the terms height, wavelength, and period as used in discussing waves.
3. Briefly explain the changes that occur in waves as they approach shore.
4. Define the following terms: crest, trough, fetch, swell, and rip current.

INVESTIGATION

MAKING WAVES

PURPOSE: To produce and study waves made by wind.

MATERIALS:

pan	drinking straw
water	food coloring

PROCEDURE:

A. Fill the pan with water to a depth of about one centimeter. Let the water sit for a few minutes.

B. Aim the straw so that when you blow through it gently, the air will go across the surface of the water from the center of the pan. See Fig. 16-16.

Fig. 16-16

C. Blow through the straw and observe the waves.
   1. Describe the waves produced.
   2. What is the *fetch* in this case?

D. While disturbing the water as little as possible, place a drop of food coloring on the surface of the water near the center of the pan.
   3. What happens to the food coloring?

E. Aim the straw at the center of the pan and blow through it gently. As the waves move on the surface, observe what happens to the food coloring.
   4. What happens to the food coloring?
   5. What do you think the food coloring represents?
   6. Which move faster, the waves or the food coloring?

F. Now blow one puff of air down from directly above the center of the pan. Observe what happens when the wave hits the side of the pan.
   7. What did you observe?

G. Increase the force at which you blow the puff of air.
   8. What is the name of the part of the wave you are watching as it moves? How did the wave change?

H. Time the puffs of air so that as the wave returns to the center of the pan, you produce another wave. In this manner only one wave at a time is seen as it goes to the side and reflects back to the center of the pan.
   9. Describe the wavelength of this wave. Remember, the wavelength is the distance the wave traveled by the time another is generated.

CONCLUSIONS:

1. How are waves produced by wind?
2. How are the currents made?
3. Where did the energy come from to make the waves and currents?

16–4. Giant Waves

At the end of this section you will be able to:

- Explain how a *tsunami* is formed.
- Describe the cycle of high and low tides each day.
- Explain the cause of tides.

The destruction shown in Fig. 16–17 was brought about by a few huge waves on May 26, 1983. Not far from this location, a school bus had just brought 43 school children to the beach for a picnic. All the children and their bus were swept out to sea. In 1896, on the same Japanese island, a few similar waves swept away 27,000 people and 10,000 houses. In this section you will learn how such huge waves are caused.

Fig. 16–17 Tsunami destruction can be severe.

KILLER WAVES

Imagine what would happen in a swimming pool if the bottom were suddenly pushed up or dropped down. All the water in the pool would move. Something similar happens when there is a sudden change on the sea floor. During an earthquake, an underwater volcanic eruption, or a mudslide, a large part of the sea floor may rise or fall. Such disturbances on the sea floor cause a very large wave, called a **tsunami** (soo-**nahm**-ee). *Tsunami* is Japanese for "harbor wave." The people of Japan have had many experiences with these waves.

Tsunami A giant wave caused by a sudden disturbance on the sea floor.

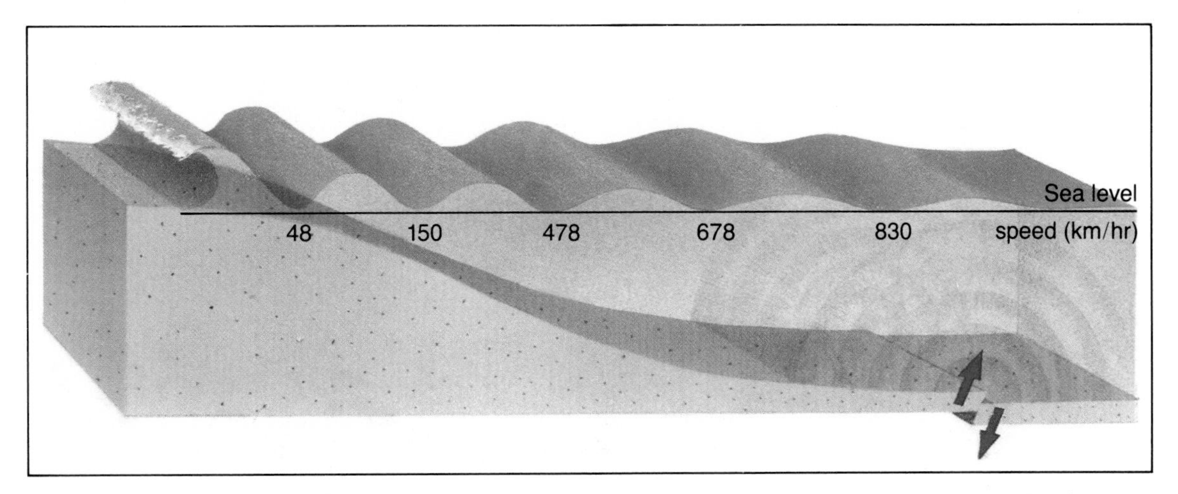

Fig. 16–18 The birth of a tsunami.

Tsunamis are very long waves. In the open sea, they often have wavelengths of about 200 kilometers. But heights of tsunami waves are usually less than one meter. A ship in the open ocean would hardly notice tsunamis passing by. However, tsunamis move at speeds up to 960 kilometers an hour. When these giant waves approach land, their heights can increase to 30 meters. See Fig. 16–18. When tsunamis hit a coastline, serious flooding can occur. Coastal towns can be heavily damaged. Tsunamis occur in groups of five to ten waves and about 5 to 20 minutes apart. Thus people should not return to the shore until the last of the waves has hit.

A warning system has been set up for the Pacific Ocean, where tsunamis are most common. When an earthquake that might cause a tsunami occurs, people living in coastal regions are warned by radio. Although tsunamis move at high speeds, earthquake waves travel much faster. Thus, with tsunamis, people usually can be given several hours' warning to move away from the shore to higher ground.

TIDES

Most people along coasts do not experience the long waves called tsunamis. There is, however, a long wave with which they are familiar. This large wave is called a **tide.** It makes the sea level rise and fall regularly in most places. Observation of *tides* shows that they are very long, regular waves. The crest of the wave brings a high tide. The trough of the wave brings a low tide. The period of the wave is about twelve hours. High and low tides are about six hours apart.

Tide A regular rise and fall of the water level along the shore.

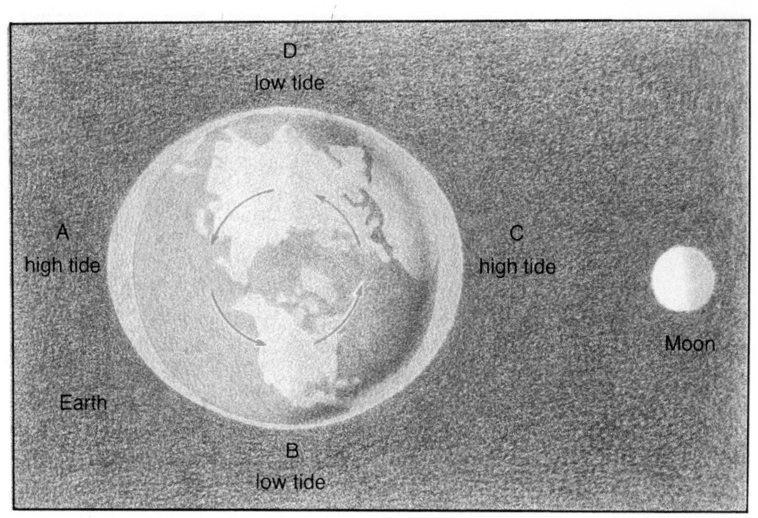

Fig. 16–19 There are two high tides and two low tides each day.

What causes such huge waves in the sea? The answer to this question involves the moon and the sun. The earth, moon, and sun all attract each other with a gravitational force. The moon's gravitational force has a greater effect on tides than does the sun's. However, when the sun's gravitational force is added to the moon's gravitational force, the tides are even higher. Because water in the sea is free to move, the moon's gravity pulls sea water into a bulge toward the moon. A similar bulge is formed on the side of the earth opposite to the moon. On that side of the earth, the moon's gravitational force on the water is much less than on the solid earth. This causes the solid earth to be pulled more than the water. Thus the water seems to rise as a tidal bulge. This means that the sea always has two bulges. One is facing toward the moon and the other is farthest from the moon.

While the earth is turning, the bulges stay lined up with the moon. In Fig. 16–19, points A and C are at high tide. As the earth rotates, points D and B will be at high tide about six hours later. When will points A and C be at high tide again? If you were measuring the tides at one point, you would observe two high tides and two low tides about every twenty-four hours. There are about six hours between each high and low tide.

The sun's gravitational force has less effect on the tides than the moon has because the sun is much farther away. However, when both the moon and sun are lined up on the same side of the earth, their gravitational pull is combined. Then unusually high tides, called *spring tides,* occur. See Fig. 16–20 (a). Spring

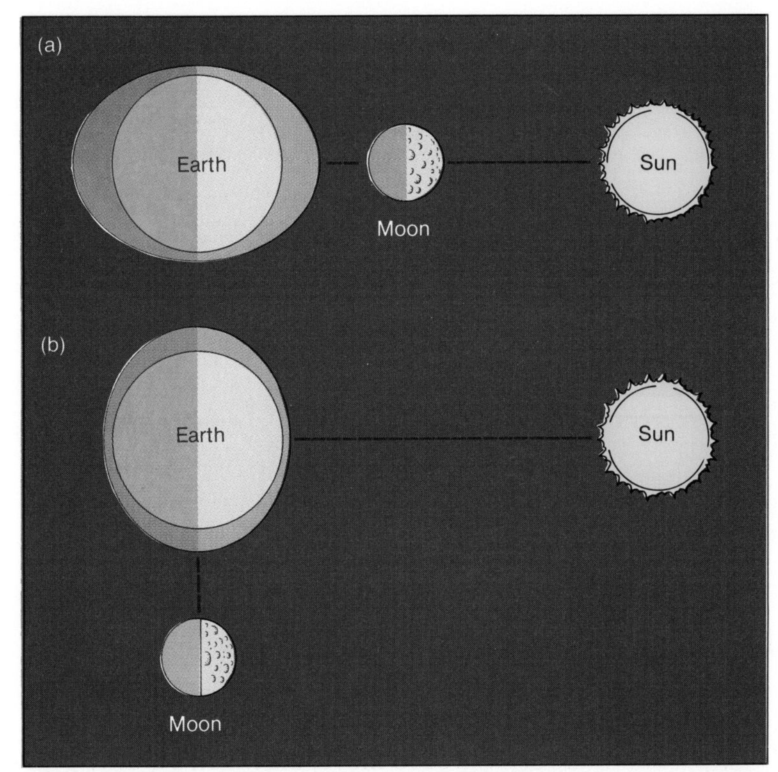

Fig. 16-20 (a) Spring tides occur when the moon, sun, and earth line up. (b) Neap tides occur when the sun and the moon are at right angles to each other.

tides occur twice each month. They do not refer to the spring season of the year. When the moon and sun are at right angles to each other, unusually low tides occur. These are called *neap tides*. See Fig. 16-20 (b). Neap tides also occur twice each month.

TIDAL EFFECTS

Not all parts of the sea are affected in the same way by the tides. The shapes of the ocean basin and the sea floor make the water respond in different ways. Each part of the sea has its own pattern of tides. For example, some places on the shore of the Gulf of Mexico have only one high and one low tide each day. In some places, the changes in water level from high to low tide are very large. The water level in the Bay of Fundy in Canada changes about ten meters. This is about the same as the height of a three-story building. See Fig. 16-21, page 450. Along straight coastlines and in the open sea, the water level does not change more than 50 centimeters. The behavior of tides at a place can only be predicted from observations made over a long period of time.

Fig. 16-21 (top) *Low tide and* (bottom) *high tide in the Bay of Fundy.*

Tidal bore A large wave that leads a tide upriver.

A single wave, called a **tidal bore,** may be found leading a tide up a river. See Fig. 16-22. In order for such waves to form, two conditions must exist. First, a large tide difference, from six to ten meters, must occur in the river's tidal basin. Second, the mouth of the river must be broad and have a gradual, sloping bottom. On the Severn River in England, the *tidal bore* is so great that surfers may ride it upstream for miles.

At places where tidal changes are large, tidal power may one day be used as a source of energy. To use tidal power, a dam would have to be built to trap water at high tide. The water

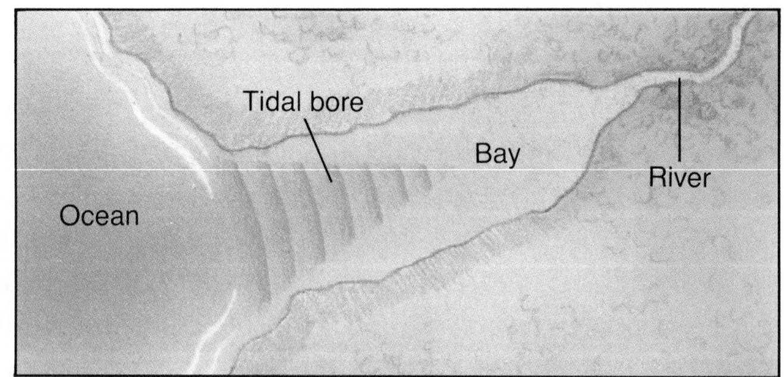

Fig. 16-22 A tidal bore.

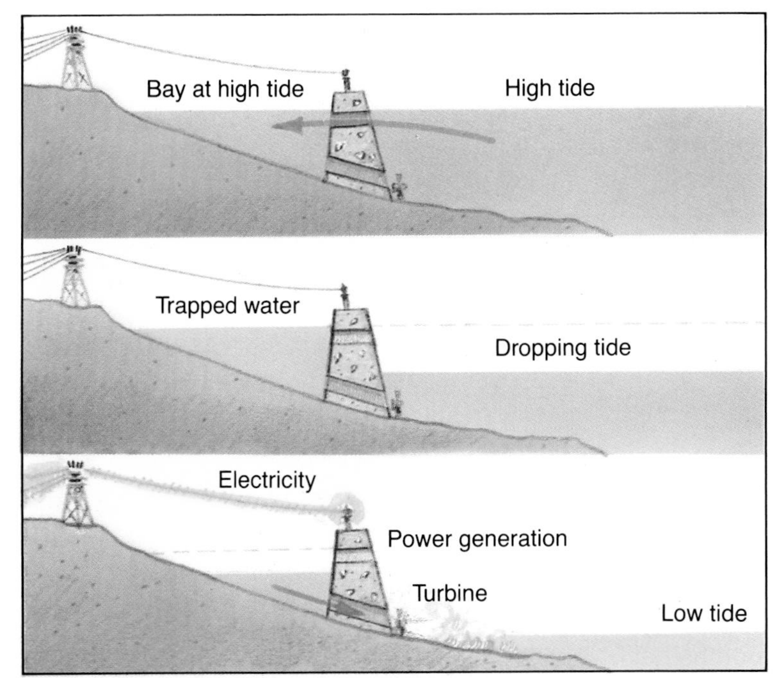

Fig. 16-23 The principle of a tidal dam.

would then be released during low tide. See Fig. 16-23. The flowing water would be used to generate electricity. An experimental tidal power plant has been constructed at the mouth of the Rance River in France.

SUMMARY

Giant waves are produced by large earth movements on the sea floor. Caused by earthquakes or volcanic eruptions, they can be dangerous to people living on the coast. In most places, the sea level changes in a regular way every day because of tides. Tides are caused mainly by the effect of the moon's gravity on the earth and its ocean waters. Tidal effects are different from place to place.

QUESTIONS

Use complete sentences to write your answers.

1. Describe several causes of tsunamis.
2. Explain why there are usually two high tides and two low tides each day.
3. The moon passed overhead on the seacoast at 9 P.M. Explain why you would expect the highest tide of the day to occur at about this time.
4. How can the behavior of tides be predicted at a place?

SKILL-BUILDING ACTIVITY

AN INFERENCE FROM THE TIDES

PURPOSE: To use water height data and other information for making an inference about the cause of tides. An inference is a conclusion drawn from observations.

MATERIALS:
graph paper   ruler
pencil

PROCEDURE:

A. A measuring tape attached to a pier piling in Boston Harbor was used to make readings of the water level. Table 16-1 shows those readings. Make a graph of the data. Label the vertical axis "Height of Water (m)" and the horizontal axis "Time (hr)."

B. The highest readings are *high tides*. The lowest readings are *low tides*. Look at the graph.
   1. How many hours apart are the high tides? How many hours apart are the low tides?

C. On January 1, the moon had passed overhead a little before the 9 P.M. measurement was taken. On December 31, the moon had passed overhead a little before 8 P.M. The 8 P.M. reading was the highest for the day.
   2. What conclusion (inference) would this suggest?

D. On Jan. 2, at 10 P.M. the moon had passed overhead a little before the measurement was made. The 10 P.M. reading was the highest for January 2.
   3. What role could the moon's gravity play here?

E. Look at the graph again. Around 9 A.M. in Boston, the moon would be on the opposite side of the earth.
   4. Is there a high tide or a low tide in Boston at that time?
   5. Of all the water in the oceans, the water at Boston, and north and south of Boston, would be the farthest from the moon at 9 A.M. Would this increase or decrease the moon's attraction?

CONCLUSIONS:
1. What causes tides?
2. Why are there two high tides in Boston in one day?

HEIGHT OF WATER IN BOSTON HARBOR ON JANUARY 1												
Time	1 A.M.	2 A.M.	3 A.M.	4 A.M.	5 A.M.	6 A.M.	7 A.M.	8 A.M.	9 A.M.	10 A.M.	11 A.M.	12 noon
Height of water (meters)	2.0	1.7	1.6	1.9	2.4	2.9	3.4	3.8	4.0	3.7	3.3	2.7
Time	1 P.M.	2 P.M.	3 P.M.	4 P.M.	5 P.M.	6 P.M.	7 P.M.	8 P.M.	9 P.M.	10 P.M.	11 P.M.	12 mid-night
Height of water (meters)	2.2	1.6	1.2	1.3	1.6	2.1	2.6	3.4	4.6	3.5	3.2	2.8

Table 16-1

CAREERS IN SCIENCE

MARINE TECHNICIAN

Marine technicians can enjoy fascinating careers that are involved with almost every aspect of the oceans. Some technicians work with oceanographers charting the ocean floor. Others work with marine biologists who are developing new sources of food from the sea. Still others participate in the undersea efforts to recover ships and their sunken treasures.

Technicians may be found on board ships or on stationary platforms in the sea. They may work near shipyards and docks or in the laboratories of a marine industry. Technicians are the people who follow through on the plans, check details, and help gather data for a particular project.

To become a technician, you'll need a two-year college degree. Courses in science, math, and drafting are helpful. Jobs are available in private industry and the government. For more information, write: American Society of Limnology and Oceanography, I.S.T. Building, University of Michigan, Ann Arbor, MI 48109.

MARINE GEOLOGIST

The world of the marine geologist is the vast ocean floor. Geologists gather samples of the sediment that lies on the ocean bottom. Then they perform experiments on these samples to find out what they contain. Marine geologists may also explore the ocean floor for oil and mineral deposits. The quality of these deposits is analyzed, and it is the marine geologist who must decide if they can be brought to the surface for human use. A person in this field might also be involved in studying the underwater crust of the earth to help understand our planet's history. Most geological research involves the use of sophisticated technologies and equipment.

To enter the field of marine geology, you must have a bachelor's degree. For high level positions, an advanced degree is also necessary. Jobs are available with oil companies, universities, and the government. For more information, write: American Geological Institute, 5202 Leesburg Pike, Falls Church, VA 22041.

CHAPTER REVIEW

VOCABULARY

On a separate piece of paper, write TRUE next to the number of each statement that is true. Next to the number of each false statement, write FALSE, then make the statement true by writing the correct term in place of the underlined incorrect term.

1. A current is a regular rise and fall of the water level in the sea.
2. A turbidity current is a current caused by the increased density of water that has mixed with sediments.
3. A fetch is a long, regular wave that moves a great distance across the ocean.
4. A swell is a wave whose top has curled forward and crashed down against the shore.
5. The tide is the distance over which wind blows steadily, producing waves.
6. A tidal bore is a large wave that leads a tide upriver.
7. Any water moving in a particular direction is called a breaker.
8. A giant wave caused by a disturbance on the sea floor is called a tsunami.

QUESTIONS

Give brief but complete answers to each of the following questions. Unless otherwise indicated, use complete sentences to write your answers.

1. Give an example of how the sun's energy causes an increase in the density of ocean water and an example of how it causes a decrease in the density of ocean water.
2. Where are the currents of cold water most likely to be found in the ocean? Where are the currents made up of warm water found?
3. Describe a well-known density current found in the Mediterranean Sea. What causes this current?
4. What effect do turbidity currents have on the ocean floor?
5. Why is the general pattern of surface ocean currents circular in the North and South Atlantic?
6. What is the Gulf Stream?
7. Name the current that causes the Gulf Stream. Which two currents does it split into in the North Atlantic?
8. What common effect is caused when the Labrador Current meets the Gulf Stream?

9. Explain how surface currents can affect climates.
10. What is the pattern of motion for water particles near the surface as a wave passes?
11. Describe what happens to the speed, wave height, and wavelength of a wave as it approaches the shore and forms a breaker.
12. Why are tsunamis called "killer waves"?
13. How is a tsunami in the open sea different from one at the shoreline?
14. What is the major cause of tides on earth?
15. Describe a usual cycle of tides during a one-day period. In your answer, include the time periods between high and low tides.

APPLYING SCIENCE

1. What would be the effect on climates near the poles and the equator if the oceans did not circulate?
2. Explain the difference in temperature of the water at the beaches in California and Florida.
3. Explain how the El Niño may affect the air temperature on the west coast of the United States.
4. Refer to Fig. 12–12, page 339. Compare the global wind patterns with the ocean currents shown in this chapter. Make a drawing to show the wind patterns and the ocean currents on a map similar to the map shown in Fig. 16–10 on page 438.

BIBLIOGRAPHY

Britton, Peter. "Waves Challenging Giant Rigs." *Popular Science,* January 1981.

Frances, Peter, and Stephen Self. "The Eruption of Krakatau." *Scientific American,* November 1983.

McKean, Ken, and Mayo Mohs. "Tracking the Killer Waves." *Discover,* August 1983.

Matthews, Samuel W. "New World of the Oceans." *National Geographic,* December 1981.

National Geographic Editors. "The Promise of Ocean Energy." *Special Report on Energy,* February 1981.

Time-Life Editors. "Restless Oceans" (Planet Earth Series). Chicago: Time-Life Books, 1984.

UNIT

5

EARTH AND SPACE

CHAPTER 17

A total solar eclipse occurs when the earth passes through the moon's shadow. In any given place, a total eclipse of the sun is likely to be seen only once in several hundred years.

THE SUN

CHAPTER GOALS

1. Describe the structure of the sun.
2. Explain how the sun changes.
3. Identify the source of the sun's energy.
4. Describe the form in which solar energy is received by the earth

17-1. Structure of the Sun

At the end of this section you will be able to:

- Name and describe the layers of the sun.
- Describe sunspots.
- Explain how disturbances on the sun may affect the earth.

On a dark, clear night in the far north, the sky begins to glow. Green and red patches of light dance across the sky. These flowing curtains of light move in strange shapes as they expand and shrink. For a while, it seems as if the sky is on fire. Then the lights fade away, leaving a faint reddish streak that also gradually fades away. These beautiful "northern and southern lights" are often seen at night near the north and south polar regions. Would you think of the sun as their cause? In this section you will see how these lights are caused when particles from the sun reach all the way to the earth.

LAYERS OF THE SUN

The sun is a huge ball of very hot gas. It would take more than a million earths to make up the same volume as the sun. The sun's density is low because it is made up of gases.

The sun is made up of several layers that blend into one another. Because all the layers are gases, there are no sharp dividing lines between them. Because the sun is so hot, the gases in the sun are mostly a mixture of bare atomic nuclei and free electrons. This matter is called *plasma*. On the earth, plasmas are found only where the temperatures are very high, such as in the flame of a welding torch.

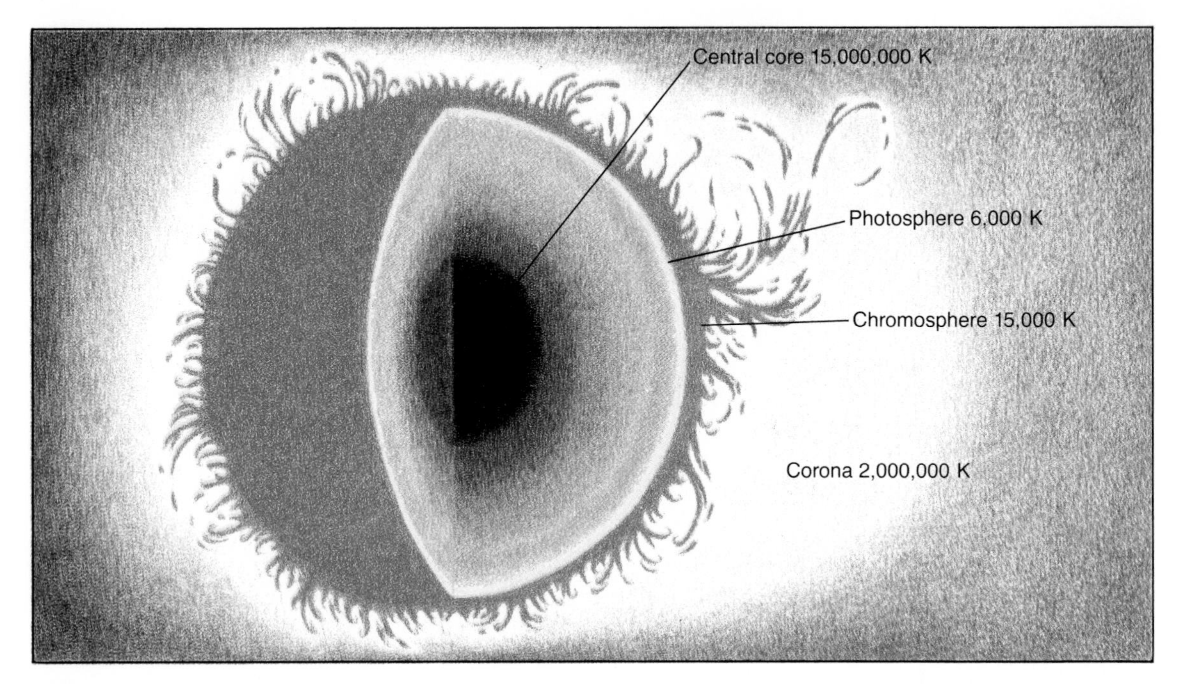

Fig. 17–1 The parts of the sun.

The sun can be divided into four layers: the core, photosphere, chromosphere, and corona. See Fig. 17–1.

1. Core. The changes that produce the sun's huge supply of energy take place in its center, or core. The temperature in the core has never been measured because it is so deep within the sun. Scientists believe that the temperature near the sun's center is at least 15,000,000 degrees on the Kelvin scale. Energy coming from this central power plant is carried outward until it reaches the sun's surface layer.

Photosphere A thin outer layer of gas in the sun, which produces visible light.

2. Photosphere. Energy from the sun's interior reaches the surface. There it causes a thin layer of gases to give off a brilliant light. This glowing layer of gas, called the **photosphere** (**fote**-oh-sfeer), produces the visible light that comes from the sun. The *photosphere* is the part of the sun that we see. CAUTION: DO NOT LOOK DIRECTLY AT THE SUN AT ANY TIME. EVEN A BRIEF LOOK CAN PERMANENTLY DAMAGE YOUR EYES. The photosphere layer has a temperature of about 6,000 degrees on the Kelvin scale.

Chromosphere The layer in the middle of the sun's atmosphere, which gives off a faint red light.

3. Chromosphere. Just above the photosphere are two layers that make up the sun's atmosphere. First is a layer, called the **chromosphere** (**kroe**-moh-sfeer), that gives off a faint red light. This light cannot be seen against the bright background of the photosphere.

4. Corona. The *chromosphere* blends into a much less dense layer of faintly glowing gas. This layer, called the **corona** (kuh-**roe**-nuh), surrounds the sun like a halo. There is no outer boundary for the *corona.* Some of the sun's material escapes completely and streams out to, and past, the earth. This stream of particles from the sun's corona makes up the *solar wind.* The parts of the sun are shown in Fig. 17-1, page 460.

Corona The outermost layer of the sun.

Fig. 17-2 The surface of the sun has a rough appearance.

SUNSPOTS

The gases that make up the sun are always moving. Energy pouring from the sun's core causes storms in the hot gases. Thus the surface of the sun appears to be rough because of these motions. When seen through a special telescope, the visible surface of the sun resembles the peel of an orange. See Fig. 17-2. The most easily seen storms on the sun's outer layers are **sunspots.** *Sunspots* are huge storms that extend from the surface of the sun into its interior. These are areas that appear dark when seen against the bright background of the photosphere. Sunspots form when a part of the sun's surface becomes cooler. This is caused by the blocking of the hot gases that constantly rise from below. The temperature inside a sunspot is about 2,000 degrees on the Kelvin scale cooler than the surrounding surface. This lower temperature makes it appear darker than the bright photosphere. A large sunspot may be five times as large as the earth's diameter. Some large groups of sunspots last for several weeks. Sometimes the same sunspots are seen for months. Small sunspots usually last for a few days to a week.

Sunspots Areas on the sun's surface that appear dark.

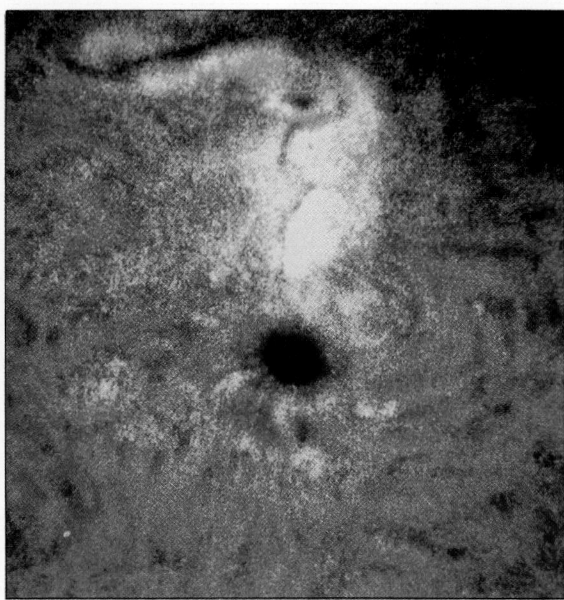

Fig. 17-3 (left) *Sunspots can usually be seen on the sun's surface.* (right) *Prominences are great clouds of glowing gas.*

Most sunspots are seen in groups close to the sun's equator. See Fig. 17-3 *(left)*. As the sun rotates, the sunspots appear to move across the face of the sun. Sunspots circle around the sun and return to their original position in about 27 days. Thus the sun appears to rotate on its axis about once every 27 days. However, because the sun is made of gas, not all of it turns at the same speed. Places near the sun's equator move faster than places near the poles. Separate spots and groups of spots may change their shapes as they rotate with the sun.

Prominence A huge cloud of gases that shoot out from areas of sunspots.

Huge clouds of bright gases, called **prominences (prom-**ih-nen-ses), shoot out from the sun's surface in the areas of sunspots. A *prominence* may extend thousands of kilometers above the sun's surface. A prominence can be seen in Fig. 17-3 *(right)*. Prominences often form a huge arch between pairs of sunspots. Sometimes giant fountains of bright gas suddenly shoot up from the sun's surface. These are solar flares. They usually appear as explosions near sunspots.

DO SUNSPOTS AFFECT THE EARTH'S WEATHER?

Observations of sunspots over many years show a general pattern when they appear. See Fig. 17-4. The number of sunspots increases every few years. After reaching a peak, the number of sunspots decreases. Then the cycle begins again. There is some evidence that the sunspot cycle may affect the

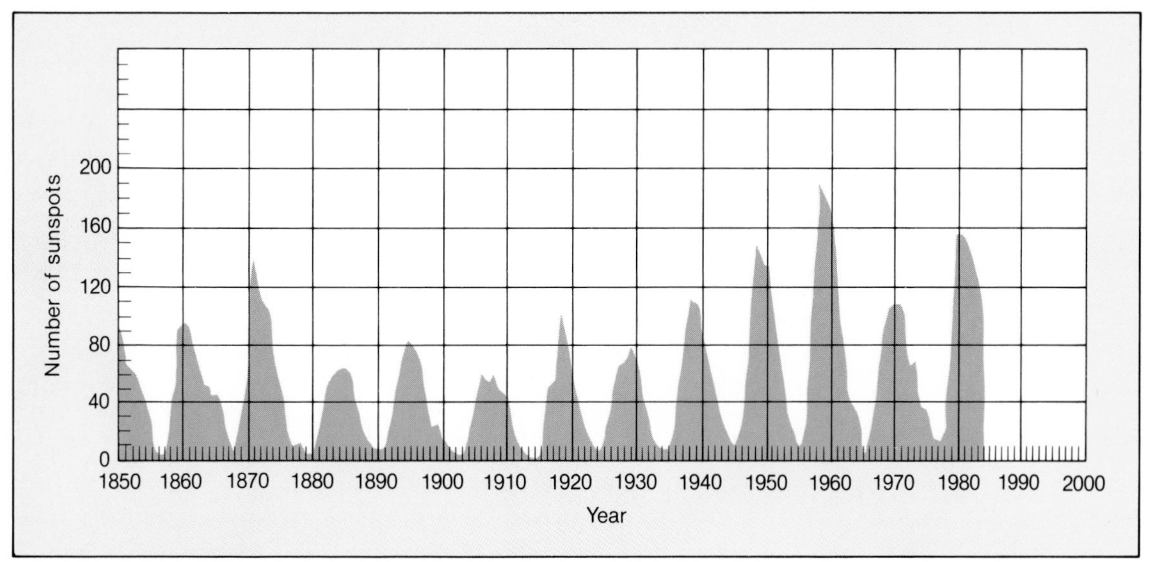

Fig. 17-4 The number of sunspots seen each year since 1850 is shown on the graph.

earth's environment. For example, the study of many tree rings shows a pattern very much like the sunspot cycle. During sunspot cycles, the space between the rings changes from the normal pattern. The distance between tree rings shows how fast the tree has grown each year. See Fig. 17-5. Each ring represents one year of growth. During years when there is abundant rain, the tree grows rapidly. The distances between rings become larger. In dry years, the tree grows less. The distances between rings are smaller. Thus the distances between the tree rings are a record of wet and dry periods. This could mean that sunspots have some effect on the earth's weather by changing the amount of rain that falls in some places.

Fig. 17-5 During sunspot cycles, the space between the rings, indicating tree growth, varies from the normal pattern.

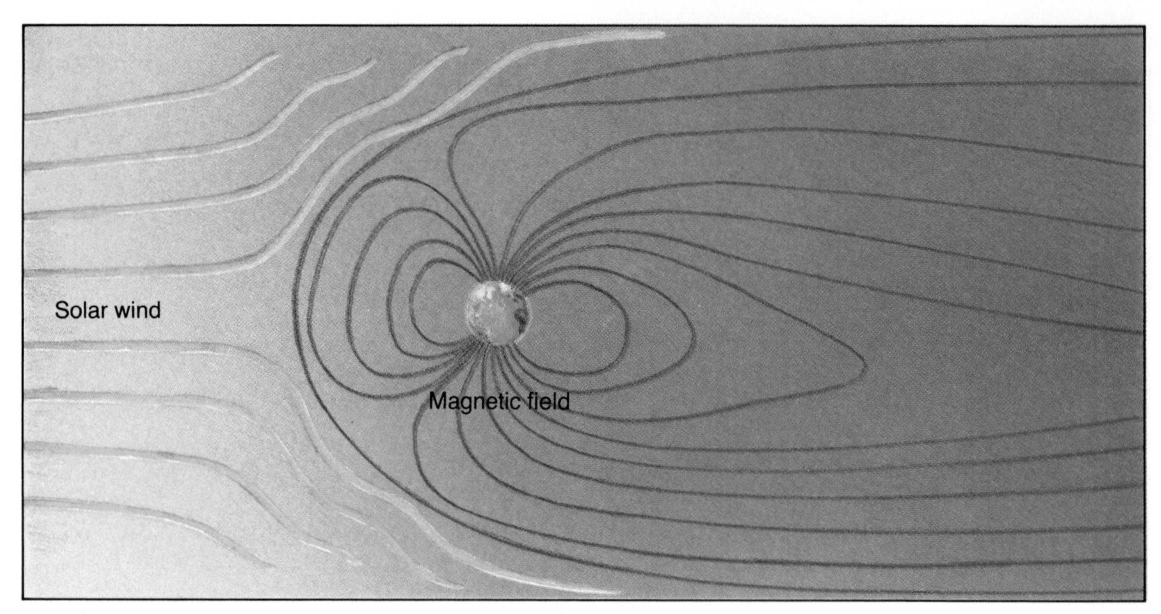

Fig. 17-6 The earth's magnetic field causes the solar wind to flow around the earth.

SOLAR WIND

The space between the planets and the sun is not empty. Solar flares may cause gases from the sun's corona to be thrown out into space. These gases are made of electrically charged particles, and they move out through the solar system as solar wind. See Fig. 17-6. The solar wind moves past the earth at speeds of 300 to 700 kilometers per second. This is much faster than the speed of a moving bullet. The solar wind could be a danger to living things. However, we have two shields that protect us, the atmosphere and a magnetic field. The earth's atmosphere is able to absorb most of the solar wind that reaches the earth. Also important is a magnetic field that surrounds the earth far out into space.

The earth acts like a giant magnet. This can be shown by holding a bar-shaped magnet so that it can turn freely. It will slowly turn and come to rest pointing north and south. A needle in a compass acts in the same way. The magnetic effect may be caused by slow movements of hot materials within the core and mantle. Like all magnets, the earth has two opposite magnetic poles, called north and south. However, these are not the same places as the geographic North and South poles, which are on the earth's axis. But compass needles point generally north and south. This is because the magnetic poles are near the geographic North and South poles. See Fig. 17-7.

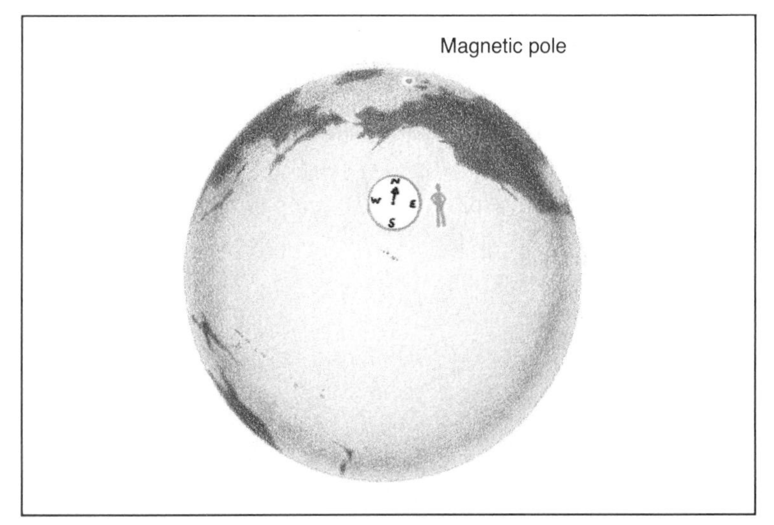

Fig. 17-7 The latitude of the North Pole is 90° N. The magnetic North Pole is at 76.2° N and 101° W. When electrically charged particles come near the earth, they are guided to the earth's magnetic poles. The particles strike the gases in our atmosphere, causing auroras.

The earth's magnetism pushes the solar wind to either side as it reaches the earth. The solar wind flows around the earth's magnetic field, like a river flowing around an island, as you can see in Fig. 17–6. Some of the charged particles pass through the earth's magnetic field and become trapped. When these charged particles strike the gases in the earth's atmosphere, light is given off. These greenish-white and blue lights are seen as **auroras,** or northern and southern lights. See Fig. 17–8. *Auroras* are most commonly seen near the poles. This is because the particles are guided by the magnetic field toward the North and South poles. Sometimes, they are seen in places closer to the equator. This appears to be caused when solar flares form on the sun's surface.

Auroras The northern and the southern lights produced when charged particles from the sun are trapped in the earth's magnetic field.

Fig. 17-8 The northern lights as seen in Alaska.

Solar flares appear quickly, like the explosion of a pool of gasoline. Prominences may suddenly rise upward from the location of the flare. Then the flare quickly fades away. However, the burst of energy from the solar flare may blow out a part of the sun's atmosphere into space. This causes a sudden increase in the solar wind. Several days after the solar flare is observed, its effects may be felt when the strong solar wind reaches the earth. The increased solar wind can cause a solar storm in the earth's atmosphere.

A solar storm can effect the earth's atmosphere in three ways. First, x-rays from the solar wind interfere with high frequency radio waves. This can cause radio blackouts and scramble signals from satellites. A second effect is caused by protons and electrons in the solar wind. These particles may push through the earth's magnetic field and produce auroras. If the solar wind is very powerful, a third effect takes the form of sudden increases in electrical power in power lines. This can produce interruption in the supply of electricity in some areas.

SUMMARY

The sun is a glowing sphere of gas made up of four layers. The energy pouring from the sun's core causes storms in the hot gases. The storms that occur on the sun's outer layers are seen as sunspots, which appear as dark areas on the bright surface. The number of sunspots increases and decreases in cycles every few years. Sunspots may have some influence on the earth's weather and thus have a possible effect on the growth of plants. A solar wind moves out from the sun through the solar system. The earth's magnetic field causes the solar wind to flow around the earth. Sometimes part of the solar wind is trapped by the earth's magnetic field and produces auroras and other effects in the earth's atmosphere.

QUESTIONS

Use complete sentences to write your answers.

1. Name and describe the four layers of the sun.
2. Describe sunspots and how they form on the sun.
3. Name two other disturbances on the sun's surface that appear related to sunspots.
4. What effect does the solar wind have on the earth?

SKILL-BUILDING ACTIVITY

DOES THE SUN ROTATE?

PURPOSE: To use problem solving skills in finding the answer to the question, "Does the sun rotate?"

MATERIALS:
paper pencil

PROCEDURE:

A. We must look at the sun from our position in space and tell whether or not it rotates. Often the scientist uses another example as a model to help find an answer to a problem.
  1. If you looked at earth from space, how could you tell that the earth rotates?
  2. Suppose you noticed that a certain feature on the earth's surface returned to view every 24 hours. What would you conclude?

B. The sun's surface is so hot that everything there exists as a gas. Usually such a surface would give no marks to watch for evidence of rotation. There are certain times, however, when gigantic sunspots can be seen to move across the sun's surface. Look at Fig. 17-9, in which the bottom photo was taken two days after the top photo.
  3. Do the sunspots appear to move together or independently?
  4. The movement shown by the photos is $\frac{1}{14}$ the distance around the sun. How long did this movement take?
  5. How many days would it take the sunspots to move around the sun?

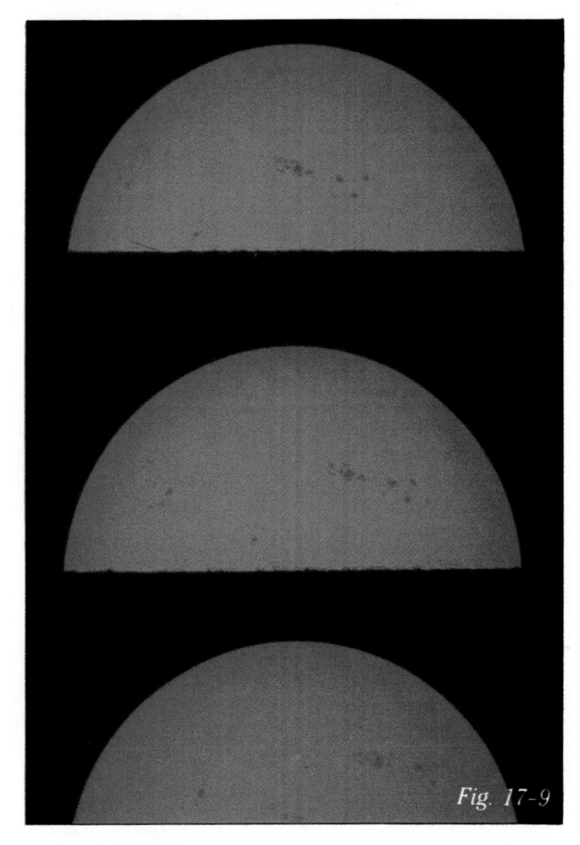

Fig. 17-9

  6. Assume the sunspots do not move on their own but are carried around by the rotation of the sun. How many days does it take for the sun to rotate once?

CONCLUSIONS:

1. How can you tell that the sun rotates?
2. Explain how you determined the length of time it takes the sun to make one complete rotation.
3. What assumption had to be made about sunspots in order to answer question 6?

17-2. The Sun's Energy

At the end of this section you will be able to:

- Identify the source of the sun's energy.
- Describe the energy that leaves the sun.
- Name the kinds of *radiant energy* received by the earth.

Each second, the sun gives off an amount of energy equal to 200 billion hydrogen bombs. How the solar furnace works remained a mystery for a very long time. However, today scientists not only understand how the sun works, but can even copy the way it makes energy. See Fig. 17-10.

Fig. 17-10 An H-bomb uses the same source of energy as the sun.

WHAT MAKES THE SUN SHINE?

Think about how much energy the sun gives off. Each second, the sun produces as much energy as the United States would use in three million years. And it has been giving off about this same amount of energy for at least the last four to five billion years. What can be the source of this marvelous power?

The most simple hypothesis would be that the sun gets its energy from fire. In other words, the sun might be hot and bright because something is being burned. However, this would be very unlikely. For example, suppose that the sun could be made of coal. Astronomers know that the sun has a mass of 1.8×10^{30} kilograms. An amount of coal equal to the mass of the sun would last only about 6,000 years. Clearly, the source of the sun's energy cannot be anything like ordinary burning.

Then, in 1835, a German scientist named Hermann von-Helmholtz developed a theory that, for a while, seemed to work. Helmholtz said that the sun's great mass caused its outer layers to be pulled in by gravity. Thus the sun was constantly shrinking. The shrinking caused the gases inside the sun to be squeezed and become heated. The shrinking of the sun would then cause it to get very hot and to shine. Helmholtz found that the sun would need to shrink only about 50 meters each year in order to produce its energy. This is a very small part of the sun's diameter of 1.4 million kilometers. However, there was one problem with this theory. Even when shrinking so slowly, the sun must have been very much larger in the past. Only 25 million years ago it would have been so big that the earth would have been inside the sun. Thus the earth could not have been older than about 25 million years. However, geologists who measure the rate of decay of radioactive elements in rocks have gathered evidence that shows the earth to be at least 4.5 billion years old. Since the earth cannot be older than the sun, this means that the sun must have been shining for at least several billion years. By the end of the 18th century, most astronomers realized that the age of the sun according to the Helmholtz's theory could not be correct. Another explanation was needed.

A clue to the mystery was provided by Albert Einstein in 1905. Einstein said that matter could be changed into energy. This principle is expressed by the famous equation: $E = mc^2$, where (E) is energy, (m) is mass, or the amount of matter changed, and (c) is the speed of light. The equation states that a small mass can be turned into a large amount of energy. Thus, the sun could produce great amounts of energy by losing some of its mass. Each second the sun changes more than 600 million tons of hydrogen nuclei into helium nuclei, with a loss of mass. Other kinds of nuclear fusion reactions also help supply the sun's energy. For example, the nuclei of carbon, nitrogen, and oxygen can be involved in a complicated reaction that also produces helium. Because the sun is so huge, it would take 100 trillion trillion years to use up its entire mass. Now scientists could build a theory to explain how the sun produces energy. Further scientific observations provided the details needed to make a complete theory.

THE SOLAR FURNACE

The sun is made up of an enormous amount of gases. Hydrogen gas makes up 74 percent of the sun's mass. Helium makes up 25 percent. The remaining 1 percent is probably made up of small amounts of all the other known chemical elements. All the matter in the sun presses down on its center with a huge force. This creates a pressure within the sun's core that is nearly one billion times the air pressure at the earth's surface. The gases near the sun's center are squeezed together so strongly that they are changed.

The high temperatures in the core cause the atoms in the gases to lose their electrons. Only the nuclei are left. The very high pressures and temperatures cause some of these atomic nuclei to join together. This process is called *fusion*. In the sun, fusion joins four hydrogen nuclei into a helium nucleus plus two hydrogen nuclei. The helium and hydrogen that are produced have a little less mass than all the four hydrogen nuclei had originally. See Fig. 17–11. The small amount of mass that disappears is changed into energy. Smaller amounts of other kinds of atomic nuclei are also changed. These nuclei also produce helium along with energy.

Each second, the sun changes about 4.2 billion kilograms of hydrogen into energy. There is enough hydrogen in the sun to supply energy at the present rate for at least five billion years.

Scientists now believe that the puzzle of the sun's energy has been solved. The answer is in the constant fusion of atomic nuclei, within the sun's core, that changes mass into energy.

Scientists have duplicated reactions similar to those that occur on the sun. One place is in the hydrogen or H-bomb. In order to reach the high temperatures needed to start the fusion reaction, an atomic bomb is used first. Another place where scientists are duplicating fusion is in fusion reactors. Here, scientists are attempting to release energy in controlled amounts. The major problem has been finding a way to contain the very high temperatures needed for fusion. At present, fusion reactors have been able to contain the high temperatures for only thousandths of a second. Someday scientists hope to find a way for producing energy from a fusion reactor continuously. This would be similar to the reactions that cause the sun to shine. The development of such a method would supply much of humankind's future need for energy.

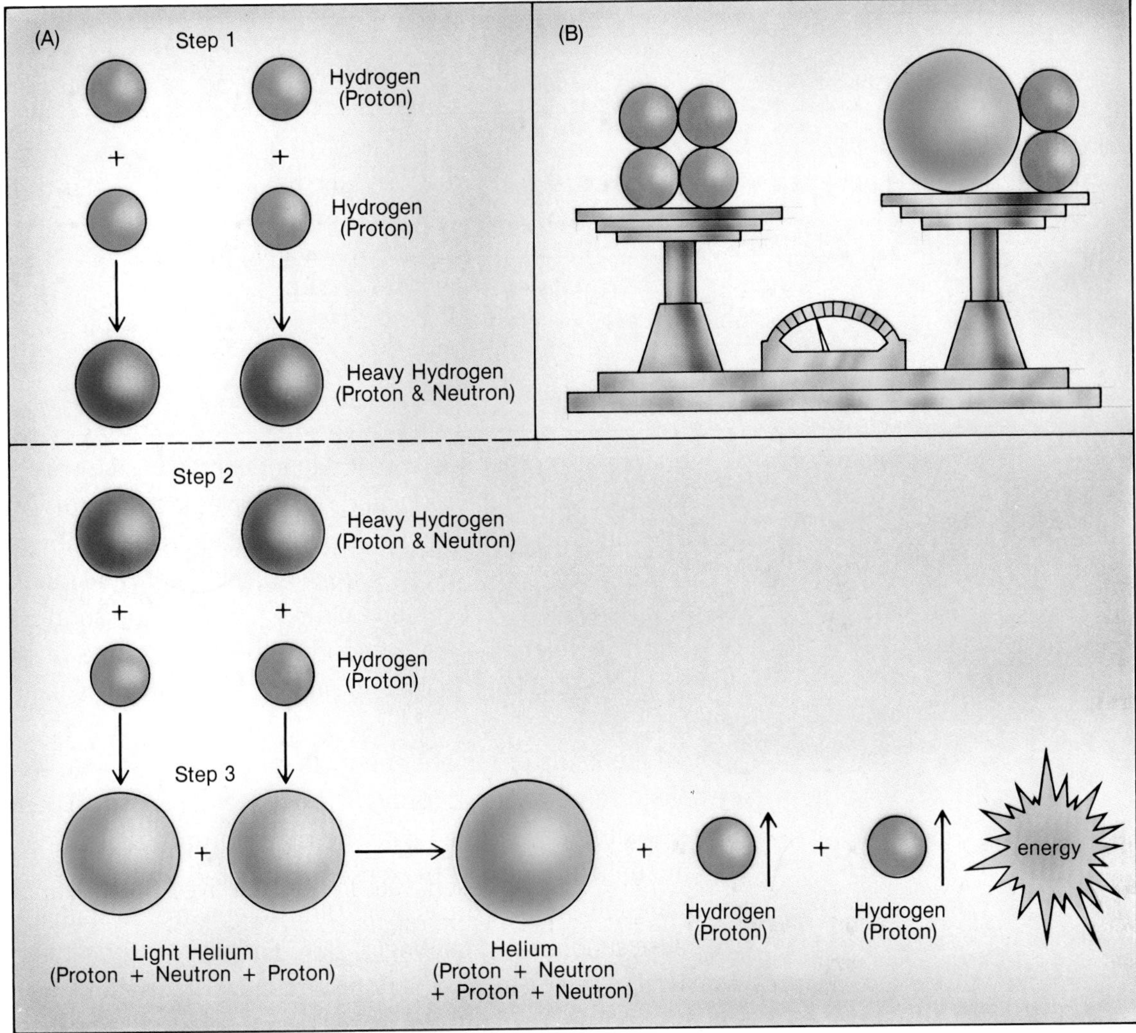

Fig. 17-11 (A) The three steps by which hydrogen nuclei combine in the sun to form helium. Step 1: Four hydrogen nuclei form two heavy hydrogen nuclei. Step 2: Each heavy hydrogen nucleus combines with another hydrogen nucleus to form a light helium. Step 3: Two light helium nuclei form a normal helium nucleus and two hydrogen nuclei. (B) At the end, the total mass is less than the total mass at the beginning.

SUNLIGHT

The processes that supply the sun's energy take place deep inside the sun. Here the temperatures are high enough for nuclear changes to occur. Energy produced inside the sun then makes its way out toward the surface. The gases near the surface absorb this energy and become hot. This causes the gases in the sun's outer layers to give off *radiant energy*. Radiant energy is energy that travels through space in the form of waves. All of the many forms of radiant energy can move through space at the very high speed of 300,000 kilometers per second. Radiant energy can travel from the sun to the earth in about eight minutes.

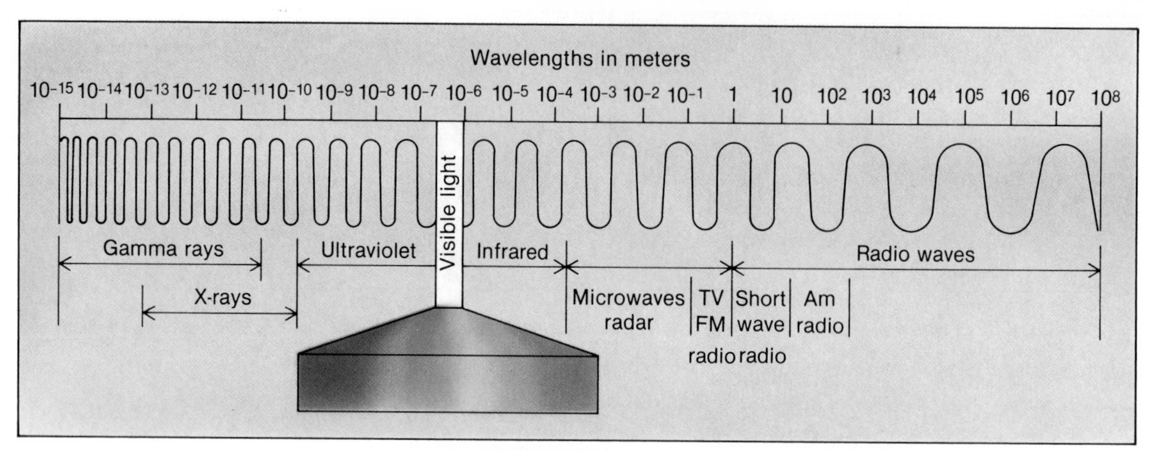

Fig. 17–12 The spectrum of radiant energy.

There are also other sources of radiant energy on the earth. For example, radio, television, and radar waves are all forms of radiant energy. The different kinds of radiant energy are shown in Fig. 17–12. The longest waves are radio, television, and radar waves. Infrared, visible light, and ultraviolet waves are shorter. Very short wavelengths are found in X-rays and gamma rays. Notice that visible light makes up only a small part of radiant energy.

Of all the kinds of radiant energy that make up sunlight, we are able to sense only two forms. The eye sees a very small part of the spectrum as visible light. The other form of radiant energy in sunlight that can be sensed is infrared waves. These waves are felt on the skin as heat. Ultraviolet waves in sunlight can cause sunburn and tanning of skin. Too much exposure to ultraviolet waves can also seriously damage your skin. Prolonged exposure to ultraviolet waves over a period of years has also been linked with skin cancer.

Most of the radiant energy leaving the sun is in the form of visible light and infrared rays. Together they make up more than 90 percent of the energy sent out by the sun. Almost all of the remaining 10 percent is made up of ultraviolet rays, X-rays, and gamma rays. Fortunately for all the living things on earth, not every kind of radiant energy given off by the sun reaches the earth's surface in large amounts. The gases in the upper parts of the earth's atmosphere absorb X-rays, gamma rays, and most of the ultraviolet rays. For example, ozone, which is a form of oxygen, is responsible for blocking out most of the ultraviolet rays. For this reason, the small amount of ozone gas in the atmosphere is very important. See Fig. 17–13.

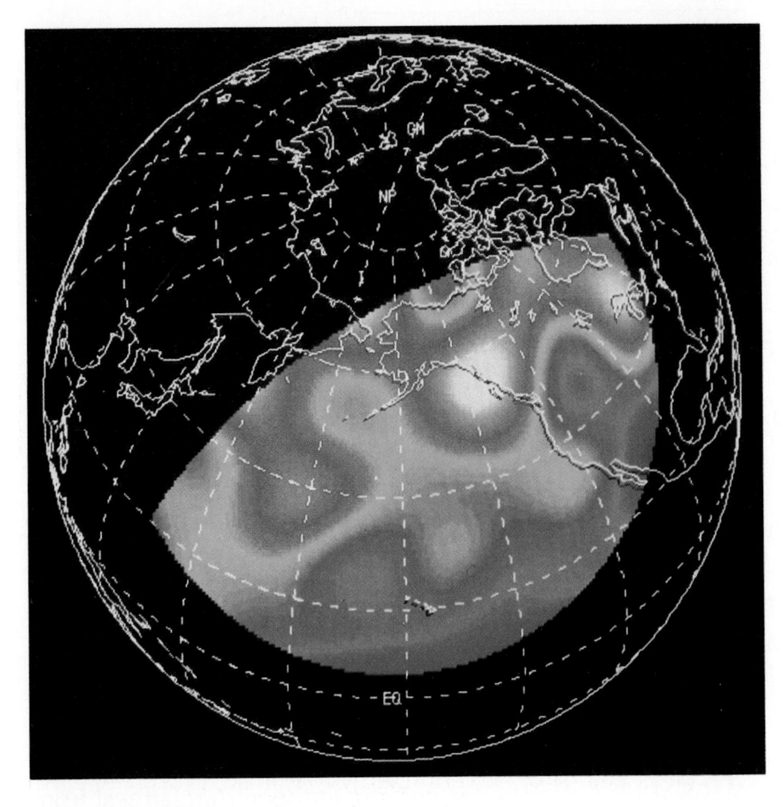

Fig. 17-13 Ozone gas on earth filters out some of the sun's harmful radiation.

SUMMARY

The secret of the sun's energy lies deep within its core. The high temperature and pressure cause separate hydrogen nuclei to join together. In this process, a small amount of the mass of the hydrogen is changed into energy. Energy from the core of the sun moves out into space. By using its hydrogen as fuel, the sun will be able to produce energy for billions of years. The earth receives this radiant energy from the sun, mostly in the form of visible light and infrared waves.

QUESTIONS

Use complete sentences to write your answers.

1. What is the source of the sun's energy?
2. What is radiant energy?
3. Name and give the percentages of the kinds of radiant energy sent out by the sun.
4. Which kinds of radiant energy are absorbed by the earth's atmosphere?

INVESTIGATION

SEPARATING LIGHT

PURPOSE: To separate light into its various colors.

MATERIALS:

diffraction grating	pencil
light bulb	paper

PROCEDURE:

A. A diffraction (dif-**rack**-shun) grating is a piece of plastic having thousands of parallel lines molded into its surface. The lines can easily be damaged by fingerprints and scratches. Hold the diffraction grating to your eye and look through it at the light bulb. See Fig. 17–14. CAUTION: NEVER LOOK AT THE SUN. YOU MAY DAMAGE YOUR EYES. You should be able to see colors spread out to the right or left. If not, turn the grating a quarter of a turn so that the colors are on the right and left sides.
   1. List the colors you see.
   2. Where in nature have you seen this set of colors before?

Fig. 17-14

B. The colors you see are the same colors in the same order that you would see if you looked at a rainbow. (DO NOT LOOK AT THE SUN!) This is the visible light spectrum. Look at Fig. 17–12 on page 472.
   3. Estimate the shortest wavelength given for the visible light spectrum.

C. The shortest wavelength of visible light will be the color nearest the light bulb. Use the grating to look at the light bulb.
   4. Which color is the one with the shortest wavelength?
   5. As you look farther away from the bulb, past the full spectrum, what do you see?

D. The longest wavelength of visible light will be the color in the spectrum farthest from the light bulb. Use the grating to look at the light bulb.
   6. Which color is the one with the longest wavelength?

CONCLUSIONS:

1. List the colors of the visible light spectrum, starting with the color of the shortest wavelength.
2. What is the shortest wavelength of light that can be seen? What is the longest wavelength of light that can be seen?
3. Name the kind of radiant energy that is just outside of the visible spectrum on the short wavelength end.
4. Name the kind of radiant energy that is just outside the visible spectrum on the long wavelength end.

TECHNOLOGY

EXPLORING SPACE: THE HUBBLE SPACE TELESCOPE

Called both the most important astronomical happening for the 20th century and perhaps the most difficult and demanding project ever undertaken by NASA, the Hubble space telescope, is now scheduled to be launched in 1986. The telescope was appropriately named after the American astronomer Edwin P. Hubble (1889–1953). Hubble's research demonstrated the existence of nebulae outside our galaxy. His work, resulting in what is now called Hubble's Law, led to an estimate of the size of the universe. He is credited with having begun the study of the universe beyond our galaxy. The Hubble space telescope will continue and extend this research. We will see what we only thought to exist before.

This massive instrument, which weighs 11,339 kilograms and is 13.4 meters long and 4.3 meters in diameter, will be placed in an orbit 515 kilometers above the earth. From that point, the telescope will be looking at objects that are 14 billion light years away. It will actually be looking back in time at stars just forming! With the information sent back by this instrument, scientists hope to determine how our galaxy was formed. The images from space will be collected with the mirror of the telescope, approximately 2.4 meters in diameter. Light will be fed into five instruments, including two new cameras—one a wide-field planetary camera, the other a faint-object camera. The wide-field camera is expected to be the most used instrument. Each instrument and camera has been designed especially for the space telescope. Each component must be tested thoroughly to make certain that it will withstand the force and vibrations of lift-off as well as the low temperature of space. Since the objects being viewed are so far away from the telescope, it will take about 24 hours to collect a single complete image. For that 24 hours, there can be no movement of the device, or the image will be distorted.

The Hubble space telescope is expected to operate for fifteen years. To make this possible, the space telescope has been designed so that the instruments and other parts can be repaired or replaced by astronauts, or be brought back to earth by a shuttle for major repairs. The telescope will be visited every five years for this purpose.

When finally launched and placed in orbit, the Hubble telescope is certain to have an effect on the way we look at the universe. Many astronomers believe that the most important discoveries will be unexpected, just as the discovery of the moons of Jupiter was unexpected by Galileo when he first aimed a telescope at the planets. We know it will give us information about 27th and 28th magnitude stars and the formation of galaxies. Scientists also expect to learn more about quasars and black holes. And astrophysicists may find data that could verify one of their GUTs, Grand Unified Theories. GUTs are theories which try to explain the various particles of matter and their behaviors with one overall hypothesis. What other questions do you think they might be able to answer by seeing into the outer reaches of space?

¡COMPUTE!

SCIENCE CONCEPT

The sun is vital to our existence. It is essential to the process of photosynthesis, which provides oxygen and food. It provides heat and, of course, it provides light. Because of the earth's revolutions, the sun's light and heat affect different parts of the globe differently. As you learned in the beginning of this text, the earth's tilt causes a change in the direction of the sun's perpendicular rays. This process creates the change in seasons. Although you can't see or feel these motions, you can observe them in changes in shadows. At any given time, a shadow's length will differ as you move north or south. It will also differ from day to day at the same location. Studying data on the length of shadows in different places at different times gives us information about the earth's movement around the sun.

COMPUTER INPUT

If a cycle of events occurs on a regular basis, it can be written in a mathematical form. Once information is in that form, it can be entered into a computer. The computer is designed to perform many calculations quickly. This is especially helpful when you need to make the same calculation over and over, changing just one or two parts of the mathematical formula each time. You could program all the factors into the computer and come out with the figures in a matter of seconds. In this case, we will program the computer to produce data about different shadow lengths based on the mathematical relationships between the earth and the sun.

WHAT TO DO

Before entering Program Shadow Moves, record your answers to the following questions on a separate piece of paper.

A. You are standing at 23.5 degrees north latitude, looking at an upright stick ten centimeters in length.
   1. On what day is the stick's shadow the longest? On what day is it the shortest?
   2. In what months does the shadow get longer? In what months does it get shorter?

B. Would your answers to the preceding questions be different if you were standing at 40 degrees north latitude or at 60 degrees north latitude?

Enter Program Shadow Moves. The computer will ask you for the latitude of your location and the length of the stick. (This program will perform the correct mathematical calculations for latitudes between 23.5 and 60 degrees north of the equator.) When these are entered, the computer will calculate the length of the shadow at that latitude on the 21st of each month. Each time you run the program, record the data. (See the sample data chart.) Compare the computer's data output with your responses to the questions above. Record the data for at least five different latitudes. Plot your data in graphs to see the changes in shadow length more easily. Try to draw conclusions about the relationship between the different variables involved. For instance, if you keep stick length and time (month and day) the same, what happens to shadow length as you go north? If you keep location (latitude) and stick length the same, what happens as you go from winter to summer?

SAMPLE DATA CHART

Latitude	Month	Shadow Length
23.5° N	Jan.	
	Feb.	
	Mar.	
	Etc.	

THE SUN: PROGRAM SHADOW MOVES

GLOSSARY

GOTO	A command in BASIC that instructs the computer to return to a previous statement or to go to another part of the program and follow the directions there.
PERIPHERALS	Computer accessories. A peripheral might be a printer, a joystick, the TV screen, or the keyboard. Everything except the actual electronic circuits are peripherals.

PROGRAM

```
100  REM SHADOW MOVES
110  DIM A(12), M$(12)
120  FOR X = 1 TO 12
130  READ M$(X), A(X)
140  NEXT
150  INPUT "WHAT IS THE LATITUDE
     (23.5 - 66)? ";LA$
160  LA = VAL(LA$)
170  IF LA < 23.5 OR LA > 66 THEN
     GOTO 150
180  INPUT "HOW LONG IS THE STICK?
     ";ST
190  PRINT "MONTH"; TAB(15)"SHADOW
     LENGTH"
200  FOR X = 1 TO 12
210  A = 90 - (LA + A(X))
220  AD = TAN (ABS (A / 57.3))
230  PRINT M$(X); TAB(15) INT((ST / AD)
     * 10) / 10
240  NEXT
250  PRINT
260  INPUT "REPEAT Y/N" A$
270  IF A$ = "Y" GOTO 150
275  END
300  DATA  JAN 21, 15.9, FEB 21, 7.8,
     MAR 21, 0
310  DATA  APR 21, -7.8, MAY 21,
     -15.9, JUN 21, -23.5
320  DATA  JUL 21, -15.9, AUG 21, -7.8,
     SEP 21, 0
330  DATA  OCT 21, 7.8, NOV 21, 15.9,
     DEC 21, 23.5
```

PROGRAM NOTES

REM statements, written into the program, are useful since they alert the user to the purposes of each section. In Program Shadow Moves, there is only one REM statement. Several others could be added to clarify the steps. Can you write REM statements to enter before lines 150, 179, and 300 that would let the user know what kind of step is coming up next?

BITS OF INFORMATION

Computer learning takes many forms. For instance, a company in Palo Alto, California, has designed a system called Dial-A-Drill. Subscribers to the system call, using a pushbutton phone, and a computerized voice asks questions in math or spelling. Answers are given by pressing the buttons on the phone. Dial-A-Drill keeps track of your right and wrong answers.

CHAPTER REVIEW

VOCABULARY REVIEW

On a separate piece of paper, match the number of the statement with the term that best completes it. Use each term only once.

photosphere	corona	aurora	solar flare
chromosphere	sunspot	prominence	radiant energy

1. The ______ is the outermost layer of the sun.
2. A dark, cooler area on the sun's surface is called a(n) ______.
3. The kind of energy given off by the sun that travels through space as waves is called ______.
4. A large arc of glowing gas seen on the sun's surface is known as a solar ______.
5. The ______ is the layer of the sun that we can see.
6. The layer of the sun that gives off a reddish glow is called the ______.
7. A(n) ______ may cause an interference of radio communications on earth.
8. The northern or southern lights are called a(n) ______.

QUESTIONS

Give brief but complete answers to the following questions. Unless otherwise indicated, use complete sentences to write your answers.

1. Compare the size, mass, and density of the sun with that of the earth.
2. What kind of matter is found in the sun?
3. Which layer of the sun produces the visible light?
4. Describe the corona of the sun.
5. Why should you never look directly at the sun?
6. What relationship exists between tree rings and sunspots?
7. How are we protected from the dangers of the solar wind?
8. In what ways does a solar storm affect the earth?
9. Explain why the sun could not be making its energy by a common burning process such as burning of coal.
10. Describe the fusion process by which the sun makes its energy.
11. Of what importance is the equation $E = mc^2$ in the way the sun makes its energy?
12. What causes the sun to give off radiant energy?
13. What kinds of radiant energy are commonly made on earth?

14. Identify the kinds of radiant energy, found in the spectrum shown in Fig. 17–12, page 472, that we can either see or feel.

APPLYING SCIENCE

1. Cut a two-centimeter-square hole in a piece of cardboard. Cover the hole by taping a piece of aluminum foil over it. With a sharp needle, punch a neat hole in the aluminum near its center. Use this device to cast an image of the sun onto a piece of white paper. CAUTION: NEVER LOOK DIRECTLY AT THE SUN. EYE DAMAGE MAY RESULT! Shading the paper by placing it in a cardboard box will make the image show up better. Adjust the distance between pinhole and paper screen to give as large an image as possible. Locate any visible sunspots and make a drawing of them. Watch the sunspots for several days and note any changes. Calculate the period of rotation of the sun based on your observations.
2. One kilogram of wood will supply about 12 million (1.2×10^7) joules of energy if burned. A joule is a unit of energy. Using $E = mc^2$, calculate the amount of energy given off if the one kilogram were converted completely into energy ($c = 3 \times 10^8$ m/sec, 300 million m/s). Compare this to the amount of evergy given off when one kilogram of wood is burned.
3. In direct sunlight, use a fine spray of water from a hose nozzle to produce a rainbow. Move the spray with respect to the sun to make the best rainbow. Make a drawing to show the location of the nozzle, sun, and rainbow. Relate the arrangement of the rainbow colors to the spectrum shown in Fig. 17–12.
4. The equation "speed = wavelength × frequency" relates these three properties of waves. Find out the frequency (Hz/s) of your favorite radio station and use the equation to calculate the wavelength of the radio waves. Use the speed of radiant energy given in the text. Your answer will be in kilometers. Convert it to meters and relate this to the wavelengths given in the spectrum of Fig. 17–12.

BIBLIOGRAPHY

French, Bevin M., and S. P. Maran. *A Meeting with the Universe*. Pasadena, CA: Planetary Society, 1984.

Kidder, Tracy. "A Blemished Sun?" *Science 81*, July/August 1981.

Lampton, Christopher. *The Sun*. New York: Franklin Watts, 1982.

Willamson, Samuel J., and Herman Z. Cummins. *Light and Color in Nature and Art*. New York: Wiley, 1983.

CHAPTER

18

The moon is seen as it enters the earth's shadow (right) *and as it leaves the shadow. The next lunar eclipse visible in parts of North America will occur August 16, 1989.*

SATELLITES OF THE SUN

CHAPTER GOALS

1. Compare the characteristics of the inner planets with those of the outer planets.
2. List and describe the smaller members of the solar system.
3. Describe the moon's features and its history.
4. Explain the motions of the earth–moon system as it revolves around the sun.

18–1. The Inner Planets

At the end of this section you will be able to:

- ☐ Explain why the solar system contains two general kinds of planets.
- ☐ Compare the sizes of the three planets closest to the earth.
- ☐ Compare the conditions on the surfaces of Mercury, Venus, Earth, and Mars.

Suppose that observers from outer space have been watching the earth since it was formed. For more than four billion years, they would not have seen anything leave the earth. Then suddenly, starting about thirty years ago, they would have noticed that the people of the earth began to send off tiny spacecraft. At first, the spacecraft only flew around the earth. Then they landed on the moon. Later, the observers would have seen other spacecraft fly to the earth's neighboring planets. The observers might conclude that the people of the earth had begun a new age of exploring.

THE SOLAR SYSTEM

The solar system is made up of the sun and billions of smaller bodies. Each of these bodies follows an orbit around the sun. The orbits are not perfect circles. Instead, the orbits have the shape of an **ellipse** (eh-**lips**). An *ellipse* is like a circle that has been flattened into an oval shape. An egg has the shape of an ellipse. However, many of the orbits, such as the earth's, are only slightly elliptical.

Ellipse The oval shape of the orbits followed by members of the solar system.

Fig. 18-1 The planets compared with each other and the sun.

The bodies of the solar system move in orbits. They are held by the force of gravity from the huge mass of the sun. In fact, the sun makes up more than 99 percent of the solar system's mass. The sun is the controlling body of the solar system. The nine planets are the best known and, in most cases, the largest other bodies of the solar system. The planets can be separated into two major groups, the inner and outer planets. The inner group of planets includes Mercury, Venus, Earth, and Mars. These planets are the smaller ones, consisting of solid rock with metal cores. The four outer planets include Jupiter, Saturn, Uranus, and Neptune. These are giant planets. They are made of ices and gases, with small solid cores. See Fig. 18-1. Pluto is the farthest planet from the sun. It is very small and seems to be different from all the other planets.

Scientists believe that the way the solar system was formed explains the differences between the inner and outer planets. The inner planets are made up mostly of solid rock and metals. They were formed close to the sun's heat and had their lighter materials driven off. Materials such as stone and metals were not easily driven off. On the other hand, the outer planets are made up mostly of ices and gases. They were formed farthest from the sun's heat and held on to the lighter materials. Also, the large masses and strong gravity of the outer planets help to hold the lighter gases. The materials that formed the ices and gases now make up most of the outer planets. Since the earth is one of the small inner planets, its history must be similar to that of its closest neighbors. Thus the spacecraft that explore these planets give scientists a better understanding of the earth. We will take a closer look at each of the planets that formed in the same way that the earth did.

Fig. 18-2 The surface of Mercury photographed from a spacecraft.

MERCURY

Mercury (**mur**-kyuh-ree) is slightly bigger than the moon but much smaller than the earth. Mercury is not easily seen in the night sky because it is so close to the sun. However, just after the sun sets at night, Mercury may be seen near the horizon. At dawn, it rises shortly before the sun comes into view. The earth's atmosphere makes it difficult to see Mercury this low in the sky. During the day, the brightness of the sun makes observations difficult. However, spacecraft have flown near Mercury, sending back pictures and information.

The surface of Mercury is covered with craters. See Fig. 18-2. When the solar system was very young, it was filled with smaller bodies left over from the time the planets formed. Many of these bodies struck the planets, leaving craters as scars. On Mercury, these craters have remained. Mercury has no atmosphere, and thus there is no weather to cause erosion. The photographs of Mercury's surface also show many giant curving cliffs that extend hundreds of kilometers, cutting across the craters. Some of the cliffs are as much as three kilometers high. They were caused by the wrinkling of Mercury's surface when the planet was cooling and shrinking. Early in its history, Mercury also went through a period of volcanic activity. This has produced large lava plains that cover about one-quarter of its surface.

Mercury rotates very slowly on its axis. A day on Mercury is equal in length to 59 earth days. Mercury's long days, combined with its closeness to the sun and lack of an atmosphere, cause its surface to become very hot. Temperatures soar to around 400°C during the long period of daylight. On the dark side, the temperature drops to about −180°C.

Mercury seems to be a planet that has hardly changed for the past four billion years. It may show how the surface of the earth could have looked in its early history.

VENUS

Venus (**vee**-nus) comes closer to the earth than any other planet. Venus is covered by a thick layer of clouds. See Fig. 18-3 *(left)*. The clouds cause it to reflect sunlight. Thus Venus is seen as a brilliant object in the night sky. It is seen just after sunset or before sunrise. Venus spins very slowly on its axis. Venus rotates in a direction opposite to the direction it moves in its orbit. It takes 243 earth days for Venus to complete one entire rotation. Since it takes 225 days for Venus to orbit the sun, there is less than one Venus day in a Venus year.

For a long time Venus was called earth's twin. Both planets are similar in size, mass, and density. However, the atmospheres of the two planets are very different. The atmosphere of Venus is made up almost entirely of carbon dioxide. It is 90 times denser than the earth's atmosphere. This causes the atmospheric pressure on the surface of Venus to be much greater than on the earth. It is about the same as the pressure in our oceans at a depth of about one kilometer. This very heavy atmosphere traps heat because of the greenhouse effect. Short-wave radiation passes through the atmosphere and is absorbed by the surface. When the radiation is given off by the surface, it leaves as long-wave radiation, which cannot pass out of the atmosphere. The atmosphere also spreads the heat evenly over the surface of the planet. The temperature difference between the sunlight and dark sides of the planet or between its equator and its poles is only a few degrees. Temperatures on the surface

Fig. 18-3 (left) *A close-up photograph of Venus taken from a spacecraft shows its heavy layer of clouds.* (right) *A radar map of the surface of Venus. Light-colored areas are highlands.*

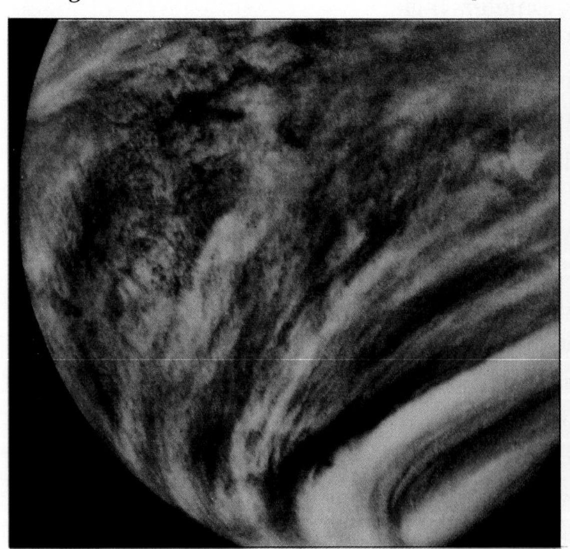

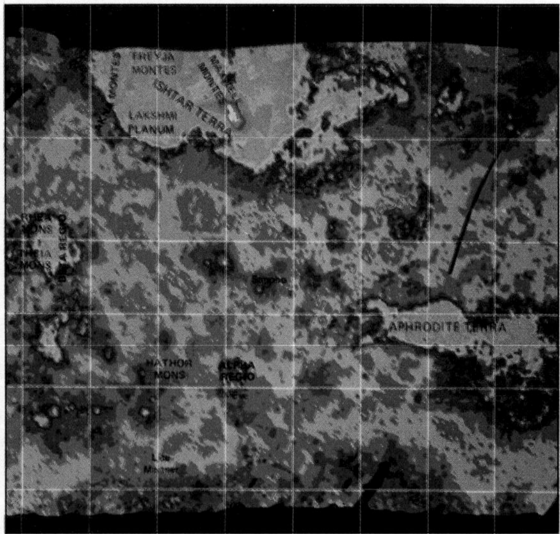

are about 475°C. This is about as hot as the temperature used to grill a steak outdoors.

Until spacecraft were able to visit Venus, little was known about this planet because of its cloud cover. Now crewless space vehicles have examined its surface by radar and landed on its surface. It was discovered that the surface of Venus had high mountains, valleys and canyons. See Fig. 18-3 *(right)*. Many of these features were the result of volcanic activity that probably also created the heavy atmosphere of Venus. There was no water found on Venus.

In the past, Venus may have had much less carbon dioxide in its atmosphere. With conditions like that, its surface temperature would have been more like that of the earth. However, once the carbon dioxide starts to build up in the atmosphere, the trapped heat causes more carbon dioxide to be released. This, in turn, causes more heat to be trapped. Eventually, a thick cloud layer covers a very hot surface. Small changes in a planet's atmosphere can set off a chain of events that may cause a great change in the climate. The conditions on Venus should make us very careful about releasing large amounts of carbon dioxide into the earth's atmosphere.

MARS

Mars is a smaller planet than the earth. It is about half as large in diameter, with only about one-tenth (0.1) the mass of earth. It probably does not have a heavy core like the earth. Mars turns once on its axis in a little over 24 hours. A visitor from earth would feel very much at home with the length of day and night on Mars. Its axis is tilted at nearly the same angle as the earth's. Thus, Mars also has four seasons. However, it takes Mars nearly two earth years to complete its orbit around the sun. Thus each of its seasons is six months long.

At the surface, the atmospheric pressure on Mars is only about two-hundredths (0.02) that of the earth's atmosphere. The Martian atmosphere is made up mostly of carbon dioxide, with about 3 percent nitrogen. But it is very thin, and therefore cannot trap much heat from the sun. As a result, the surface of Mars is always very cold. Even during the summer, the temperature near the Martian equator probably does not get above 0°C. Like the earth, Mars has polar icecaps of frozen water. Nighttime temperatures near the poles approach −200°C.

Fig. 18-4 The largest volcano on Mars is called Olympus Mons.

Many forces helped to shape the surface of Mars. There are many craters that were formed during its early history. Large areas of the planet have been flooded with lava, forming lava plains. Huge volcanic cones are found on Mars' surface. The largest is about 600 kilometers across and 25 kilometers high. See Fig. 18-4.

Scientists were surprised to find evidence that Mars once had running water. Photographs of the Martian surface show many channels and other features that could only have been made by running water. See Fig. 18-5. However, the surface of Mars is now dry and the atmosphere contains no water vapor. Where has the water gone? Much of it is locked up in the polar icecaps. There may also be a thick layer of frozen water beneath the planet's surface. There also seems to be frozen water in the soil particles just below the surface.

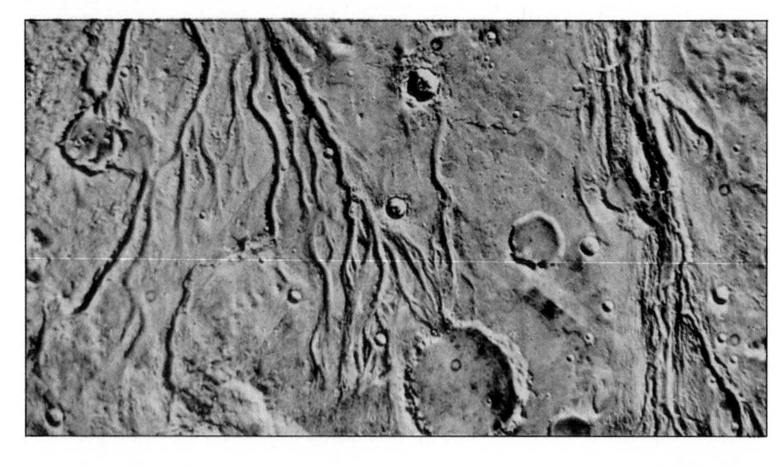

Fig. 18-5 An example of the channels on the surface of Mars that were made by running water.

Windblown material has also played a part in forming the Martian surface. Huge dust storms sweep across the planet during its spring season. The winds reach much higher speeds on Mars than on the earth. Thus the materials carried by the wind cause much more erosion of the rock than on the earth. Large deposits of wind-carried material are very common all over Mars.

In 1976, two United States spacecraft landed on Mars' surface. Those spacecraft carried small laboratories that tested the soil for life. The experiments, however, revealed no sure sign of any kind of life.

Many kinds of evidence show that the climate of Mars was once much different than it is today. At one time, millions or billions of years ago, this planet must have had a warmer climate, running rivers, and a denser atmosphere. It must have been a very different kind of planet from the freeze-dried Mars we are now observing. What caused this change on Mars is not known. Perhaps the same events that caused the earth to pass through its cold periods have also affected Mars. Further exploration of Mars may lead to a better understanding of the causes of all planetary climates.

SUMMARY

The solar system contains two kinds of planets, the solid, earthlike planets, and planets that are very large and are made mostly of ice and gas. The planets closest to the sun—Mercury, Venus, Earth, and Mars—are called the inner planets. Mercury lacks an atmosphere and is covered with craters. Venus has very high temperatures caused by its heavy covering of clouds. Mars now has a cold, dry surface with many kinds of features.

QUESTIONS

Use complete sentences to write your answers.

1. Why are the inner planets so different from the outer planets of the solar system?
2. Compare the sizes of Mercury and Venus to Earth.
3. List conditions for Mercury, Venus, Earth, and Mars. Include atmosphere, surface features, temperature (°C), length of day (Earth days).

INVESTIGATION

PLANET ORBIT MODELS

PURPOSE: To draw scale models of the inner planets' orbits.

MATERIALS:

plain paper, 2 straight pins, cardboard, string, pencil, ruler

PROCEDURE:

A. Place a sheet of paper over the cardboard. Draw a small circle near the center of the paper. Through the center of the circle, press a pin through the cardboard to anchor it well. Label the circle SUN.

B. Look at Fig. 18-6 (a). You will use the method shown in the figure to draw the orbits of Mars, Earth, Venus, and Mercury, in that order, on the same sheet of paper.

Planet	Distance Between pins (cm)	Distance Around loop (cm)
Mars	1.7	19.9
Earth	0.2	12.1
Venus	0.1	8.8
Mercury	1.0	6.5

scale: 1.0 cm = 25 million km

The following table shows how far apart to set the second pin from the sun pin, and how large a loop to make.

C. When making the orbits, do not move the pin labeled SUN. Move only the second pin the proper distance each time. Move it in a straight line so the holes it makes are in a line.

D. To make the string loop, cut a piece of string several centimeters longer than the distance around the loop. Bring the ends of the string together and tie a knot one-half the distance shown in the table. For example, for Mars, cut a piece of string about 22 centimeters long. Bring the ends together and tie a knot at 9.9 centimeters. Check the knot as you tighten it. See Fig. 18-6 (b).

E. Keep the string tight as you move the pencil while making the orbit. You may wish to practice once or twice on a separate sheet of paper. Draw the orbits carefully. Label each orbit with its planet's name as soon as you draw it.

CONCLUSION:

1. Describe the shapes of the orbits of the inner planets.

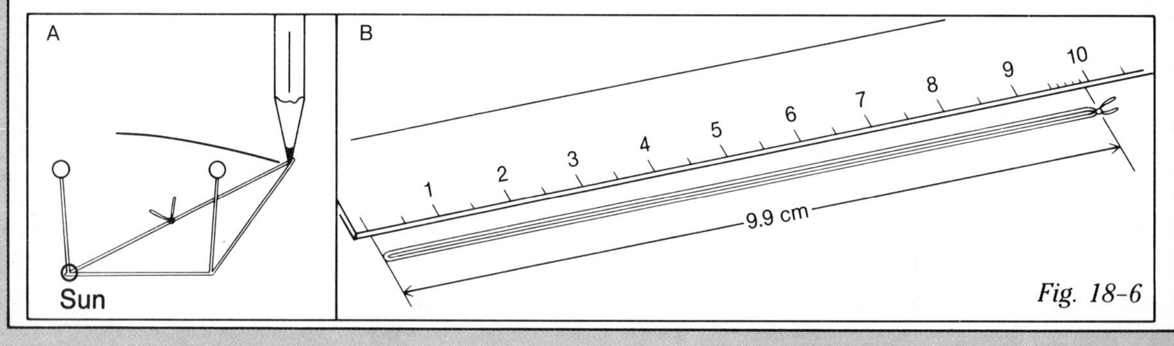

Fig. 18-6

18–2. Outer Planets

At the end of this section you will be able to:

- ☐ Describe the main features of Jupiter, Saturn, Uranus, and Neptune.
- ☐ Compare Jupiter's moons with the inner planets.
- ☐ Explain how Pluto is different from the other planets.

Five hundred years ago, frail sailing ships set out on exploratory voyages. Their voyages changed the world. Similarly, in modern times, spacecraft have been sent out to explore the outer parts of the solar system. See Fig. 18–7. What these spacecraft have discovered has been as much of a surprise as the discovery of the New World was in the 15th century.

Fig. 18–7 (left) *An unmanned spacecraft.* (right) *A sixteenth century sailing ship.*

JUPITER

Jupiter (**jou**-pih-ter) was the greatest and strongest among the ancient Roman gods. It also is the giant among the planets. Jupiter has almost 2.5 times the mass of all the other planets put together. Its diameter is 11 times the diameter of the earth. Although it would take more than 1,000 earths to equal Jupiter's volume, Jupiter has only 318 times the earth's mass. This is because the density of Jupiter is less than one-third that of the earth. It is also more than five times farther from the sun than the earth and therefore receives far less heat. Strangely, Jupiter gives off more heat than it receives from the sun. Some of this heat is left over from the time the planet was formed. More heat is also produced by the planet's slow shrinking that squeezes and heats the inside of Jupiter. Like the earth, Jupiter is also surrounded by a magnetic field. Charged particles from the sun are trapped in this magnetic field. This belt of trapped particles makes up a danger zone around Jupiter. Spacecraft approaching the planet must protect their instruments from the effects of these particles.

The surface of Jupiter is covered with a cloud layer that is separated into colored bands. See Fig. 18-8. The clouds are made up of about 88 percent hydrogen and 11 percent helium gases. There are also small amounts of water, ammonia, methane, and other compounds. The hydrogen and helium, being transparent, are not visible. The color in the clouds comes from different kinds of colored compounds that are found in lower layers of the cloud covering. Jupiter probably does not have a solid surface. Inside, the gases that make up its atmosphere simply become thicker and denser. Finally the dense atmosphere blends into an ocean made up mainly of liquid hydrogen. Near the center of Jupiter, the pressure is so great that the liquid hydrogen becomes similar to a metal. It is this liquid metallic hydrogen within Jupiter that causes its powerful magnetic field. Jupiter probably has a small rocky core at its center.

Fig. 18-8 The upper cloud layer of Jupiter is made up of colored bands.

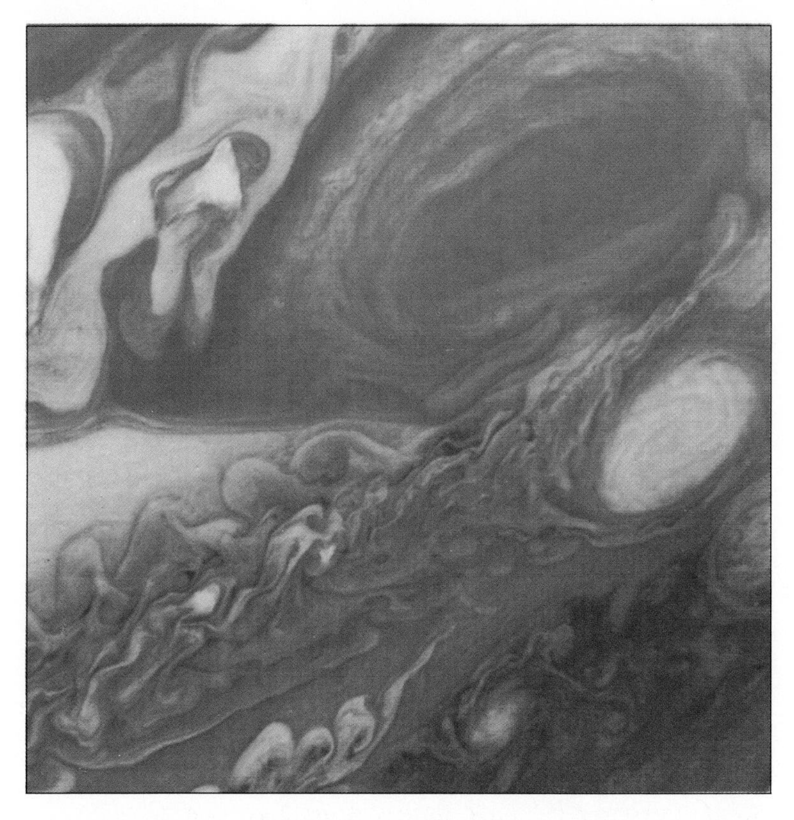

Fig. 18-9 Winds cause the swirls and currents seen in the colored bands on Jupiter.

It takes this giant planet nearly 12 earth years to complete its orbit around the sun. But Jupiter rotates very rapidly. One day on Jupiter is less than ten hours long. The rapid rotation of Jupiter causes its outer cloud layers to separate into colored bands. Heat coming from the inside causes the clouds to have high and low pressure regions just as in the earth's atmosphere. Jupiter's rapid rotation causes these highs and lows to wrap completely around the planet. Strong winds flow between the pressure belts, causing swirls and twisting motions in the gases. See Fig. 18-9. Storms similar to hurricanes develop in the atmosphere from time to time and appear as colored spots. One such storm is called the Great Red Spot. It has been observed in the same place on Jupiter for over three centuries.

THE MOONS OF JUPITER

In 1610, using a telescope he had constructed, the Italian scientist Galileo discovered that Jupiter had four large moons. These four satellites are called the Galilean moons of Jupiter. They are named, in order of increasing distance from the planet, *Io* (**eye**-oh), *Europa* (yoo-**roh**-puh), *Ganymede* (**gan**-ih-meed), and *Callisto* (kuh-**lis**-to). Each of these satellites is larger than the planet Mercury. See Fig. 18–10. Each moon is quite different from the huge planet it orbits. Io, Jupiter's closest moon, has a bright yellow-to-orange-colored surface that, surprisingly, is free of craters. It has active volcanoes whose lava flows may have covered craters and given it a fairly smooth surface. The next satellite, when moving away from Jupiter, is Europa. It is about the size of the earth's moon. Europa has a light yellow-colored surface that is covered by dark cracks. There are also large deposits of ice on Europa's surface. The two outermost satellites, Ganymede and Callisto, are similar in appearance. Their surfaces are covered with a mixture of rock and ice marked by many craters. Jupiter has at least 12 other satellites that are much smaller than the four Galilean moons. Around Jupiter, there is also a faint ring that is made up of very small particles.

The Galilean satellites of Jupiter share many characteristics with the earthlike inner planets. Future exploration of these moons may provide better understanding of the working of planets like the earth.

Fig. 18–10 Photographs of the Galilean moons of Jupiter. Clockwise, beginning at the bottom left: Io, Europa, Ganymede, and Callisto.

Fig. 18-11 The colored bands in Saturn's upper cloud layers can be seen in this photograph.

SATURN

Saturn (**sat**-urn) is the next planet beyond Jupiter. It is the least dense of all the planets. Its density is less than that of water. If there could be a large enough ocean, Saturn would float in it. This means that Saturn, like Jupiter, is rich with the light element hydrogen. Also, like Jupiter, Saturn is believed to have a small, rocky core surrounded by a thick layer of liquid hydrogen. It produces more heat than it receives from the sun. Saturn rotates very rapidly, once every 10 hours and 40 minutes. This causes its outer cloud layers to separate into bands. Photographs of Saturn taken from spacecraft show bands of pale yellow, golden brown, and reddish brown colors as shown in Fig. 18–11.

Saturn's most observable feature is its great system of rings. See Fig. 18–12. Saturn's rings are made up of many small pieces of ice or substances coated with ice. The size of the chunks of ice material ranges from a few centimeters across to several meters. The rings are divided into three large bands, each of which is made of many smaller rings. These are similar to the rings on a phonograph record.

Fig. 18-12 Saturn's rings are hundreds of separate ringlets, like grooves in a record album.

URANUS AND NEPTUNE

If you were on Uranus (**yoor**-uh-nus) or Neptune (**nep**-toon), the sun would look so far away that it would be hard for you to tell it from other stars. Thus both Uranus and Neptune are very cold. Each has a diameter about four times that of the

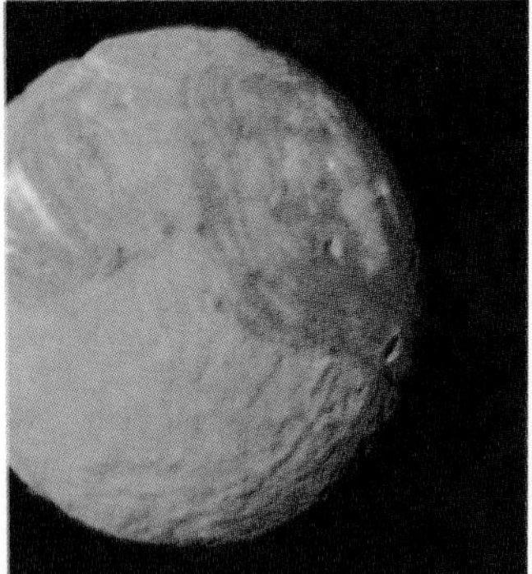

Uranus (top) and its satellite Miranda (bottom).

earth. Each has a small, solid core surrounded by a thick layer of ice and liquid hydrogen. The planets have a pale greenish or blue-green color with only very faint bands like Jupiter and Saturn. Uranus and Neptune take about 17 hours to complete one rotation. Uranus rotates in an unusual way because it is "tipped over." Its north pole points almost directly toward the sun, rather than up and down as with the other planets. Uranus has five large satellites and ten smaller ones. It also has at least eleven rings similar to those of Saturn. Close-up photos from the Voyager 2 spacecraft have shown that Uranus' rings are made of dark boulder-size rocks.

PLUTO

There are reasons to believe that Pluto did not form the same way as any other planet. First, Pluto is a very small planet. Its diameter seems to be about equal to the distance between Los Angeles and Washington, D.C. This makes Pluto the smallest planet. Second, Pluto's orbit is different from the orbits of the other planets. While the orbits of other planets lie in nearly the same plane, Pluto follows a strange, tilted orbit that is the most elliptical of all the planetary orbits. Third, Pluto has a very low density and seems to be made up of frozen methane and other ices. Such a composition is unlike that of any other planet. Answers to many of the questions about Pluto must await some future time when spacecraft can finally reach the very edge of the solar system.

SUMMARY

The outer planets are made up mostly of liquids, ices, and gases. Jupiter, Saturn, Uranus, and Neptune have a small, rocky core surrounded by a thick layer of liquid hydrogen, ice, and clouds. Pluto is different from the other outer planets and is believed to have formed differently from them.

QUESTIONS

Use complete sentences to write your answers.

1. Compare the main features of Jupiter, Saturn, Neptune, and Uranus.
2. Name and describe the four largest moons of Jupiter.
3. How does Pluto differ from the other outer planets?

INVESTIGATION

A SCALE MODEL OF THE SOLAR SYSTEM

PURPOSE: To draw models of the size of each planet in the solar system and, using the same scale, to get an idea of the distances planets are from the sun.

MATERIALS:

unlined paper, pencil, pencil compass, ruler

PROCEDURE:

A. It is difficult to get a good idea of the relative sizes of the sun and its system of planets. Since they are all very large, we will use circles as scale models to represent them. The scale that we will use is 1 cm = 1,000 km.

1. The sun has a diameter of 1,400,000 km. Using the scale, what diameter circle would represent the sun?

B. Copy the table below. Complete the third column by finding the scale diameter of each planet.

Name	Diameter (km)	Scaled Diameter (cm)
Mercury	5,100	
Venus	13,000	
Earth	13,000	
Mars	6,900	
Jupiter	140,000	
Saturn	120,000	
Uranus	53,000	
Neptune	50,000	
Pluto	6,000 (?)	

C. On one side of an unlined sheet of paper draw a circle to scale for each planet, using a pencil compass. The circles may overlap. Label each circle with the name of the planet it represents.

2. Name four large planets that are similar in size.
3. Name five small planets that are similar in size.

D. Copy the table below. Complete the third column by finding the scale distance each planet is from the sun. Use the same scale as for planet size.

Name	Distance from Sun (km)	Scale Distance from Sun (cm)
Mercury	58,000,000	
Venus	110,000,000	
Earth	150,000,000	
Mars	230,000,000	
Jupiter	780,000,000	
Saturn	1,400,000,000	
Uranus	2,900,000,000	
Neptune	4,500,000,000	
Pluto	5,900,000,000	

E. Look at your scale distances and answer the following questions.

4. Is it possible to draw a scale distance model of the solar system inside the classroom using this scale?
5. What is the scale distance to Saturn in kilometers?

CONCLUSION:

1. Why do commercial models of the solar system use one scale for the sizes of the sun and planets and a different scale for the distances of the planets?

18-3. Smaller Members of the Solar System

At the end of this section you will be able to:

- Explain how a belt of *asteroids* may have been formed as part of the solar system.
- Compare the properties of *meteors* and *comets* with other smaller members of the solar system.
- Describe two ways in which moons may have formed around planets.

On June 30, 1908, a mysterious explosion occured in Siberia. People asleep in their tents 80 kilometers away were blown up into the air. In a nearby town, windows were broken. The explosion was detected as far away as England. The explosion was caused by an unknown object from space. The solar system contains many bodies that float between the planets.

ASTEROIDS

Between the orbits of Mars and Jupiter is a wide region of the solar system that does not contain a planet. In this zone are found a huge number of lumps and boulders. These objects are are called **asteroids.** See Fig. 18-13. The total number of *as-*

Asteroid A smaller body in the solar system, usually found in an orbit between Mars and Jupiter.

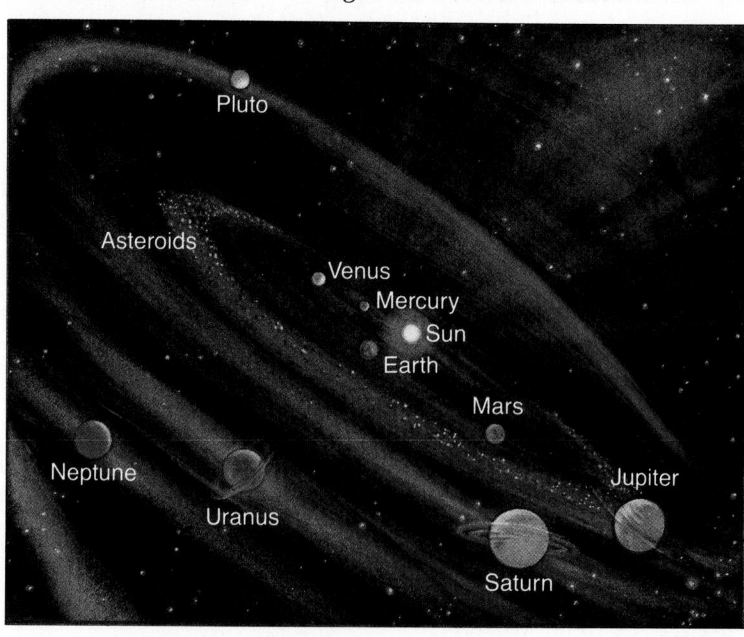

Fig. 18-13 The belt of asteroids lies between the orbits of Mars and Jupiter. The orbits of the planets are not shown with the correct size in this diagram.

teroids is not known. Only the larger ones reflect enough sunlight to be seen through telescopes. About 2,000 asteroids have been discovered. An asteroid is said to have been discovered when it has been photographed several times and its orbit is determined. See Fig. 18–14. Thousands more have appeared in photographs, but no one has spent the time necessary to find their orbits. About 230 asteroids are known to have diameters greater than 100 kilometers. The largest, called *Ceres* (**seer**-eez), has a diameter of about 1,000 kilometers. There are probably thousands more with diameters of a few kilometers or less. All the asteroids put together probably would not equal the mass of a small planet. Ceres alone makes up about 30 percent of the total mass of all the known asteroids.

Fig. 18–14 Asteroids appear as streaks in telescopic photographs.

At one time, scientists thought that the asteroids might be the wreckage of a small planet. This planet was supposed to have been torn apart by Jupiter's tremendous force of gravity. However, scientists now know that there is not enough material in the asteroids to make a planet. They now think that most of the asteroids were once part of about 50 large bodies that formed along with the planets. But Jupiter's gravity caused them to start crashing into each other. These collisions broke them into the pieces that are found today. Only a few of the original larger bodies, such as Ceres, remain. Many smaller asteroids seem to be bits of matter left over as larger bodies grew in the young solar system. Asteroids probably existed in other parts of the solar system. But then the planets acted like giant vacuum cleaners that swept up these small objects. The large distance between Mars and Jupiter allowed the leftover material to remain as the belt of asteroids.

Most of the asteroids follow orbits that keep them in the region between Mars and Jupiter. However, sometimes an asteroid passes close enough to Mars or Jupiter to be pulled into a new orbit. This causes some asteroids to follow orbits that carry them to other parts of the solar system. For example, 28 asteroids follow paths that cause them to cross the orbit of the earth as they move around the sun. Some of these can come close to the earth and could collide with this planet. However, such collisions with asteroids are very rare. It is believed that it was an asteroid that caused the mysterious explosion in Siberia. The only very recent crater known to exist on the earth is found in Arizona. See Fig. 18–15.

Fig. 18–15 This crater in Arizona was made when a meteorite struck the earth about 30,000 years ago.

METEORS

Meteor A bright streak of light in the sky, which is caused by rock material burning up in the atmosphere.

While collision with large objects is not common, many of the very small objects in the solar system often collide with the earth. Most of these objects are very small pieces of rock. They usually have a mass of less than one gram. When they reach the earth's atmosphere, friction causes most of them to get so hot that they burn up at altitudes near 100 kilometers. The bright streaks they create in the sky are called **meteors** (**meet**-ee-orz), or "shooting stars." Some *meteors* look like huge balls of fire. Sometimes one of the objects is so large that parts of it may reach the earth's surface. These chunks of material that reach the earth's surface are called *meteorites*. They usually can be identified by marks on their surfaces. These marks are caused by the heat produced by friction as they pass through the atmosphere. About 90 percent of the meteorites found are made of stone similar to ordinary rocks on the earth. The others are made mostly of iron mixed with nickel. Most meteorites seem to be asteroids that have wandered out of their usual place in the solar system. Some meteorites found on the earth have not changed since the solar system formed. Thus scientists are able to learn much about the early history of the solar system from their studies of meteorites.

COMETS

Comet An icy body that comes from the edge of the solar system and moves toward the sun.

Scientists believe that the entire solar system is surrounded by a distant cloud. This cloud is made up of what have been called "dirty snowballs." Chunks of ice, rock, and dust slowly move around the sun at distances so great that it takes the sun's light more than a year to travel there. This cloud was probably formed along with the rest of the solar system about five billion years ago. Because it is so far away, we would not even know this cloud was there. However, sometimes the gravity of a passing star pulls one of the clumps of ice and rock from the cloud. It may begin moving toward the sun. The icy body then becomes a **comet.** Each *comet* follows a different path as it moves in toward the sun.

Comets often are affected by the gravity of Jupiter and Saturn. For example, a comet can be pulled by Jupiter's gravity into an orbit that takes it close to the sun. When this happens, the comet may become a member of the group of comets that are seen from the earth at different times.

A comet that follows an orbit bringing it near the sun can become one of the most unusual objects we ever see in the sky. When a comet passes the orbit of Mars, it may begin to grow a long tail. The sun's heat and the solar wind cause material to glow and stream away from the comet. See Fig. 18–16. A comet's tail always points away from the sun, like long hair streaming away from the wind. See Fig. 18–17. Some comets develop two tails, whereas others may hardly have a visible tail.

Some comets can be seen from the earth at regular times. For example, Halley's comet has been seen about every 76 years for many centuries. However, each time a comet makes a passage around the sun it loses some of its materials. Thus a comet usually becomes smaller and dimmer after each appearance until it finally disappears. A comet may also break into smaller parts that continue to follow the original orbit. Meteor showers are seen when these pieces collide with the earth's

Fig. 18–16 Solar wind and heat are responsible for the long tail that is characteristic of many comets.

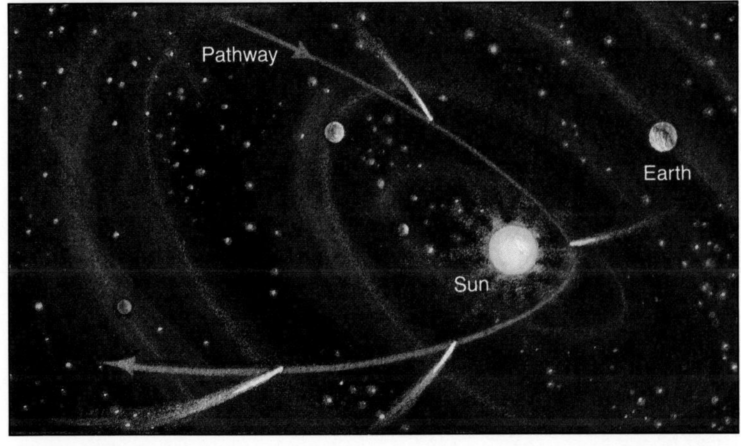

Fig. 18–17 The tail of a comet always points away from the sun.

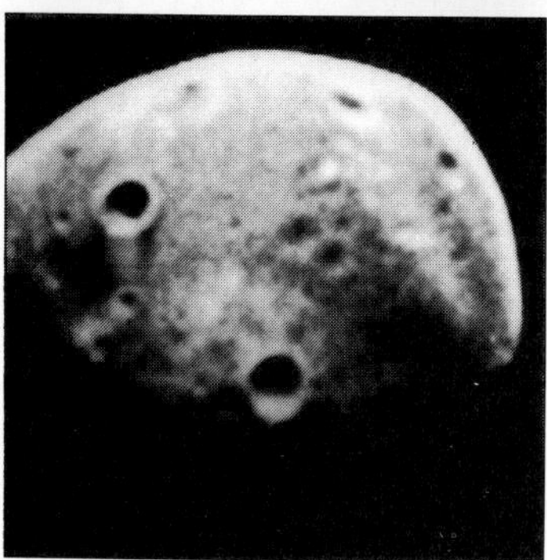

Fig. 18–18 Mars has two tiny, irregularly shaped moons. (top) *Phobos, the larger, inner moon.* (bottom) *Deimos, the smaller, outer moon.*

atmosphere. At times during the year, as many as 50 to 100 meteors are seen in an hour. Such meteor showers occur when the earth moves through one of the groups of bodies that were once part of comets. Fairly heavy meteor showers can be seen each year near the dates of May 4 to 6, August 10 to 14, and October 19 to 23. Most of these meteor showers are heaviest after midnight.

MOONS

Other kinds of bodies found in the solar system are moons. Planets may come to have one or more satellites, or moons, in two ways. First, a planet's moons can be formed along with the planet. For example, the large moons of Jupiter seem to have been formed like a miniature solar system around the planet as the planet itself formed. The inner moons have rocky interiors and the outer moons have icy interiors. Saturn also has a similar system of moons, with at least 17 satellites. The largest moon of Saturn, called *Titan* (**tie**-tun), is the only satellite in our solar system to have an atmosphere.

A second way that a planet may come to have moons is by the capture of large asteroids. For example, the two satellites of Mars, called *Phobos* (**foh**-bohs) and *Deimos* (**dee**-mohs), are asteroids that are held by Mars' gravity. See Fig. 18–18.

SUMMARY

The most numerous of the smaller bodies in the solar system are asteroids, which are found in a region between the orbits of Mars and Jupiter. Icy bodies from the most distant parts of the solar system may become visible as comets when they begin to follow orbits that bring them close to the sun. Meteors are objects that sometimes collide with the earth but are usually burned up in the atmosphere. A planet's moons may be formed along with the planet or captured at a later time.

QUESTIONS

Use complete sentences to write your answers.

1. What explanation is given by scientists for the presence of the asteroid belt?
2. In what ways do comets differ from meteors?
3. Describe two ways moons may have formed around planets.

INVESTIGATION

THE PATHS OF ASTEROIDS AND COMETS

PURPOSE: To draw the paths of several asteroids and a comet, and to compare these with the paths of planets from the Investigation in Section 18–1.

MATERIALS:

plain paper	pencil
cardboard	string
2 straight pins	ruler

PROCEDURE:

A. The procedure you will use is the same as in the Investigation of Section 18–1. Place a sheet of paper over the cardboard and stick one pin near the center of the paper, anchoring it firmly. Draw a small circle around the pin and label it SUN. You will not move this pin when making the asteroid orbits.

B. Use the method shown in Fig. 18–19 to draw the orbits of the asteroids on one side of your paper and the orbit of the comet on the other side. The comets have a much larger path that goes from very near the sun to the outer boundary of the solar system. The comet path cannot be drawn on the same scale as the asteroids.

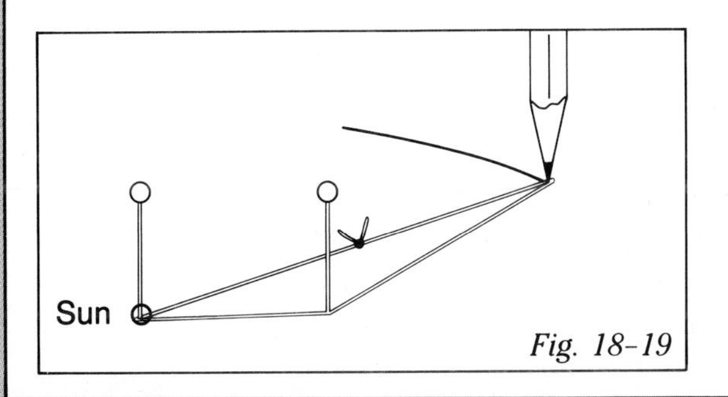

Fig. 18–19

C. Cut one piece of string at a time, making each several centimeters longer than the distance around the loop given in the table below. Bring the ends of the string together and tie a knot one-half the distance shown in the table. For example, for the typical asteroid, the distance around the loop is 6.4 centimeters. Cut a piece about 9 centimeters long, and tie the ends together, making a loop 3.2 centimeters long. See Fig. 18–6(b) on page 488.

Object	Distance Between Pins (cm)	Distance Around Loop (cm)
Typical asteroid	0.6	6.4
Apollo	2.5	7.0
Hildago	11.0	28.0
Halley's Comet	10.4	21.2

CONCLUSIONS:

1. Which orbit is most like the orbits of the planets drawn in the Investigation of Section 18–1?
2. Most asteroids follow a typical path. Which asteroid has a path that is much closer to the sun than that of the typical asteroid?
3. Which asteroid has a path that takes it much farther from the sun than the typical asteroid?
4. Compare the path of Halley's comet to the path of a typical asteroid.

18-4. The Moon

At the end of this section you will be able to:

- Compare the surface of the moon with the surface of the earth.
- Give a brief description of the moon's history.
- Describe three theories that may account for the origin of the moon.

One accomplishment of the space program has been travel to the moon. See Fig. 18-20. Some scientists predict that, by the early 21st century, humans will be able to live on the moon. Could you be one of the first "natives" on the moon?

Fig. 18-20 The panel instruments of an Apollo spacecraft with the moon in the background.

THE SURFACE OF THE MOON

The moon is the earth's nearest neighbor in space. It is so close that you can easily see the main features of the moon's surface. With a pair of binoculars, you can see rugged mountain ranges, wide level plains, and hundreds of large and small craters. These features stand out because there is no atmosphere to hide them. However, it was not until astronauts landed on the moon, beginning in 1969, that the actual conditions of earth's closest neighbor became known.

In some ways, the moon resembles the earth. Both have a crust and a mantle. The moon may also have a core like the earth, although it is probably much smaller. The inside of the moon may also be hot like that of the earth.

However, the astronauts who visited the moon were more impressed by the differences between the surface of the moon and that of the earth than by their similarities. First, there is no atmosphere on the moon. But instruments have been able to measure small amounts of the gases argon, neon, and helium. Also, there is no water on the moon. No sign of life has been found on the moon. It appears that the only living things ever to have been on the moon were the astronauts from earth. During the day the temperature on the moon's surface rises to above 100°C. During the night the temperature drops to below −100°C. The moon has no atmosphere to shield it from the sun's heat or help hold the heat during the night.

The moon is smaller than the earth. Its mass is also much less than the earth's mass. Because of this, the force of gravity felt on its surface is only one-sixth that felt on earth. Thus the astronauts on the moon's surface were able to jump higher and move by leaps and bounds, as if they had springs on their feet. They fell in slow motion and landed softly.

The surface features of the moon have not been changed by wind and water. Many of the features that can be seen today on the moon have been there since its early history. The moon is not free from all change, however. The lack of an atmosphere allows the surface to be hit constantly by small meteorites and solar wind particles. This breaks up some of the rocks on the surface. On the moon, the small pieces of rock are not washed away or blown away as they would be on earth. As a result, a fine dust covers almost all the moon's surface. See Fig. 18-21. The astronauts were surprised to find that the moon dust is a dark gray color. It looks tan from the earth.

Fig. 18-21 Moon dust is stirred up by this astronaut on the moon.

HISTORY OF THE MOON

The rocks brought back from the moon are surprisingly old. All these moon rocks are more than three billion years old. However, almost all the rock on the earth's surface is younger than that. Changes that take place on the earth produce young rock, which covers the older rock. On the moon, there seem to have been no major changes for three billion years.

Evidence from the moon rocks seems to support the following outline of the moon's history. The moon came into being at the same time as the earth. This was about 4.6 billion years ago. At first, its outer layers were hot and molten. Then the heavy

Fig. 18-22 (left) *Features on the moon's surface can be seen with the eye alone. Tyco can be seen as the large crater near the bottom.* (right) *Craters cover much of the moon's surface.*

materials settled downward. The lighter materials rose to the surface. A crust formed as the outer parts cooled and hardened. During its earliest years, the moon had many volcanoes. However, the volcanoes stopped erupting about three billion years ago. Also, during this early period, the moon was struck many times by huge chunks of material that were flying through space. These collisions tore great basins in the moon's surface. Volcanic eruptions later filled these low places with lava. These lava-filled basins now appear as dark-colored plains on the moon's surface. Early astronomers thought that these dark areas were seas and called them **maria** (**mar**-ee-ah), which is Latin for "seas." Most of the rest of the moon's surface is made up of light-colored highlands. The highlands are the remains of the moon's original crust, which were not covered by the lava flow. The dark *maria* and the light highlands are easily seen when you look at the full moon. See Fig. 18-22 *(left).*

Maria Lava plains that are seen as dark areas on the moon's surface.

The highlands of the moon are marked by thousands of craters. See Fig. 18-22 *(right).* These craters are the scars left by meteorites of many different sizes that crashed into the moon's surface throughout its history. A few of them are several hundred kilometers in diameter.

Some craters have streaks of light-colored material leading out from their edges. These streaks are called **rays.** The *rays* consist of material that was thrown out of the crater when it was formed. One large crater, named Tycho (**tie**-ko), has a large system of rays. Tycho is easily seen with the aid of binoculars when the moon is full. See Fig. 18-22 *(left).*

Rays Light-colored streaks that lead out from some of the craters on the moon.

WHERE DID THE MOON COME FROM?

Three theories have been put forward to explain the origin of the moon.

1. Daughter theory. The *daughter theory* suggests that the moon was once part of the earth. It says that the rotation of the earth and the gravity of the sun produced a bulge in the earth. Then, long ago, a part of the earth broke away. It flew into space, and became the moon.

2. Sister planet theory. The *sister planet theory* says that the earth and moon are twin planets that formed at the same time.

3. Capture theory. The third theory is called the *capture theory*. It suggests that the moon was formed in some other part of the solar system and was "captured" by the earth's gravity.

It is unlikely that the moon was once part of the earth, as the daughter theory suggests. The theory is unlikely because there are too many differences in the chemical makeup of the two bodies. The capture theory does not agree with the principles of gravity. According to these principles, if the earth captured the moon, the orbit of the moon would be different than it is today. The sister theory seems the most likely to be true. However, not all scientists agree with this theory. The information gained from exploring the moon has not settled the question about the moon's origin.

SUMMARY

The surface of the moon is very different from the earth. This is because the moon does not have an atmosphere. The rocks of the moon have changed very little since its early history. They show that the moon's surface was once hot and molten, and covered with volcanoes. It is believed that the moon formed along with the earth.

QUESTIONS

Use complete sentences to write your answers.

1. Compare the conditions on the moon's surface with the conditions found on the earth's surface.
2. Outline the moon's history as supported by a study of its rocks.
3. Describe three theories that answer the question, "Where did the moon come from?"

INVESTIGATION

SURVIVAL ON THE MOON

PURPOSE: To find out what is important for survival on the moon.

MATERIALS:
paper pencil

PROCEDURE:

A. You and two members of your crew are returning to the base ship on the sunlit side of the moon after carrying out a 72-hour exploration trip. Your small rocket craft has crash-landed about 300 kilometers from the base ship. You and the crew need to reach the base ship. In addition to your spacesuits, your crew was able to remove the following items from the rocket craft:

4 packages of food concentrate
20 m nylon rope
1 portable heating unit
1 magnetic compass
1 box of matches
1 first-aid kit
2 50-kg tanks of oxygen
20 L of water
1 star chart
1 case of dehydrated milk
1 solar-powered radio set
3 signal flares
1 large piece of insulating fabric
1 flashlight
2 45-caliber pistols, loaded.

B. Using what you know about the moon, rate each item in the above list according to how important it would be in getting you back to the base ship. List the most important first, the least important last. Number them 1 through 15. Answer the following:
1. Which three items were the most important? Explain.
2. Which items would be useless? Explain your answer.

C. Compare your list with the one supplied by your teacher. Astronauts would list the items in this order.

D. To score your list against the astronauts' list, do the following:
3. Beside each item on your list, place the number that represents the difference between your ranking and the astronauts' ranking. For example if you listed oxygen first, you would write 0 in front of oxygen on your list. If you had listed it third, then you would write 2, and so on.
4. After placing a score beside each item on your list, add up the individual scores to get a total. Compare your score with those of other students.
5. What is your total score?
6. The lower your total score, the closer you came to surviving the return trip to the base ship. How did your chance of surviving compare to other students' chances?

CONCLUSIONS:
1. What does the moon lack that humans need for survival?
2. What materials would you need to survive on the moon?

18–5. The Earth and Moon

At the end of this section you will be able to:

- ☐ Describe the motions of the earth–moon system.
- ☐ Relate each *phase* of the moon to a position in its orbit around the earth.
- ☐ Explain how *eclipses* of the sun and moon occur.

High above the earth's surface the space shuttle orbiter releases a space telescope. See Fig. 18–23. The telescope will remain in an orbit around the earth while the space shuttle returns to earth. From its position high above the earth's atmosphere, the space telescope can see deep into space. The space telescope is one example of many different kinds of human-made satellites that have been developed to orbit the earth.

Fig. 18–23 The space shuttle is a useful tool for releasing and repairing satellites and for conducting experiments in space.

THE EARTH–MOON SYSTEM

A **satellite** is a body that moves in an orbit around a much larger body. For example, the space telescope is a *satellite* of the earth because its mass is very small compared to the mass of the earth. Is the moon also a satellite of the earth? The moon is often called the earth's natural satellite. This is because the mass of the earth is so different from that of the moon that the moon could be called a satellite of the earth. However, rather

Satellite A body that moves in an orbit around a larger body.

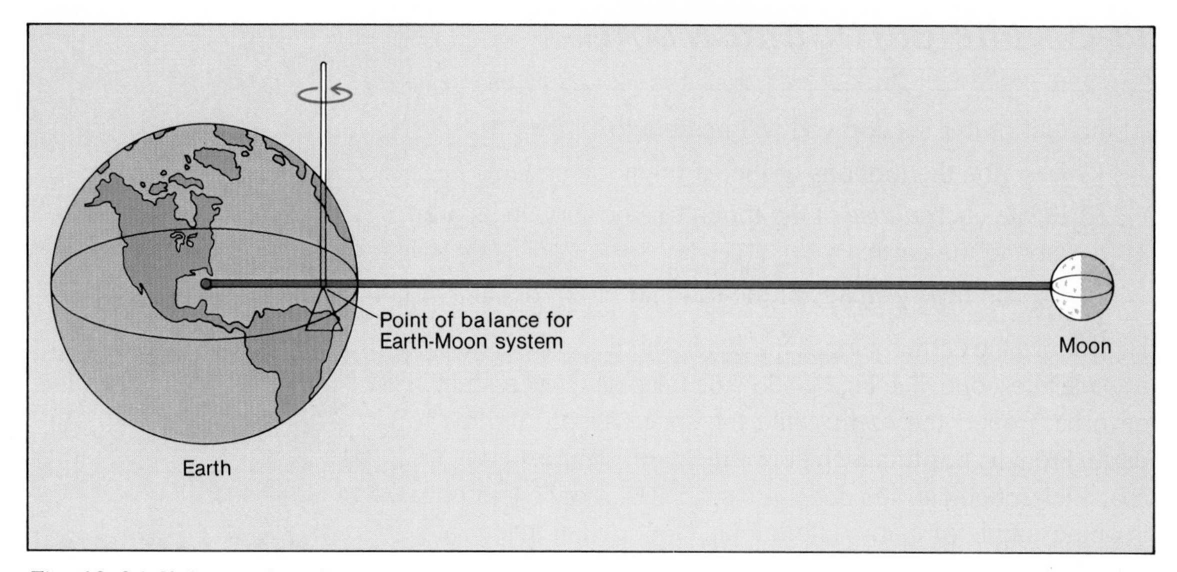

Fig. 18–24 If the earth and the moon were connected, the balance point of the system would be within the earth.

than the moon being a satellite of the earth, it and the earth make up a system that is like a pair of planets moving around the sun together. If the earth and moon were connected by a solid bar, the system would balance at a place inside the earth. See Fig. 18–24. Instead of the earth moving in a smooth orbit with the moon circling the earth, the earth and moon together orbit the sun. At the same time, each turns around the balance

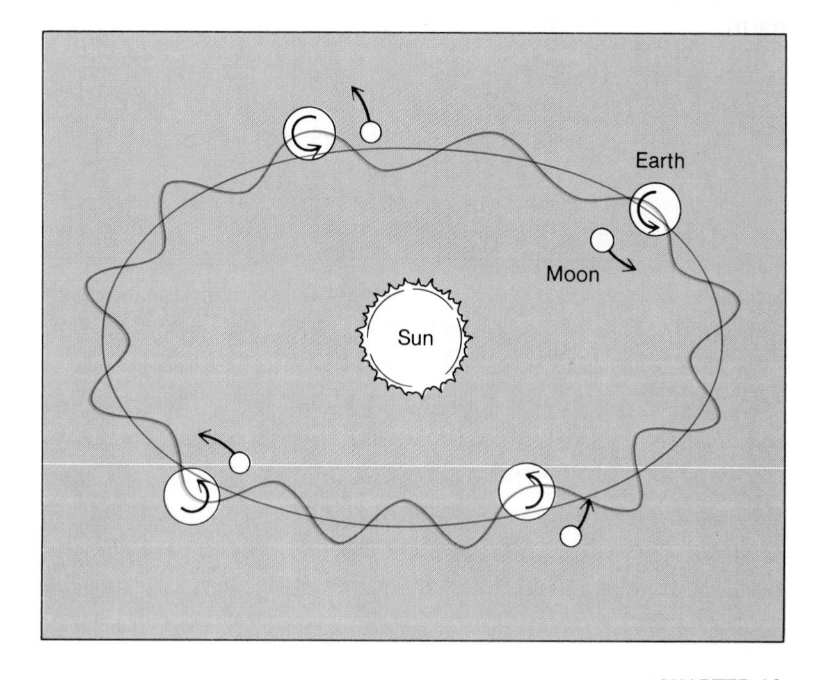

Fig. 18–25 The earth and the moon move around the sun as a single system.

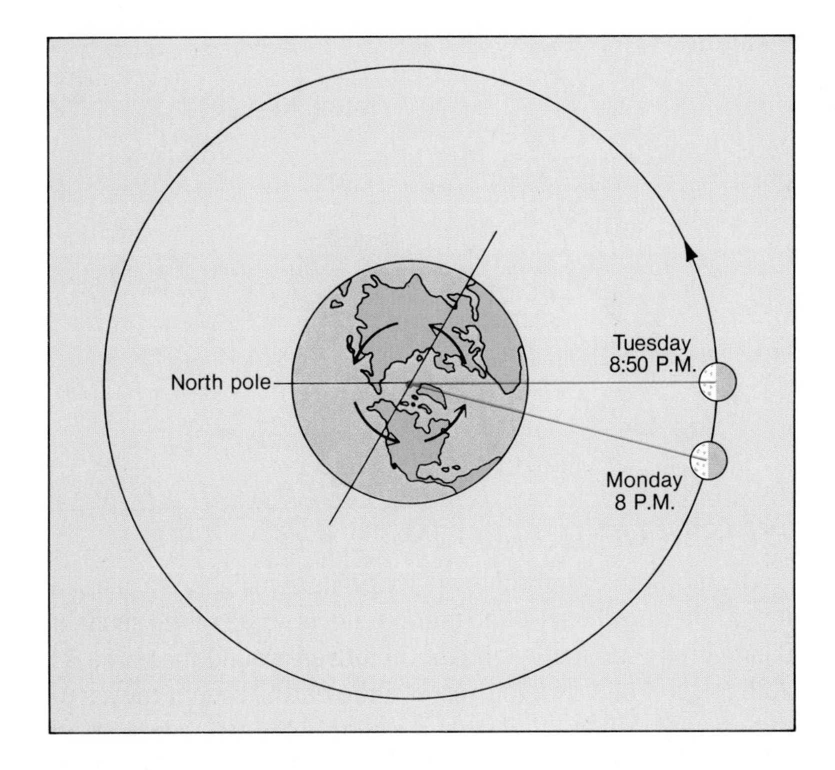

Fig. 18–26 The moon moves ahead in its orbit each day, so a place on the earth must rotate about 50 minutes longer each day to bring the moon back into view.

point. See Fig. 18–25. However, when the moon is seen from the earth, it looks as if the moon is circling the earth. For this reason, the moon is usually said to move in an orbit around the earth.

The moon makes one complete trip around the earth every 27 1/3 days. The moon also turns on its axis once every 27 1/3 days as it moves around the earth. As a result, the same side of the moon always faces the earth.

We can see the moon appear on the eastern horizon every day. This is because the earth's eastward rotation carries us around until the moon can be seen. The moon appears to rise in the east, then move across the sky, and set in the west. If the moon did move around the earth, it would always rise and set at the same time of day. However, the moon does move around the earth in an eastward direction. Thus the earth must turn a little farther each day to "catch up" with the moon and bring it back into view again. It takes about 50 minutes each day for the earth to rotate the extra amount needed to "catch up" with the moon. This means that the moon rises about 50 minutes later each day. See Fig. 18–26.

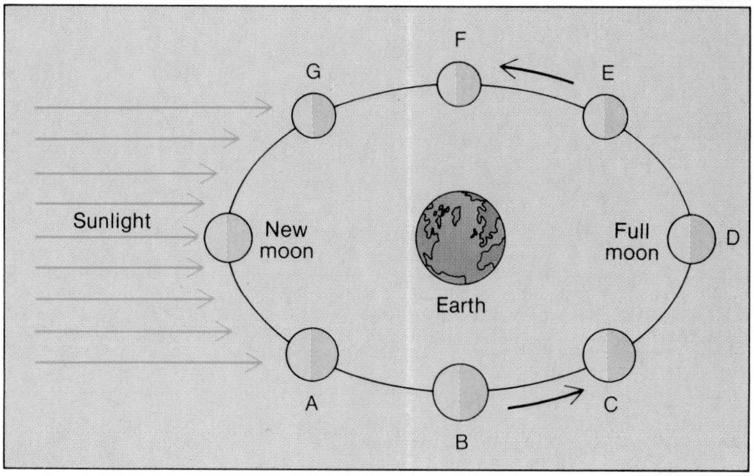

Fig. 18–27 The moon moves constantly between the sun and the earth, then away from the sun.

PHASES OF THE MOON

The moon does not shine or give off its own light. We can see the moon only because the moon reflects some of the sunlight falling on its surface. As the moon moves around the earth, its path takes it between the sun and the earth. Then it moves to the opposite side of the earth, away from the sun, as shown in Fig. 18–27. The position of the moon, the earth, and the sun determines how the moon looks to us as we see it from the earth. The different amounts of the lighted and dark parts of the moon seen from the earth are called the **moon's phases** (**faze**-uhz). See Fig. 18–28.

Moon phase The changing appearance of the moon as seen from the earth.

When the moon is between the earth and the sun, its dark side is facing the earth. This *moon phase* is called the *new moon*. If you were standing on the dark side of the moon during the new moon phase, you would see the sunlit side of the earth. As the moon moves in its orbit, a small part of the lighted side can be seen from the earth. This is the *crescent* (**kres**-ent) phase. When the moon has moved one-quarter of the way through its orbit, one-half of its bright side is seen from the earth. This phase is called the *first quarter*. After the first quarter, comes the football-shaped *gibbous* (**gib**-us) phase. After this phase, the amount of sunlit surface increases until all of the sunlit side of the moon can be seen from the earth. The fully-lit side of the moon is called the *full moon*. As the moon continues its orbit, the sunlit part of the moon begins to move out of view. The phases following full moon are gibbous, last quarter, crescent, and then another new moon. It takes 29 1/2 days to go through one complete cycle of moon phases. For

Fig. 18-28 The phases of the moon (left to right and top to bottom): *crescent, quarter, gibbous, full.*

example, from one full moon to the next is 29 1/2 days. However, the time it takes the moon to make one revolution around the earth is 27 1/3 days. The reason is that the earth also moves around the sun. Thus when the moon makes one revolution around the earth in 27 1/3 days, the earth has moved farther around the sun. The moon must move in its orbit about two more days or 29 1/2 days to return to its original phase.

THE MOON AND THE CALENDAR

In the past, people used phases of the moon to measure the passage of time. The calendar month originally came from the 29 1/2-day cycle of moon phases. A year, or the time it takes for the earth to make one complete orbit of the sun, is about 365 1/4 days. But, in a 365-1/4 day year, there are more than twelve 29 1/2 day cycles of moon phases. Since 365 1/4 cannot be divided evenly by 29 1/2, Western societies do not use a calendar based on the phases of the moon. Western calendars use a system of 12 months that add up to 365 days. To make up the extra quarter day of each year, a whole day is added every fourth year. In this way, we keep our calendar in step with the earth's movement around the sun.

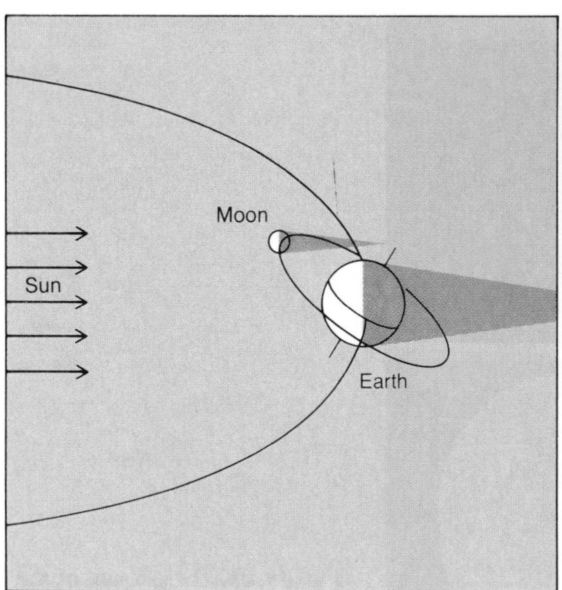

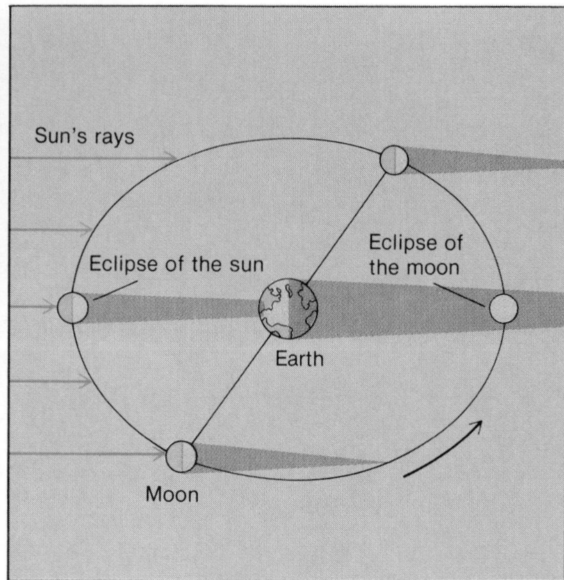

Fig. 18–29 (left) *The moon's tilting orbit makes its shadow usually miss the earth. The orbit is tilted less than is shown.* (right) *The moon's shadow causes a solar eclipse. The earth's shadow causes a lunar eclipse.*

ECLIPSES

Both the earth and the moon cast long shadows in space. At the new moon, the moon is between the earth and the sun. You might wonder why the shadow from the moon does not fall on the earth every new moon phase. The answer is found in the way the moon's orbit is slightly tilted in relation to the earth's orbit around the sun. See Fig. 18–29 *(left)*. Usually, the moon crosses too high or low between the earth and the sun for its shadow to fall on the earth. However, as the earth and the moon move together around the sun, the moon sometimes crosses directly between the sun and the earth. See Fig. 18–29 *(right)*. When this happens, an **eclipse** of the sun occurs. Only the tip of the moon's shadow falls on the earth's surface. Thus only a small part of the world is able to see a total *eclipse* of the sun. See Fig. 18–30. The shadow moves so quickly that the eclipse lasts only a few minutes at any one place. In the outer part of the shadow, where the light is not completely blocked out, only a part of the sun is seen to be covered. This is called a *partial eclipse*. Because the moon follows an elliptical orbit around the earth, it is not always the same distance away. During some eclipses, the moon is too far from the earth for its shadow to reach the earth's surface. This causes an *annular eclipse*. Annular means "ring." During an annular eclipse, the outer edges of the sun are still visible, forming a ring.

Eclipse The earth or the moon passing through the shadow of the other.

Fig. 18–30 A solar eclipse.

Fig. 18-31 The beginning of an eclipse of the moon.

An eclipse of the moon occurs when the moon passes through the earth's shadow. See Fig. 18-31. Eclipses of the moon are as rare as eclipses of the sun. However, unlike a solar eclipse, an eclipse of the moon is seen by everyone on the dark side of the earth.

SUMMARY

Both the earth and the moon move around a balance point located below the earth's surface. Together they make up the earth-moon system. As the moon orbits the earth, its appearance changes as it passes through its phases. Sometimes eclipses are produced when the shadow of the moon falls on the earth or when the moon moves through the earth's shadow.

QUESTIONS

Use complete sentences to write your answers.

1. Starting with a moonrise, describe the moon's movement in the sky. Explain why the moon will rise 50 minutes later each day.
2. List the eight phases of the moon, in the order in which they occur, starting with the new moon.
3. During which phase is the moon between the sun and the earth?
4. During which phase is the earth between the sun and the moon?
5. What causes an eclipse of the sun?
6. What causes an eclipse of the moon?

INVESTIGATION

THE PHASES OF THE MOON

PURPOSE: To demonstrate the phases of the moon using a model.

MATERIALS:

ball, one-half painted black
chalkboard
pencil
paper

PROCEDURE:

A. The ball represents the moon; your head, the earth; and a drawing on the chalkboard, the sun. Stand by your desk with your back to the sun.

B. Turn the unpainted side of the ball toward the sun. See Fig. 18–32 *(left)*. Hold it at eye level in front of you.

1. What phase of the moon does this represent?
2. On a sheet of paper, draw a circle. Make your circle look like the moon model as you see it.

Fig. 18–32

C. Turn your body clockwise one-quarter of a turn (90 degrees). See Fig. 18–32 *(right)*. Again hold the moon model at eye level with the light side toward the sun.

3. What phase of the moon is this?
4. On your paper, make a drawing of the moon model as you see it. With your pencil, shade in the part that is dark.

D. Turn another quarter clockwise. You should now be facing the sun. Hold the moon at eye level with the light side toward the sun.

5. What phase of the moon does this represent?
6. On your paper, draw a circle and shade it in to look like the model as you see it.

E. Turn one-eighth turn counterclockwise. Again turn the light half of the moon toward the sun.

7. What phase of the moon is this?
8. Make a drawing of what the moon model looks like.

F. Return to the position in step D. If you have the time, try continuing this same process until you have returned to your original position. Try to find a position that represents the crescent phase that occurs just after new moon and compare it to the drawing you made for the answer to question 8. Find positions that demonstrate the gibbous moon phases.

CONCLUSION:

1. Make a drawing similar to Fig. 18–27 on page 510 showing the positions of the moon, earth, and sun that you demonstrated in this activity. Label each moon position with its proper phase.

TECHNOLOGY

SAVING PLANET EARTH

Suppose a comet were on a collision course with earth. Could anything be done? At a conference held by NASA in 1981, scientists and weapons experts discussed this possibility and the methods to be used to save the planet.

Is this a fantasy space game? No, but new findings, especially those whose meaning is unclear, present new questions. In this case, careful study, combining the work of paleontologists and geochronologists, has suggested that mass extinctions have occurred about every 26 to 30 million years. A deposit of clay, rich in an element usually seen only in meteorites, was found worldwide. This layer separates rocks that contain dinosaur fossils from rocks that do not.

What could have happened during that period to eliminate the dinosaur? Whatever it was also produced the rich layer of clay and occurred every 26 to 30 million years. Some scientists believe it takes a period of approximately 26 million years for our solar system to travel through the crowded central portion of the Milky Way. Perhaps, scientists hypothesize, an asteroid crashed into the earth at that point. The tremendous explosion that resulted could have thrown so much dust and so many fragments of both the asteroid and earth into the atmosphere that sunlight could not reach the earth. The earth became cold, plants died, and the dinosaurs starved or froze to death. Eventually, this dust settled out of the atmosphere, forming the worldwide layer of clay. Once the sun could get through the atmosphere, life continued. If this happened once, perhaps it could happen again. And if the earth were to be bombarded with comets, the atmosphere would again be filled with dust and other particles. Sunlight would be blocked and the earth would cool. Would some life forms again disappear? At the 1981 NASA conference, it was agreed that this need not happen, not if the new technology of space were combined with the peaceful use of rocketry. With enough forewarning, a missile could be launched to meet the oncoming asteroid at a distance far removed from earth. In fact, weeks before the orbit of the comet could send it crashing into the earth, a computer-directed missile could be fired toward the comet. Last-minute adjustments could be made from earth to keep the missile on target. Scientists and weapons experts agree that it would not be necessary to use nuclear weapons nor to hit the asteroid directly. The explosive force itself behind the comet would actually be enough to push it into another orbit.

This scene may never be played out. It is possible that data from further research will disprove the theories concerning the mass extinctions of certain plants and animals. Part of planning for the future, however, is anticipating as many possibilities as are imaginable. In this case, the combined research efforts of scientists interested in the earth's past and those interested in its future could help prevent a possible disaster. In any case their further research will seek a greater understanding of the earth's relationship to other bodies in space.

¡COMPUTE!

SCIENCE INPUT

Our knowledge of the solar system changes as the technology of space telescopes and space travel becomes more and more advanced. In the earliest times, astronomical theories depended on observations from earth. Today, computerized images of the planets have been sent from space vehicles launched precisely for research purposes. Through all these scientific and engineering efforts, we have learned a great deal about the characteristics of these satellites of the sun and the entire solar system. Scientists hope that further research will push us past the limits of our own galaxy where we may gain an understanding of the structure of the universe. Program Satellite will help you to review the data presented in your textbook chapters about our solar system's planets.

COMPUTER INPUT

A computer can perform a number of functions, that is, a number of tasks, as you have already seen. Program Planets uses the computer's ability to select random numbers to design a four-question quiz for you (see Chapter 13 Compute! for the definition of random). Program Planet will ask questions, tell you when you're correct, and at the end of the quiz, tell you your final score. Other programs in the Compute! section have used the computer's ability for mathematical calculations. More complicated programming might involve computer graphics, to create images or graphs or very sophisticated mathematical computations.

Not every computer can perform all possible computer functions. What the computer can do depends on its power, or memory capacity. When talking about computers, users may refer to an 8K RAM or a 16K RAM machine. This information describes the amount of storage a computer has. (See Chapter 8 Compute! for a definition of "bits" and "bytes.") 8K means that the particular computer can manipulate and store approximately 8,000 bytes of information. Different computer functions require different amounts of storage capacity. Therefore, if you were buying software, it would be important to know how much storage it needed. A program needing 16K RAM will not run on your 8K RAM computer.

WHAT TO DO

On a separate piece of paper, make a chart similar to the one below, listing all terms and their definitions. The definitions should be short and simple. Study the chart before entering the program. When you run the program, it will produce a quiz of four questions using the characteristics of the planets you input. Lines 220 to 320 are for data statements. Your terms and definitions are data. Be sure to type the statement in the correct form. An example is given in line 220.

The program will tell you the correct answer if your answer is wrong. After four questions it will give you a score and tell you how many questions out of four you've answered correctly. Run the program at least five times, recording the terms you've missed so you know the ones you need to study.

DATA CHART

Sun Satellite	Characteristic
Mercury	closest to the sun
Jupiter	has 6 satellites
Venus	covered by a thick layer of clouds
Mars	believed to have been like earth
etc.	. . .

SUN SATELLITES: PROGRAM PLANETS

GLOSSARY

RAM — Random Access Memory. That part of the computer's memory in which data can be input, stored, changed, and deleted.

ROM — Read Only Memory. That part of the computer's memory that may not be changed or erased.

PROGRAM

```
100  REM PLANETS
110  DIM T$(10), D$(10)
120  FOR X = 1 TO 10: READ
     T$(X),D$(X): NEXT X:K=0
130  FOR Q = 1 TO 4: LET N = INT (10 *
     RND (1) + 1)
140  PRINT : PRINT "WHICH SUN
     SATELLITE HAS THIS
     CHARACTERISTIC?"
150  PRINT "--> ";D$(N): INPUT "-->
     ";A$
160  IF A$ = T$(N) THEN PRINT "YOU'RE
     CORRECT!"
170  IF A$ = T$(N) THEN K = K + 1
180  IF A$ < > T$(N) THEN PRINT "THE
     CORRECT ANSWER IS:  ";T$(N)
190  FOR T = 1 TO 2000: NEXT T: NEXT
     Q
200  PRINT "--> END OF QUIZ"
210  PRINT "--> ";K;" OF 4 ANSWERS
     WERE CORRECT"
220  DATA   MERCURY, CLOSEST TO
     THE SUN
310  DATA
320  END
```

PROGRAM NOTES

This program is written to be used with ten data statements. New words and definitions can, however, replace the old ones at any time. The program can, in that way, be used for more than ten planet characteristics or for any other set of words and their definitions. If you wanted to add more data statements, you would change line 110, which tells how many spaces of memory are being saved for data. You would also change line 140, of course, if you had changed the subject of your terms and definitions.

BITS OF INFORMATION

Space exploration depends heavily on the use of computers. The technology of space is so complicated that, without the tremendous capacity of the computer to do many kinds of calculations at the same time and quickly, it would be impossible to organize a space mission. Computers also permit two-way input into space vehicles. On a space mission it is not only the astronauts who pilot the spacecraft. The scientific and engineering ground crew can also assist through communication with the spacecraft's computer. This provides an additional safety factor.

Satellites also are in two-way communication with earth. They transmit data about conditions on earth and in the atmosphere. NASA has developed a suitcase-size, computerized rescue kit based on satellite communications. If, for example, you were lost at sea, you would open the kit and transmit a signal to a nearby communications satellite. The satellite would then transmit your signal to a station on land.

CHAPTER REVIEW

VOCABULARY

On a separate piece of paper, match the number of each statement with the term that best completes it. Use each term only once.

meteor	gibbous	meteorite	new moon
ray	eclipse	maria	asteroid
comet	sister theory	moon phase	ellipse

1. A smaller body of the solar system found in orbit between Mars and Jupiter is called a(n) ______.
2. ______ is the changing appearance of the moon as seen from the earth.
3. Lava plains on the moon are known as ______.
4. A(n) ______ is an icy body that comes from the edge of the solar system.
5. A(n) ______ results when the earth or the moon passes through the shadow of the other.
6. The ______ phase occurs when the moon is between the sun and the earth.
7. A(n) ______ is a bright streak of light in the sky.
8. A natural object that survives its plunge through the earth's atmosphere is known as a(n) ______.
9. The ______ moon phase occurs just before or after the full moon.
10. The ______ is an explanation of the origin of the moon that seems most likely true.
11. A light-colored streak leading out from a crater on the moon is called a(n) ______.
12. A(n) ______ is the name for the shape of the path followed by a solar system member.

QUESTIONS

Give brief but complete answers to each of the following questions. Unless otherwise indicated, use complete sentences to write your answers.

1. Why is the sun called the controlling body of the solar system?
2. Compare the two general kinds of planets in the solar system and how they formed.
3. Which one of the inner planets is the smallest?
4. Which of the inner planets have an atmosphere? What effect has this had on their surfaces that is not true of planets that have no atmosphere?

5. What is the source of the extra energy that Jupiter gives off that it doesn't receive from the sun?
6. Describe the makeup of Jupiter, starting with the outer cloud layer and going in to the center of the planet.
7. List the four Galilean moons of Jupiter in order from innermost to outermost.
8. Describe the rings of Saturn.
9. What unusual feature does Uranus have?
10. Where is the asteroid belt located?
11. What are comets made of and where do they come from?
12. Describe two kinds of meteorites that reach the earth's surface.
13. What explanation is given for the way in which Mars got its moons?
14. What surface features can you see on the moon with a pair of binoculars?
15. Why is a large part of the moon's surface covered by a fine dust?
16. Describe the theory that seems to best explain how the moon was made.
17. Why do we always see the same side of the moon?
18. What is a partial eclipse of the sun?

APPLYING SCIENCE

1. Using a pair of binoculars, or a small telescope, locate Jupiter in the night sky. Look for the four Galilean moons. The binoculars will need to be held steady against a fence post or similar stable object to keep it from vibrating. Make a drawing of the location of the moons of Jupiter that you can see.
2. Using a pair of binoculars, or a small telescope, look at the moon. Make a drawing of the part of the moon that is visible. Label any of the surface features that you can identify.

BIBLIOGRAPHY

Branley, Franklyn M. *Halley's Comet 1986.* New York: Dutton, 1983.

Chapman, Clark R. *Planets of Rock and Ice.* Pasadena, CA: Planetary Society, 1984.

Gore, Rick. "The Planets, Between Fire and Ice." *National Geographic,* January, 1985.

Kowal, Charles. "The Chiron Mystery." *Omni,* June 1983.

Miller and Hartman. *The Grand Tour.* New York: Workman, 1981.

Time-Life Editors. *Solar System* (Planet Earth Series). Time-Life Books, 1983.

CHAPTER

19

Stars looking like bright streaks circle the North Star above two observatories. Observatories are used to gather and read many different kinds of information carried in starlight.

STARS AND GALAXIES

CHAPTER GOALS

1. Describe a way to locate stars seen in the night sky.
2. List ways that the distance between the earth and a star can be measured.
3. Identify the main source of scientific information about stars.
4. Trace the steps in the life history of a typical star.

19–1. Observing Stars

At the end of this section you will be able to:

- Explain one way *constellations* can be useful.
- Describe ways of gaining information about stars.
- List two ways of finding the distance between the earth and a star.

There are more stars in space than all the drops of water in the oceans or all the grains of sand on the beaches of the earth. Except for the sun, the stars are also so distant from the earth that even the largest and most powerful telescopes show them only as pinpoints of light. Since the stars are so far away, it might seem difficult to learn about them. But information about the stars is carried in their light. By reading this information, scientists have learned what stars are made of, how large they are, and how they are moving in space.

STAR PATTERNS

Long ago people thought that stars were lights on a huge dome that surrounded the earth. However, we now know that each star is actually a sun.

On a clear, dark night, about 2,000 stars can be seen. With a little practice, you can begin to see that there are patterns among the stars. You can begin to recognize the pictures that people have used for thousands of years to identify these patterns. According to this old way of studying the stars, the particular star patterns can be seen as forming the outlines of people,

Fig. 19-1 This is a photograph of the constellation Orion. Can you see the figure of a hunter?

Constellation A group of stars in a particular pattern.

animals, or familiar objects. A group of stars in such an arrangement is called a **constellation** (kon-stuh-**lay**-shun). Most *constellations* are named after animals or ancient gods and heroes. Often it is hard to see how a group of stars makes an outline of something. See Figs. 19-1 and 19-2. At times, it may be easier to recognize a constellation as a baseball diamond rather than a horse with wings. However, no matter how you identify them, it is useful to be familiar with some of the con-

Fig. 19-2 Orion the hunter is depicted here with the prominent stars shown. The three bright stars forming the belt are easily found in the winter sky.

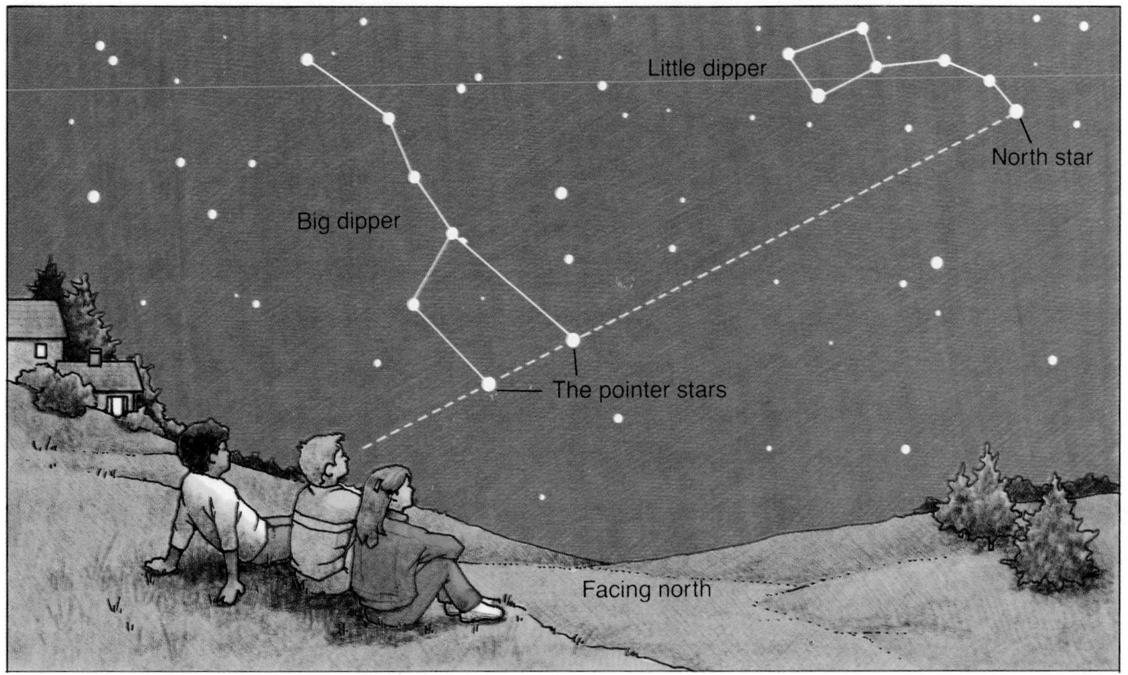

Fig. 19-3 The two stars in the bowl of the Big Dipper point to Polaris, the North Star.

stellations. They can be used as guides, or maps, to find certain stars in the sky. Once you become familiar with a few constellations, it is easy to find other constellations or individual stars.

The group of stars known as the Big Dipper is an example of how to use constellations as guides. The Big Dipper is part of a constellation called *Ursa Major*, or Big Bear. This constellation is easily seen in the northern sky all during the year in the Northern Hemisphere. Two stars forming the side of the bowl of the Big Dipper can be used as pointers to find the North Star. A line drawn through these pointer stars and going about five times the distance between them leads to the star called *Polaris* (poh-**lar**-us), or North Star. See Fig. 19-3. Polaris is the star at the end of the handle of the Little Dipper, which is part of the constellation called *Ursa Minor*, or Little Bear. A camera aimed at Polaris with the shutter left open for several hours will produce a photograph similar to Fig. 19-4. All stars except Polaris will show as a circular streak on the film. Polaris is almost directly above the North Pole. Thus all the other stars seem to circle around it as the earth rotates on its axis.

Fig. 19-4 The stars seem to circle in the sky as a result of the earth's rotation.

When we look at the sky, we see the stars in constellations as if they were located on a kind of ceiling over the earth. For example, the stars in the Big Dipper all seem to be at the same

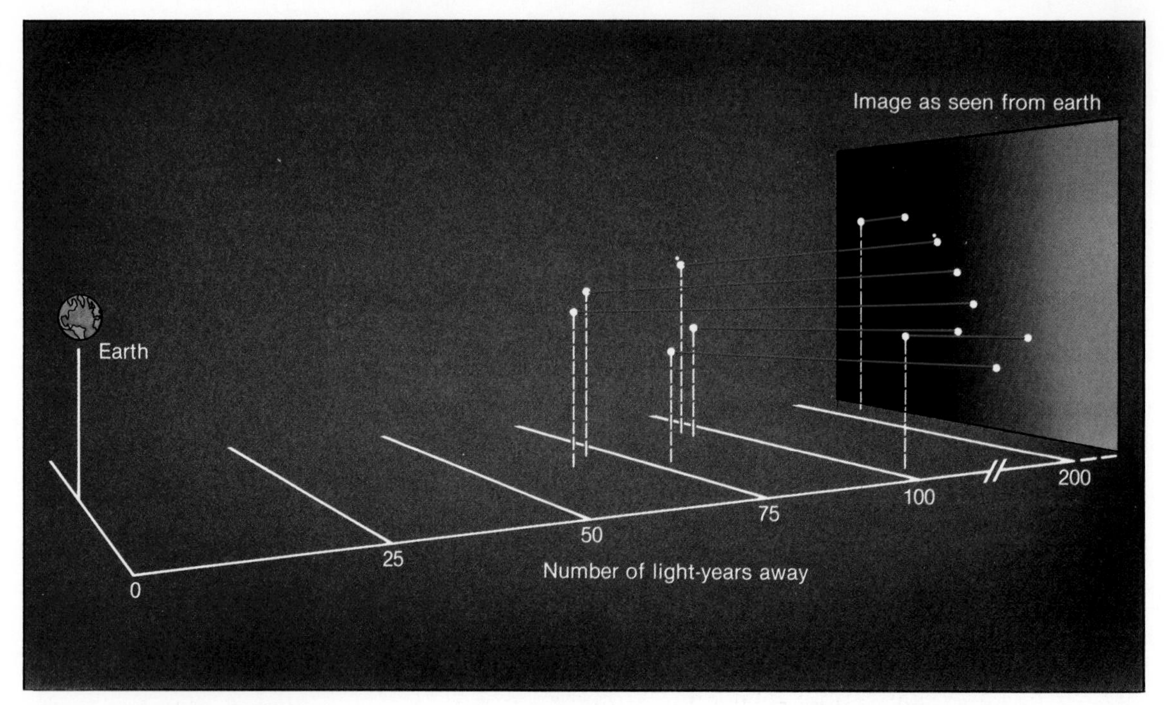

Fig. 19–5 The stars that make up the Big Dipper are at different distances from the earth.

distance from the earth. This is so because of their great distance. But the stars are actually widely scattered at different distances from us. See Fig. 19–5.

You may see little connection between the appearances of the constellations and their names. This is because the names were invented by people in a different society thousands of years ago.

STARLIGHT

Every star is able to send some kind of message to the earth. The message is in the form of radiant energy, which can travel through space. For example, when you look at a star, you can see some of its radiant energy in the form of visible light. Stars also give off invisible forms of radiant energy. There are radio waves, X-rays, ultraviolet, infrared, and gamma rays. *Astronomers* study the stars by capturing some of their radiant energy with telescopes.

Some kinds of telescopes can be used on the earth's surface. The most familiar are the telescopes that use lenses or special mirrors. These telescopes collect and magnify the visible light from stars. Radio telescopes capture and study radio waves that reach the earth's surface. Infrared rays can also be detected by ground-based telescopes. However, X-rays, ultraviolet rays, and gamma rays from stars are mostly blocked by the earth's atmosphere. Thus the telescopes that capture these forms of energy must do their observing from satellites or space vehicles high above the atmosphere.

All of the kinds of radiant energy sent out by stars carry information about the stars' atoms. The light from a star can be compared to music that is made up of sounds produced by different instruments. Each instrument adds its own sound to the music. In the same way, the different atoms in a star add a characteristic part to the energy given off by the star. Scientists study starlight by first collecting it with a telescope. Then a special instrument on the telescope, called a *spectroscope,* separates the light into its characteristic parts. These parts are then recorded on photographic film or in computers for later study.

By studying the records made by spectroscopes, astronomers are able to learn about what kinds of atoms are in stars. All scientific knowledge about stars comes from the messages that can be read in starlight.

HOW FAR AWAY IS A STAR?

When looking at the night sky, you can see that not all stars have the same brightness. Actually, some of the brightest objects seen are planets. Venus, Mars, Jupiter, and Saturn are easily seen without a telescope. Planets can be sorted out from stars by observing their positions from night to night. Since stars are much farther away from the earth than the planets, stars appear to keep the same position among other stars. However, planets are much closer. Thus they appear to move against the background of stars.

Some close stars actually do seem to move very slightly when observed every six months. As the earth moves in its orbit, these stars seem to shift their positions in relation to the more distant stars. This is similar to the way a nearby tree appears to move against the distant background when viewed from a moving car. See Fig. 19–6. Trees that are farther away appear to move less against the distant background. This apparent motion can be used to give an idea of how far away the trees are.

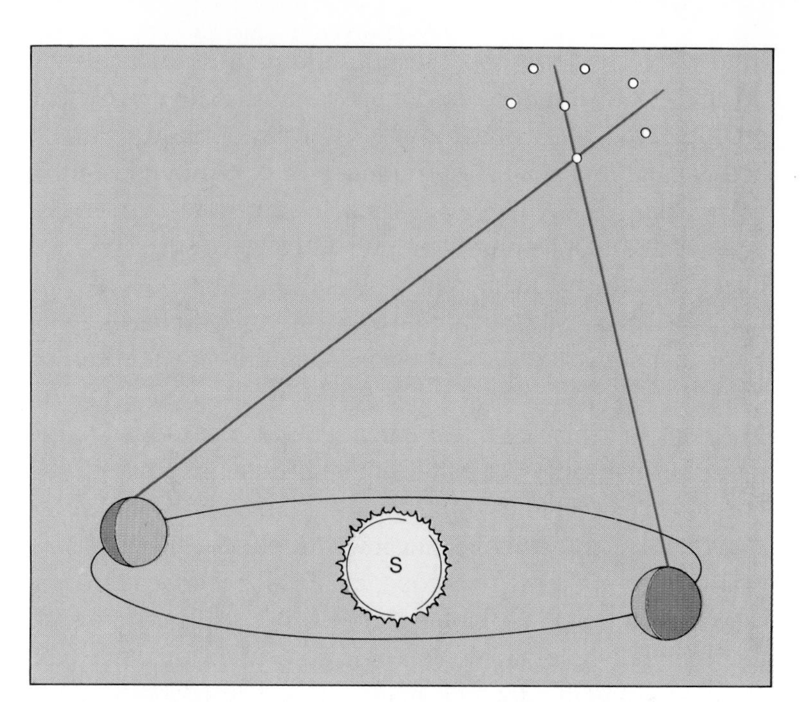

Fig. 19-6 Close stars seem to shift their position when seen from opposite sides of the earth's orbit.

Parallax The apparent shift of position of a star when seen from different points in the earth's orbit.

Light-year The distance light travels in one year.

This apparent motion is called **parallax** (**par**-uh-laks). By using *parallax*, the distances to nearby stars can be measured.

At first astronomers used kilometers to measure how far from earth objects in space were. But as they began to measure the distances to stars, the numbers became too large to use easily. For example, the distance of the closest star to the earth, other than the sun, is 4×10^{13} (40 trillion) kilometers. So astronomers decided to use another unit, the **light-year,** to measure large distances. A *light-year* is the distance light travels in one year. Light travels about 300,000 kilometers per second. To figure out how far light would travel in one year, you have to find out how many seconds there are in one year. You can do this by multiplying 60 (the number of seconds in one minute) by 60 (the number of minutes in one hour) by 24 (the number of hours in one day) by 365 (the number of days in one year). This gives you the number of seconds in one year. The number of seconds in one year can then be multiplied by the speed of light to give more than 9×10^{12}, or 9 trillion, kilometers. Thus it is easier to say that the nearest star is 4 light-years away than to write out the number of kilometers.

Scientists use other methods to measure the distances to stars that are far away. One method makes use of the brightness of the star. Have you ever judged the distance to a street light by how bright it looked? If you have, you were estimating distance using brightness. The actual amount of light a star sends out is called the **absolute magnitude** (**ab**-suh-loot **mag**-nih-tood). Stars have different *absolute magnitudes*. Some are 10,000 times brighter than the sun. Others are only ten-thousandths (0.0001) as bright as the sun. Once the absolute magnitude of a star is known, its distance can be found by comparing its absolute magnitude to its brightness as actually seen from the earth. You might judge the distance to a street light by making the same kind of comparison.

Absolute magnitude A measure of the actual brightness of a star.

Astronomers have different ways of finding the absolute magnitudes of stars. One of the best ways makes use of a kind of star whose brightness changes during periods of one day or up to many days. The length of time it takes for these stars to change their brightness is related to their absolute magnitudes. Once the absolute magnitude of these stars is found, the distance to them, and others close to them, can be found. This is done by comparing their brightness, as actually seen from the earth, to their absolute magnitudes. For example, a star that has a high absolute magnitude may appear dim. This means that the star is probably very far away.

SUMMARY

Anyone who observes the night sky notices that there are patterns in the arrangement of the stars. These patterns are now used as guides to locate the positions of stars. Detailed information about stars comes mainly from the energy that they send out into space. Observing closer stars from different points in the earth's orbit can show their distance from the earth. Stars farther away show a change in brightness that can be used to determine their distance from earth.

QUESTIONS

Use complete sentences to write your answers.

1. Why is knowing the location of some constellations useful?
2. How is information gained about stars?
3. What various kinds of scientific information can we get from starlight?
4. List two methods used to measure the distance from earth to stars. Which one can be used only for close stars?

INVESTIGATION

THE PARALLAX EFFECT

PURPOSE: To find out how the parallax effect helps us judge distance.

MATERIALS:

pencil
chalk
chalkboard
modeling clay
ruler

PROCEDURE:

A. Hold your pencil at arms length with the eraser end up. Close your right eye and, using only your left eye, line up the pencil with some object on the far wall of the room. Without moving the pencil or your head, open your right eye and close your left eye.
   1. Describe what happened to the position of the pencil compared to the object on the far wall.

B. What happened in step A is called the "parallax effect." It is the basis of our ability to tell distances with just our eyes. Hold the pencil again as in step A. While keeping your right eye closed, move your head to the right. Do not move the pencil.
   2. In what way is the effect like that in step A?

C. Copy the following table.

Distance of Pencil from Eye (cm)	First Eye (Number)	Second Eye (Number)
30		
60		
90		

D. Draw 10 vertical lines on the chalkboard 10 cm apart. From left to right, number the lines from 1 to 10.

E. To see the parallax effect, one must observe a near object (pencil) against a distant background (chalkboard) by looking at it from two different locations (each eye). Place the pencil, pointed end up, on the table between you and the lines drawn on the chalkboard or wall. Anchor the pencil in clay so it does not move. Hold your head level, 30 centimeters from the pencil. Using one eye only, sight the point of the pencil so it lines up with one of the numbered lines on the chalkboard or wall. Then, without moving your head, open the other eye and close the first eye. Note the number with which the point of the pencil now lines up. Estimate fractions between numbers. Record these two numbers in your table.

F. Move your head 60 centimeters from the pencil and repeat step E. Record the numbers in your table.

G. Move your head 90 centimeters from the pencil and repeat step E. Record the numbers in your table.

CONCLUSIONS:

1. How did moving farther from the pencil affect the parallax?
2. Our eyes judge distances to objects automatically until the objects are far from us. How could we use parallax to judge distances that are too far for just our eyes to judge? (Hint: See step B.)
3. What conditions are necessary to be able to judge distance by parallax?

19–2. The Life of a Star

At the end of this section you will be able to:

- Explain why stars have different colors.
- Relate the color of a star to its absolute magnitude and surface temperature.
- List the stages in the life history of a typical star.

Circling high above the atmosphere, the space telescope will have a much clearer view of the universe than telescopes on the ground. See Fig. 19–7. Outside the atmosphere, the space telescope will probe seven times deeper into space than before.

Fig. 19–7 The orbiting space telescope can see objects ten times more clearly and fifty times farther away than telescopes on the ground.

STAR COLORS

The messages carried by the light from stars have shown scientists that there are many kinds of stars. Some stars are larger than the earth's orbit around the sun. Others are smaller than the earth. There are stars made of gas thinner than air and others that are harder than diamond. One kind of star repeatedly blows up like a balloon, then shrinks, and blows up again, in an endless cycle. One difference among stars is their color. They appear at night in a wide variety of colors. This is because the color of starlight is related to the temperature on the star's surface. For example, you have probably seen how an iron poker held in a fire changes color as it gets hot. As a piece of iron is heated, it first becomes red. After further heating, the iron glows with a yellow light. Finally, it becomes a brilliant blue-white. In the same way, the color of a star shows its surface temperature.

The coolest stars have surface temperatures of about 3,000 K or higher. Astronomers have discovered that there is also a relationship between the star's surface temperature and its brightness, or absolute magnitude. Generally, the higher the temperature of the star, the higher the absolute magnitude.

TYPES OF STARS

When stars are plotted on a chart according to their surface temperature and absolute magnitude, they fall into three main groups. See Fig. 19–8. Most stars fall into a narrow band on the chart. The band begins with hot stars at the upper left and ends with cool stars at the lower right. Since most known stars fall into this group, they are called the *main sequence* (**see**-kwens) stars. A second group of stars appears at the upper right of the chart. These are bright stars even though their red color shows low surface temperatures. Their brightness results from their great size. The stars in this group are called *red giants*. One red giant is so large that, if it were placed at the center of the solar system, its surface would reach past Mars. A third group of stars appears at the bottom of the chart. These are called *white*

Fig. 19–8 Stars are grouped in this chart according to their surface temperature and absolute magnitude.

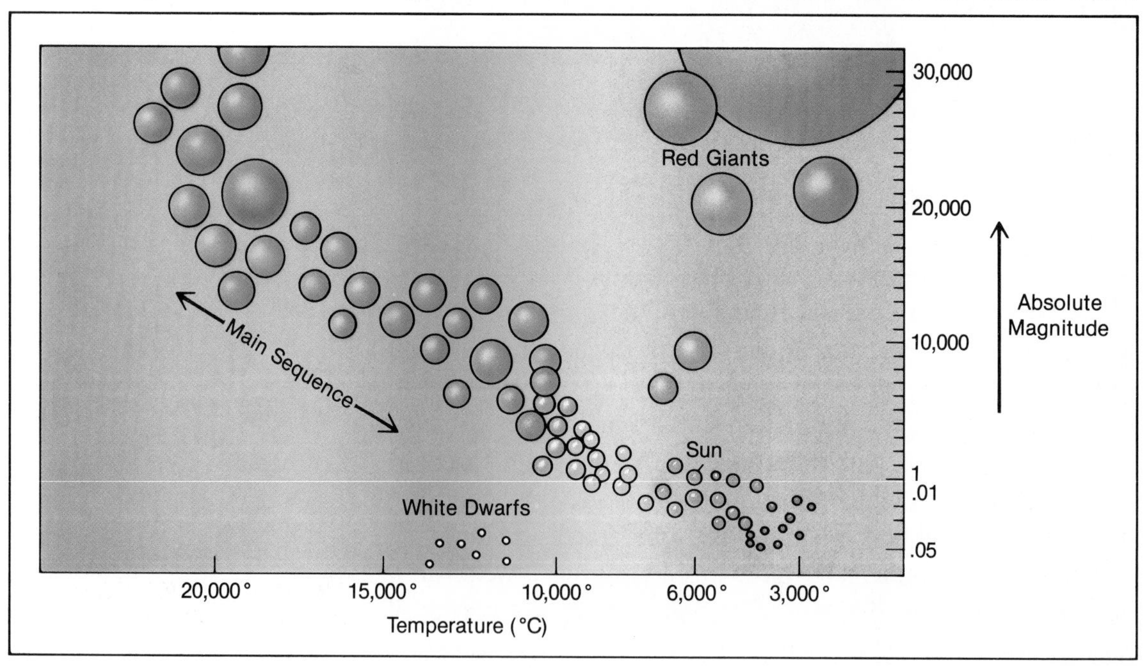

dwarfs. They are very hot and produce white light. But they are very small and therefore not very bright. Some white dwarfs are only the size of the earth.

LIFE OF STARS

Most stars have life histories that are measured in billions of years. The groups of stars shown on the chart in Fig. 19-8 have helped astronomers piece together the life history of stars. The stars in each group are believed to represent different stages in the life history of typical stars. A star begins as a cloud of gas and dust. Such clouds are common in the space between stars. Many such clouds have been observed. See Fig. 19-9. The force of gravity pulls the clouds of dust and gas together. As the center of the cloud shrinks, it becomes hot. Finally, the temperature rises high enough to begin the process of fusion. In fusion, hydrogen nuclei are brought together to make helium nuclei, giving off energy in the process. The star begins the first stage of life as its interior furnace begins to supply energy. The heat produced prevents the star from shrinking any more. The star becomes a main sequence star. It shines with a brilliant blue-white light.

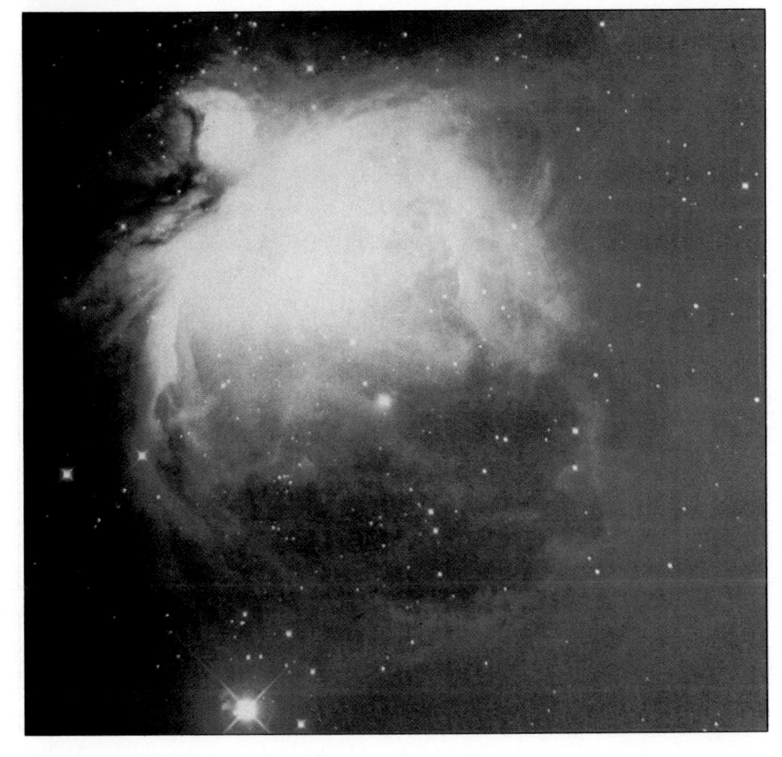

Fig. 19-9 A cloud of gas and dust seen in the constellation Orion.

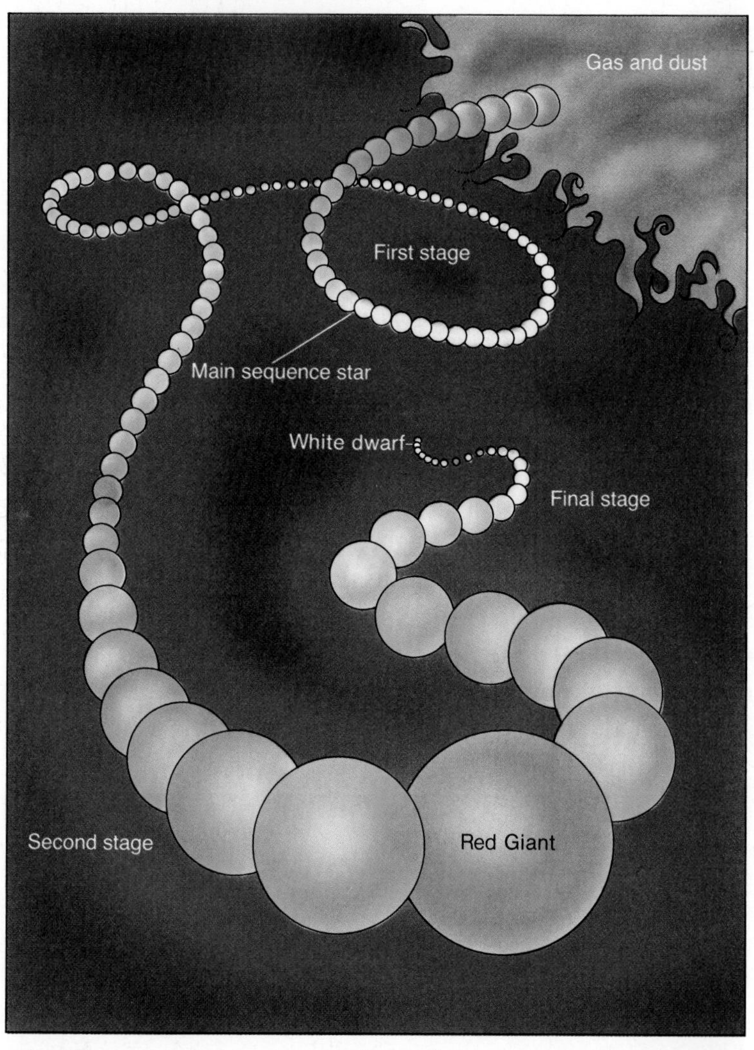

Fig. 19–10 Stages in the life history of a typical star.

In this stage the young star has a large supply of hydrogen. See Fig. 19–10. For a long period of time, it shines with a steady light. The star will be in this stage longer than in any other stage in its life. Thus, most known stars are in the main sequence stage. The sun, for example, is a main sequence star.

Eventually, a star will have used up nearly all of its hydrogen. Then its gravity starts to make it shrink. This makes the interior even hotter. It becomes possible for the helium nuclei to join to make heavier elements. These reactions give a new burst of energy to the star. It begins the second stage in its life history. See Fig. 19–10. The star expands to giant size. But its large size causes the surface temperature to drop. It becomes one of the red giants seen on the chart in Fig. 19–8.

When the helium fuel is also used up, the star shrinks again. Now the star enters its third and final stage. See Fig. 19–10. Its smaller size allows more energy to reach its surface, and it shines with a white light. A white dwarf seen on the chart in Fig. 19–8, page 530, has been formed. White dwarfs are very old stars that have largely used up their supply of nuclear fuel. They slowly grow colder until they burn out. They finally die as black dwarfs, like the ashes from a fire.

The larger the size of a star at birth, the faster it goes through the three stages. This is because the force of gravity is much greater. Gravity causes the star to shrink faster. For example, the star Sirius (**seer**-ee-us) is 2.5 times the mass of the sun. It is going through its life stages about 15 times faster than the sun. The sun probably has about five billion years to go before it will become a red giant. At that time it will expand so much that the solar system will be destroyed.

UNUSUAL STARS

Very large stars have a different life story. Stars having more than about four times the mass of the sun use up their hydrogen fuel very quickly. These larger stars explode at the time when a smaller star would only be through half its life. The explosion can become as bright as 10,000 suns. Such a flare-up in the sky is called a **supernova** (soo-pur-**noe**-vuh). Among the stars seen from the earth, a *supernova* is observed about once every 140 years. These stars collapse after the explosion. The great force crushes the atoms, leaving only neutrons. The result is a *neutron star*. A neutron star is about 15 kilometers in diameter and is very dense. One spoonful of its matter would weigh more than 9×10^7 million metric tons on the earth. If the earth were squeezed down to the same density, it would fit into a baseball park.

Supernova A large star that suddenly explodes.

Neutron stars are also known as *pulsars*. At first, scientists thought that pulsars were an alien radio source. They give off rapid pulses or bursts of radio waves, light, and X-rays. A neutron star can spin on its axis as many as thirty times each second. A source of energy in its surface sends out a stream of radiation in a pattern similar to that of the rotating light in a lighthouse. When the earth is in the path of one of these streams of radiation, radio telescopes can pick up the signals as rapid pulses.

Astronomers believe that some very large stars may collapse with such force that they pass beyond the stage of being a neutron star. They become *black holes*. A black hole is an object so small and dense that its gravity will allow nothing, not even light to escape from it. Because no energy of any kind can come from a black hole, there is no way to observe it directly. If you tried to take a flash picture of a black hole the light would only disappear down the hole. However. astronomers believe that they may be able to locate some black holes by the effect they have on nearby objects. Black holes will swallow anything that comes near. For example, a black hole close to a star would strip matter off its neighbor. The material lost from the star would then give off energy such as X-rays as it disappeared down the black hole. Powerful X-ray sources are believed to be the result of black holes taking in matter from nearby stars.

One of the most likely possibilities for a black hole is Cygnus X–1. It was the first X-ray source discovered in the constellation Cygnus. Cygnus X–1 is about six to eight times the mass of the sun and about 8,000 light-years from the earth. Scientists continue to search for other possible black holes.

SUMMARY

The color of a star is determined by its surface temperature. Red stars are coolest, while blue stars are the hottest. Sorting out stars according to their color and brightness is the key to learning all about their life histories. Most stars move through several stages, during which they change in color and size. They end up as small, black dwarfs. Large stars have a different fate. They may end up as a very small dense object with unusual properties.

QUESTIONS

Use complete sentences to write your answers.

1. What determines the color of a star?
2. What two properties of stars help us learn their relative age?
3. Starting right after a star is formed from gas and dust, list three changes it goes through before it becomes a black dwarf.

SKILL-BUILDING ACTIVITY

STAR COLORS AND TEMPERATURES

PURPOSE: To use an experiment to infer why a star's color shows its temperature.

MATERIALS:

masking tape
plastic ruler (30 cm)
No. 30 uncoated steel wire (3 m)
1.5 V flashlight bulb
miniature socket
flashlight battery (size D)
black paper
grating

PROCEDURE:

A. Cover each edge of the plastic ruler with masking tape. Wind about 3 meters of the wire around the ruler so the turns do not touch. Tape each end of the coil in place, leaving about five centimeters free on one end. Put the bulb in the miniature socket.

B. Tape one socket wire to the free end of the coil. Tape the other wire to the base of the flashlight battery. Put the setup on the black paper. See Fig. 19–11 *(top)*.

C. Like stars, electric light bulbs radiate a continuous spectrum. You will control the amount of current flow and the temperature of the bulb by touching the center post of the battery to different places on the coil of wire. See Fig. 19–11. Try this now.

1. When does the bulb glow brightest?

D. From the brightest light, in about five moves find where the bulb just glows. At the same time, your partner will watch the light through a grating that breaks up the spectrum of light. To block stray light, cup your hand around your eye.

2. Are all the color bands visible when the light is brightest? Which colors seem to fade most?
3. Looking closely at the bulb, describe the spectrum when the bulb is dim.

E. Betelgeuse (**bet**-'l-jooz), Rigel (**rye**-jul), and Capella (kuh-**pell**-uh) are three stars in the winter constellation Orion (oh-**rye**-un). They appear bright red, bluish, and yellow-white, respectively, as shown in the bar graphs. See Fig. 19–11 *(bottom)*.

4. Which star gives the brightest violet-blue (v-b) color band?
5. Which star has a yellow-orange band brighter than its other colors?

CONCLUSIONS:

1. Based on the information you obtained from the light-bulb experiment, which star in step E is the hottest star?
2. Why is Capella yellow-white?

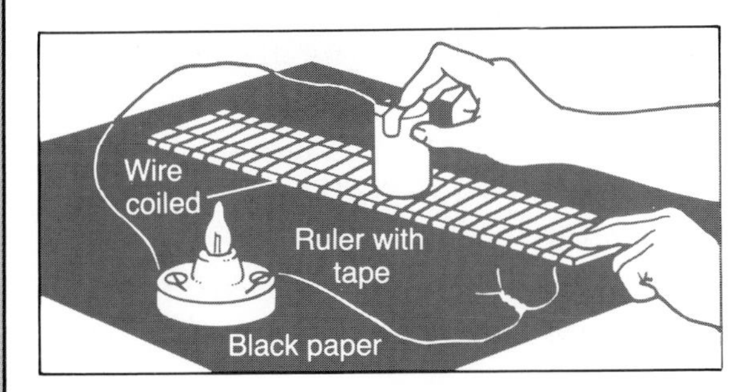

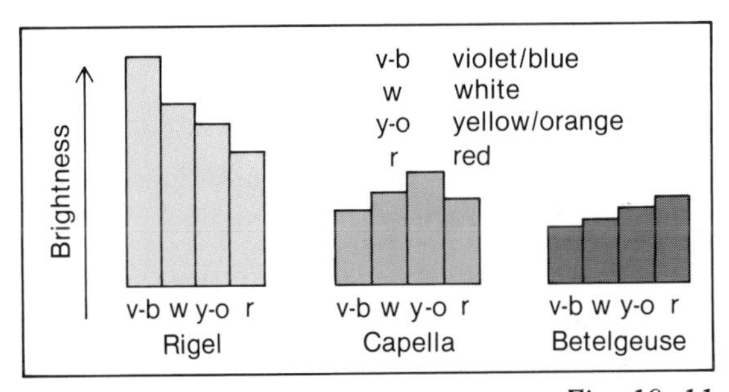

Fig. 19–11

19–3. Galaxies

At the end of this section you will be able to:

- Explain what a galaxy is.
- Describe the structure of the Milky Way galaxy.
- Give three theories for the origin of the universe.

On a clear moonless night you can see a bright, thin cloud stretching across the sky. See Fig. 19–12. This cloud of light actually consists of billions of stars. It is so large that it would take light 100,000 years to travel from one edge to the other.

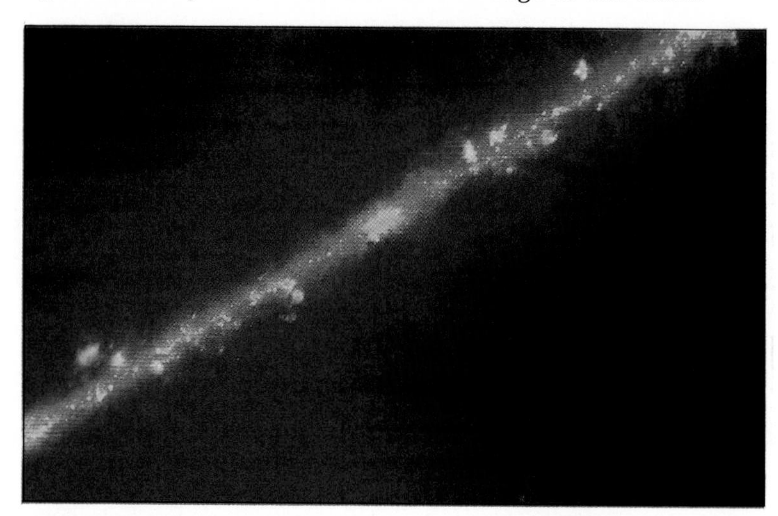

Fig. 19–12 The Milky Way as seen by the Infrared Astronomy Satellite (IRAS).

THE MILKY WAY

Milky Way A faint band of light seen spread across the night sky, produced by the stars in our galaxy.

Galaxy A huge system of stars held together by gravity.

The faint band of light we see on any clear, dark night is called the **Milky Way.** With binoculars or a small telescope, the *Milky Way* can be seen to be made up of thousands of faint stars. The combined light of these stars makes the Milky Way appear to be a continuous streak across the sky. Why are so many stars seen in one part of the sky?

The sun and all other stars seen from the earth are part of a huge system of stars called a **galaxy.** If the *galaxy* were the size of North America, the stars would be like tiny specs, smaller than a thousandth of a centimeter, and 150 meters apart. The galaxy to which the sun belongs contains more than 100 billion stars. The stars are held in a galaxy by their gravity, which tends to pull the stars toward each other. The galaxy to which the sun belongs is usually called the Milky Way galaxy.

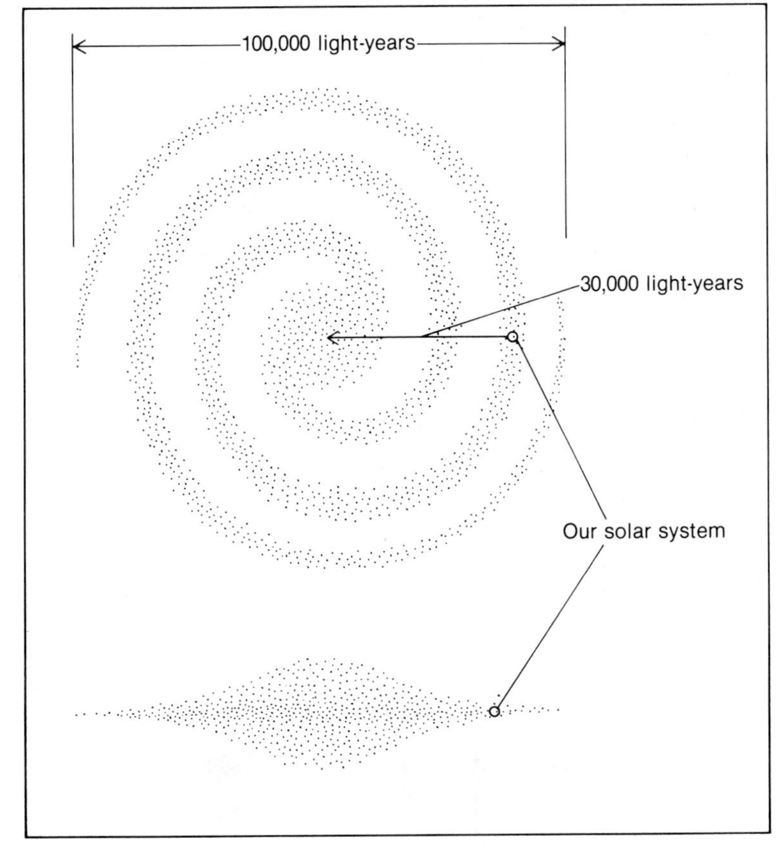

Fig. 19–13 Our solar system is near the edge of our galaxy, the Milky Way.

The galaxy is shaped like a disk with a bulge at its center. The solar system is located closer to the edge than to the center of the galaxy. Thus, looking toward the galaxy's center, we see many more stars than when looking elsewhere. See Fig. 19–13.

OUR GALAXY

On a very clear, dark night, if you look at the constellation of Andromeda, it is possible to see a small, fuzzy patch of light among the stars. A telescope shows that this is a galaxy about 2.2 million light-years away. See Fig. 19–14, page 538. It has a spiral shape that makes it look like a pinwheel. Scientists think that the Milky Way galaxy would also look like this if we could see it from far away.

Our galaxy is more than 100,000 light-years from edge to edge. It has a thick center that is about 10,000 light-years across. Surrounding the center is a flat disk that is divided into spiral arms. The sun is located in one of these spiral arms about 30,000 light-years from the center, as can be seen in Fig. 19–13.

Fig. 19-14 The Andromeda Galaxy.

The Milky Way galaxy is made up of about 200 billion stars. Most of these stars are located in the flat disk made up of the spiral arms. To us, the stars appear to be crowded together in the sky. However, this is a result of our looking at the stars many thousands of light-years into space. This makes the stars appear to be stacked together. Actually, the stars are very far apart. To get some idea of relative distances, imagine how far from each other a few baseballs would be if you scattered them across the entire United States. However, if seen from the moon, the baseballs would appear to be grouped together. There are 60 known stars within 17 light-years of the sun. The closest is 4.3 light-years away. Between the stars there are large amounts of gas and dust.

All the stars in the Milky Way galaxy revolve around its center, in the same way that the planets revolve around the sun. The sun takes about 230 million years to make one revolution around the galaxy.

THE UNIVERSE

Telescopes that look deep into space beyond our own galaxy show that the universe contains billions of other galaxies. Just as stars are seen to differ from each other, not all galaxies are alike. In addition to the spiral shape of the Milky Way and Andromeda galaxies, some galaxies are shaped like footballs or like spheres. Other galaxies have no general shape. All galaxies that can be observed have one thing in common. They are all moving away from each other at a very high speed. Scientists can tell that the galaxies are moving apart because of the information that can be read from their light. There is more red than there would be if the galaxies were not moving or were moving toward the earth. This is called the *red shift* of the starlight. The red shift in the light from the galaxies is caused when the light waves are spread out as the source moves away. You may have heard the same effect with sound waves that come from a moving source. For example, the sound of a siren from a moving car changes as the car moves past you. The sound waves are crowded together as the car approaches then spread apart as the car moves away. In the same way, light waves from a distant galaxy can show that it is moving away.

The red shift seems to show that the galaxies that are farthest away are moving at the highest speeds. If this is true, then the entire universe must be expanding. To get a clearer sense of this, think of the galaxies as dots painted on a balloon. The balloon expands as it is blown up. The dots are seen to move away from each other. In the same way, if the universe is expanding, then all of the galaxies will be seen to be moving away from each other.

One of the most puzzling of all objects seen by astronomers is the **quasar** (**kway**-zahr). *Quasars* seem to be young galaxies that are still forming. The amount of energy quasars produce is much greater than the energy given off by ordinary galaxies. Thus quasars are bright enough to be seen at great distances. Many quasars seem to be 10 to 15 billion light-years away. The light from quasars shows a very large red shift, indicating that they are moving away at high speeds.

Quasar A small, starlike object, very far from the earth, that produces a huge amount of energy.

HOW DID THE UNIVERSE BEGIN?

The idea that the universe is expanding has led scientists to the *Big Bang* theory. This theory says that at one time the entire universe was concentrated in a small volume. Then, about 15 to 20 billion years ago, there was a tremendous explosion that caused the universe to begin expanding. About one million years after the Big Bang, huge clouds of hydrogen and helium began to form. The clouds were the beginning of the galaxies. In time, stars started to grow in the galaxies. Within about 10 billion years after the Big Bang, all of the galaxies that are now seen had formed. Today, we see the results of that Big Bang as a rapidly expanding universe with galaxies moving away from us. Scientists have also found that all space is filled with weak radio waves. These radio waves are believed to be left by the great explosion that was the Big Bang.

Other theories have been developed to explain the universe. For example, the *steady-state* theory says that new matter is always being created in space. As the expanding universe carries the old stars away, new ones are created to replace them. Thus, according to the steady-state theory, the universe is in a state of balance, with no beginning or end. Another theory is the *oscillating universe* theory. *Oscillating* means "to swing back and forth." The oscillating universe theory says that gravity will finally stop the outward movement of the galaxies caused by the Big Bang. Then they will move back together, forming a hot mass, until another Big Bang sends them flying out again. This is repeated over and over, so the universe will continue to expand and shrink forever.

At the present time, most astronomers believe that the steady-state and oscillating universe theories cannot be proven. There is evidence that the universe started with an explosion and has continued to expand since that time. The way the universe may have begun and how it might end is a question that scientists still are trying to answer.

SUMMARY

Like most stars, the sun is part of a galaxy. The galaxy to which the sun belongs is shaped something like a saucer, with spiral arms reaching out from a central bulge. We see evidence of our galaxy in the form of a band of stars crossing the sky. The universe contains billions of other galaxies that are apparently moving apart as the universe expands.

QUESTIONS

Use complete sentences to write your answers.

1. What is a galaxy?
2. What causes the Milky Way we see in the night sky?
3. Describe the size and shape of the Milky Way galaxy.
4. Explain three theories for the origin of the universe.

TECHNOLOGY

SPACELAB

Spacelab 1 flew for ten days in November 1983. In that time, it produced information to fill 280,000 volumes of an encyclopedia. It will probably take ten years to analyze all the data. *Spacelab 1* was an important first because it not only took scientific experiments into space but also took the scientists with them. It was significant too as an international research effort joining the forces of the European Space Agency and NASA.

The specialists on *Spacelab 1* carried out 72 experiments designed by scientists in 11 countries. These included investigations into how the biological clocks of living things are affected by zero gravity; how zero gravity affects the human immunological system; and how space sickness is caused. *Spacelab 1* carried the Imaging Spectrometric Observatory (ISO), a set of unique instruments for studying the atmosphere.

Spacelab produced a new kind of scientist—one who is competent to perform experiments in biology, physics, chemistry, and astronomy in a very limited period of time. *Spacelab* produced a new space traveler, the payload specialist, who is not one of the crew. The mission also proved that people in space were extremely valuable. As equipment broke down, the scientists in space could repair it. A camera designed to produce extremely clear photographs of 2,590-square-kilometer regions of the earth jammed. This could have led to a major disappointment, since up to that time only 40 percent of the earth had been mapped well enough for resource planning. What *Spacelab* could do in weeks would normally take from 10 to 20 years of expensive photographing from planes. What happened? The camera was repaired on board and produced 1,000 important photographs.

Why are *Spacelab* experiments so important? Space can provide conditions that do not normally occur on earth, making some scientists and manufacturers wonder if certain things that would be impossible to make on earth might be built or developed in space. For example, one experiment was designed to test the possibility of manufacturing single, perfect crystals of silicon in the weightlessness of space. These crystals could be used to make faster and less expensive computer chips. On earth, under the influence of gravity, melting silicon is affected by currents of heat. The particles do not always cool evenly, producing impurities. Videotapes from *Spacelab* showed no evidence of heat currents on board. Styrofoam spheres, more uniform than any on earth, were produced. More research is needed to know whether mineral crystals produced in space are so free of defects that it is worth the tremendous cost of manufacturing in orbit. Orbiting space stations may someday produce synthetic minerals that must now be mined on earth. *Spacelab* is scheduled for four more flights before 1988. The scientific community awaits these opportunities for new experiments. Each of these flights offers another chance for increased international cooperation in scientific research.

¡COMPUTE!

SCIENCE INPUT

We can measure the distance to some nearby stars if we know the parallax angle. You learned in this chapter that the parallax is the apparent shift of position of a star when seen from different points in the earth's orbit. The parallax angle of a star is the angle it makes across the sky as seen from one of the points in the earth's orbit.

Using what is called the method of trigonometric parallax, we can calculate the distance to nearby stars. The formula is:

$$\underset{\text{(light years)}}{\text{DISTANCE TO STARS}} = \frac{3.26}{\underset{\text{(seconds)}}{\text{PARALLAX ANGLE}}}$$

Parallax angles used to be measured with photographs taken six months apart. Now new devices called measuring engines enable scientists to measure stars five times more distant than could be determined previously.

COMPUTER INPUT

The computer's ability to do mathematics quickly and accurately is a great help in doing long calculations. This ability represents the immense progress made in mathematics because of new technology. One of the first tools used to assist in the solution of mathematical problems was the abacus. The slide rule, invented in 1630, was a wonderful tool for mathematicians, engineers, and scientists because it was small, portable, and required only the application of human energy. Attempts to develop mechanical calculators, powered by hand, began in the 17th century. The first practical device for calculation, however, wasn't invented until 1889. It only did multiplication. The electronic revolution brought us the hand calculator, which gave us speed as well as convenience in mathematical work. The full range of the computer's abilities, though, can make the hand calculator seem like an ancient tool.

In this exercise, the speed of a computer's mathematical capabilities will be matched against your own and those of a hand calculator.

WHAT TO DO

You will need paper and pencil, a watch or clock with a secondhand, hand calculator, and a computer. On a separate piece of paper, copy the data chart below.

Using the formula given above and the data for the parallax angle of some nearby stars, calculate with pencil and paper the distance from the earth to those stars. Do the same calculations using a hand calculator. Record the time it took to do each of these tasks.

Enter Program Parallax. (Remember to check your manual for the correct command for a clear screen statement.) Save the program on tape or disk and then run it. Record the time it took for the computer to do the calculations.

DATA CHART

Star's Name	Parallax Angle	Distance
Sirius	.379	
Rigel	.005	
Alpha Centauri	.760	
Altair	.198	
Barnard's Star	.544	
Procyon	.294	
Wolf 359	.402	
Beta Cassiopeia	.073	

STAR DISTANCES: PROGRAM PARALLAX

GLOSSARY

DISK DRIVE — That section of the hardware that "reads" the program stored on disk. Some software is in cassette form, not on disk. In that case a drive is not necessary.

PIXEL — A picture element. A pixel is a point or tiny box on the screen. The screen is composed of vertical and horizontal pixels.

RESOLUTION — The number of pixels on a TV screen or monitor. High resolution means there are a large number of pixels. Low resolution means there are fewer. The higher the resolution, the sharper the image or display.

PROGRAM

```
100  REM PARALLAX
105  REM REPLACE THIS STATEMENT
     WITH CLEAR SCREEN STATEMENT
110  INPUT "WHAT IS THE PARALLAX
     ANGLE? ";P$
120  P = VAL(P$)
130  IF P = 0 THEN GOTO 110
135  REM FORMULA FOR PARALLAX
140  D = 3.26/P
150  PRINT "STAR DISTANCE = ";D;"
     LIGHT YEARS"
160  END
```

PROGRAM NOTES

When calculating your time, be sure to record the time it took to enter the program and to correct any errors. After comparing calculation times among the three methods, add the entry and correction time to the computer's time. How does that change the differences among the three methods? Do you think a computer is always the most efficient method for mathematical calculation? In what situations might you choose to use a hand calculator instead of a computer?

BITS OF INFORMATION

You don't have to be a seasoned programmer to become involved in the computer world. Many young adults have even won computer contests without programming. Two 17-year-old students from Harbor Spring, Michigan, won a five-day, all-expense trip to Washington, D.C., and $500 in cash. Their idea, which they carried out, was to give senior citizens computer lessons. Another 11-year-old from Dubuque, Iowa, won $4,000 from a food company for her idea for a nutrition program. If you are interested in computer contests, keep watch for news through your school computer club, computer magazines, computer retail stores, and manufacturers' advertisements.

Your local library may have a computer newsboard. If it doesn't, you might ask the librarian about the possibility of adding such a bulletin board, to announce computer events that are open to people in your community or state.

CHAPTER REVIEW

VOCABULARY

On a separate piece of paper, match each term with the number of the phrase that best explains it. Use each term only once.

absolute magnitude	black hole	main sequence	supernova
light-year	Ursa Major	Milky Way	constellation
parallax	quasar	Polaris	galaxy

1. Commonly called the North Star.
2. Any particular pattern of stars.
3. A small, starlike object, very far from the earth, that produces huge amounts of energy.
4. A unit used in measuring distances to stars.
5. The shift in position of a star when seen from different parts of the earth's orbit.
6. A measure of the actual brightness of a star.
7. Thought to be the result of the collapse of a very large star.
8. Used to help locate the North Star.
9. A large star that suddenly explodes.
10. The group to which most stars belong.
11. A faint band of light running across the sky.
12. A huge system of stars.

QUESTIONS

Give brief but complete answers to each of the following questions. Unless otherwise indicated, use complete sentences to write your answers.

1. How did most of the constellations get their names?
2. Describe how you would locate the North Star.
3. To what constellation does the North Star belong?
4. List three kinds of radiant energy that telescopes on the earth can receive.
5. Where must telescopes be placed to receive all of the different kinds of radiant energy given off by stars?
6. Describe two kinds of scientific information astronomers get from starlight.
7. How can you tell whether an object is a star or a planet?
8. What method is used to measure the distance from earth to faraway stars?

9. Why can't a red star be identified as a red giant, considering only its surface temperature?
10. Describe how stars are born, according to scientific theory.
11. According to theory, how does a star become a red giant?
12. How is a supernova related to a neutron star?
13. Why do we see the Milky Way as a faint band of light running across the sky?
14. Locate and describe a galaxy that astronomers believe looks like the Milky Way galaxy.
15. What evidence do scientists have for believing that the universe is presently expanding?

APPLYING SCIENCE

1. Observe the night sky and locate the North Star. Extend a slightly bent line from the pointer stars of the Big Dipper beyond the North Star until you see a constellation of five stars that form a *W*. This constellation is Cassiopeia. Draw a sketch to show the relative positions of the Big Dipper, North Star, and Cassiopeia.
2. Look in a newspaper, an almanac, or the current issue of *Sky and Telescope* to find which planets are visible. Locate any one of the planets in the night sky. View it for several nights at the same time and explain how you know it to be a planet.
3. Explain why the moon appears to move with you as you ride along in a car, while the trees, street lights, buildings, etc., all move past you.

BIBLIOGRAPHY

Couper, Heather, and Terence Murtach. *Heavens Above: A Beginner's Guide to the Universe.* New York: Watts, 1981.

Gold, Michael. "The Cosmos Through Infrared Eyes." *Science 84,* March 1984.

Kunitzen, Paul. "How We Got Our Arabic Star Names." *Sky and Telescope,* January 1983.

Menzel, Donald H., and Hay M. Pasachoff. *A Field Guide to the Stars and Planets.* Boston: Houghton Mifflin, 1983.

National Geographic Editors. *Our Universe.* Washington, DC: National Geographic Society, 1980.

Whitney, Charles. *Whitney's Star Finder.* New York: Knopf, 1981.

GLOSSARY

A

Abrade To wear away.
Absolute magnitude A measure of the actual brightness of a star.
Abyssal plain A flat, wide deposit of sediments that covers the deeper parts of the ocean basin floor.
Air mass A large body of air that has taken on the temperature and humidity of a part of the earth's surface.
Anemometer An instrument used to measure wind speed.
Anticline Rock layers folded upward.
Anticyclone A large mass of air spinning out of a high pressure area.
Asteroid A smaller body in the solar system, usually found in an orbit between Mars and Jupiter.
Asthenosphere The partly molten upper layer of the mantle.
Astronomy The branch of earth science that studies the universe.
Atmosphere The layer of gases that make up the outside of the earth.
Atomic particles Small particles that make up atoms. Protons, neutrons, and electrons are three kinds of atomic particles.
Auroras The northern and southern lights produced when charged particles from the sun are trapped in the earth's magnetic field.
Autumnal equinox The time of the year, about September 22, when the earth's axis is tilted neither toward nor away from the sun.
Axis The imaginary line through the center of the earth on which the earth rotates.

B

Barometer An instrument used to measure atmospheric pressure.
Batholiths Very large bodies of igneous rock that are formed underground.
Beaches Parts of the shoreline covered with sand or small rocks.
Biomass Plant materials that are used to produce energy.
Breaker A wave whose top has curled forward and crashed down against the shore.

C

Cavern A series of underground passages made when water dissolves limestone.
Celsius The temperature scale used most in science. It is based on the freezing point of water, 0°C, and the boiling point of water, 100°C.
Channel The path of a stream of water.
Chemical weathering The ways rock breaks down by changing some of the materials in the rock.
Chromosphere The middle layer in the sun's atmosphere which gives off a faint red light.
Climate The average weather at a particular place.
Comet An icy body that comes from the edge of the solar system.
Compound A substance made up of two or more kinds of atoms joined together.
Condensation The process by which a gas is changed into a liquid.
Conservation Saving the natural resources by controlling how they are used.
Constellation A group of stars in a particular pattern.
Continental glacier A sheet or mass of ice that covers a large area of land.
Continental rise A gently sloping area at the base of the continental slope.
Continental shelf The flat submerged edge of a continent.
Continental slope The sloping part of the sea floor that marks the boundary between the sea floor and continental shelf.
Contour line A line drawn on a topographic map joining points having the same elevation.
Convection The movement of material caused by differences in its temperature.
Convection Movement of air caused by temperature differences.
Convection cell A complete circle of moving air caused by temperature differences.
Coral A small animal that lives in great numbers in the warm shallow parts of the sea.
Core The center layer of the earth made up of an inner and an outer part having a radius of about 3,500 km.

Coriolis effect Bending of the paths of moving objects as a result of the earth's rotation.
Corona The outermost layer of the sun.
Creep A slow downhill movement of loose rock or soil.
Crosscutting The principle that says faults or magma flows cutting through other rocks must be younger than the rocks they cut.
Crust The thin, solid outer layer of the earth.
Crystal A mineral form with a definite shape that results from the arrangement of its atoms or similar units.
Current Water moving in a particular direction.
Cyclone A kind of storm that is formed along the polar front.

D

Delta A deposit formed at the mouth of a river or stream.
Density The amount of mass in a given volume of a substance.
Dew point The temperature at which air becomes saturated with water vapor.
Diastrophism Movement of solid rock in the crust.
Differentiation The process by which planets became separated into layers.
Dike Body of igneous rock formed underground in cracks that cut across layers of existing rock.
Dome mountains Mountains formed when a part of the crust is lifted by rising magma.
Dune A deposit of wind-carried materials, usually sand.

E

Eclipse The earth or moon passing through the shadow of the other.
Electron shell An area where a certain number of electrons move at a definite distance from the nucleus.
Element A substance made up of only one kind of atom.
Ellipse The oval shape of the orbits followed by members of the solar system.
Epicenter The point on the earth's surface directly above the focus of an earthquake.
Era A time period in the earth's history. Each era is separated by major changes in the crust and changes in living things.
Erosion All the processes that cause pieces of weathered rock to be carried away.
Evaporation The process of water changing from a liquid to a gas at normal temperatures.
Exfoliation The peeling off of scales or flakes or rock as a result of weathering.
Experiment A setup to test a particular hypothesis.

F

Fault The result of movement of rock along either side of a crack in the earth's crust.
Fault-block mountains Mountains formed when large blocks of rock are tilted over.
Fetch The distance over which the wind blows steadily producing a group of waves.
Fission The splitting of the nuclei of certain kinds of atoms.
Focus The point of origin of an earthquake.
Folding Bending of rocks under steady pressure without breaking.
Fossil Any evidence of or the remains of a plant or animal that lived in the past.
Fossil fuel Mineral fuel found in the earth's crust.
Front The boundary separating two air masses.
Fusion The joining together of nuclei of small atoms to produce larger atoms with the release of energy.

G

Galaxy A huge system of stars held together by gravity.
Gem An attractive or rare mineral substance that is hard enough to be cut and polished.
Geologic column An arrangement showing rock layers in the order in which they were formed.
Geology The branch of earth science that studies the solid part of the earth.
Geothermal power Energy obtained from heat in the earth's crust.
Glacier A large body of moving ice and snow.
Greenhouse effect The name given to the method by which the atmosphere traps energy from the sun.
Ground water The water that soaks into the ground, filling openings between rocks and soil particles.
Guyot A submerged volcano whose top has been flattened by wave erosion before it sank below the surface.

H

Half-life The length of time taken for half a given amount of a radioactive substance to change into a new substance.

Horizon Any separate layer of soil.

Humidity The amount of water vapor in the air.

Humus Material in soil that comes from remains of plants and animals.

Hurricane A small but intense cyclonic storm formed over warm parts of the sea.

Hydrocarbons The principal substances found in petroleum. They contain only hydrogen and carbon.

Hydroelectric power Electricity produced when water moves downhill.

Hydrosphere All the bodies of water on the earth's surface.

Hypothesis A statement that explains a pattern in observations.

I

Igneous rock Rock formed when molten material cools.

Index fossils The remains of a plant or animal that lived during only a small part of the earth's history.

Interface A zone where any of the three parts of the earth (the lithosphere, hydrosphere, and atmosphere) come together.

Invertebrates Animals without backbones.

Ion An atom that has become electrically charged as a result of the loss or gain of electrons.

Ionosphere A region of the upper atmosphere containing many electrically charged particles.

Isostasy The state of balance between different parts of the lightweight crust as it floats on the dense mantle.

Isotherm A line on a map joining locations that record the same temperature.

J

Joint A crack in solid rock with no movement along the sides of the fracture.

K

Key bed A rock layer that is easily recognized and found over a large area.

Kilogram An SI unit of mass equal to 1,000 grams.

L

Landslide A sudden movement of large amounts of loose material down a slope.

Latitude The distance in degrees north or south of the equator.

Lava Magma that has flowed out onto the earth's surface.

Law of superposition The principle that, in a group of rock layers, the top layer is generally the youngest and the bottom layer the oldest.

Legend Information on a map that explains the meaning of each symbol used on a map.

Light-year The distance light travels in one year.

Liter An SI unit of volume. A liter is equal to 1.05 quarts.

Lithosphere The earth's outer layer averaging about 100 km in thickness.

Loess A deposit of windblown dust.

Longitude The distance in degrees east or west of the prime meridian.

Longshore current Movement of water parallel with the shoreline.

M

Magma Molten rock beneath the earth's surface.

Mantle The layer of earth below the crust having a thickness of about 2,870 km.

Map projection The curved surface of the earth shown on the flat surface of a map.

Maria Lava plains that are seen as dark areas on the moon's surface.

Meander A wide curve in the channel of an old river.

Measurement An observation made by counting something.

Mesosphere The atmospheric layer above the stratosphere extending from about 50 km to 85 km, where temperature decreases with increasing altitude.

Metamorphic rock Rock that has changed by the action of heat or pressure without melting.

Meteor A bright streak of light in the sky caused by rock material burning up in the atmosphere.

Meteorology The branch of earth science that studies the atmosphere.

Meter The basic SI unit of length. A meter is equal to 39.4 inches.

Mid-ocean ridge A chain of underwater mountains found on the floor of the oceans.

Milky Way A faint band of light seen spread

across the night sky produced by the stars in our galaxy.

Mineral A solid, naturally forming substance that is found in the earth and that always has the same properties.

Mineral deposit A body of ore.

Molecule The smallest part of a substance with the properties of that substance.

Moon phase The changing appearance of the moon as seen from the earth.

Moraine A layer of till.

N

Natural resource Any of the substances from the earth that can be used in some way.

Nodules Lumps of sediment found on the sea floor that contain valuable minerals.

Nuclear energy Energy that is produced by change in the nuclei of atoms.

Nucleus The central part of an atom that contains the protons and neutrons.

O

Observations Any information that we gather by using our senses.

Oceanography The branch of earth science that studies the oceans and the ocean floor.

Ooze A kind of sea floor sediment formed from remains of living things.

Ore A rock that contains useful minerals and can be mined at a profit.

Organic Coming from any living thing.

P

Parallax The apparent shift of position of a star when seen from different parts of the earth's orbit.

Permeability A measure of the speed with which water can move between the pore spaces in a rock.

Petrify To replace the material in the remains of a living thing with hardened mineral matter.

Petroleum A fossil fuel that is usually a liquid, that comes from the decay of ancient life in warm, shallow seas.

Photosphere A thin outer layer of gas in the sun which produces visible light.

Physical weathering The ways rock break into smaller pieces without any change in the materials in the rock.

Plankton Small plants and animals that float near the surface of the sea.

Plate tectonics The theory that the earth's crust is made up of moving plates.

Plateau Large raised areas of level land.

Pluton Any large intruded rock mass formed from magma.

Polar front The boundary where cold polar air meets warm air.

Pollution The adding of harmful substances to the environment.

Porosity The amount of the open spaces or pores in soil or rock.

Precipitation Water that falls from clouds, usually as rain or snow.

Prominence A huge cloud of gases that shoot out from areas of sunspots.

Q

Quasar A small object very far away which produces a huge amount of energy.

R

Radiant energy A form of energy that can travel through space.

Radioactive Describes a substance that changes over a period of time into a completely new substance.

Rays Light-colored streaks leading out from some craters on the moon.

Recycling The process of recovering a natural resource and using it again.

Red clay A kind of sediment found on the floor of most of the open sea.

Relative age A method of telling if one thing is older than another. It does not give the exact age.

Relative humidity The amount of water vapor in the air compared to the amount of water vapor the air could hold at that temperature.

Rift valley A deep, narrow valley formed where the earth's crust separates.

Rock A piece of the earth that is usually made up of two or more minerals mixed together.

Rock cycle The endless process by which rocks are formed, destroyed, and formed again in the earth's crust.

Runoff Water running downhill over the land surfaces.

S

Salinity The number of grams of dissolved salts in 100 g of sea water.

Satellite A body that moves in an orbit around a larger body.

Scale The relationship between the distance on a map and the actual distance on the earth's surface.

Scientific method A guide scientists use in solving problems.

Scientific theory A general explanation that includes many hypotheses that have been tested and found to be correct.

Sea cave A hollow space formed by waves at the base of a sea cliff.

Sea cliffs The steep faces of rocks eroded by waves.

Sea stack A column of rock left offshore as sea cliffs are eroded.

Seamount A submerged volcano rising from the ocean floor.

Sediment Any substance that settles out of water.

Sedimentary rock A kind of rock made when a layer of sediment becomes solid.

Seismic waves The vibrations caused by movement along a fault.

Seismograph An instrument that records seismic waves.

Shoreline The place where the land meets water.

Sill Body of igneous rock formed underground in cracks that run parallel with layers of existing rock.

Slump The sudden downward movement of a block of rock or soil.

Solar cell A device that is used to change sunlight directly into electricity.

Solar system The sun with its nine planets and smaller bodies that move around it.

Spit A long, narrow deposit of sand formed where a shoreline changes direction.

Standard Time The time based on one meridian and used for a time zone.

Stratosphere The atmospheric layer above the troposphere, between about 20 km and 50 km, showing a temperature increase with increasing altitude.

Subduction Where one plate is forced beneath another plate.

Submarine canyon A deep valley cut into a continental shelf.

Summer solstice The time of the year, about June 22, when the earth's axis is tilted most toward the sun.

Sunspots Areas on the sun's surface that appear dark.

Supernova A large star that suddenly explodes.

Swells Long, regular waves that move great distances across the oceans.

Syncline Rock layers folded downward.

T

Talus A pile of broken rock that gathers at the bottom of a cliff or steep slope.

Terrace A platform beneath the water at the base of a sea cliff.

Thermal pollution Release of heat into streams, lakes, oceans, or the atmosphere.

Thermocline A zone of rapid temperature change beneath the sea surface.

Thermosphere The layer of the atmosphere above the mesosphere, extending from about 85 km to about 600 km, where temperature increases with increasing altitude.

Tidal bore A large wave that leads a tide upriver.

Tide A regular rise and fall of water level in the sea.

Till Material deposited directly from a glacier.

Topographic map A map that shows the shape of the land surface.

Transpiration The process by which plants take up water through their roots and release water vapor from their leaves.

Trench A deep valley on the ocean floor found along the edges of some ocean plates.

Tributary A stream that flows into a river.

Troposphere The most dense layer of the atmosphere. It lies closest to the earth's surface.

Tsunami A giant wave caused by a sudden disturbance on the sea floor.

Turbidity current A current that is caused by the increased density of water as a result of mixing with sediments.

U

Unconformity A boundary separating younger rock layers from older layers that were exposed to erosion.

Uniformitarianism The principle that the same processes acting on the earth today also acted on the earth in the past.

Upwelling The rise of water to the sea surface.

V

Valley glacier A river of ice in a mountain valley.

Vernal equinox The time of the year, about March 22, when the earth's axis is tilted neither toward nor away from the sun.

Vertebrates Animals that have backbones.

Volcano A place where lava reaches the surface and builds a cone or other surface feature.

W

Water cycle The processes by which water leaves the oceans, is spread through the atmosphere, falls over the land, and runs back to the sea.

Water table The level below the ground surface where all open spaces are filled with water.

Watershed The land area that drains into a stream or river.

Weathering All the processes that break rock into smaller pieces.

Windbreak A barrier that causes wind to move more slowly.

Winter solstice The time of the year, about December 22, when the earth's axis is tilted most away from the sun.

APPENDIX A: THE SI (METRIC) SYSTEM

METRIC SYSTEM PREFIXES	
Greater than 1:	**Less than 1:**
kilo (k) = 1,000 hecto (h) = 100 deka (da) = 10	deci (d) = 0.1 centi (c) = 0.01 milli (m) = 0.001 micro (μ) = 0.000 001

METRIC-ENGLISH EQUIVALENTS
1 km = 0.621 miles
1 m = 39.37 inches
1 cm = 0.394 inches
1 mm = 0.0394 inches
1 kg = 2.2046 pounds
1 g = 0.035 ounce
1 L = 1.06 quart
1 mL = 0.00106 quart

COMMONLY USED UNITS		
	Unit	**Can be compared to**
Length	kilometer (km) 1000 m	The distance covered by walking for twelve minutes.
	meter (m)	The height of a door knob.
	centimeter (cm) 0.01 m	The thickness of a bread slice.
	millimeter (mm) 0.001 m	The thickness of a dime.
Mass	kilogram (kg) 1000 g	The mass of this book.
	gram (g) 0.001 kg	The mass of a paper clip.
Volume	liter (L) 1000 mL	Four times the volume of one cup.
	milliliter (mL) 0.001 L	The volume occupied by two nickels.

APPENDIX B: CHARACTERISTICS OF PLANETS

Planet	Average Distance From Sun (millions of km)	Time Taken to Go Once Around Orbit	Period of Rotation	Mass (Earth = 1)	Number of Satellites
Mercury	57.9	88.0 days	59 days	0.055	0
Venus	108.2	225.0 days	243 days	0.815	0
Earth	149.6	365.25 days	24 hrs	1.0	1
Mars	227.9	687.0 days	24 hrs 37 min	0.108	2
Jupiter	778.3	11.9 yrs	9 hrs 54 min	318.0	16
Saturn	1427.0	29.46 yrs	10 hrs 40 min	95.2	17
Uranus	2870.0	84.0 yrs	17 hrs	15.0	15
Neptune	504.0	165.0 yrs	17 hrs 54 min	17.0	2
Pluto	5900.0	248.0 yrs	6 days 9 hrs	0.002	1

APPENDIX C: WEATHER MAP SYMBOLS

A weather map is made by placing symbols and numbers at locations on a map. The symbols and numbers represent the weather observed at those locations. A set of lines (isobars) connects the places that have the same atmospheric pressure. Heavier lines with solid triangles or half-circles on them show fronts. The symbols for fronts are:

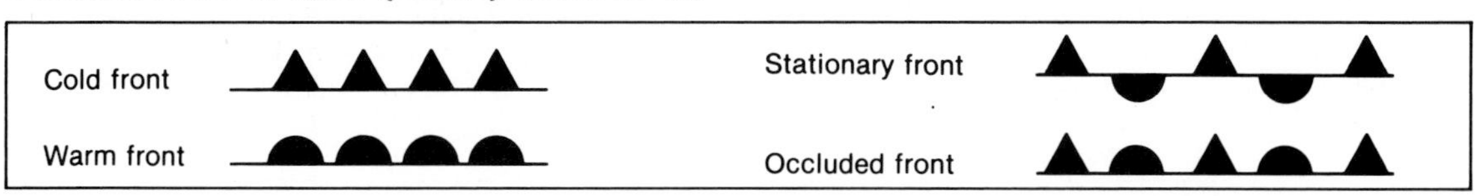

Observations taken at individual locations are plotted on a small circle on the map where the readings were taken. These circles along with symbols describing the weather conditions are called *station models*. Some symbols used to make station models are:

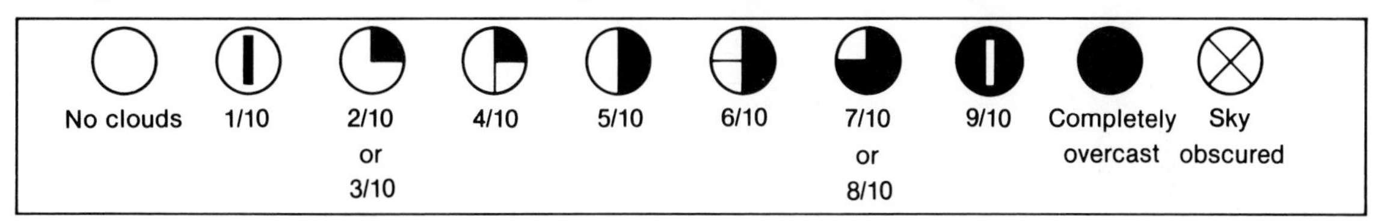

1. Cloud cover

North

Northeast

Southeast

2. Wind direction

3. Wind speed

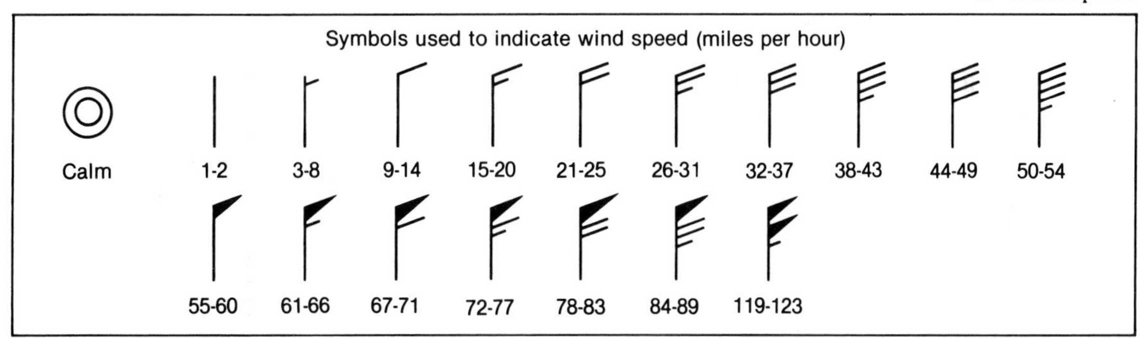

For example, a basic station model might be:

Wind from northwest at 15 m.p.h., sky one-half covered with clouds.

In addition to the symbols shown above, station models also include other numbers and symbols.

An example of a complete station model is shown on right.

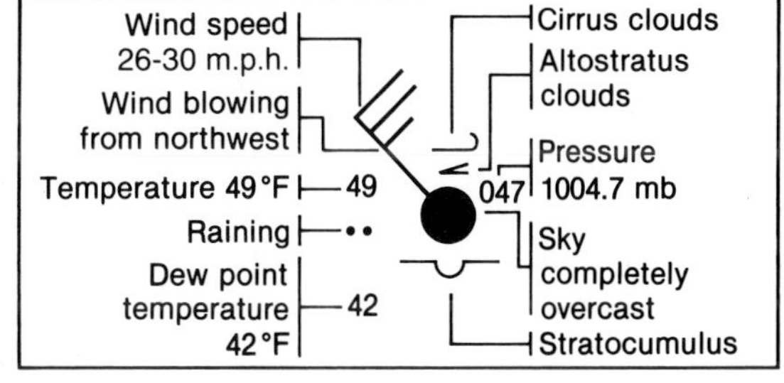

*mb = millibars

APPENDIX D: PROPERTIES OF SOME MINERALS COMMONLY FOUND IN ROCKS

LIGHT-COLORED MINERALS ABUNDANT IN IGNEOUS ROCKS				
Mineral	Hardness	Cleavage or Fracture	Color	Streak
Quartz	7	Fractures with curving lines on broken surface	Colorless or smoky gray to pink	None
Feldspar	6	Cleavage in two directions, nearly 90°	Usually white to pink	None
Mica (Muscovite)	2–2.5	Cleavage in one direction; splits into thin sheets	Colorless to slightly brown or greenish	None
DARK-COLORED MINERALS ABUNDANT IN IGNEOUS ROCKS				
Mineral	Hardness	Cleavage or Fracture	Color	Streak
Mica (Biotite)	2–2.5	Cleavage in one direction; splits into thin sheets	Dark brown to black	None
Chlorite	1–2.5	Cleavage in one direction; may show uneven fracture	Pale green to brown, rose and yellow	Light green
Hornblende	5–6	Cleavage two ways, at 56° and 124°	Black to greenish-black	Gray-green
Augite	5–6	Cleavage two ways at about 90°	Dark green to black	Gray-green
Olivine	6.5–7	Uneven fracture with curved lines on broken surface	Olive green to brown	None
LIGHT-COLORED MINERALS COMMONLY FOUND IN SEDIMENTARY ROCKS				
Mineral	Hardness	Cleavage or Fracture	Color	Streak
Calcite	2–3	Cleavage in three directions at greater than 90°	Colorless	None
Dolomite	3.5–4	Cleavage in three directions at greater than 90°	White or various colors	White
Halite	2.5	Three cleavages at 90°	Colorless to light gray	None
Gypsum	2	Good cleavage in one direction	Colorless to gray-white	None

DARK-COLORED MINERALS COMMONLY FOUND IN SEDIMENTARY ROCKS				
Mineral	**Hardness**	**Cleavage or Fracture**	**Color**	**Streak**
Hematite	5.5–6.5	Uneven fracture	Reddish-brown to black	Brown-red
Magnetite	6	Irregular fracture	Black; metallic luster (attracted to a magnet)	Black
Limonite	1–2.5	Earthy fracture	Yellow-brown or dark brown to black	Yellow-brown
LIGHT-COLORED MINERALS COMMONLY FOUND IN METAMORPHIC ROCKS				
Mineral	**Hardness**	**Cleavage or Fracture**	**Color**	**Streak**
Chrysotile	2.5–4	Splintery fracture; may separate into fibers (asbestos)	Greenish-gray	None
Talc	1	Cleavage in one direction making thin flakes	White to gray; greasy feel	White
DARK-COLORED MINERALS COMMONLY FOUND IN METAMORPHIC ROCKS				
Mineral	**Hardness**	**Cleavage or Fracture**	**Color**	**Streak**
Corundum	9	Irregular fracture	Usually brown	None
Epidote	6–7	Cleavage in two directions, distinct in one direction poor in the second; fracture uneven	Green to dark brown or black	None
Garnet	6.5–7.5	Irregular fracture	Red and brown	None
MINERALS COMMONLY FOUND IN VEINS IN ROCKS				
Mineral	**Hardness**	**Cleavage or Fracture**	**Color**	**Streak**
Pyrite	6–6.5	Uneven fracture	Brass yellow	Black
Galena	2.5	Cleavage in three directions at 90°	Silver-gray	Black
Sphalerite	3.5–4	Cleavage in six directions at 60°	Usually dark brown to black	Yellow-brown
Chalcopyrite	3.5–4	Uneven fracture	Yellow-brown metallic luster	Green-black

INDEX

NOTE: Page numbers in **boldface** type refer to illustrations and those in *italic* type refer to definitions.

A

B

C

G

H

I

J

K

P

Q

R

U

V

W

X

Y

Z

PHOTO CREDITS

Chapter 1: p. xii, United States Geological Survey; p. 2 (tl) G. Gualco/U.P./ Bruce Coleman; (tr) Walter Rawlings/ Robert Harding Assoc.; p. 3 Nicholas DeVore III/ Bruce Coleman; p. 4 (tr) Stephen J. Krasemann/ D.K.R. Photo; (cr) Bill Curtsinger/ Photo Researchers; p. 5 Dan McCoy/ Rainbow; p. 7 (cl) James Sugar/ Black Star; (cr) John Marshall; p. 8 John de Visser/ Black Star; p. 9 Ralph Perry/ Black Star; p. 10 Ralph Perry/ Black Star; p. 13 Mark Tuschman/ Discover Magazine; p. 14 Eric Kroll/ Taurus Photos; p. 15 James Sugar/ Black Star.

UNIT 1: pp. 26–27 Peter Turner/ Image Bank

Chapter 2: p. 28 NASA/ Photo Researchers; p. 31 (tl) Focus on Sports; (tr) Focus on Sports; p. 32 HRW Photo by Richard Haynes; p. 33 NASA; p. 38 NASA Science Photo Library Int'l/ Taurus Photos; p. 41 HRW Photo by Russell Dian; p. 45 HRW Photo by Russell Dian; p. 48 David Falconer/ Bruce Coleman; p. 52 (tr) Dennis di Cicco/ Sky & Telescope Magazine; p. 54 Robert Harding Picture Library; p. 61 (tl) Charles Henneghien/ Bruce Coleman; (tr) HRW Photo by Richard Haynes.

Chapter 3: p. 64 S. J. Krasemann/ Peter Arnold; p. 66 (tr) HRW Photo by Richard Haynes; p. 69 (tl) Woods Hole Oceanographic Institution; p. 72 Meffert/Stern/ Black Star; p. 76 (t) Galen Rowell/ Peter Arnold; p. 77 NASA/ Jet Propulsion Laboratory; p. 79 Department of Geological Sciences/ Cornell University.

Chapter 4: p. 86 GOLT/ Photo Researchers; p. 90 (t) Marshall Lockman/ Black Star; p. 92 Gene Daniels/ Black Star; p. 95 Ken Sakamoto/ Black Star; p. 97 Snead/ Bruce Coleman; p. 99 (tc) Gary Rosenquist/ Earth Images; (bc) Gary Rosenquist/ Earth Images; (bl) Gary Rosenquist/ Earth Images; p. 100 Peter Arnold/ Peter Arnold, Inc.; p. 101 (tl) W.E. Ruth/ Bruce Coleman; (tc) Robert Harding Picture Library; (tr) United States Geological Survey; p. 103 Robert P. Carr/ Bruce Coleman; p. 105 W.H. Hodge/ Peter Arnold; p. 106 (tl) Manuel Rodriguez; (tc) Michael Gallagher/ Bruce Coleman; (tr) Russ Kinne/ Photo Researchers; p. 107 (tl) Manuel Rodriguez; (tc) Grant Heilman; (tr) Bob & Clara Calhoun/ Bruce Coleman; p. 109 (tl) Grafton M. Smith/ Image Bank; (tr) E. R. Degginger/ Bruce Coleman.

Chapter 5: p. 112 Manuel Rodriguez; p. 114 (t) Bart Bartholomew/ Black Star; (b) William L. Ramsey; p. 116 (t) Joy Spurr/ Bruce Coleman; (b) Manuel Rodriguez; p. 117 (cl) Michael Markew/ Bruce Coleman; (br) Stephen J. Krasemann/ D.R.K.; p. 119 U.S.D.A./ Soil Conservation; p. 120 Manuel Rodriguez; p. 121 (tl) J. Callahan; (tr) Jacques Janeoux/ Peter Arnold; p. 123 (c) Jack W. Pykinga/ Bruce Coleman; p. 127 (tl) Manuel Rodriguez; (tr) Stephen J. Krasemann/Peter Arnold; p. 128 Grant Heilman; p. 129 (tr) Stephen Fuller/ Peter Arnold; p. 131 NASA; p. 133 Orlana Sentinel Star/ Black Star; p. 136 Peter Arnold; p. 137 (tr) Harvey Lloyd/ Peter Arnold; p. 139 Adolahe/ Southern Light; p. 141 (tl) Paul Shambroom/ Photo Researchers; (tr) Michael Heron/ Woodfin Camp.

Chapter 6: p. 146 Tom Bean/ Tom Stack; p. 148 (tl) Steve Coombs/ Photo Researchers; (tr) Manuel Rodriguez; p. 149 (tl) George Holton/ Photo Researchers; p. 150 (tl) Leland Brown/ Photo Researchers; p. 156 Sabine Weiss/ Photo Researchers; p. 157 (bl) Robert P. Carr/ Bruce Coleman; (bc) Lillian N. Bolstad/ Peter Arnold; (br) Sund/ Image Bank; p. 158 (br) Grant Heilman; p. 159 Loraine Wilson/ Robert Harding Picture Library; p. 160 (tr) Baron Wolman/ Woodfin Camp & Assoc.; (br) Robert Frerck/ Woodfin Camp & Assoc.; p. 161 (tl) Hank Morgan/ Rainbow; (tr) HRW Photo by Ken Lax; p. 162 (bl) Jack Fields/ Photo Researchers; (br) John S. Shelton; p. 163 (tr) Bruce Roberts/ Photo Researchers; p. 164 HRW Photo by Richard Haynes; p. 165 David Madison/ Bruce Coleman.

Chapter 7: p. 168 David Stone/ Rainbow; p. 170 Robert Harding Picture Library; p. 172 (tr) Manuel Rodriguez; (cr) Manuel Rodriguez; p. 173 (bl) G.R. Roberts/ Photo Researchers; p. 174 William Ramsey; p. 177 John Reader/ National Geographic Society; p. 178 (cr) William Ramsey; p. 179 (tl) Tass/ Sovfoto; (bl) George Roos/ Peter Arnold; p. 180 Photo Works/ Photo Researchers; p. 181 (tr) Stephen J. Krasemann/ D.R.K. Photo; (bl) Runk/ Schoenberger/ Grant Heilman/ N. Museum/ Franklin & Marshall College; p. 182 (tl) Manuel Rodriguez; p. 185 J. Fennell/ Bruce Coleman; p. 186 HRW Photo by Ken Karp; p. 187 (tr) Grant Heilman; p. 188 Runk/Schoenberger/ Grant Heilman; p. 205 (tl) John Curtis/ Taurus Photos; (tr) Walter H. Hodge/ Peter Arnold.

UNIT 2: pp. 210–211, Michael Freeman/ Bruce Coleman

Chapter 8: p. 212 Krafft-Explorer/ Photo Researchers; p. 220 Novosti/ Image Bank; p. 221 (cl) Photo Researchers; (bl) Karen Tompkins/ Tom Stack; (bc) Karen Tomkins/ Tom Stack; (br) Ft. Worth Museum of Science & History/ Susan Gibler/ Tom Stack; p. 222 Alfred Pasieka/ Taurus Photos; p. 223 Runk/ Schoenberger/ Grant Heilman/ Franklin & Marshall College; p. 226 Granger; p. 227 (tl) Bob Gossington/ Tom Stack; (cr) Manuel Rodriguez; p. 228 (tl) M. Liacos; (tr) B.M. Shaub; p. 229 (t) HRW Photo by Richard Haynes; (bl) J. Fennell/ Bruce Coleman; (br) Adrian Davies/ Bruce Coleman; p. 231 Larry Dale Gordon/ Image Bank; p. 233 Gabe Palmer/ Image Bank.

Chapter 9: p. 238 Alan Pitcairn/ Grant Heilman; p. 240 NASA/ Black Star; p. 241 (t) John Running; (b) E.R. Degginger/ Bruce Coleman; p. 242 Fred Ihrt/ Image Bank; p. 243 (t) Shelly Grossman/ Woodfin Camp; (c) Robert P. Carr/ Bruce Coleman; p. 244 Bob & Clara Calhoun/ Bruce Coleman; p. 247 Michael G. Borum/ Image Bank; p. 248 Keith Gunner/ Bruce Coleman; p. 249 Keith & Donna Dannen/ Photo Researchers; p. 250 (tl) Leo Touchet/ Woodfin Camp; (tr) Walter Rawlings/ Robert Harding Picture Library; (c) Jack Baker/ Image Bank; p. 251 (t) Bob McKeever/ Tom Stack; (b) David W. Hamilton/ Image Bank; p. 252 (tl) William Felger/ Grant Heilman; p. 255 Nicholas de Vore III/ Bruce Coleman; p. 256 Dan McCoy/ Rainbow; p. 257 (cl) David Krasnor/ Photo Researchers; (cr) Stephen J. Krasemann/ Photo Researchers; (bl) Art Kane/ Image Bank; (br) Budd Titlow/Naturegraphics/ Tom Stack; p. 261 (tl) Dan McCoy/ Rainbow; (tr) Dan McCoy/ Rainbow.

Chapter 10: p. 266 Bob & Clara Calhoun/ Bruce Coleman; p. 269 (b) Nicholas DeVore III/ Bruce Coleman; p. 271 Nicholas DeVore III/ Bruce Coleman; p. 272 Nicholas DeVore III/ Bruce Coleman; p. 274 Jeff Foott/ Bruce Coleman; p. 275 Dr. George Gerster/ Photo Researchers; p. 276 (bl) Woods Hole Oceanographic Institute; p. 281 Mark Sherman/ Bruce Coleman; p. 282 (tl) Mark Sherman/ Bruce Coleman; p. 286 Katherine S. Thomas/ Bruce Coleman; p. 288 Wendell Metzen/ Bruce Coleman; p. 289 (cl) Horst Schafer/ Peter Arnold; (cr) Paola Kock/ Photo Researchers; (bl) Robert Whitney; (br) Robert Whitney; p. 290 Marc Solomon/ Image Bank; p. 292 Dennis Hogan/ Tom Stack; p. 294 (br) Ron Sherman/ Bruce Coleman; p. 297 Peter Arnold; p. 298 (br) Jonathan T. Wright/ Bruce Coleman; p. 299 George Gerster/ Photo Researchers; p. 300 Terence Moore/ Woodfin Camp; p. 301 George Gerster/ Photo Researchers; p. 303 Ph. Charliat/ Photo Researchers.

UNIT 3: pp. 306–307 Stephen J. Krasemann/ D.R.K. Photo

Chapter 11: p. 308 Wards Science/Science Source/ Photo Researchers; p. 310 (t) William Thompson/ Earth Images; p. 312 Alfred Pasieka/ Taurus Photos; p. 313 (cl) Robert P. Carr/ Bruce Coleman; p. 315 HRW Photo by Richard Haynes; p. 316 J. Serafin/ Peter Arnold; p. 317 (tl) HRW Photo by Richard Haynes; p. 318 (tr) Jules Bucher/ Photo Researchers; p. 323 (tl) Michal Heron/ Woodfin Camp; (tr) NASA/ Science Source/ Photo Researchers.

Chapter 12: p. 328 Bjorn Bolstad/ Photo Researchers; p. 330 (tl) Hanson Carroll/ Peter Arnold; p. 333 (br) Bill Evans; p. 342 Norman Owen Tomalin/ Bruce Coleman; p. 345 (tl) Thomas W. Putney/ Delaware Stock Photos; (tr) Dan McCoy/Rainbow; p. 347 (tl) L.L.T. Rhodes/ Taurus Photos; p. 347 (tr) Eric Kroll/ Taurus Photos.

Chapter 13: p. 350 John Deeks/ Photo Researchers; p. 352 (tl) HRW Photo by Russell Dian; (tr) Jonathan Wright/ Bruce Coleman; p. 353 (tl) Dave Baird/ Tom Stack; (tr) Russ Kinne/ Photo Researchers; p. 354 (tl) Tom Stack/ Tom Stack & Assoc.; (tc) Hans Namuth/ Photo Researchers; (tr) Dale Jorgenson/ Tom Stack; p. 356 Norman Owen Tomalin/ Bruce Coleman; p. 357 HRW Photo by Richard Haynes; p. 358 Richard Weiss/ Peter Arnold; p. 365 (tr) Brian Parker/ Tom Stack; p. 367 (tr) Shostal; p. 368 (t) NOAA; (bl) Doug Millar/ Photo Researchers; p. 369 (tl) Henry Lansford/ Photo Researchers; (tr) Spencer Swanger/ Tom Stack; p. 370 Frederic Lewis; p. 373 Joel Gordon/ D.P.I.; p. 375 (tl) Joe Munroe/ Photo Researchers; (bl) Wards Science/ Science Source/ Photo Researchers; p. 379 NOAA.

UNIT 4: pp. 384–385 Guido Alberto Rossi/ Image Bank

Chapter 14: p. 386 Bob Evans/ Peter Arnold; p. 388 Wards Science/ Science Source/ Photo Researchers; p. 395 Peter Arnold; p. 397 Vivian M. Peevers/ Peter Arnold; p. 399 Runk Schoenberger/ Grant Heilman; p. 401 Dan McCoy/ Rainbow; p. 403 Sullivan & Rogers/ Bruce Coleman.

Chapter 15: David Doubilet; p. 411 E.R. Degginger/ Bruce Coleman; p. 416 D. Brewster/ Bruce Coleman; p. 417 NASA; p. 419 (t) Tom McHugh/ Steinhart Aquarium/ Photo Researchers; (b) Frank W. Lane/ Bruce Coleman; p. 420 W.H./ Image Bank; p. 421 (t) George Whiteley/ Photo Researchers; (b) Dan McCoy/ Rainbow; p. 422 (tr) Shell Oil Company; p. 425 (tl) Christina Thompson/ Woodfin Camp; (tr) Jim Pickerell.

Chapter 16: p. 428 Bob Evans/ Peter Arnold; p. 433 HRW Photo by Richard Haynes; p. 434 Larry Mulvehill/ Photo Researchers; p. 436 NOAA; p. 441 John Bryson/ From RG/ Photo Researchers; p. 444 Ron Church/ Photo Researchers; p. 445 HRW Photo by Richard Haynes; p. 446 Kyodo Photo Services; p. 450 (tr & cr) Clyde H. Smith/ Peter Arnold; (bl) Weston Kemp; p. 453 (tl) Eric Kroll/ Taurus Photos; (tr) Jack Fields/ Photo Researchers.

UNIT 5: pp. 456–457 Rainbow

Chapter 17: p. 458 Dan McCoy/ Rainbow; p. 461 NASA/ Science Source/ Photo Researchers; p. 462 (l) Ward's Scientific/ Photo Researchers; (r) G.C. Fuller/ S.P.L./ Photo Researchers; p. 463 (bl) Michael P. Gadomski/ Bruce Coleman; p. 465 (bl) Jack Finch/S.P.L./ Photo Researchers; p. 467 Ward's Science/ Science Source/ Photo Researchers; p. 468 U.S. Navy/ S.P.L./ Photo Researchers; p. 473 NASA/ Science Source/ Photo Researchers; p. 474 HRW Photo by Richard Haynes; p. 475 NASA/ Robert Harding Picture Library.

Chapter 18: p. 480 R. Royer/ Science Photo Library/ Photo Researchers; p. 483 Astron Society of the Pacific/ Science Source/ Photo Researchers; p. 484 (l) NASA; (r) NASA; p. 486 (t) NASA; (b) Astron Society of the Pacific/ Science Source/ Photo Researchers; p. 490 NASA; p. 491 NASA; p. 492 NASA; p. 493 (tl) NASA; (br) NASA/ A.M.N.H.; p. 497 (tr) Lowell Observatory; (br) Litton Industries; p. 499 (cl) NASA/ A.M.N.H.; p. 500 (tl & bl) NASA/ National Space Science Data Center; p. 502 Rainbow; p. 503 NASA; p. 504 (tl) Dennis di Cicco; (tr) NASA; p. 507 NASA; p. 511 (tl) Allen Green/ Photo Researchers; (tr) Jules Bacher/ Photo Researchers; (cl) Dennis di Cicco; (cr) Russ Kinne/ Photo Researchers; (bl) Jules Bacher/ Photo Researchers; (bc) Wards Science/ Science Source/ Photo Researchers; (br) Russ Kinne/ Photo Researchers; p. 512 (bl) Tersch Observatories; p. 513 Wards Science/ Science Source/ Photo Researchers; p. 514 HRW Photos by Richard Haynes; p. 515 Clyde H. Smith/ Peter Arnold.

Chapter 19: p. 520 James Sugar/ Black Star; p. 522 (tr) Wards Science/ Science Source/ Photo Researchers; p. 523 (br) Bill Belknap/ Photo Researchers; p. 531 Ronald Royer/ Photo Researchers; p. 536 NASA; p. 538 Hale Observatories/ Science Source/ Photo Researchers; p. 541 NASA.